IET MATERIALS, CIRCUITS AND DEVICES SERIES 77

Integrated Optics

Other volumes in this series

Volume 2	**Analogue IC Design: The current-mode approach** C. Toumazou, F.J. Lidgey and D.G. Haigh (Editors)
Volume 3	**Analogue–Digital ASICs: Circuit techniques, design tools and applications** R.S. Soin, F. Maloberti and J. France (Editors)
Volume 4	**Algorithmic and Knowledge-Based CAD for VLSI** G.E. Taylor and G. Russell (Editors)
Volume 5	**Switched Currents: An analogue technique for digital technology** C. Toumazou, J.B.C. Hughes and N.C. Battersby (Editors)
Volume 6	**High-Frequency Circuit Engineering** F. Nibler *et al.*
Volume 8	**Low-Power High-Frequency Microelectronics: A unified approach** G. Machado (Editor)
Volume 9	**VLSI Testing: Digital and mixed analogue/digital techniques** S.L. Hurst
Volume 10	**Distributed Feedback Semiconductor Lasers** J.E. Carroll, J.E.A. Whiteaway and R.G.S. Plumb
Volume 11	**Selected Topics in Advanced Solid State and Fibre Optic Sensors** S.M. Vaezi-Nejad (Editor)
Volume 12	**Strained Silicon Heterostructures: Materials and devices** C.K. Maiti, N.B. Chakrabarti and S.K. Ray
Volume 13	**RFIC and MMIC Design and Technology** I.D. Robertson and S. Lucyzyn (Editors)
Volume 14	**Design of High Frequency Integrated Analogue Filters** Y. Sun (Editor)
Volume 15	**Foundations of Digital Signal Processing: Theory, algorithms and hardware design** P. Gaydecki
Volume 16	**Wireless Communications Circuits and Systems** Y. Sun (Editor)
Volume 17	**The Switching Function: Analysis of power electronic circuits** C. Marouchos
Volume 18	**System on Chip: Next generation electronics** B. Al-Hashimi (Editor)
Volume 19	**Test and Diagnosis of Analogue, Mixed-Signal and RF Integrated Circuits: The system on chip approach** Y. Sun (Editor)
Volume 20	**Low Power and Low Voltage Circuit Design with the FGMOS Transistor** E. Rodriguez-Villegas
Volume 21	**Technology Computer Aided Design for Si, SiGe and GaAs Integrated Circuits** C.K. Maiti and G.A. Armstrong
Volume 22	**Nanotechnologies** M. Wautelet *et al.*
Volume 23	**Understandable Electric Circuits** M. Wang
Volume 24	**Fundamentals of Electromagnetic Levitation: Engineering sustainability through efficiency** A.J. Sangster
Volume 25	**Optical MEMS for Chemical Analysis and Biomedicine** H. Jiang (Editor)
Volume 26	**High Speed Data Converters** A.M.A. Ali
Volume 27	**Nano-Scaled Semiconductor Devices** E.A. Gutiérrez-D (Editor)
Volume 28	**Security and Privacy for Big Data, Cloud Computing and Applications** L. Wang, W. Ren, K.R. Choo and F. Xhafa (Editors)
Volume 29	**Nano-CMOS and Post-CMOS Electronics: Devices and modelling** S.P. Mohanty and A. Srivastava
Volume 30	**Nano-CMOS and Post-CMOS Electronics: Circuits and design** S.P. Mohanty and A. Srivastava
Volume 32	**Oscillator Circuits: Frontiers in design, analysis and applications** Y. Nishio (Editor)
Volume 33	**High Frequency MOSFET Gate Drivers** Z. Zhang and Y. Liu
Volume 34	**RF and Microwave Module Level Design and Integration** M. Almalkawi
Volume 35	**Design of Terahertz CMOS Integrated Circuits for High-Speed Wireless Communication** M. Fujishima and S. Amakawa
Volume 38	**System Design with Memristor Technologies** L. Guckert and E.E. Swartzlander Jr.
Volume 39	**Functionality-Enhanced Devices: An alternative to Moore's law** P.-E. Gaillardon (Editor)
Volume 40	***Digitally Enhanced Mixed Signal Systems*** C. Jabbour, P. Desgreys and D. Dallett (Editors)

Volume 43	**Negative Group Delay Devices: From concepts to applications** B. Ravelo (Editor)
Volume 45	**Characterisation and Control of Defects in Semiconductors** F. Tuomisto (Editor)
Volume 47	**Understandable Electric Circuits: Key concepts. 2nd Edition** M. Wang
Volume 48	**Gyrators, Simulated Inductors and Related Immittances: Realizations and applications** R. Senani, D.R. Bhaskar, V.K. Singh and A.K. Singh
Volume 49	**Advanced Technologies for Next Generation integrated Circuits** A. Srivastava and S. Mohanty (Editors)
Volume 51	**Modelling Methodologies in Analogue Integrated Circuit Design** G. Dundar and M.B. Yelten (Editors)
Volume 53	**VLSI Architectures for Future Video Coding** M. Martina (Editor)
Volume 54	**Advances in High-Power Fiber and Diode Laser Engineering** I. Divliansky (Editor)
Volume 55	**Hardware Architectures for Deep Learning** M. Daneshtalab and M. Modarressi
Volume 57	**Cross-Layer Reliability of Computing Systems** G. Di Natale, A. Bosio, R. Canal, S. Di Carlo and D. Gizopoulos (Editors)
Volume 58	**Magnetorheological Materials and Their Applications** S. Choi and W. Li (Editors)
Volume 59	**Analysis and Design of CMOS Clocking Circuits for Low Phase Noise** W. Bae and D.K. Jeong
Volume 60	**IP Core Protection and Hardware-Assisted Security for Consumer Electronics** A. Sengupta and S. Mohanty
Volume 64	**Phase-Locked Frequency Generation and Clocking: Architectures and circuits for modem wireless and wireline systems** W. Rhee (Editor)
Volume 65	**MEMS Resonator Filters** R.M. Patrikar (Editor)
Volume 66	**Frontiers in Hardware Security and Trust: Theory, design and practice** C.H. Chang and Y. Cao (Editors)
Volume 67	**Frontiers in Securing IP Cores; Forensic detective control and obfuscation techniques** A. Sengupta
Volume 68	**High Quality Liquid Crystal Displays and Smart Devices: Vol. 1 and Vol. 2** S. Ishihara, S. Kobayashi and Y. Ukai (Editors)
Volume 69	**Fibre Bragg Gratings in Harsh and Space Environments: Principles and applications** B. Aïssa, E.I. Haddad, R.V. Kruzelecky and W.R. Jamroz
Volume 70	**Self-Healing Materials: From fundamental concepts to advanced space and electronics applications, 2nd Edition** B. Aïssa, E.I. Haddad, R.V. Kruzelecky and W.R. Jamroz
Volume 71	**Radio Frequency and Microwave Power Amplifiers: Vol. 1 and Vol. 2** A. Grebennikov (Editor)
Volume 72	**Tensorial Analysis of Networks (TAN) Modelling for PCB Signal Integrity and EMC Analysis** B. Ravelo and Z. Xu (Editors)
Volume 73	**VLSI and Post-CMOS Electronics Volume 1: VLSI and post-CMOS electronics and Volume 2: Materials, devices and interconnects** R. Dhiman and R. Chandel (Editors)

Integrated Optics

Volume 1: Modeling, material platforms and fabrication techniques

Edited by
Giancarlo C. Righini and Maurizio Ferrari

The Institution of Engineering and Technology

Published by The Institution of Engineering and Technology, London, United Kingdom

The Institution of Engineering and Technology is registered as a Charity in England & Wales (no. 211014) and Scotland (no. SC038698).

First published 2020

The Institution of Engineering and Technology
Michael Faraday House
Six Hills Way, Stevenage
Herts, SG1 2AY, United Kingdom

www.theiet.org

British Library Cataloguing in Publication Data
A catalogue record for this product is available from the British Library

ISBN 978-1-83953-341-9 (Hardback)
ISBN 978-1-83953-342-6 (PDF)

Typeset in India by MPS Limited

Dedicated to two exceptional women: Prof. Vera Russo, who 50 years ago got me involved in integrated optics, and my wife, Marta, who for as many years went with me with her love and great patience.

– Giancarlo C. Righini

To Profs. Georges Boulon, Eugene Duval, Andre Monteil, and all the friends of the PCML laboratory in Lyon, where I started to develop my passionate interest for glass photonics.

– Maurizio Ferrari

Contents

About the editors

Giancarlo C. Righini is former director of the Enrico Fermi Center and of the National Department on Materials and Devices, National Research Council of Italy (CNR). He was also research director at the Nello Carrara Institute of Applied Physics, CNR, and vice-president of the International Commission for Optics. His research interests concern fiber and integrated optics, glass materials, and micro-resonators. He has published over 500 research papers. He is fellow of OSA, SPIE and Italian Physical Society (SIF), and also founding member and fellow of European Optical Society (EOS) and Italian Society of Optics and Photonics (SIOF).

Maurizio Ferrari is the director of research with the Institute for Photonics and Nanotechnologies, CNR, Italy, where he is also head of the CSMFO – Caratterizzazione e Sviluppo di Materiali per la Fotonica e Optoelettronica Lab and the Institute for Photonics and Nanotechnologies (IFN-CNR) Trento unit. He is coauthor of more than 400 publications in international journals, of several book chapters, and he is involved in numerous national and international projects concerning glass photonics. He is a OSA and SPIE fellow.

Preface

The birth of integrated optics is commonly associated with the publication, in 1969, of the article by Stewart E. Miller titled "Integrated Optics: An Introduction". In reality, as discussed in more detail in Chapter 1, Volume 1, of this book, the confinement of light in thin-film structures had been experimentally demonstrated in earlier works, in 1963, and its potential for new optical devices and applications had already been imagined. It was in 1968, for instance, that Shubert and Harris brought up the idea of realizing an integrated data processor based on two-dimensional refractive lenses and thin-film modulators.

Whatever it be, it is sure that the 1960s have represented a decade of extraordinary advances for optics, where the fundaments of the modern photonics were laid. Begun with the invention of laser, those years saw the birth of semiconductor laser industry, the seeds of the fiber optic telecommunications, and the first envisioning of compact, lightweight multifunctional guided-wave optical devices.

A long way has been travelled since those early works, and great advances have been made in both the materials and the technologies for integrated optics. As an example, let us just refer to the case of glassy materials: since the very beginning, low-loss optical waveguides have been produced by using oxide glasses. Commercial optical glasses, e.g., borosilicates, were first exploited using both deposition processes (mostly, RF-sputtering) and diffusion processes, either at low energy (ion exchange) or at high energy (ion implantation). Even soda–lime glasses (commercial microscope slides) were frequently used in the laboratories for the preliminary tests. Such waveguides were purely passive, so the researchers started to search for a proper approach permitting to add functionalities to glass. Excellent results were obtained, for instance, with the doping of glasses with semiconductor particles, to enhance nonlinear properties, or with rare earth ions, leading to the development of integrated optical amplifiers and lasers. By the way, the laser cavity could be produced by direct UV writing of gratings in glasses containing photorefractive oxides, like germanium or tin oxides. The subsequent efforts to optimize the rare-earth-doped glasses and the devices based on them have led to the use of transparent glass-ceramics, which combine the properties of the homogeneous matrix with those of the nanocrystals embedded in it. Then, after consideration that the transmission window of oxide glasses was not suitable for important applications in the mid-infrared wavelength range, the research moved toward non-oxide glasses, and chalcogenide glass thin films have proven to be very good for integrated optical devices operating up to 10 μm. More recently, the same chalcogenide glasses have been deposited onto plastic substrates, demonstrating the potential of flexible and stretchable glass-integrated photonic devices. At the same time, also in response to requests by the microelectronics industry, major glass companies have developed ultrathin flexible glasses, thus opening the way to a monolithic glass flexible photonics.

Analogous developments have occurred for other materials and technologies, so that nowadays various material platforms (some of them are fully compatible with complementary metal-oxide semiconductor (CMOS) fabrication processes) are available for the fabrication of on-chip photonic devices.

Due to these exceptional advances in integrated optics, we considered that it could be worth to collect in a new book some novel scientific contributions from both top and emerging researchers, with the aim of—on one side—illustrating some fundaments to newcomers in this field and—on the other—presenting to the experts a state of the art and some recent achievements. The vastity of the field, of course, prevented us from providing an exhaustive review; a hard decision we had to take, for instance, was to leave out details of semiconductor integrated optics. Besides this, the collected material was massive, and the publisher considered more convenient to split the 25 chapters of the book into two volumes.

Volume 1 begins with a short summary of the chronological development of integrated optics in the last 50 years. Then, since the development of high-performance integrated optical circuits providing complex functionalities requires an accurate modeling of their components, the following three chapters discuss numerical and analytical tools to model waveguides, devices, and nanophotonic metasurfaces, with particular attention to nonlinear effects.

The second section takes most of Volume 1, discussing material platforms and fabrication techniques. Two introductive chapters (Chapters 7 and 8) present the fundaments of thin-film deposition methods, by a physical and a chemical route, respectively. Five chapters review the fabrication methods and the applications in integrated optics of different materials, rare-earth-doped glasses (Chapter 5), lithium niobate (Chapter 6), photorefractive glasses, crystals, and polymers (Chapter 9), liquid crystals (Chapter 10), and silicon nitride (Chapter 11). Fabrication of channel waveguides and integrated optical structures is then analyzed in detail in two more chapters, dealing with femtosecond laser writing and with ion beam techniques (Chapters 12 and 13, respectively).

Volume 2 begins with a section devoted to optical (Chapter 1) and structural (Chapter 2) characterization techniques of materials and surfaces; Chapter 3 outlines the novel information that can be obtained by performing spectroscopic measurements in the THz spectral domain, a crucial step to acquire the knowledge of material properties at these frequencies, in view of a future THz photonics.

Finally, the second and last section of Volume 2 includes a group of chapters more related to guided-wave structures, devices, and applications. It begins with a discussion of plasmonic nanostructures and waveguides (Chapter 4) and continues with a review of the growth or deposition processes of dielectric crystalline thin films; the liquid-phase epitaxy process of rare-earth-doped fluoride and oxide crystalline films is described in detail, together with their applications, especially in the laser field (Chapter 5).

The two chapters that follow deal with different aspects of whispering gallery mode microcavities, namely, with the integration of open-ring microresonators and with the use of microspherical resonators for sensing (Chapters 6 and 7, respectively).

The applications are at the core of the last five chapters: nonlinear phenomena in proton-exchanged lithium niobate waveguides and their applications to classical and quantum optics are described in Chapter 8. Technologies and modeling challenges of next-generation long-wavelength infrared detector arrays are the subject of Chapter 9, whereas Chapter 10 presents an in-depth analysis of arrayed waveguide gratings for telecom and spectroscopic applications. The status of integrated quantum photonics, namely, of integrated systems where the photons themselves act as carriers of quantum information, is reviewed in Chapter 11; so far, most quantum photonics has not yet achieved monolithic integration, but much work is being done and progress is fast.

The conclusive chapter (Chapter 12) in some way goes back to the fundaments of integrated optics and to the need of designing and optimizing each component of the circuit: with today's complex functionalities required, genetic algorithms and neural networks may become necessary. As an example, an optical reservoir computer is presented, which is used as a programmable optical signal information processing device for a sensing application.

We are aware that some excellent books on integrated optics are available, but none very recent; so, we sincerely hope that the readers will appreciate our effort to provide an updated overview of integrated optics. We apologize for the incompleteness, even if almost unavoidable when dealing with a so-wide R&D area, and for any omission, possible but certainly unintentional.

Last but not least, we want to thank all the authors of the chapters for their excellent work and precious collaboration. We also wish to remember Marc De Micheli, who passed away abruptly in July 2019, when still working on his contribution, and to thank his friend and colleague Pascal Baldi, who completed the chapter on nonlinear integrated optics in lithium niobate. Finally, thanks are due to the editorial staff of The IET, especially Olivia Wilkins and Sarah Lynch, for their full support.

Giancarlo C. Righini
Maurizio Ferrari

Chapter 1

1969–2019: 50 years of integrated optics

Giancarlo C. Righini[1] *and Stefano Pelli*[1]

There is no unique definition of integrated optics. Usually, the term *integrated optics* describes a technology which permits one to construct optical devices or circuits constituted by several optical components which are connected by optical waveguides and able to perform some more or less complex functions. In a broader sense, integrated optics includes the whole research area aimed at exploiting guided-wave techniques (excluding optical fibres) to design and fabricate novel or advanced optical devices. What is generally agreed is that the birth of integrated optics is conventionally associated with the publication, in 1969, of the paper by Miller titled 'Integrated Optics: An Introduction' [1], even if a few earlier papers had already described the confinement of light in thin-film structures and considered its potential for new optical devices. A long journey has been made since then, which has led to the development of new materials, fabrication techniques and integration platforms; in parallel, new applications have been conceived and demonstrated.

The other chapters in this book are providing a detailed presentation of the fundamental principles and the recent advances in fabrication techniques, optical materials and devices, besides illustrating some of the many applications. The present chapter, therefore, does not aim at recalling the continuous progress occurred in the field of integrated optics in the past 50 years but focuses on the beginning and the (current) end of the journey, namely on the early times of optical waveguides and on some appealing novelties of the last years.

1.1 1969: the birthyear of integrated optics

The publication of the September 1969 issue of *The Bell System Technical Journal (BSTJ)* was crucial to launch a new field of research in optics. In fact, even if some work on optical waveguides had already been done in the previous years, that issue presented an organic body of articles, with the vision presented by Miller [1] and the theoretical analyses by Marcatili, Miller and Goell [2–5]. It represented a novel starting point for several researchers in optics worldwide.

[1]Nello Carrara Institute of Applied Physics, National Research Council of Italy (IFAC CNR), Firenze, Italy

As a curiosity, and to frame this great step forward in a more general technological context, one can remember that 1969 was also the year of the first man landed on the Moon on Apollo 11 mission (July 20–21) and of the first-ever communication between two computers. The latter event occurred in California on October 29, through ARPANET (Advanced Research Projects Agency Network). Remaining at Bell Labs, other two notable advances were made in that same year, in computer science and optoelectronics, respectively. A Bell Labs team, including Ken Thompson and Dennis Ritchie, who had been working as a partner of the MULTICS (Multiplexed Information and Computing Service) project and was trying to find an alternative to it, in August 1969 implemented a self-hosting computer operating system which later became UNIX (the name was actually given well into the 1970s) [6]. The exceptional achievement in optoelectronics was due to the pioneering work done at Bell Labs by George Smith and Willard Boyle, who invented the charge-coupled device in 1969 (even if the first description of the device and its experimental verification were published in April 1970) [7,8]. Forty years later, that invention earned them the 2009 Nobel Prize in Physics.

1.1.1 What there was before?

The search for an effective mean of transmitting information over long distances without being hampered by the environmental conditions, first of all by rain and water vapour, had started well before the discovery of laser. Thus, in the 1950s there was an emerging interest in looking at the possibility of using a microwave waveguide system for long-haul communication in place of the coaxial cable or radio relay systems used at that time. An early paper by Barlow [9] already discussed the potential of microwave waveguides for multichannel communication; to quote his 1947 statement, 'in this matter ... we are taking a peep into the future, but the project is well worthy of close attention'. Thus, a few years later, there was quite a general consensus that the guided propagation inside metallic pipes could provide a means to obtain a tremendous capacity of radio and TV channels. Two papers, among several others, may be chosen as a representative of that field of research: 'Waveguide as a Communication Medium' by Miller [10] and 'Characteristics of Waveguides for Long-Distance Transmission' by Karbowiak and Solymar [11]. In his detailed article, which included transmission experiments at 9 GHz in a copper pipe of inner diameter 4.73″ ($\approx$12 cm) and 500-ft long ($\approx$152 m), Miller concluded that the combination of the solid pipe with mode filters was attractive as a communication medium [10]. It was also envisaged that some signal regeneration would be necessary at repeaters spaced by some 25 mi ($\approx$40 km). A few years later, in 1961, Karbowiak and Solymar also considered special waveguides, namely dielectric-coated and helical waveguides, and ring or disc structures, and concluded that using pulse code modulation and regenerative repeaters, spaced from 10 up to 40 mi ($\approx$16–64 km), a very advanced system 'of unequally capacity', carrying several hundreds of television channels, could be implemented [11]. It is interesting to note that the authors only had a doubt that such 'enormous' communication capacities would ever be required!

The first demonstration of a laser source in 1960 by Maiman [12] paved the way to the design and development of optical components performing the same functions as the then-available microwave components. Quite soon, it became clear that optical fibres had the capability of easily winning the competition with metal pipes for long-distance high-capacity communication. Analogously, the existing parallel-plate microwave waveguides found a corresponding optical component, and the very first published images of optical modes in a dielectric waveguide appeared in a short note by Kaplan in 1963 [13]. Starting from the formula expressing the minimum thickness of the slab symmetrical waveguide to support the propagation of the mode of order m

$$d \geq D_m = \frac{m}{\sqrt{k_1 - k_2}} \frac{\lambda}{2}, \tag{1.1}$$

a four-mode waveguide was implemented by using two optically flat BK-7 glass plates spaced 32-μm apart and 77-mm long, and by filling the inner space with a mixture of benzene and chlorobenzene. Transverse magnetic (TM) modes were excited by a He–Ne laser with polarization perpendicular to the slab plane, and their number was controlled through temperature control, acting on the refractive index of the mixture.

Thus, the year 1963 may perhaps be considered the real birthyear of integrated optics; as a matter of fact, a few more papers were published in that year concerning the presence of a dielectric waveguide effect in another important optoelectronic structure: the laser diode p–n junction [14–18]. The first contribution was published by Yariv *et al.* [14,15], who calculated that such a junction, being a 'sandwich' structure constituted by a core depletion layer with dielectric constant higher than in the p and n regions, could support both transverse electric (TE)-guided and TM-guided waves. Moreover, they produced an experimental confirmation [16] by photographing the front surface of a cooled forward-bias GaAs p–n junction: a bright region was visible, instead of the nearly uniform light expected in the absence of the waveguide effect. Waveguide trapping was also observed by Ashkin *et al.* [17] in a GaP junction at room temperature; they, however, found that guiding was distinguishable only when the junction had a reverse bias. Accordingly, they said that the result of Bond *et al.* [16], concerning GaAs junctions in forward bias, could be explained only by postulating a much larger trapping angle than 'anything observed' in GaP. Numerical calculations of the propagation constants in an active dielectric slab were performed by Hall *et al.* [18], comparing the results with the parameters of GaAs laser diodes.

The experimental work reported by Kaplan [13] had been carried out at Wheeler Laboratories. These Labs, located in Great Neck, NY, had been founded in 1946 by Wheeler to develop microwave circuits and antennas for missile systems tracking and guidance radar. There, from 1964 to 1967, a pioneering activity, also thanks to a NASA contract, was performed on guided-wave optical components. The activity led to the demonstration of various types of slab waveguides, such as a structure with glass core in liquid cladding, one with liquid core in between two polished glass plates, and an all-solid structure with core and cladding glasses [19–21]. Figure 1.1 shows the images, taken at a microscope focused on the exit edge, of the modes supported by solid-core liquid-cladding waveguides constituted by fused quartz plates immersed in cineole ($C_{10}H_{18}O$; n_D = 1.457@20°C). The photo on the left shows the single mode (TE_0) of an Ultrasil quartz plate 0.06-mm thick, whereas on the right, the photo of the TE_{03} modes supported by a fused quartz plate 0.13-mm thick, polished on purpose by Dell Optics Co.

In the same period, namely January to November 1964 [19], liquid-core, solid-cladding waveguides were also implemented. Figure 1.2 shows the experimental apparatus to test these dielectric-slab waveguides, consisting of two 77-mm plates of Schlieren quality borosilicate crown glass ($k = 2.2952$), suspended in chlorobenzene; the inside surfaces of the plates were optically flat. The spacing between the plates was adjusted by means of a differential screw, in the range of ≈35–47 μm. The refractive index of the liquid core could be adjusted by varying the temperature. Much care was taken to guarantee the parallelism of the plates: after adjustment, the plates were parallel to within 0.01 mrad.

Various components, including directional couplers, modulators, laser and detector, were also studied and experimentally demonstrated. For instance, using nitrobenzene as the liquid core, experimental tests of amplitude, phase and polarization modulators were performed [19]. Later, the solid-core liquid-cladding configuration was used with the goal of demonstrating a waveguide laser [20]. In this case, the core consisted of a thin Corning-0580 Nd-glass, 77 mm × 25 mm × 0.048 mm, with both ends aluminized for 4% transmission, immersed in a temperature-controlled dichlorobenzene solution. A sketch of the experimental laser is shown in Figure 1.3. The waveguide and an FX-38A flash lamp were located at the conjugate foci of an elliptical reflector. Due to the low quality factor of this cavity, the predicted threshold energy was of the order of 60 J for single-mode operation; tests were made at lower pump energies to avoid damaging the Nd glass, and laser action was not observed [20].

The technique of proton irradiation to modify the refractive index of the glass was also proven at Wheeler Labs to represent an effective approach to the

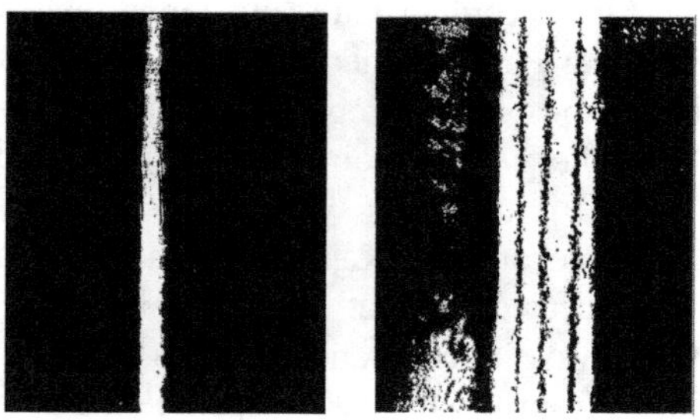

Figure 1.1 Photographs of some guided modes in an Ultrasil quartz plate (core) immersed in cineole (cladding): TE_0 mode (left) and TE_{03} (right). Adapted from Schineller et al. [19]

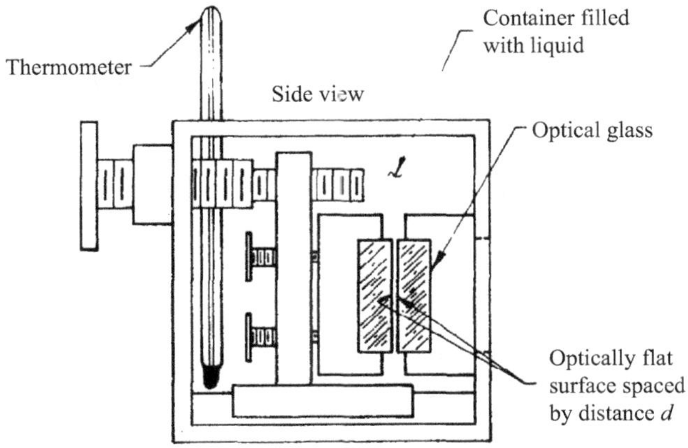

Figure 1.2 Sketch of the experimental apparatus to test liquid-core solid-cladding optical waveguides. The container, with glass windows, is filled with the liquid that acts as waveguide core. The thickness of the waveguide, i.e. the spacing between the two glass flats, is adjusted by the differential screw (one turn of the screw changes the spacing by 10 μm), whereas the refractive index of the core may be varied by changing the temperature of the liquid (the heather is not shown in the figure). The smaller differential screws are used to adjust the parallelism of the cladding glass plates. Reproduced from Schineller et al. [19]

fabrication of optical waveguides, with high potential for the implementation of waveguides with a complex shape and of large arrays of waveguides [20–23].

The following years saw the development of analytical and numerical tools to describe the optical guided propagation in thin-film structures [24–26]. It is worthwhile to highlight the fact that Osterberg and Smith [25,26] introduced the use of prism couplers to couple the light from a source into and out a glass thin film, either homogeneous or inhomogeneous (i.e. having a higher refractive index close to the surface than in the bulk glass; the authors referred to Pilkington float glass and found that the surface in contact with molten tin had a refractive index higher than the opposite surface, with a difference in the range of 0.0003–0.002). It was necessary, however, to wait a few years more to find a correct and complete analysis, with experimental proofs, of the mechanism of prism coupling in/out a thin-film waveguide [27–30].

To complete this brief overview of the activities preceding the fabulous year 1969, it is necessary to draw the attention to two papers published in 1966 and 1968 by Anderson *et al.* [31,32] and especially to the paper by Shubert and Harris in 1968 [33]. The former papers were among the earliest to highlight the importance of the photolithographic process and thin-film techniques for the fabrication of dielectric waveguides. It can just be noted that the authors used the term 'infrared wavelengths' to refer to the band which is now currently defined 'near-infrared';

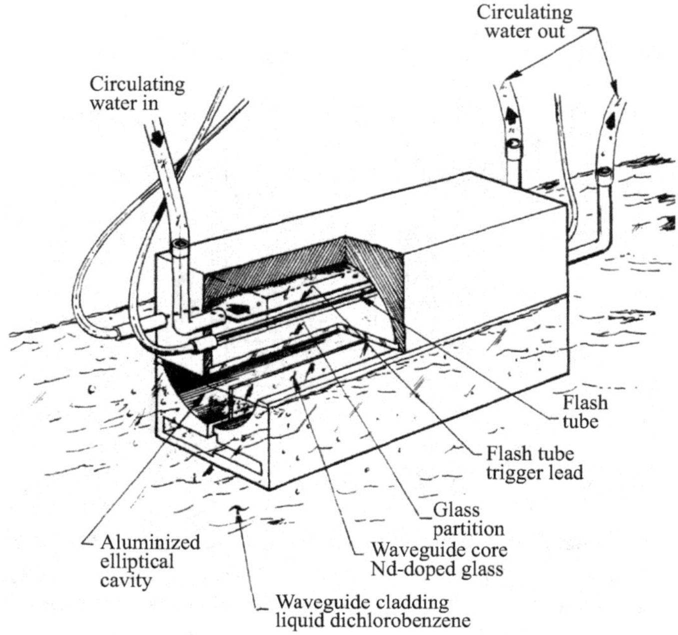

Figure 1.3 Sketch of the proposed solid-core liquid-cladding laser. The case is divided into two sections, separated by a glass window; water is circulated through the sealed upper half where the flashtube is located, whereas the open bottom half, containing the Nd-doped core, is immersed in a temperature-controlled container of dichlorobenzene. Reproduced from Schineller et al. [20]

they were in fact testing a GaAs slab waveguide with a He–Ne laser emitting at 1.15 μm [32]. In the latter paper, Shubert and Harris analysed the waveguide modes and the beam excitation mechanism in a thin film and achieved experimental confirmation of the guided modes in a structure consisting of a 1-μm (or thicker) epoxy core compressed between two glass flats, at the same time complaining that there was 'considerable difficulty in fabricating thin films that are adequate for *integrated optical* applications'. Besides using for the first time, at least at our knowledge, the term 'integrated optical' in the proper meaning of waveguide-based circuit, the real breakthrough of that paper was the very first vision of a full thin-film optical guided-wave device [33]. As a matter of fact, taking the inspiration from the concurrent development of coherent optical processing systems [34], supported by the availability of laser sources, Shubert and Harris suggested the possibility of implementing an integrated data processor utilizing two-dimensional (2D) refractive lenses and modulators constituted either by photochemical materials or electronically controllable devices. In addition to the design where the components, including the detector, were embedded in the guiding film, they also considered a partial integration, with the components embedded in the device's substrate. By the way, it can be noted that their vision motivated, in the 1970s and afterwards, several papers on the design and fabrication of waveguide lenses [35–48] as well as of integrated optical processors [49–52].

To give a more complete picture of the pioneering period of planar guided-wave optics, it may be worthy to add that very little was known in the Western countries in the 1960s–70s about the ongoing research in that area in the Soviet Union. Some information became available to a broad audience quite later [53], letting us know, for instance, about articles published in 1967 in Russian on the propagation properties of an anisotropic or asymmetric dielectric waveguide [54,55]. Only in 1991, a journal in English devoted to fibre and integrated optics appeared, published by the Academy of Sciences of USSR in cooperation with the Institute of Physics Publishing Ltd. (IoP), with the aim of providing 'Soviet scientists with extensive opportunities for making their work accessible abroad' [56]. Soviet Lightwave Communications had a good start [57], but unfortunately journal subscriptions soon collapsed, and in 1994 IoP ceased to publish it.

Situation was even worse as to the contemporary knowledge about Chinese research in this field. Due also to the Great Proletarian Cultural Revolution (1966–76), little activity was performed in Chinese Universities and very few publications (less than a dozen) on integrated optics appeared in that period, only written in Chinese. Among them, two review papers on integrated optics were published in 1973 and 1974 [58–60]. Nowadays, of course, situation has completely changed, and most of the top-level papers by Chinese authors are published in international journals, even if several domestic journals cover the optics- and photonics-related topics.

1.1.2 The activity at Bell Labs

Bell Labs, officially Bell Telephone Laboratories, Incorporated, were founded in 1925 and were primarily a development organization, even if very many innovations and discoveries must be ascribed to the physicists, material scientists, engineers in their research departments. Quite obviously, the long-distance transmission of information was a subject of interest there since the very beginning. As remembered by Kaminow in a review article [61], in the 1960s Stewart Miller was the director of a laboratory of Bell Labs at the Crawford Hill Lab in Holmdel, NJ, where research was continuing on millimetre waveguide communication systems, already proposed in the 1950s [10]. At the same time, in the Bell Labs in Murray Hill, NJ, some 40 km apart, a large effort was devoted to the development of the early lasers. Schawlow was a member of the technical staff, while Townes was a professor at Columbia University with a consulting contract to Bell Labs. As a result, in 1958 they published a fundamental paper on laser [62] and filed the US patent application 2,929,922 on 'Masers and Maser Communication Systems', which was assigned to Bell Labs in March 1960. Also Javan *et al.* were working in Murray Hill and at the beginning of 1961 they demonstrated the first gas laser, using an He–Ne mixture, capable of coherently emitting at some lines in the wavelength range from 1.1 to 1.2 μm [63]. The research on optical communications was also starting to be pushed ahead [64].

In the meantime, integrated electronics was moving very fast. The idea of how to make a whole circuit (transistors, diodes, resistors and capacitors) in a single silicon chip was conceived in 1958, independently, by Jack Kilby at Texas Instruments and Robert Noyce at Fairchild Electronics. In 1961, both the companies introduced their first commercial integrated circuits (ICs) which were followed in 1962 by the products of Signetics (later bought by Philips and incorporated in Philips Semiconductors) and Motorola. In Japan, the trial production of ICs started in the early 1960s, and by the mid-1960s the mass production begun, especially for application in computers. It was in 1965 that Gordon Moore, then director of R&D at Fairchild, predicted a doubling of the number of components per chip every year in the period 1965–75 [65]. Such prediction, later modified in doubling the number of transistors about every 2 years, was the basis of what is now known as Moore's law.

In such a scenario, it was to be expected that a strong attention would be addressed to the optical components which could be complementary to the fibre communication systems. The 1969 September issue of the *BSTJ*, collecting the papers by Miller *et al.* [1–5], was in a sense the focus of all the pioneering research on thin-film optical waveguides and was pivotal in bringing this new R&D area to front and in clearly outlining the potential of guided-wave planar components for the development of advanced, compact and rugged optical devices, thus opening the way to the wide area that is currently labelled as integrated photonics. Of course, not all the proposed ways of implementing integrated optical devices later proved to be feasible or convenient. As an example, referring to Miller's paper [1], while the proposals of an integrated resonator utilizing grating mirrors (Figure 1.4 (a)) and of a frequency selective filter based on a directional coupler (Figure 1.4(b)) have been widely converted into practice, the scheme suggested to realize an integrated laser based on Nd ions as active material and an electroluminescent material (such as doped zinc sulphide) as a pump (Figure 1.4(c)) was overcome by the later availability of laser diodes.

1.2 The first two decades: 1969–89

The availability of easy and low-cost fabrication processes is critical for the development and the success of a novel research area. Accordingly, the first years after the naming of integrated optics saw several experimental demonstrations of different waveguide fabrication technologies, most often trying to improve the results presented in the pioneer publications. The early focus was on glass materials, which had the advantage of full compatibility with optical fibres and provided an easy and low-cost tool to develop technological processes that later could also be adapted to other optical materials [66,67]. The reader is also referred to Chapter 5 for a close-up of a specific area of glass integrated optics, namely rare-earth-doped glass and glass-ceramic waveguides [68]. Let us give here some details on one of

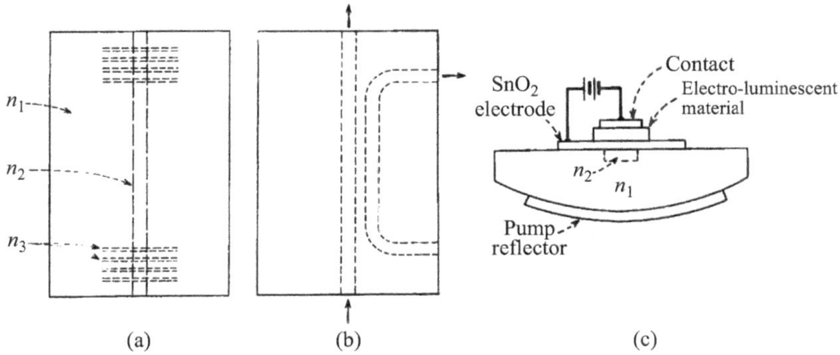

Figure 1.4 Some integrated optical (IO) elements suggested in Miller's paper [1]: (a) IO resonator made by a waveguide and two Bragg grating mirrors; (b) IO directional coupler and (c) possible schemes of an IO laser based on Nd-doped material, with an electroluminescent material pump and a spherical reflector. Adapted from Miller [1]. Courtesy of Nokia Corporation and AT&T Archives © 1969 Nokia Bell Labs. All Rights Reserved

the simplest and most effective fabrication techniques, namely ion-exchange, before briefly overviewing the development of the other materials and techniques.

1.2.1 Ion-exchange technique

One of the first areas explored for planar waveguide fabrication was that of ion-exchange in glass [69–72]. On one side, there had been the observation by Osterberg and Smith [26] of the guiding capability of the surface layer of floating glass, where the refractive index is increased due to the tin from the melt entering into the glass. On the other side, the chemical strengthening of glass through ion-exchange process [73], already known since the beginning of twentieth century, was attracting in the 1950s–60s a great attention for industrial processes, as witnessed by the number of patents filed [74]. Moreover, in 1968, Nippon Sheet Glass announced the discovery, in cooperation with NEC Corporation, of an innovative method, based on ion exchange, to vary in a parabolic shape the central and peripheral refractive indices of a glass fibre, with the aim of reducing the spreading of the envelope of a propagating optical pulse and thus increasing the fibre transmission capacity [75]. Ten more years were necessary, however, to introduce these fibres in the market under the product name of SELFOC® [76]. Almost in parallel, at Bell Labs, in 1969, Na^+–Li^+ exchange was used to produce a controlled change of refractive index in a rod glass, for potential low-resolution imaging devices [77].

It was, therefore, to be expected that soon the process would be applied to the routine fabrication of planar optical waveguides: in 1972, Izawa and Nagakome used electrically induced ion migration to diffuse Tl^+ ions into a borosilicate crown glass containing alkali ions K^+ and Na^+ [78]. By using a second diffusion step

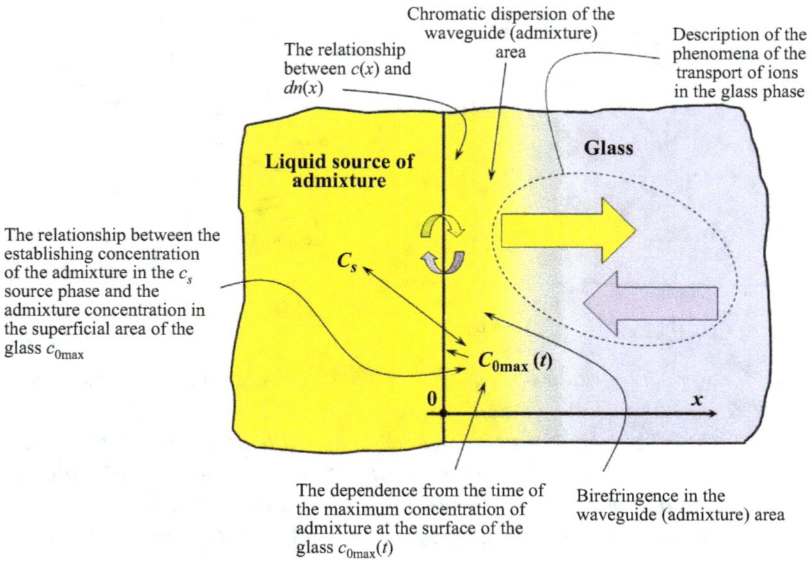

Figure 1.5 Sketch of the ion-exchange process at the interface between a glass and a liquid source of substituting ions. Reproduced from Rogoziński [72] under Creative Commons Attribution License

(exchanging K^+ and Na^+ for Tl^+), they produced buried waveguides, with propagation losses estimated 'sufficiently smaller' than 0.1 dB/cm. A purely thermal exchange process of K^+, Tl^+, Ag^+ and Na^+ was demonstrated in 1973 by Giallorenzi *et al.*, who also measured losses of less than 0.1 dB/cm in selected samples [79]. It may be noted that these low values of propagation losses at 0.6-µm wavelength in ion-exchanged waveguides remained unbeaten for long. One year later, in 1974, Saruwatari and Izawa demonstrated the first planar glass waveguide laser using Tl^+ exchange in an Nd-doped borosilicate glass. The guide was multimode, 4-mm long, and the laser had pulsed operation [80].

Figure 1.5 provides a graphical summary of the ion-exchange process for the fabrication of planar optical waveguides, showing (in section) the interface between the admixture (salt melt containing the ions that must enter the glass) and the glass containing the alkali ions to be substituted [72]. Assuming, for simplicity, that the process is binary, the admixture contains a single exchange ion A and the glass a single alkali ion B; the important parameters to describe the process are c_s, the concentration of A in the ion's source (admixture); $c_{0\max}(t)$, the concentration of A at the surface of glass (which must reach an equilibrium value) and $c_A(x)$, the concentration of A inside the glass (which also depends on time). If we also assume that in the process a number density ΔN of ions B are substituted with an identical ΔN of ions A, by using the Lorenz–Lorenz equation, one finds that the maximum change $\Delta n_0 = n_0 - n_g$ of the refractive index (n_0 and n_g are the refractive index at the exchanged glass surface and the refractive index of the glass substrate, respectively) is linearly proportional to ΔN, according to the expression [71]:

$$\Delta n = \frac{2\pi\left(n_g^2 + 2\right)}{3n_g}(\alpha_A - \alpha_B)\Delta N, \tag{1.2}$$

where α_A and α_B are the ionic polarizability volumes. In practice, the maximum index changes achievable when substituting the Na^+ ions in the glass substrate are in the range of 0.01–0.2 (see Table 1.1). The A ions are generally supplied as nitrates, and the table indicates the melting temperature T_{min} of the nitrate and the maximum temperature T_{max}, where chemical stability degrades, and decomposition reactions may occur. When using nitrates, however, it is suggested to use temperatures lower than 400°C, to avoid chemical attack of the glass surface by the NO_3 anions at higher temperatures. It is also possible to use eutectic mixtures of different salts to work at a lower melting temperature. An example is the exchange of copper ions from $CuSO_4$:Na_2SO_4 and $CuSO_4$:K_2SO_4 eutectic baths at temperatures in the range of 530°C–585°C [54], whereas the melting temperatures of the single salts are 1,069°C, 884°C and 110°C for potassium, calcium and copper sulphate, respectively. A rich bibliography on ion exchange in glass, with focus on copper exchange, is present in a PhD thesis [81].

The major advantage of ion exchange is the simplicity of the technique, requiring only a furnace, a container of the nitrate salt to be melt and a proper holder of the sample. Figure 1.6 shows a sketch of the apparatus for thermal ion exchange, with a glass sample holder; in alternative, stainless steel holders are frequently used. The constancy of the melt temperature must be controlled by a thermocouple [72]. The apparatus becomes slightly more complex when one wants to apply an electric field to the melt/substrate system to exploit field-assisted ion exchange. In cases of Ag^+–Na^+ or Cu^+–Na^+, in particular, an alternative to the melt, which may also be aggressive with respect to the glass surface, is to use a

Table 1.1 Most commonly used ions for waveguide fabrication in glass, ordered according to the increasing value of the refractive index change

Ion	Ionic radius (Å)	Nitrate – T_{min} (°C)	T_{max} (°C)	Δn_0 (s)
Na^+	0.95	307	380	–
K^+	1.33	334	410	0.01
Li^+	0.65	264	600	0.012
Rb^+	1.49	310	370	0.015–0.02
Cu^+	0.96	430*	–	0.015–0.06
Cs^+	1.65	414	510	0.04
Ag^+	1.26	212	444	0.1
Tl^+	1.49	206	430	0.1–0.2

Reference is made to ion exchange from a nitrate salt melt, except Cu^+: for copper, CuCl (*) is often used [81].

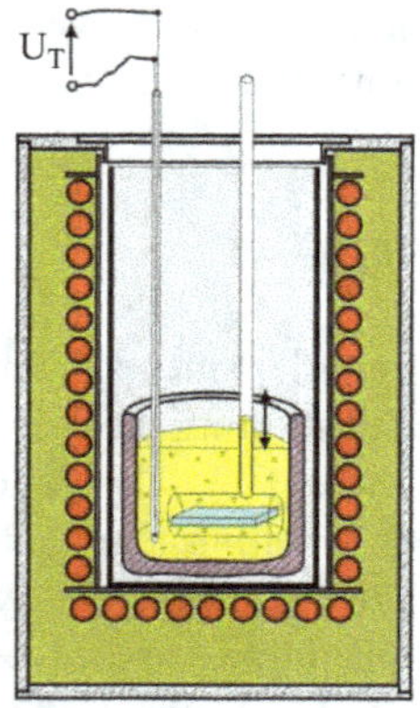

Figure 1.6 Sketch of the cross section of a simple electric furnace for thermal ion exchange. The molten salt is in the crucible, made in glass, ceramic or stainless steel. In the figure, the glass substrate is placed into a glass holder. To monitor the temperature, a thermocouple inside a silica glass tube is immersed in the crucible. During the process, the thermocouple voltage U_T is continuously recorded. Reproduced from [72] under Creative Commons Attribution License

metal film as a source of substituting ions. The Ag or Cu film is deposited onto the glass: by heating and applying a voltage across the glass, the film oxidizes and the so-created Ag^+ or Cu^+ ions migrate into the glass. In this case, there is a further advantage: the creation of channel waveguides is simple, and the process results more industrially compatible and suitable for mass production.

A couple of interesting reviews of the advances occurred in the first two decades of integrated optics concerning modelling and performing of the ion-exchange process, and its use for the design and fabrication of integrated optical components in glass, were published in the late 1980s [82–85].

1.2.2 Other materials and technologies

The use of ion implantation to create an optical waveguide in silica glass through the local modification of the refractive index had been suggested and tested by Schineller *et al.* [21–23]. Its application, however, appeared more interesting for the fabrication of guiding structures in semiconductor materials; thus, in the 1970s, there were papers reporting optical waveguides in GaAs [86,87], GaP [88], ZnTe [89], as well as a review assessing pros and cons of ion implantation, to conclude that this technology is readily compatible with integrated optics and may constitute a major tool in the fabrication of the new devices [90]. However, whereas the first demonstration of a simple device fabricated by proton implant in GaAs, namely a directional coupler, was published in 1973, it was necessary to wait until 1989 to see the realization of a planar waveguide laser in Nd-doped crystal [91]. A 6-mm deep waveguide was formed in a 10 mm $\times$ 5 mm Nd:YAG crystal, 1-mm thick, by

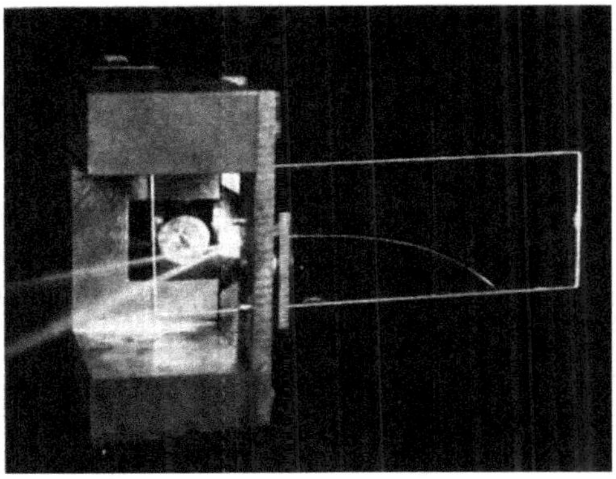

Figure 1.7 Photographic image of one of the first ridge glass waveguides produced by RF sputtering of a borosilicate glass. Reproduced from Goell and Standley [93]. Courtesy of Nokia Corporation and AT&T Archives © 1969 Nokia Bell Labs. All Rights Reserved

implanting He^+ ions at 77K, with ion energies up to 2.8 MeV at a total dose of 7×10^{16} ions/cm^2. The light confinement was due to the barrier, of lower refractive index, created by the ions at the end of their track. The laser cavity was obtained by butting two plane mirrors against the polished crystal end faces; an output power of 1 mW at 1.06 μm was achieved, with a low slope efficiency of about 1.7% respect to the absorbed power, which can be explained by the lack of structure's optimization [91]. An updated overview of the optical waveguides produced by ion beams is presented in Chapter 13 [92].

Another physical deposition method used since the very beginning to fabricate thin-film glass waveguides was radio-frequency (RF) sputtering. It was first suggested by Goell and Standley, from Bell Labs, who produced film waveguides with propagation loss below 1 dB/cm by sputtering barium borosilicate Corning 7059 glass [93]. They also realized ridge waveguides, about 0.3-μm thick and 20-μm wide, from the same films by using backs puttering with a quartz fibre as shadow mask: a photo of a curved waveguide is shown in Figure 1.7 [93]. Since then, RF sputtering was rapidly adopted in several research labs [94–101] and has always been widely used, also thanks to its capability of adapting to different materials and of tailoring the optical properties of the deposited film by changing the gas in the vacuum chamber [100–102].

Among the earlier integrated optical components fabricated by RF sputtering, one can mention waveguide branches [103], grating filters [104,105], an integrated electro-optical modulator fabricated by sputtering a ferroelectric $LiTaO_3$ film on top of a 7059-glass substrate [106] and waveguide amplifiers [107,108]. The range of materials of interest for integrated optics and deposited by RF sputtering soon broadened, including single crystal and epitaxial films [109–113]. More details on the different sputtering processes, namely direct current (DC), RF and magnetron, as well as a recap on other physical deposition processes, such as vacuum evaporation, pulsed laser deposition and molecular beam epitaxy, are given in Chapter 7 [114].

Looking at the references given so far, it is quite clear that the first significant efforts for the development of the new field of integrated optics were done in the United States, but very soon there were growing activities worldwide. Additionally, in 1974, it was already possible to have a good idea of what was going on in Europe and Japan: the French company Thomson-CSF decided to devote an issue of its technical journal to optical communications, thus testifying the existing strong industrial interest for the overall field of fibre and integrated optics [115]. Almost at the same time, Suematsu published an interesting review on the progress of integrated optics in Japan [116]. In the same period, in USSR, too, there were many ongoing researches on the fabrication of optical waveguides, even with original approaches [117], on corrugated structures in thin-film waveguides [118,119] and on non-linear effects in guided-wave structures [120]. A review on optical phenomena in thin-film waveguides was also published in 1974 [121].

A critical issue for the operation of all the optical waveguides is the coupling of light from the source and to another guide or a detector. Although it is clear that the only rugged and efficient solution for hybrid integrated optical circuits and devices is constituted by butt coupling to the laser diode or detector through the polished end facets, it is also true that sometimes the source may be an external laser (e.g. in the characterization tests in the laboratory). In the latter case, prism coupling is very effective [27–30]; several papers, however, have investigated the characteristics of gratings in thin films and their use as in/out couplers [104,105,118,119,122–124]. A more complex solution is sketched in Figure 1.8, where output coupling is achieved by the use of a random coupling region in conjunction with a holographic element; this technique permits highly selective coupling, with more than 30 dB discrimination against unwanted modes [124].

It should not be forgotten that in the 1980s, another technology emerged for the fabrication of glass planar optical waveguides, namely sol–gel. Although sol–gel, as an alternative method for the preparation of glasses and ceramics at considerably lower temperatures than conventional melting process, had been known since the late nineteenth century, it was only at the beginning of the 1980s [125], also with the organization in Padova, Italy, of the first International Workshop on Glasses and Glass Ceramics from Gels [126], that the actual modern development of sol–gel science begun [127]. Soon, sol–gel thin films were tested as optical waveguides, and silica–titania glass films appeared of particular interest, due to the possibility of tuning their refractive index by changing the relative content of silicon dioxide and titanium dioxide [128–135]. Among the first devices fabricated in silica–titania sol–gel films, we can recall the sensors developed by the group of Lukosz exploiting the properties of waveguide gratings embossed in the films [128]. The sol–gel films can also be easily doped with semiconductor nanoparticles [132] for non-linear functionality, or with dyes [133] and rare earths [134,135] for optical amplification and lasing.

From all these publications, it clearly emerges that, after a first phase where passive waveguides and components were developing, the interest was moving to

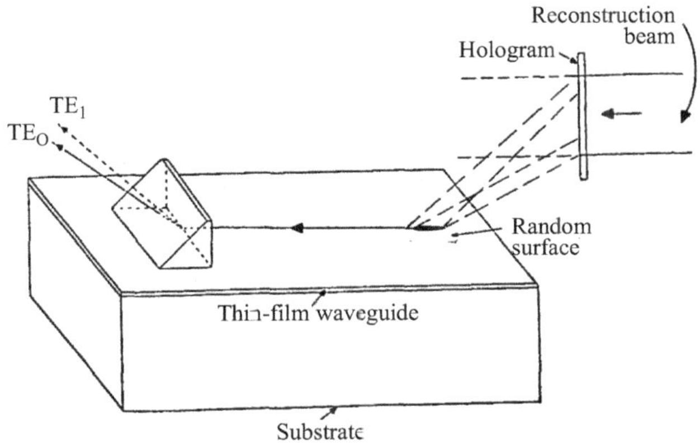

Figure 1.8 Representation of a holographic input coupler, capable of selective mode excitation. The modes are coupled out of the waveguide by a conventional prism coupler. Reproduced, with permission, from Reference [124]

electrically controllable devices, such as optical modulators, and active devices, i.e. optical amplifiers and lasers. Correspondingly, the focus of material scientists was moving from amorphous materials, like silica, multicomponent glass and silicon nitride, to ferroelectric crystals and semiconductors, with the corresponding fabrication technologies. Thus, the Czochralski growth of the crystals and Ti:indiffusion, Li-outdiffusion and proton exchange to produce higher index layers were the fundamental processes for the fabrication of waveguides and modulators in lithium niobate and lithium tantalate, even if other approaches were pursued, too, and early results were already published in the early mid-1970s [136–143]. Figure 1.9 shows the photo of a phase modulator realized in Li-outdiffused $LiNbO_3$ and the cross section of the device: the waveguide is 10-mm long and 100-μm wide, and the distance of electrodes is the same as the guide width, i.e. 100 μm. Due to the poor overlap of the modulating field and the guided mode (see Figure 1.9(b)), the performance of the modulator was not very good: the power needed was 0.3 mW/MHz/rad for a bandwidth of 2 GHz [139].

A review on waveguides and micro-waveguides in lithium niobate was published in a recent book [144], where one can also find a more general review on optical waveguides [145]. A deeper discussion of integrated optics in lithium niobate is provided in Chapter 6 [146], whereas non-linear phenomena and devices in lithium niobate are discussed in Chapter 8, Volume 2 [147].

Remaining in the area of crystalline materials, it is quite obvious that III–V semiconductors, and to a less extent II–VI and IV–VI semiconductors, have always played a fundamental role in the development of integrated optics, where the laser source is, in most cases, constituted by a laser diode and the monolithic integration (MI) of the source with the passive or tuneable circuit components has always been a priority target [148–158].

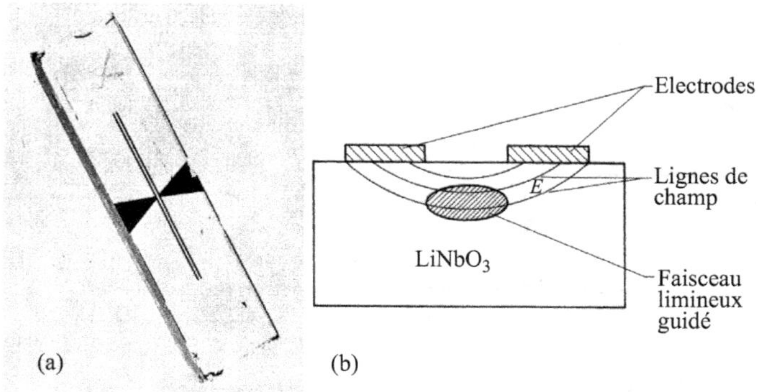

Figure 1.9 Waveguide phase modulator fabricated on a $LiNbO_3$ substrate: (a) photo of the device, where the electrodes are clearly visible and (b) cross section of the modulator, showing the overlap of the electric field between the electrodes and the buried Li-outdiffused waveguide. Adapted from Papuchon [139]

This brief description of the fundamental materials and processes whose development started in the 1970s should have included many more items, both on the materials' side (e.g. silicon nitride, polymeric materials, liquid crystals, semiconductor-doped and rare-earth-doped glasses) and on the fabrication technology side (e.g. plasma-enhanced chemical vapour deposition, flame hydrolysis deposition, spray pyrolysis, sol–gel deposition and ultraviolet or femtosecond (fs)-laser writing). Due to the limited space, however, and to avoid excessive overlap, the interested reader is referred to Chapter 8 for details on chemical deposition techniques [159], Chapter 9 for photorefractive waveguides [160], Chapter 10 for the applications of liquid crystals in integrated optics [161], Chapter 11 for silicon-nitride-integrated optics [162] and Chapter 12 for fs-laser writing of integrated optical structures in glass [163]. Many books and conference proceedings may provide a broad overview on advanced materials and technologies at the very beginning or in the first decade of the 2000s, and only very few can be mentioned here [164–168].

1.3 Fifty years later: 2019–20

Many advances in materials and technologies have occurred in the 50 years past the article by Miller [1], but it is a bit hard to classify one of them as really disruptive. Most of the advances in integrated optics, or integrated photonics, as it is more common to label this field of R&D in the current millennium, were due to improvements of what already existed, rather than to new discoveries. A big step forwards, however, has been made in the fabrication platforms, based on different materials: high-index glasses, polymers, silicon nitride, lithium niobate and semiconductors. The most spectacular advance and an effective breakthrough has been made in InP and silicon photonics: InP technology allows us to integrate all the photonic components, including lasers, optical amplifiers and high-performance modulators [150,154,155,169–171], whereas Si photonics lacks two crucial building blocks, lasers and optical amplifiers. This disadvantage, however, is largely compensated by the possibility of using Complementary metal–oxide–semiconductor (CMOS)-like fabrication processes, which allow us to achieve a dense integration of photonic components, hence unprecedented efficiency and functionality. The capability of producing large volumes of Si-based photonic IC (PICs) at low cost has made photonics competitive with electronics, both in performance and cost [172–175]. For more details, an updated overview of silicon photonics is available in two very recent books [176,177].

1.3.1 Recent issues in silicon photonics and photonic–electronic integration

It is the high refractive index contrast permitted by silicon technology that has allowed for a submicrometric size of waveguides and tight bends, leading to the dense packing of optical functions in a chip. The design of a PIC may therefore be quite complex, and the use of CMOS process has made the design of PICs to evolve in a similar way to electronics, through the use of process design kits (PDKs), namely of information packages containing sufficient information (e.g. geometries and models) to design a chip that can effectively be fabricated in a foundry. Nowadays, PICs, as the electronic ICs, are mostly manufactured in a small number of manufacturing and prototyping facilities ('foundries' or 'fabs'), serving a much larger community of designers. Figure 1.10 illustrates the design flow when using a PDK as the main interface between the designer and the foundry [173].

A PDK for PICs is very similar to those for traditional CMOS, which contain information on the fabrication process, design rules, mask layers, models for simulation and predefined components. How to move from a photonic–electronic logical circuit to the real device depends very much on the fabrication technology used to combine the electronic and photonic circuit elements. Figure 1.11 presents different integration methods [173]. The last solution (Figure 1.11(e)), i.e. the MI, where photonic and electronic components lie side-by-side on a same chip, appears the most attractive and simplifies the design process. MI, however, still implies a number of boundary conditions and challenges, such as the thermal management, which may have different impact on the performance of photonic or electronic components. On top of the chart, the interconnection by wire bonding (WB – Figure 1.11(a)) would reduce the complexity of thermal management and of off-chip connections but would be the worst solution with regard to interconnect electrical parasitics, connection density and reusability of photonic–electronic cells.

Another approach, with higher connection density and lower design complexity, is represented by the heterogeneous 2.5D technology (more properly called interposer technology), where a large Si bare die (the interposer) is designed specifically to accommodate smaller dies: each of them may be a standard 2D die, one

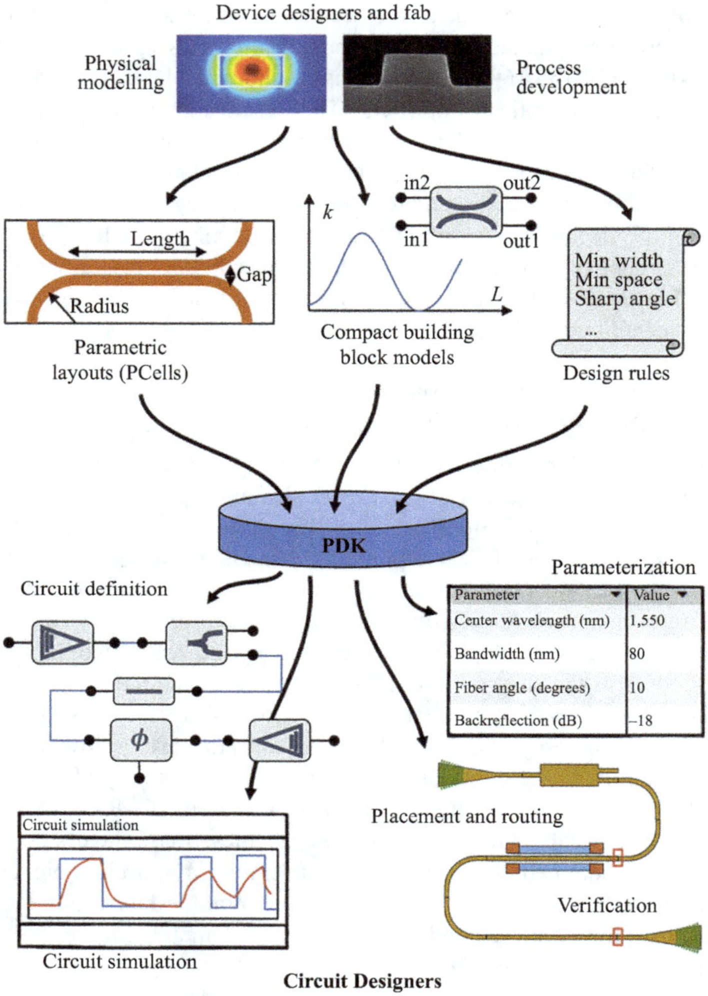

Figure 1.10 Sketch of the design flow of a PIC. The PDK must contain the descriptions of the building blocks (layout and circuit models) and the design rules of the fabrication process. It constitutes the interface between the circuit designer and the device designer and the foundry ('fab'). Reproduced, with permission, from Reference [173]

custom-built for 2.5D assembly, or a 3D IC (IP – Figure 1.11(b)). Finally, with intermediate characteristics, one can use flip-chip integration through micropillars or microbumps (FC – Figure 1.11(c)), or 3D stacking (3D – Figure 1.11(d)); in these two cases, the dense electrical interconnects between PIC and IC can give rise to electrical crosstalk, and thermal effects from the IC can significantly affect the PIC performance.

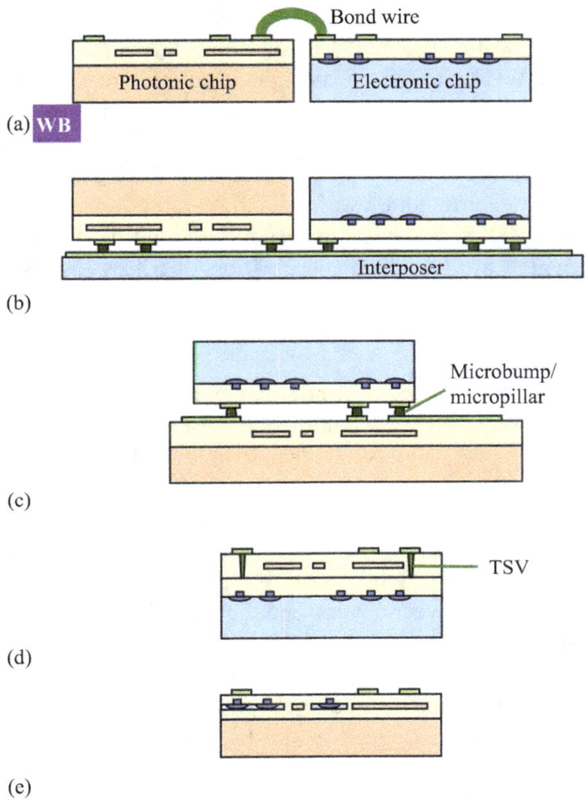

Figure 1.11 Different schemes for photonic–electronic integration: (a) WB – side-by-side wire bonding, (b) IP – 2.5D integration on interposer, (c) FC – 3D flip-chip/micropillar integration, (d) 3D – 3D stacking with through-silicon vias and (e) MI – monolithic front-end integration. Reproduced, with permission, from Reference [173]

On the photonic side, issues to be faced concern the modal dispersion and nonlinearity. Submicrometric Si waveguides usually support two guided modes, for the quasi-TE polarization and for the quasi-TM polarization, which have propagation properties very different and very wavelength dependent. Moreover, due to the strong confinement of light, the optical power densities may be so high that nonlinear effects are no longer negligible, possibly implying signal cross-coupling, signal distortion and spectral broadening.

1.3.2 Towards full 3D integrated photonics

With the increasing complexity and device density of PICs, efficient waveguide crossing components (WGXs) are indispensable in many network topologies and have therefore became key building blocks, requiring the optimization of their performance. Staying in a planar layout, WGX designs based on adiabatic aperture widening are large, while resonant designs allow a compact footprint but have narrow bandwidth. WGXs based on multimode interference, even if with multimode behaviour, have a number of attractive features. A device of this type, fabricated in a CMOS-compatible process using 248 nm lithography, with only one patterning step, exhibited loss down to 0.18 ± 0.03 dB/crossing and crosstalk of -41 ± 2 dB, uniform across an 8″ wafer [178].

Ultralow-loss WGX arrays based on a periodic multimode structure have also been designed, which are based on the creation of a low-loss Bloch wave in a matched periodic structure [179]. A loss of 0.04 dB on average, equal to theoretical design efficiency, and crosstalk suppression over 35 dB in a CMOS-compatible geometry were demonstrated.

To make a further step, however, non-planar circuit topologies become inevitable; in fact, a large-scale PIC may comprise tens of thousands of crossings, so that even very low losses of the single WGX lead to an overall detrimental effect on the PIC performance. In fact, even if the in-plane WGX loss is about 15–40 mdB/crossing, the resulting total optical loss for a PIC can easily reach some 10 dB, together with possibly untenable levels of crosstalk. One approach is to develop a multilayer photonic platform, where closely spaced waveguide layers allow an efficient transfer of light between layers, and the most distant waveguide layers can be used for ultralow-loss waveguide crossings [180].

A recent proposal, which offers the capability of very low crosstalk (less than -75 dB) and to bridge series of parallel waveguides, is based on 3D-printed freeform polymer structures, realized in situ by direct-write laser lithography and two-photon polymerization [181]. Figure 1.12 shows the concept of the optical waveguide overpass (WOP) and details of a demonstrator device, fabricated on a silicon-on-insulator (SOI) wafer having a 220-nm-thick device and a 2-μm-thick buried oxide. The demonstrator included a 4×4 switch-and-select circuit that consists of four 1×4 switches at the input and four 4×1 switches at the output. The footprint of a single WOP amounts to approximately $15 \times 160\ \mu m^2$, including two 50-μm-long tapers for coupling the WOP to the SiP waveguides [181].

The move from in-plane (2D) integrated optics to 3D PICs has been made easier by the continuous evolution of the lithographic techniques and in particular by the excimer or fs-laser writing, either additive (mostly based on two-photon polymerization) or subtractive (laser ablation), besides fs-laser inducing of local structural modifications inside transparent materials. At this regard, an interesting paper demonstrated the capability of combining both two-photon polymerization and fs-laser ablation, thus improving the nanofabrication efficiency and enabling the fabrication of complex 3D micro-structures/nano-structures [182].

The target of a full 3D integrated photonic circuit, however, was already pursued at the end of the 1990s using polymers, reactive ion etching, shadow and grey-level photolithography [183], but new efforts have been made in the last decade, following two main approaches: multilayer structures [180,184,185] and laser micro/nanofabrication [182,186–189]. An interesting perspective was discussed in a recent paper [185], where it was shown that, by the use of thin-film technology

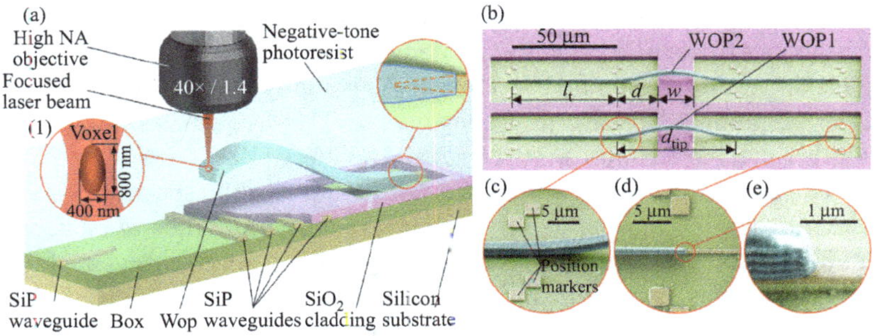

Figure 1.12 Concept and realization of waveguide overpasses (WOPs) in the Si-photonics platform: (a) a WOP is laser-written into a liquid negative photoresist deposited onto the PIC. The lithographic spatial resolution is determined by the size of the volumetric pixel (voxel) resulting from two-photon polymerization, shown in inset (1) and coupling tapers in the WOP and the waveguide are indicated in inset (2); (b) scanning electron microscope (SEM) image of the WOP (colours added by image processing), with close-ups of different parts of the WOP shown in images (c)–(e). Reproduced, with permission, from Reference [181]

and hybrid waveguide design, lithium niobate and silicon photonics could be integrated in future 3D microsystems, which could therefore exploit the superior electro-optical and non-linear optical properties of lithium niobate. The search for a fabrication technology enabling the incorporation of highly non-linear single crystals in amorphous materials has also led to appealing developments in glass-based photonic platforms. As an example, patterning of microscopic crystals inside glass by fs-laser irradiation was used by Zhu *et al.* to add non-linear properties to rare-earth-doped glasses [190], and by Stone *et al.* to grow uniform single crystals in an $LaBGeO_5$ glass and to write crystal-in-glass waveguides and Y-junctions [188].

Laser microfabrication, moreover, has the advantage of making possible the customization, almost on the fly, of any circuit structure, be it for testing different design approaches or to adapt any device to specific application needs. For such a purpose, recently, other fabrication approaches based on additive manufacturing (AM) have also been developed. Their limit, so far, is that devices fabricated this way are still intended to be used at infrared (IR), or even microwave wavelengths, for the obvious reason that in these spectral regions the printing resolution required is lower than in the visible or near-IR window and therefore more easily attainable. A further present limitation of this technology is related to the dimensions of printing area, usually smaller than 20 cm in diameter (8″ wafer), and even less when high resolution is required. Currently, however, there is a fast-developing trend; for instance, Weidenbach *et al.*, using a commercial 3D printer, fabricated a passive IO device in a TOPAS® polymer, operating at the low edge of the THz band and exhibiting losses of about 6.3 dB/m and 10-mm critical bending radius [191]. At higher frequencies, Yang *et al.* demonstrated losses as low as 0.9 dB/m at 0.75 THz in a VeroWhitePlus^{TM} polymer Kagome-type waveguide [192], and Pandey *et al.* produced plasmonic waveguides at THz frequencies, implementing, among other passive structures, 3D bends and splitters [193].

In the last years, experimental results in the visible spectrum were also achieved by additional manufacturing; as an example, Bollgruen *et al.* fabricated waveguide structures (about 145-µm wide) by ink-jet printing several layers of acrylate-based resins on polymer methacrylate, with losses were about 1.4 dB/cm at 785 nm [194], whereas Udofia and Zhou produced a series of waveguides (still rather wide, more than 154 µm) and passive devices (e.g. conformal splitters and combiners) with a custom system based on a soft and stretchable material (a polymer gel) [195]. Obviously, these structures are heavily multimodal, but in many ways this situation replicates what happened with the first optical waveguides, as recalled in Section 1.1.1.

1.3.3 Flexible photonics

The past decade has seen the development of flexible, conformable and wearable multifunctional electronic devices, which offer ruggedness, portability, light weight and low cost of production, making them more appealing, in some application areas, than conventional devices on rigid substrates. There has been, in particular, a growing adoption of these flexible devices for solar power harnessing, healthcare and consumer electronics applications. For more information on the materials and the fabrication technologies of flexible electronics, the interested reader is referred to some very recent books [196–199].

According to a recent report, the flexible electronics market is expected to grow from USD 23.92 billion in 2018 up to USD 40.37 billion by 2023, at a compound annual growth rate of 11% between 2018 and 2023 [200]. An example in front of every one of the spreading of flexible electronics is given by the number of foldable phones already on the market in January 2020 (e.g. Samsung Galaxy Z Flip, Motorola Razr, Samsung Galaxy Fold and Royole FlexPai), with more expected to come throughout the year. It is to be noted that the Samsung Galaxy Z Flip is the first to feature a foldable glass display.

Following this trend, a similar effort could be expected to develop flexible photonic devices. Organic materials would appear as the most convenient platform, due to their properties of mechanical flexibility, low cost and large-scale manufacturing; unfortunately, they also present some cons, such as recrystallization, thermal degradation and oxidation. Some excellent results in the so-called soft photonics have already been obtained; many optical polymers, in fact, exhibit low optical loss (down to 0.01 dB/cm in the 1.3–1.5-µm wavelength range) and can be easily processed. Polymer waveguides can be fabricated using conventional lithograph, direct writing, micro-moulding or imprint/embossing. The solution printing technique, in particular, has a high potential for the large-scale manufacture and integration of micro- and nano-devices. It was successfully applied to the

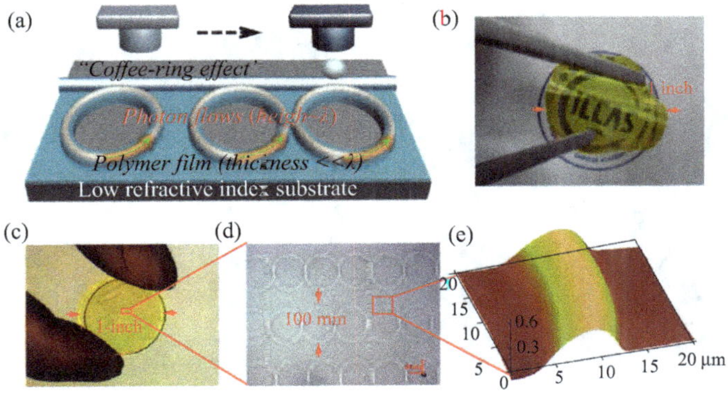

Figure 1.13 Design and fabrication method of a wafer-scale organic printed photonic chip: (a) a thin-film spin-coated from a polymer solution is locally dissolved by printing solvent droplets, which produce various microscale waveguide structures; (b) a free-standing flexible photonic chip (doped with a yellow dye) peeled off from the substrate; (c) image of large-scale optical structures on a 1″ wafer; (d) microscope image of printed microring chains, showing uniform size and well-defined pattern and (e) AFM image of a typical printed structure showing the smoothness and height with respect to the surrounding film. Reproduced from Zhang et al. [201] under Creative Commons License

fabrication of waveguides, microresonators, microlaser and more complex structures, such as an add-drop filter constituted by three input/output printed 1D waveguide ports and two microring processing modules [201]. Figure 1.13 summarizes this solution-processed fabrication technique for soft photonic circuits.

Inorganic materials, however, remain the long-term choice for the development of stable and high-performance devices; a first step along this direction was made with optoelectronic devices, such as flexible light-emitting diodes, photodetectors and solar cells [202]. Hybrid flexible circuits, namely composed by an inorganic photonic circuit onto an organic substrate, are feasible, and various solutions have already been proposed and demonstrated [203–211]. The transfer-and-bond method, for example, permits to transfer silicon photonics onto plastic substrates; as a proof of concept, Mach–Zehnder interferometers and microresonators were first fabricated in a standard SOI wafer and then, through a suitable process, transferred onto a flexible polydimethylsiloxane (PDMS) film, still preserving optical functionality and tuneability [203]. Figure 1.14 shows how a highly integrated and functional Si photonic circuit can be transferred from a standard wafer substrate to a flexible plastic substrate, yet retaining its optical performance.

Xu *et al.* reported on silicon nanomembrane (SiNM)-based devices transferred by a modified transfer printing method [204]. In their work, SiNM waveguides

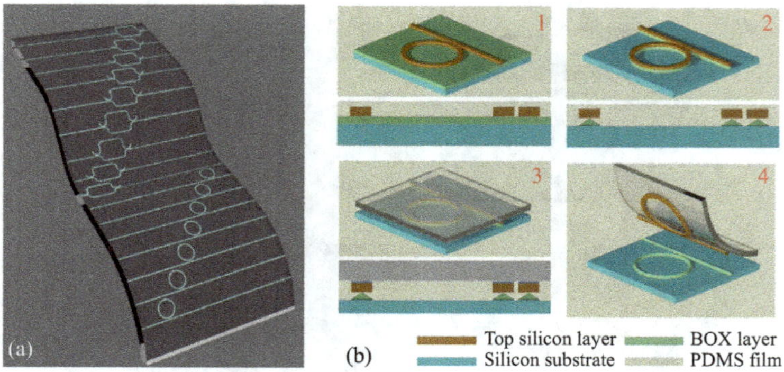

Figure 1.14 Schematic of transfer and bond fabrication of flexible silicon photonic circuits: (a) illustration of Si PIC on a flexible plastic substrate and (b) fabrication procedures to transfer and bond Si PICs from a wafer substrate to a flexible PDMS substrate. Reproduced, with modifications, from Chen et al. [203] under Creative Common License

230-nm thick, 30-μm wide and up to 5.7-cm long were transferred to flexible Kapton films with high transfer yield and exhibited propagation loss ~1.1 dB/cm.

Another approach which has emerged recently is based on the use of chalcogenide glass waveguides [205,206]. Bulk glasses in the As_2S_3 binary system or $Ge_{23}Sb_7S_{70}$ ternary alloy were prepared by melt-quenching process and were then used as sources for thermal evaporation of films with thickness around 400 nm on top of polyimide or PDMS substrates. Waveguides and photonic components may be patterned with a single step by using thermal imprinting through an elastomer mask. The relevant advantages of chalcogenide glasses must be ascribed to their low deposition temperature, tuneability of their refractive index (from 2 to 3.5, depending on composition) and very low propagation loss (less than 1 dB/m) in the band around 1.5 μm.

Many real applications, however, require that the devices are not only bendable but also stretchable: this requirement imposes further constraints on the materials and structures. Starting from the chalcogenide glass photonic platform, a solution strategy was adopted, which combines interconnecting waveguides with photonic components realized in stiffened parts of the substrate, so to minimize shape deformation of these components. In practice, high-index chalcogenide glasses and low-index epoxy polymers were used, both monolithically integrated on PDMS elastomer substrates [206].

On the other hand, plastic substrates have the same limitations as all organic devices, mostly related to ageing and temperature stability. There has been, therefore, much interest to move to a flexible glass substrate and, possibly, to an all-glass platform. The latter approach, which would avoid complicated pattern transfer, has been successfully investigated by Li *et al.* exploiting the properties of chalcogenide glasses, which also allow high-index-contrast structures [205]. In the other case, it was electronics, again, to open the way, pushing for the development of ultra-thin flexible glasses, suitable for displays and cellular phones' covers [211]. Thus, one has now the possibility of choosing among various flexible glasses, produced by the major glass companies worldwide, with thickness in the range of 20–200 μm. Various processes may be used for the device fabrication, from pattern transfer, by the consolidated electronic technologies [212], to fs-laser direct writing [213]; ion exchange or RF sputtering deposition [214] would also be suitable for waveguide and passive components' fabrication.

1.4 Conclusions

Many very significant advances in materials and technologies have occurred in more than 50 years since the first demonstrations of confinement of light in thin films and the naming of integrated optics for the planar optical guided-wave circuits. The start of integrated optics, however, was not fast. It may be worth recalling that the first integrated optical circuit, which just consisted of a laser and a modulator, was reported only in 1987 [215], i.e. almost 20 years after the visionary proposal by Miller, and almost 30 years after the implementation of the first integrated electronic circuit, containing several transistors in a silicon substrate [216,217].

The main aim of the present chapter has been to recall the pioneering efforts of many researchers to model and implement waveguides, couplers and modulators, i.e. the building blocks of what will become the PIC. It implied also the search for suitable materials and process technologies: the history of this research area has seen the developments of growing complexity, from the oxide glass platform to nitrides, chalcogenide glasses, ferroelectric crystals and semiconductors.

Even traditional material systems, such as glasses, have evolved towards more complex systems (e.g. glass ceramics) and more functionalities, thanks to the easiness of introducing into the amorphous matrix active rare-earth ions [218] and nanoparticles of various types [219]. Polymeric materials, on the other hand, are very attractive for their huge potential of enabling full 3D microfabrication of PICs, also through the use of AM. It is true that so far only very few demonstrations of multimode optical waveguides made by AM have been reported, and a few more at microwave frequencies, but this situation is not much dissimilar from that of waveguides in the 1960s, described earlier in this chapter; thus, in a similar way, one can expect future technological developments, e.g. towards fully integrated THz circuits [220].

In the meantime, the pursuit of electronics led to a breakthrough in the photonic integration capabilities when the possibility of using CMOS-compatible processes was demonstrated. Achieving a higher level of integration of photonic components and of PICs with electronic ICs poses new challenges to researchers, despite the fact that excellent results had already been obtained. As an example, a record result reported in 2014 by a group at Infinera, Sunnyvale, concerned the demonstration of a 40-wavelength, dual-polarization, InP-based transmitter PIC with 80 quadrature phase shift keying (QPSK) Mach–Zehnder modulators, capable of 2.25 Tbit/s by employing over 1,700 monolithically integrated functions [221]. Other relevant results, published in the first months of 2020, confirm that this is a strategic area not only for research but also for industrial development [222–226].

Finally, a current challenge comes from the goal of developing flexible and stretchable PICs. Even in this case, very promising results have been published, but many difficulties must still be overcome. The development of stretchable devices, in particular, is quite complex, requiring collaborative research of experts in micro- and nano-photonics, materials' growth, interface science and micromechanics.

In conclusion, we can say that integrated optics went beyond the landmark of 50 years in a very good shape, and its close heir, integrated photonics, keeps going and sees a bright future in many application areas, from telecommunication to sensing, to biology and medicine, to space research.

References

[1] Miller S.E. 'Integrated optics: an introduction'. *Bell Syst. Tech. J.* 1969; **48**(7):2059–2069.

[2] Marcatili E.A.J. 'Dielectric rectangular waveguide and directional coupler for integrated optics'. *Bell Syst. Tech. J.* 1969; **48**(7):2071–2102.

[3] Marcatili E.A.J. 'Bends in optical dielectric guides'. *Bell Syst. Tech. J.* 1969; **48**(7):2103–2132.

[4] Marcatili E.A.J. and Miller S.E. 'Improved relations describing directional control in electromagnetic wave guidance'. *Bell Syst. Tech. J.* 1969; **48**(7): 2161–2188.

[5] Goell J.E. 'A circular-harmonic computer analysis of rectangular dielectric waveguides. *Bell Syst. Tech. J.* 1969; **48**(7):2133–2160.

[6] Ritchie D.M. 'The UNIX system: The evolution of the UNIX time-sharing system'. *Bell Labs Tech. J.* 1984; **63**(8):1577–1593.

[7] Boyle W.S. and Smith G.E. 'Charge coupled semiconductor devices'. *Bell Syst. Tech. J.* 1970; **49**(4):587–593.

[8] Amelio G.F., Tompsett M.F. and Smith G.E. 'Experimental verification of the charge coupled device concept'. *Bell Syst. Tech. J.* 1970; **49**(4):593–600.

[9] Barlow H.M. 'The exploitation of micro-waves for trunk wave-guide multi-channel communications'. *J. Br. Inst. Radio Eng.* 1947; **7**:251–257.

[10] Miller S.E. 'Waveguide as a communication medium'. *Bell Syst. Tech. J.* 1954; **33**(6):1209–1265.

[11] Karbowiak A.E. and Solymar L. 'Characteristics of waveguides for long-distance transmission'. *J. Res., D: Radio Propag.* 1961; **65D**(1):75–88.

[12] Maiman T.H. 'Stimulated optical radiation in ruby'. *Nature* 1960; **187**: 493–494.

[13] Kaplan R.A. 'Optical waveguide of macroscopic dimensions in single-mode operation'. *Proc. IEEE* 1963; **51**(8):1144–1145.

[14] Yariv A., and Leite R.C.C. 'Dielectric-waveguide mode of light propagation in p-n junctions'. *Appl. Phys. Lett.* 1963; **2**(3):55–57.
[15] Yariv A. and Leite R.C.C. 'Erratum: dielectric-waveguide mode of light propagation in *p-n* junctions'. *Appl. Phys. Lett.* 1963; **2**(8):161.
[16] Bond W.L., Cohen B.G., Leite R.C.C. and Yariv A. 'Observation of the dielectric waveguide mode of light propagation in *p-n* junctions'. *Appl. Phys. Lett.* 1963; **2**(3):57–59.
[17] Ashkin A. and Gershenzon M. 'Reflection and guiding of light at *p-n* junctions'. *J. Appl. Phys.* 1963; **34**:2116–2119.
[18] Hall R.N. and Olechna D.J. 'Wave propagation in an active dielectric slab'. *J. Appl. Phys.* 1963; **34**:2565–2566.
[19] Schineller E.R., Heinemann H.M., Wilmot D.W. and Redlien H.W. Development of Macroscopic Waveguide and Waveguide Components for Optical Systems. Final Report, 28 Jan.–28 Nov. 1964. NASA Contractor Report CR-332, January 1966.
[20] Schineller E.R. and Wilmot D.W. Development of Optical Waveguides and Components – Progress, February 1965 to March 1966. NASA Contractor Report CR-76893, April 1966.
[21] Schineller E.R. Summary of the Development of Optical Waveguides and Components. NASA Contractor Report CR-860, August 1967.
[22] Wilmot D.W. and Schineller E.R. 'Optical waveguides formed by proton irradiation of fused silica'. *J. Opt. Soc. Am.* 1966; **56**(10):1434 (ThE17).
[23] Schineller E.R., Flam R.P. and Wilmot D.W. 'Optical waveguides formed by proton irradiation of fused silica'. *J. Opt. Soc. Am.* 1968; **58**(9):1171–1176.
[24] Kane J. and Osterberg H. 'Optical characteristics of planar guided modes'. *J. Opt. Soc. Am.* 1964; **54**(3):347–352.
[25] Osterberg H. and Smith L.W. 'Transmission of optical energy along surfaces: Part I, Homogeneous media'. *J. Opt. Soc. Am.* 1964; **54**(9):1073–1078.
[26] Osterberg H. and Smith L.W. 'Transmission of optical energy along surfaces: Part II, Inhomogeneous media'. *J. Opt. Soc. Am.* 1964; **54**(9): 1078–1084.
[27] Tien P.K., Ulrich R. and Martin R.J. 'Modes of propagating light waves in thin deposited semiconductor films'. *Appl. Phys. Lett.* 1969; **14**(9):291–294.
[28] Harris J.H., Shubert R. and Polky J.N. 'Beam coupling to films'. *J. Opt. Soc. Am.* 1970; **60**(8):1007–1016.
[29] Tien P.K. and Ulrich R. 'Theory of prism-film coupler and thin-film light guides'. *J. Opt. Soc. Am.* 1970; **60**(10):1325–1337.
[30] Midwinter J.H. and Zernike F. 'Experimental studies of evanescent wave coupling into a thin-film waveguide'. *Appl. Phys. Lett.* 1970; **16**(5): 198–200.
[31] Anderson D.B. and Shaw C.B. 'Dielectric waveguide for infrared wavelengths'. *Proceedings of G-MTT International Microwave Symposium*; Detroit, USA, May 1968. (IEEE, 1968) pp. 331–336. DOI: 10.1109/GMTT.1968.1123447.
[32] Anderson D.B. and August R.R. 'Applications of microphotolithography to millimeter and infrared devices'. *Proc. IEEE* 1966; **54**(4):657–661.

[33] Shubert R. and Harris J.H. 'Optical surface waves on thin films and their application to integrated data processors'. *IEEE Trans. Microwave Theory Tech.* 1968; **MTT-16**(12):1048–1054.

[34] see, for instance: Vander Lugt A. 'Signal detection by complex spatial filtering'. *IEEE Trans. Inf. Theory* 1964; **IT-10**(2):139–145.

[35] Shubert R. and Harris J.H. 'Optical guided-wave focusing and diffraction'. *J. Opt. Soc. Am.* 1971; **61**:154–161.

[36] Ulrich R. and Martin R.J. 'Geometrical optics in thin-film light guides'. *Appl. Opt.* 1971; **10**:2077–2085.

[37] Righini G.C., Russo V., Sottini S. and Toraldo di Francia G. 'Geodesic lenses for guided optical waves'. *Appl. Opt.* 1973; **12**:1477–1483.

[38] Zernike F. 'Luneburg lens for optical waveguide use'. *Opt. Commun.* 1974; **12**:379–381.

[39] Spiller E. and Harper J.S. 'High resolution lenses for optical waveguides'. *Appl. Opt.* 1974; **13**:2105–2108.

[40] Verber C.M., Vahey D.W. and Wood V.E., 'Focal properties of geodesic waveguide lenses'. *Appl. Phys. Lett.* 1976; **28**:514–516.

[41] Anderson D.B., Davies R.L., Boyd, J.T. and August R.R. 'Comparison of optical waveguide lens technologies'. *IEEE J. Quantum Electron.* 1977; **QE-13**:275–282.

[42] Chen B.-U., Marom E. and Lee A. 'Geodesic lenses in single-mode $LiNbO_3$ waveguides'. *Appl. Phys. Lett.* 1977; **31**:263–265.

[43] Sottini S., Russo V. and Righini G.C. 'General solution of the problem of perfect geodesic lenses for integrated optics'. *J Opt. Soc. Am.* 1979; **69**(9): 1248–1254.

[44] Valette S., Marque A. and Mottier, P. 'High-performance integrated Fresnel lenses on oxidised silicon substrate'. *Electron. Lett.* 1982; **18**:13–15.

[45] Zang, D.Y. 'Waveguide optical planar lenses in $LiNbO_3$ – theory and experiment'. *Opt. Commun.* 1983; **47**:248–250.

[46] Pitt C.W., Skinner J.D. and Trotter G.R. 'Computer simulation of thin film lenses'. *Opt. Commun.* 1985; **53**:87–90.

[47] Righini G.C. and Molesini G. 'Design of optical-waveguide homogeneous refracting lenses'. *Appl. Opt.* 1988; **27**; 4193–4199.

[48] Righini G.C., Belli G., Varasi M. and Vannucci A. 'Waveguide Fresnel lenses for integrated optical processors'. *Proc. SPIE* 1990; **1177**: 209–215.

[49] Hamilton M.C., Wille D.A. and Miceli W.J. 'An integrated optical R.F. spectrum analyzer'. *Opt. Eng.* 1977; **16**:475–478.

[50] Anderson D.B., Boyd J.T., Hamilton M.C. and August R.R. 'An integrated optical approach to the Fourier transform'. *IEEE J. Quantum Electron.* 1977; **QE-13**:268–274.

[51] Righini G.C., Russo V. and Sottini S. 'Signal processing in integrated optics employing geodesic lenses'. *Proc. SPIE* 1979; **164**:20–26.

[52] Tsai C.S. 'Integrated-optical device modules in $LiNbO_3$ for computing and signal processing'. *J. Mod. Opt.* 1988; **35**:965–967.

[53] Goncharenko A.M., Deryugin L.N., Prokhorov A.M. and Shipulo G.P. 'Development of integrated optics in the USSR'. *J. Appl. Spectr.* 1978; **29**(6):1435–1444.

[54] Goncharenko, A.M. 'Electromagnetic properties of a planar anisotropic waveguide'. *Zh. Tekh. Fiz.* 1967; **37**, 822.

[55] Deryugin L.N., Marchuk A.N. and Sotin V.E. 'Properties of planar asymmetric dielectric waveguides on a dielectric backing'. *Izv. Vyssh. Uchebn. Zaved., Ser. Radioelektron.* 1967; **10**(2):134.

[56] Semenov A.S. 'Journal Soviet Lightwave Communications'. *Sov. J. Quantum Electron.* 1990; **20**(9):1151.

[57] Semenov A.S. 'Soviet lightwave communications – the first year of publication'. *Sov. J. Quantum Electron.* 1990; **22**(6):572–573.

[58] Chen B. private communication.

[59] Yu R. 'Integrated optics'. *Chin. J. Lumin.* 1973; **10**:244–256 (in Chinese; title translated by Chen B.).

[60] Han Q. 'Integrated optics and film waveguides'. *Physics* 1974; **5**:279–289 (in Chinese; title translated by Chen B.).

[61] Kaminov I.P. 'Optical integrated circuits: a personal perspective'. *J. Lightwave Tech.* 2008; **26**(9):994–996.

[62] Schawlow A.L. and Townes C.H. 'Infrared and optical masers'. *Phys. Rev.* 1958; **112**(6):1940–1949.

[63] Javan A., Bennett W.R., Jr. and Herriott D.R. 'Population inversion and continuous optical maser oscillation in a gas discharge containing a He-Ne mixture'. *Phys. Rev. Lett.* 1961; **6**(3):106–110.

[64] Kompfner R. 'Optics at bell laboratories – optical communications'. *Appl. Opt.* 1972; **11**(11):2412–2425.

[65] Moore G.E. 'Cramming more components onto integrated circuits'. *Electronics* 1965; **38**(8):114–116. (Reprinted in *IEEE SSCS Newsl.* 2006; **11**(3): 33–35).

[66] Najafi S.I. *Introduction to Glass Integrated Optics.* (Norwood, MA, USA, Artech House, 1992).

[67] Righini G.C. and Chiappini A. 'Glass optical waveguides: a review of fabrication techniques'. *Opt. Eng.* 2014; **53**(7): 071819, 1–14.

[68] Tran T.N.L., Zur L., Chiasera A., *et al.* 'Chapter 5: Rare-earth-doped glasses and glass-ceramics for integrated optics' in Righini G.C. and Ferrari M. (eds.). *Integrated Optics – From Fundamentals to Recent Advances.* (London, The IET, 2020).

[69] Righini G.C. and Pelli S. 'Ion exchange in glass: a mature technology for photonic devices'. *Proc. SPIE* 2001; **4453**, 93–99.

[70] Mazzoldi P. and Sada C. 'A trip in the history and evolution of ion-exchange process'. *Mater. Sci. Eng. B* 2008; **149**(225):112–117.

[71] Tervonen A., West B.R. and Honkanen S. 'Ion-exchanged glass waveguide technology: a review'. *Opt. Eng.* 2011; **50**(7): 071107, 1–15.

[72] Rogoziński R. 'Chapter 7: Ion exchange in glass – the changes of glass refraction' in Kilislioglu A. (ed.). *Ion Exchange Technologies.* (London, Intech Open,

2012) (pdf available from https://www.intechopen.com/books/ion-exchange-technologies/ion-exchange-in-glass-the-changes-of-glass-refraction (last accessed on February 1, 2020)).

[73] Varshneya A.K. 'Chemical strengthening of glass: lessons learned and yet to be learned'. *Int. J. Appl. Glass Sci.* 2010; **1**(2):131–142.

[74] see, for instance Hood H.P. and Stookey S.D. 'Method of making a glass article of high mechanical strength and article made thereby'. *U.S. Patent N. 2,779*,136; 1955. Weber N. 'Strengthened glass article and method of producing same'. *U.S. Patent* N. 3,218.220; 1965. Chisholm R.S., Sleighter G. E., Ernsberger F.M. 'Method of strengthening glass by ion exchange and article made therefrom'. *U.S. Patent* N. 3,287,201; 1966.

[75] Uchida T., Furukawa M., Kitano I, Koizumi K. and Matsumura H. 'A light-focusing fiber guide'. *IEEE J. Quantum Electron.* 1969; **5**(6):331–331.

[76] SELFOC® – The World First Self-Focusing Lens (available from Nippon Sheet Glass – Story 9; http://100th.nsg.com/story/09/) (last accessed on January 30, 2020).

[77] Pearson A.D., French W.C. and Rawson E.C. 'Preparation of a light focusing glass rod by ion-exchange techniques'. *Appl. Phys. Lett.* 1969; **15**(2):76–77.

[78] Izawa T. and Nakagome H. 'Optical waveguide formed by electrically induced migration of ions in glass plates'. *Appl. Phys. Lett.* 1972; **21**(12): 584–586.

[79] Giallorenzi T.G., West E.J., Kirk R., Ginther R. and Andrews R.A. 'Optical waveguides formed by thermal migration of ions in glass'. *Appl. Opt.* 1973; **12**(6):1240–1245.

[80] Saruwatari M. and Izawa T. 'Nd-glass laser with three-dimensional optical waveguide'. *Appl. Phys. Lett.* 1974; **24**(12):603–605.

[81] Lehky J.M. 'Optical waveguides with copper film ion-exchange in glass'. PhD Thesis, 2006; Swiss Federal Institute of Technology, Zurich (available from https://doi.org/10.3929/ethz-a-005203774) (last accessed on January 30, 2020).

[82] Findakly T. 'Glass waveguides by ion exchange: a review'. *Opt. Eng.* 1985; **24**(2):244–250.

[83] Ramaswamy R.V. and Srivastava R. 'Ion-exchanged glass waveguides: a review'. *J. Lightwave Tech.* 1988; **6**(6):984–1002.

[84] Righini G.C. 'Passive integrated optical components'. *J. Mod. Opt.* 1988; **35**(6): 847–848; and articles therein.

[85] Ramaswamy R.V. and Srivastava R. 'Recent advances in ion-exchanged optical waveguides and components'. *J. Mod. Opt.* 1988; **35**(6):1049–1067.

[86] Garmire E., Stoll H., Yariv A. and Hunsperger R.G. 'Optical waveguiding in proton-implanted GaAs'. *Appl. Phys. Lett.* 1972; **21**(3):87.

[87] Somekh S., Garmire E., Yariv A., Garvin H.L. and Hunsperger R.G. 'Channel optical waveguide directional couplers'. *Appl. Phys. Lett.* 1973; **22**(1):46.

[88] Barnoski M., Hunsperger R.G., Wilson R. and Tangonan G. 'Proton-implanted GaP optical waveguide'. *J. Appl. Phys.* 1973; **44**(4), 1925.

[89] Valette S., Labrunie G. and Lizet J. 'Optical waveguides in ion-implanted ZnTe'. *J. Appl. Phys.* 1975; **46**(6), 2731.
[90] Townsend P.D. 'Ion implantation and integrated optics'. *J. Phys. E: Sci. Instrum.* 1977; **10**(3):197–203.
[91] Chandler P.J., Field S.J., Hanna D.C., *et al.* 'Ion-implanted Nd:YAG planar waveguide laser'. *Electron. Lett.* 1989; **25**(15):985–986.
[92] Chen F. 'Chapter 13: Optical waveguides produced by ion beams' in Righini G.C. and Ferrari M. (eds.). *Integrated Optics – From Fundamentals to Recent Advances*. (London, The IET, 2020).
[93] Goell J.E. and Standley R.D. 'Sputtered glass waveguide for integrated optical circuits'. *Bell Syst. Tech. J.* 1969; **48**(10):3445–3448.
[94] Fujimori M., Sasaki M. and Honda M. 'Tantalum pentoxide thin films for light guide'. *Fujitsu Sci. Tech. J.* 1972; **18**:177–189.
[95] Goell J.E. 'Barium silicate films for integrated optical circuits'. *Appl. Opt.* 1973; **12**:737–742.
[96] Pitt C.W. 'Sputtered-glass optical waveguides'. *Electron. Lett.* 1973; **9**(17): 401–403.
[97] Deitch R.H., West E.J., Giallorenzi T.G. and Weller J.F. 'Sputtered thin films for integrated optics'. *Appl. Opt.* 1974; **13**:712–715.
[98] Cheng Y.C. and Festwood W.D. 'Losses in tantalum pentoxide waveguides'. *J. Electron. Mater.* 1974; **3**:37–50.
[99] Yin Z.Y. and Garside B.K. 'Low-loss GeO_2 optical waveguide fabrication using low deposition rate rf sputtering'. *Appl. Opt.* 1982; **21**:4324–4328.
[100] Ingrey S.J., Westwood W.D., Cheng Y.C. and Wei J. 'Variable refractive index and birefringent waveguides by sputtering tantalum in O_2–N_2 mixtures'. *Appl. Opt.* 1975; **14**:2194–2198.
[101] Paulson W.M., Hickernell F.S. and Davis R.L. 'Effects of deposition parameters on optical loss for rf-sputtered Ta_2O_5 and Si_3N_4 waveguides'. *J. Vac. Sci. Tech.* 1979; **16**:307.
[102] Aagard R.L. 'Optical waveguide characteristics of reactive dc-sputtered niobium pentoxide films'. *Appl. Phys. Lett.* 1975; **27**:605.
[103] Yajima H. 'Dielectric thin-film optical branching waveguide'. *Appl. Phys. Lett.* 1973; **22**:647.
[104] Flanders D.C., Kogelnik H., Schmidt R.V. and Shank C.V. 'Grating filters for thin-film optical waveguides'. *Appl. Phys. Lett.* 1974; **24**(4):194.
[105] Hong C.S., Shellan J.B., Livanos A.C., Yariv A. and Katzir A. 'Broad-band grating filters for thin-film optical waveguides'. *Appl. Phys. Lett.* 1977; **31**:276.
[106] Gia Russo D.P. and Kumar C.S. 'Sputtered ferroelectric thin-film electro-optic modulator'. *Appl. Phys. Lett.* 1973; **23**:229.
[107] Yajima H., Kawase S., Kawase Y. and Sekimoto Y. 'Amplification of 1.06 μm using a Nd:glass thin-film waveguide'. *Appl. Phys. Lett.* 1972; **21**:407–409.
[108] Chen B.-U. and Tang C.L. 'Nd-glass thin-film waveguide: an active medium for Nd thin-film laser'. *Appl. Phys. Lett.* 1976; **28**(8):435–437.

[109] Takada S., Ohnishi M., Hayakawa H. and Mikoshiba N. 'Optical waveguides of single-crystal $LiNbO_3$ film deposited by rf sputtering'. *Appl. Phys. Lett.* 1974; **24**:490.
[110] Ishitani A. and Kimura M. 'Single-crystal $Sr_2Nb_2O_7$ film optical waveguide deposited by rf sputtering'. *Appl. Phys. Lett.* 1976; **29**(5):289.
[111] Okada A. 'Optical waveguiding properties of sputtered $LiNbO_3$ single crystal thin films on $LiTaO_3$ substrates'. *Ferroelectrics* 1976; **14**(1): 739–742.
[112] Shiosaki T., Ohnishi S., Murakami Y. and Kawabata A. 'High rate epitaxial growth of ZnO films on sapphire by planar magnetron rf sputtering system'. *J. Cryst. Growth* 1978; **45**(12):346–349.
[113] Wu M.-S., Azuma A., Shiosaki T. and Kawabata A. 'Low-loss epitaxial ZnO optical waveguides on sapphire by rf magnetron sputtering'. *J. Appl. Phys.* 1987; **62**:2482.
[114] Chiasera A. 'Chapter 7: Thin-film deposition: physical techniques' in Righini G.C. and Ferrari M. (eds.). *Integrated Optics. Recent Advances and Future Prospects*. (London, The IET, 2020).
[115] Special Issue on Optical Communications, Revue Technique Thomson-CSF 1974; **6**(4).
[116] Suematsu Y. 'The progress of integrated optics in Japan'. *IEEE Trans. Microwave Theory Tech.* 1975; **23**(1):16–20.
[117] Anikin V.I. and Gorobets A.P. 'Investigation of planar waveguides prepared for integrated optics systems by solid-state diffusion'. *Sov. J. Quantum Electron.* 1975; **5**(7):787–90.
[118] Kiselev V.A. 'Diffraction coupling of radiation into a thin-film waveguide'. *Sov. J. Quantum Electron.* 1975; **4**(7):872.
[119] Zlenko A.A., Kiselev V.A., Prokhorov A.M., Spikhal'skii A.A. and Sychugov V.A. 'Emission of surface light waves from a corrugated part of a thin-film waveguide'. *Sov. J. Quantum Electron.* 1975; **4**(7):839–42.
[120] Reutov A.T. and Tarashchenko P.P. 'Generation of a sum-frequency wave in a microscopic optically nonlinear waveguide made of lithium niobate'. *Sov. J. Quantum Electron.* 1975; **4**(7):910–911.
[121] Zolotov E.M., Kiselev V.A. and Sychugov V.A. 'Optical phenomena in thin-film waveguides'. *Sov. Phys. Usp.* 1974; **17**(1):64.
[122] Hsu H.P. and Chang W.S.C. 'The efficiency of grating coupling of two thin-film waveguides'. *Opt. Quantum Electron.* 1975; **7**:459–464.
[123] Ghizoni C., Chen B.-U. and Tang C. 'Theory and experiments on grating couplers for thin-film waveguides'. *IEEE J. Quantum Electron.* 1976; **12**(2): 69–73.
[124] Soares O.D.D. and Ash E.A. 'Holographic selective mode coupling to integrated optical waveguides'. *Electron. Lett.* 1976; **12**(10):239–239.
[125] Mukherjee S.P. 'Sol-gel processes in glass science and technology'. *Int. J. Non Equilibr. Process* 1980; **42**(1):477–88.
[126] Gottardi V. (ed.) 'Preface. Proceedings of the international workshop on glasses and glass ceramics from gels'. *J. Non-Cryst. Solids* 1982; **48**(1): vii.

[127] Brinker C.J. and Scherer G.W. *Sol-Gel Science: The Physics and Chemistry of Sol-Gel Processing*. (Boston, USA, Academic Press, 1990) 908 p.

[128] Tiefenthaler K., Briguet V., Buser E., Horisberger M. and Lukosz W. 'Preparation of planar optical SiO_2-TiO_2 and $LiNbO_3$ waveguides with a dip coating method and an embossing technique for fabricating grating couplers and channel waveguides'. *Proc. SPIE* 1983; **401**:165–73.

[129] Fabes B.D., Taylor D.J., Weisenbach L., *et al.* 'Laser processing of channel waveguide structures in sol-gel coatings'. *Proc. SPIE* 1990; **1328**: 319–28.

[130] Guglielmi M., Colombo P., Mancinelli Degli Esposti L., Righini G.C. and Pelli S. 'Planar and strip optical waveguides by sol-gel method and laser densification'. *Proc. SPIE* 1991; **1513**:44–9.

[131] Brusatin G., Guglielmi M., Innocenzi P., *et al.* 'Microstructural and optical properties of sol-gel silica-titania waveguides'. *J. Non-Cryst. Solids* 1997; **220**(2):202–9.

[132] Guglielmi M., Martucci A., Righini G.C. and Pelli S. 'CdS- and PbS-doped silica-titania optical waveguides'. *Proc. SPIE* 1994; **2288**:174–182.

[133] Yanagi H., Hishiki T., Tobitani T., Otomo A. and Mashiko S. 'Thin film lasing from a dye-doped silica/titania composite'. *Chem. Phys. Lett.* 1998; **292**(3):332–338.

[134] Orignac X., Barbier D., Min Du X., Almeida R.M., McCarthy O. and Yeatman E. 'Sol–gel silica/titania-on-silicon Er/Yb-doped waveguides for optical amplification at 1.5 μm'. *Opt. Mater.* 1999; **12**(1):1–18.

[135] Zampedri L., Tosello C., Rossi F., *et al.* 'Erbium-activated monolithic silica xerogels and silica-titania planar waveguides: optical and spectroscopic characterization'. *Proc. SPIE* 2001; **4282**:200–209.

[136] Shintaro Miyazawa S. 'Growth of $LiNbO_3$ single-crystal film for optical waveguides'. *Appl. Phys. Lett.* 1973; **23**(4):198.

[137] Kaminow I.P. and Carruthers J.R. 'Optical waveguiding layers in $LiNbO_3$ and $LiTaO_3$'. *Appl. Phys. Lett.* 1973; **22**(7):326.

[138] Carruthers J.R., Kaminow I.P. and Stulz L.W. 'Diffusion kinetics and optical waveguiding properties of outdiffused layers in lithium niobate and lithium tantalate'. *Appl. Opt.* 1974; **13**(10):2333–2342.

[139] Papuchon M. 'Modulateurs électrooptique de lumière guidée'. *Rev. Tech. Thomson-CSF* 1974; **6**(4):983–1023.

[140] Schmidt R.V. and Kaminow I.P. 'Metal-diffused optical waveguides in $LiNbO_3$'. *Appl. Phys. Lett.* 1974; **25**(8):458–460.

[141] Phillips, W. and Hammer, J.M. 'Formation of lithium niobate-tantalate waveguides'. *J. Electr. Mat.* 1975; 4(3):549–565.

[142] Minakata M., Noda J. and Uchida N. '$LiNb_xTa_{1-x}O_3$ optical waveguiding layer in $LiTaO_3$ fabricated by diffusion of $LiNbO_3$ sputtered film'. *Appl. Phys. Lett.* 1975; **26**(7):395.

[143] Pan'kin V.G., Petukhova S.N., Yu Pchelkin V., Shashkin V.V. and Shvarts N.L. 'Investigation of waveguides formed by diffusion of titanium in Y-cut lithium metaniobate'. *Sov. J. Quantum Electron.* 1978; **8**(2):177.

[144] Courjal N., Bernal M.-P., Caspar A., *et al.* 'Lithium niobate optical waveguides and microwaveguides' in You K.Y. (ed.). *Emerging Waveguide Technology*. (London, Intech Open, 2018) pp. 153–174.
[145] Selvaraja S.K. and Sethi P. 'Review on optical waveguides' in You K.Y. (ed.). *Emerging Waveguide Technology* (London, Intech Open, 2018) pp. 95–129.
[146] Sada C. 'Chapter 6: Lithium niobate integrated optics' in Righini G.C. and Ferrari M. (eds.). *Integrated Optics – From Fundamentals to Recent Advances*. (London, The IET, 2020).
[147] Baldi P. and De Micheli M. 'Chapter 8, Volume 2: Nonlinear integrated optics in proton-exchanged lithium niobate waveguides and applications to classical and quantum optics' in Righini G.C. and Ferrari M. (eds.). *Integrated Optics – From Fundamentals to Recent Advances*. (London, The IET, 2020).
[148] Shelton J.C., Reinhart F.K. and Logan R.A 'Rib waveguide switches with MOS electrooptic control for monolithic integrated optics in GaAs-$Al_xGa_{1-x}As$'. *Appl. Opt.* 1978; **17**:2548–2555.
[149] Thompson G.H.B. *Physics of Semiconductor Laser Devices*. (Chichester, Wiley-Interscience, 1980) 572 p.
[150] Coldren L., Furuya K., Miller B. and Rentschler J. 'Etched mirror and groove-coupled GaInAsP/InP laser devices for integrated optics'. *IEEE J. Quantum Electron.* 1982; **18**(10):1679–1688.
[151] Benson T. 'Etched-wall bent-guide structure for integrated optics in the III-V semiconductors'. *J. Lightwave Tech.* 1984; **2**(1):31–34.
[152] Inoue H., Hiruma K., Ishida K., Asai T. and Matsumura H. 'Low loss GaAs optical waveguides'. *IEEE Trans. Electron Devices* 1985; **32**(12):2662–2668.
[153] Liao A.S.-H. and Wang S. 'Semiconductor injection lasers with a circular resonator'. *Appl. Phys. Lett.* 1980; **36**(10):801.
[154] Kawamura Y., Wakita K., Itaya Y., Yoshikuni Y. and Asahi H. 'Monolithic integration of InGaAs/InP DFB lasers and InGaAs/InAlAs MQW optical modulators'. *Electron. Lett.* 1986; **22**(5):242–243.
[155] Sakano S., Inoue H., Nakamura H., Katsuyama T. and Matsumura, H. 'InGaAsP/InP monolithic integrated circuit with lasers and an optical switch'. *Electron. Lett.* 1986; **22**(11):594–596.
[156] Dupuis R.D. 'Metalorganic chemical vapor deposition of III-V semiconductors'. *Science* 1984; **226**(4675):623–629.
[157] Krikorian E. 'Abstract: Thin films for electro-optical devices'. *J. Vac. Sci. Tech.* 1975; **12**(1):186.
[158] Smith D.L. and Pickhardt V.Y. 'Molecular beam epitaxy of II-VI compounds'. *J. Appl. Phys.* 1975; **46**(6):2366.
[159] Lukowiak A. and Borak B. 'Chapter 8: Thin film deposition: chemical techniques' in Righini G.C. and Ferrari M. (eds.). *Integrated Optics – From Fundamentals to Recent Advances*. (London, The IET, 2020).

[160] Tran L.T.N., Berneschi S., Nunzi Conti G. and Ferrari M. 'Chapter 9: Photorefractive waveguides' in Righini G.C. and Ferrari M. (eds.). *Integrated Optics – From Fundamentals to Recent Advances*. (London, The IET, 2020).

[161] Asquini R. and D'Alessandro A. 'Chapter 10: Integrated optics using liquid crystals' in Righini G.C. and Ferrari M. (eds.). *Integrated Optics – From Fundamentals to Recent Advances*. (London, The IET, 2020).

[162] Bradley J.D.B., Wang R., Frankis H.C., Bonneville D.B., Kiani K.M. and Mbonde H.M. 'Chapter 11: Silicon nitride integrated optics' in Righini G.C. and Ferrari M. (eds.). *Integrated Optics – From Fundamentals to Recent Advances*. (London, The IET, 2020).

[163] Eaton S.M., Sotillo B., Fernandez T.T., *et al.* 'Chapter 12: Femtosecond laser writing of integrated optical structures in glasses' in Righini G.C. and Ferrari M. (eds.). *Integrated Optics – From Fundamentals to Recent Advances*. (London, The IET, 2020).

[164] Righini G.C. and Najafi S.I. (eds.). *'Integrated optics devices III'*. *Proceedings SPIE N. 3620*. (Bellingham, SPIE, 1999).

[165] Righini, G.C. and Honkanen S. (eds.) *'Integrated optics devices IV'*. *Proceedings SPIE N. 3936*. (Bellingham, SPIE, 2000).

[166] Zimmermann H.K. *Integrated Silicon Optoelectronics*. (Berlin, Springer, 2010).

[167] Tong X.C. *Advanced Materials for Integrated Optical Waveguides*. (Basel, Springer International Publishing, 2014).

[168] You K.Y. *Emerging Waveguide Technology*. (London, Intech Open, 2018).

[169] Smit M., Williams K. and van der Tol J. 'Past, present, and future of InP-based photonic integration'. *APL Photonics* 2019; **4**(5):050901.

[170] Williams K.A., Trajkovic M., Rustichelli V., Lemaître F., Ambrosius H.P. M.M. and Leijtens X.J.M. 'High-speed energy-efficient InP photonic integrated circuit transceivers'. *Proc. SPIE* 2019; **10924**:1092411. (7 pp).

[171] Yao W., Smalbrugge B., Smit M.K., Williams K.A. and Wale M.J. 'A 6 × 30 Gb/s tunable transmitter PIC with low RF crosstalk from an open-access InP foundry'. *IEEE J. Sel. Top. Quantum Electron*. 2019; **25**(5):1–10.

[172] Thomson D., Zilkie A., Bowers J.E., *et al.* 'Roadmap on silicon photonics'. *J. Opt.* 2016;**18**(7):073003.

[173] Bogaerts W. and Chrostowski L. 'Silicon photonics circuit design: methods, tools and challenges'. *Laser Photonics Rev.* 2018; **12**(4): 1700237, 1–29.

[174] Shen Y., Meng X., Cheng Q., *et al.* 'Silicon photonics for extreme scale systems'. *J. Lightwave Technol.* 2019; **37**(2):245–59.

[175] Teng M., Honardoost A., Alahmadi Y., *et al.* 'Miniaturized silicon photonics devices for integrated optical signal processors'. *J. Lightwave Technol.* 2020; **38**(1):6–17.

[176] Kasper E. and Yu J. (eds.). *Silicon-Based Photonics*. (Singapore, Jenny Stanford Publishing, 2020).

[177] Lourdudoss S., Bowers J.E. and Jagadish C. (eds.). *Future Directions in Silicon Photonics*. (Cambridge, MA, USA, Academic Press, 2019).

[178] Zhang Y., Yang S., Lim A.E.-J., *et al.* 'A CMOS-compatible, low-loss, and low-crosstalk silicon waveguide crossing'. *IEEE Photonics Technol. Lett.* 2013; **25**(5):422–425.

[179] Liu Y., Shainline J.M., Zeng X. and Popović M.A. 'Ultra-low-loss CMOS-compatible waveguide crossing arrays based on multimode Bloch waves and imaginary coupling'. *Opt. Lett.* 2014; **39**(2):335.

[180] Sacher W.D., Mikkelsen J.C., Huang Y., *et al.* 'Monolithically integrated multilayer silicon nitride-on-silicon waveguide platforms for 3-D photonic circuits and devices'. *Proc. IEEE* 2018;**106**(12):2232–2245.

[181] Nesic A., Blaicher M., Hoose T., *et al.* 'Photonic-integrated circuits with non-planar topologies realized by 3D-printed waveguide overpasses'. *Opt. Express* 2019; **27**(12):17402–17425.

[182] Xiong W., Zhou Y.S., He X.N., *et al.* 'Simultaneous additive and subtractive three-dimensional nanofabrication using integrated two-photon polymerization and multiphoton ablation'. *Light Sci. Appl.* 2012; **1**(4):e6.

[183] Garner S.M., Lee S.-S., Chuyanov V., *et al.* 'Three-dimensional integrated optics using polymers'. *IEEE J. Quantum Electron.* 1999; **35**(8):1146–55.

[184] Song B., Stagarescu C., Ristic S., Behfar A. and Klamkin J. '3D integrated hybrid silicon laser'. *Opt. Express* 2016; **24**(10):10435–44.

[185] Rusing M., Weigel P.O., Zhao J. and Mookherjea S. 'Toward 3D integrated photonics including lithium niobate thin films: a bridge between electronics, radio frequency, and optical technology'. *IEEE Nanotechnol. Mag.* 2019; **13**(4):18–33.

[186] Schumann M., Bückmann T., Gruhler N., Wegener M. and Pernice W. 'Hybrid 2D–3D optical devices for integrated optics by direct laser writing'. *Light Sci. Appl.* 2014; **3**(6):e175.

[187] Gross S. and Withford M.J. 'Ultrafast-laser-inscribed 3D integrated photonics: challenges and emerging applications'. *Nanophotonics* 2015; **4**(3):332–352.

[188] Stone A., Jain H., Dierolf V., *et al.* 'Direct laser-writing of ferroelectric single-crystal waveguide architectures in glass for 3D integrated optics'. *Sci. Rep.* 2015; **5**(1):1–10.

[189] Koos C., Freude W., Randel S., *et al.* 'Hybrid photonic multi-chip integration enabled by 3D nano-printing'. *2017 IEEE Photonics Conference (IPC) Part II*. (Orlando, FL, USA, IEEE, 2017) p. 1–2.

[190] Zhu B., Dai Y., Ma H., Zhang S., Lin G. and Qiu J. 'Femtosecond laser induced space-selective precipitation of nonlinear optical crystals in rare-earth-doped glasses'. *Opt. Express* 2007; **15**(10):6069–6074.

[191] Weidenbach, M., Jahn, D., Rehn, A., *et al.* '3D printed dielectric rectangular waveguides, splitters and couplers for 120 GHz'. *Opt. Express* 2016; **24**:28968–28976.

[192] Yang, J., Zhao, J., Gong, C., *et al.* '3D printed low-loss THz waveguide based on Kagome photonic crystal structure'. *Opt. Express* 2016; **24**:22454–22460.

[193] Pandey, S., Gupta, B. and Nahata, A. 'Terahertz plasmonic waveguides created via 3D printing'. *Opt. Express* 2013, **21**:24422–24430.

[194] Bollgruen P., Wolfer T., Gleissner U., *et al.* 'Ink-jet printed optical waveguides'. *Flexible Printed Electron.* 2017; **2**:045003.

[195] Nseowo Udofia E. and Zhou W. '3D printed optics with a soft and stretchable optical material'. *Addit. Manuf.* 2020; **31**:100912.

[196] Hussain M.M. and El-Atab N. *Handbook of Flexible and Stretchable Electronics*. (Boca Raton, USA, CRC Press, 2019) 551 p.

[197] Martinez R.V. *Flexible Electronics: Fabrication and Ubiquitous Integration*. (Basel, Switzerland, MDPI, 2019) 160 p.

[198] Khanna V.K. *Flexible Electronics*. (Bristol, UK, Institute of Physics Publishing, 2019) 500 p.

[199] Samorì P. and Palermo V. *Flexible Carbon-based Electronics*. (Weinheim, Germany, Wiley VCH, 2019) 334 p.

[200] *Flexible Electronics & Circuit Market by Application, Circuit Structure, Vertical and Geography – Global Forecast to 2023*. (2018) (available from https://www.reportlinker.com/p05343329/Flexible-Electronics-Circuit-Market-by-Application-Circuit-Structure-Vertical-And-Geography-Global-Forecast-to.html (accessed on January 2020)).

[201] Zhang C., Zou C.-L., Zhao Y., *et al.* 'Organic printed photonics: From microring lasers to integrated circuits'. *Sci. Adv.* 2015; **1**:e1500257.

[202] Ma Z. and Liu D. (eds.). *Inorganic Flexible Optoelectronics Materials and Applications*. (Weinheim, Wiley-VCH, 2019).

[203] Chen Y., Li H. and Li M. 'Flexible and tunable silicon photonic circuits on plastic substrates'. *Sci. Rep.* 2012; **2**:622.

[204] Xu X., Subbaraman H., Hosseini A., Lin C.Y., Kwong D. and Chen R.T. 'Stamp printing of silicon-nanomembrane-based photonic devices onto flexible substrates with a suspended configuration'. *Opt. Lett.* 2012; **37** (6):1020–1022.

[205] Li L., Lin H.T., Qiao S.T., *et al.* 'Integrated flexible chalcogenide glass photonic devices'. *Nat. Photonics* 2014; **8**:643–649.

[206] Li L., Lin H., Qiao S., *et al.* 'Monolithically integrated stretchable photonics.' *Light Sci. Appl.* 2018; **7**:17138.

[207] Li L., Zhang P., Wang W.M., *et al.* 'Foldable and cytocompatible sol-gel TiO_2 photonics'. *Sci. Rep.* 2015; **5**:13832.

[208] Zhang R., Wang Q. and Zheng X. 'Flexible mechanochromic photonic crystals: routes to visual sensors and their mechanical properties'. *J. Mater. Chem. C* 2018; **6**(13):3182–3199.

[209] Li H., Cao Y., Wang Z. and Feng X. 'Flexible and stretchable inorganic optoelectronics'. *Opt. Mater. Express* 2019; **9**(10):4023–4049.

[210] Benali A., Claude J.-B., Granchi N., *et al.* 'Flexible photonic devices based on dielectric antennas'. *J. Phys. Photonics* 2020; **2**(1):015002.

[211] Garner S.M. *Flexible Glass: Enabling Thin, Lightweight, and Flexible Electronics*. (Hoboken, USA, John Wiley & Sons, 2017) 380 p.

[212] Linghu, C., Zhang, S., Wang, C., *et al.* 'Transfer printing techniques for flexible and stretchable inorganic electronics'. *npj Flexible Electron.* 2018; **2**:26.

[213] Huang S., Li M., Garner S.M., Li M.-J. and Chen K.P. 'Flexible photonic components in glass substrates'. *Opt. Express* 2015; **23**(17):22532–22543.

[214] Chiasera A., Sayginer O., Iacob E., *et al.* 'Flexible photonics: RF-sputtering fabrication of glass-based systems operating under mechanical deformation conditions'. *To be presented at Fiber Lasers and Glass Photonics: Materials through Applications II, SPIE Photonics Europe, Paper 11357-3.* (Strasbourg, France, 2020).

[215] Suzuki M., Noda Y., Tanaka H., Akiba S., Kushiro Y. and Isshiki H., 'Monolithic integration of InGaAsP/InP distributed feedback laser and electroabsorption modulator by vapor phase epitaxy'. *J. Lightwave Tech.* 1987; **5**(9):1277–1285.

[216] Kilby J.S. 'Miniaturized electronic circuits [US Patent No. 3,138, 743]'. *IEEE Solid-State Circuits Soc. Newsl.* 2007; **12**(2):44–54.

[217] Noyce R.N. 'Semiconductor device-and-lead structure, reprint of U.S. Patent 2,981,877 (Issued April 25, 1961. Filed July 30, 1959)'. *IEEE Solid-State Circuits Soc. Newsl.* 2007; **12**(2):34–40.

[218] Righini G.C. and Ferrari M. 'Photoluminescence of rare-earth-doped glasses'. *Riv. Nuovo Cimento* 2005; **28**(12):1–53.

[219] De Pablos-Martin A., Ferrari M., Pascual M.J. and Righini G.C. 'Glass-ceramics: a class of nanostructured materials for photonics'. *Riv. Nuovo Cimento* 2015; **38**(7–8):311–369.

[220] Pfeiffer U.R., Jain R., Grzyb J., Malz S., Hillger P. and Rodríguez-Vízquez P. 'Current status of terahertz integrated circuits – from components to systems'. *2018 IEEE BiCMOS and Compound Semiconductor Integrated Circuits and Technology Symposium (BCICTS)*. (San Diego, 2018) p. 1–7.

[221] Summers J., Vallaitis T., Evans P., *et al.* 'Monolithic InP-based coherent transmitter photonic integrated circuit with 2.25 Tbit/s capacity'. *Electron. Lett.* 2014; **50**(16):1150–1152.

[222] Shi B. and Klamkin J. 'Defect engineering for high quality InP epitaxially grown on on-axis (001) Si'. *J. Appl. Phys.* 2020; **127**(3):033102.

[223] Theurer M.A.D., Moehrle M., Sigmund A., *et al.* 'Flip-chip integration of InP to SiN photonic integrated circuits'. *J. Lightwave Technol.* 2020; (Early Access) doi: 10.1109/JLT.2020.2972065.

[224] Tran M.A., Huang D., Guo J., Komljenovic T., Morton P.A. and Bowers J. E. 'Ring-resonator based widely-tunable narrow-linewidth Si/InP integrated lasers'. *IEEE J. Sel. Top. Quantum Electron.* 2020; **26**(2):1–14.

[225] van der Tol J.J.G.M., Jiao Y., Van Engelen J.P., Pogoretskiy V., Kashi A.A. and Williams K. 'InP membrane on silicon (IMOS) photonics'. *IEEE J. Quantum Electron.* 2020; **56**(1):1–7.

[226] Blumenthal D.J. 'Photonic integration for UV to IR applications'. *APL Photonics* 2020; **5**(2):020903.

Part I

Modelling of waveguides and devices

Chapter 2
Numerical tools for integrated optical circuits design

Francesco Prudenzano[1]

The development of high-performance integrated optical circuits providing complex functionalities requires an accurate modelling of their basic components, i.e. of the constituent photonic devices and their optical interconnections, including the conventional optical waveguides. This goal requires models which must take into account optical propagation characteristics in devices and/or waveguiding structures/media, such as the dispersion with temperature and/or wavelength, non-linear effects, anisotropy, active light/matter interaction and a number of other complex aspects. Therefore, optical modelling is a preliminary and necessary approach for the fabrication of the optical components required by current and future applications in a large variety of fields as biomedicine, high-bandwidth communications, aerospace, remote-sensing systems, industrial applications.

This chapter briefly recalls some numerical techniques, many of which exploited in commercial design tools. The most largely employed numerical techniques are simply mentioned, since they are largely diffused, appreciated and well known by physicists and optical engineers for their versatility. As an example, the finite element method (FEM), the beam propagation method (BPM) and the finite-difference time domain (FDTD) are employed in commercial optical design tools used by most researchers in optics. Instead, classical approaches as coupled mode theory (CMT) and rare-earth rate equations are better recalled via examples of modelling requiring a customized and ad hoc refined approach.

2.1 Numerical tools available on market

Nowadays there is an increasing and stringent demand for optical systems in the field of data transmission, remote sensing, free-space optical communication, biomedicine, material processing, medical diagnostics, chemical and biological sensing, industry and military applications. Both integrated waveguides and fibre-optic-based photonic components constitute the key elements of such optical

[1]Dipartimento di Ingegneria Elettrica e dell'Informazione, Politecnico di Bari, Bari, Italy

systems. Continuous advances in optical components are performed, and more refined optical sources, detectors, power combiners, modulators, switches, wavelength-demultiplexing devices (WDM's), filters will enable the next optical systems to satisfy this ever-increasing demand. These optical components require an accurate modelling. Also, the conventional waveguiding structures, starting from the ridge or the channel waveguides, cannot be directly investigated by solving Maxwell's equations with exact analytical methods, and suitable numerical techniques are required. The fabrication of ever more powerful personal computers and the development in modelling techniques, such as the available on market computer-aided design (CAD) tools for modelling optical and ultra-short microwave components, play a catalyst and fundamental role in the research scenario evolution. The weight of these two factors in enabling innovative optical component production can be considered comparable with the synthesis of novel materials, i.e. the improvements in the field of material science, or the exploitation of novel fabrication technologies [1]. Every optical and microwave engineer knows well about how design tools and device modelling are strategic in (i) the advancement of conventional optical components, (ii) the design refinement of current optical components and (iii) the reduction of time consumption in exploring new production strategies. Verifying novel ideas and visions on which extremely innovative devices could be based requires ad hoc simulations, especially if the cost of a direct experimental test is not negligible. Till the early 1990s, handmade microwave and optical models were implemented by a multitude of worldwide researchers. Microwave/optical simulators were often based on the same or similar approaches and frequently employed the same routines, proving to be fundamental work instruments in order to reduce production cost. For this reason, a few years later, there was a large spreading of CADs, massively employed for a variety of actual optimization problems. The CAD employment at first became a welcome best practice and then a standard procedure in the industrial production. These commercial tools are sometime combined with in-house developed ones. Among the main objectives of the modelling, there is the electromagnetic investigation of the dielectric waveguiding structures, these being the key elements of integrated optics.

The FDTD, the FEM and the BPM are considered universal approaches. In this section, the fundamental concepts of these three methods are outlined. The FDTD approach allows one to solve Maxwell's equations directly, i.e. in the time domain. The algorithm for updating electromagnetic fields in small time steps allows one to evaluate the evolution of the electromagnetic fields. The FEM approach typically solves Maxwell's equations in the frequency domain, but time domain can be taken into account. Maxwell's equations are converted into a variational form that involves integral expressions. It is particularly efficient from the point of calculation cost. These two methods are largely employed since they practically allow the electromagnetic investigation of any waveguiding geometry, providing the calculation of the phase constant, propagation loss and electromagnetic field profile, starting from the geometrical and optical/spectroscopic parameters. Moreover, the electromagnetic investigation sometimes, in commercial codes, is related with other modules useful for the investigation of thermal or mechanical phenomena allowing a multiphysics investigation.

The FDTD approach constitutes the basis for several simulators, providing a numerical analysis which approximates the differential operators [2]. This approximation is largely used for solving various differential and integral–differential equations. The versatility of the method is related to the general idea of considering continuous space–time domain approximated by suitable spatial and temporal calculation grids. Therefore, discrete sets of points in time and space are chosen to perform the calculation. Each differentiable function $f(x)$ defined in space and time domains is substituted by an array of function samples calculated on the grid. The implementation of this idea can be performed in a variety of different ways. The derivative in a point can be obtained via a forward difference approximation, involving the evaluation of $f(x)$ and $f(x+h)$; a backward difference scheme involving $f(x)$ and $f(x-h)$; a central difference scheme, involving $f(x-h)$ and $f(x+h)$ and allowing an improved accuracy. As expected, the computation cost increases by increasing the desired accuracy. The second-order accuracy of the central difference calculation scheme requires the function calculation in the three points $x-h$, x, $x+h$. Analogously, the third-, fourth- or higher order approximation requires to consider the calculation in at least three, four or a higher number of points in addition to the nominal one, x. By considering the case of one-dimensional Maxwell's equations, the accuracy of the finite-difference scheme can be increased by four times by using half-step grids, but in this way, the number of grid points is increased by the same times. However, as is known, the number of array points for sampling both electric and magnetic fields can be halved. In fact, as a consequence of Maxwell's equation, the electric field can be calculated by the magnetic field in contiguous points. The same is similarly true for the magnetic field, which is given by the electric field in the contiguous points. By exploiting these Maxwell's equation properties, two separate systems of grids can be imagined: one only for electric field and the other only for magnetic field. The grids are shifted each other by half step, in both time and space domains. This kind of grid or mesh calculation is also known as a staggered one. Collocated grid refers to an approach in which electric and magnetic fields are defined in the same points. In particular, the fields are calculated on the staggered grid with time offsets via a leapfrog scheme, this being particularly efficient for the reduced computational cost. The FDTD method can be directly extended to three-dimensional problems, by adopting the Yee grid [3–8]. The finite difference approach solves the discretized second-order wave equation, in scalar, semi-vectorial or vectorial forms.

While the finite-difference method approximates Maxwell's equations directly, via the approximation of the differential operator curl, the FEM takes into account Maxwell's equations without any approximation. In this case, the solution space is approximated [1,2]. The solution space is obtained by subdividing the computational domain into small spatial region/patches. Different kinds of polynomials are considered to approximate the electromagnetic field solution on each patch. The trial functions are defined within each patch with unknown coefficients. The patches and the related local polynomials are called finite elements. Typical finite elements are triangles and rectangles in two-dimensional problem, and tetrahedrons and cuboids in three-dimension space. Usually the polynomials may be constant, linear, quadratic or cubic ones. The tangential continuity of the electric and magnetic field is imposed across/at the boundaries of neighbouring patches according to the electromagnetic field theory. A variational formulation is then exploited by obtaining a linear system whose solution approximates the exact solution. Therefore, the algebraic problem, allowing the electromagnetic investigation of complex geometrical shapes, is obtained [9–12].

The BPM is widely used for the investigation of propagation problems. This technique is versatile, allowing the inclusion of several physical effects, such as polarization and non-linearities. The main idea in BPM formulation is that it is possible to identify a peculiar/preferred direction of optical field propagation within the integrated optic device. This direction is generally that along which most of the optical power is transported. Conversely, in BPM approach, the power flow in the direction that is perpendicular to the preferred one can be neglected [13]. In fact, optical waveguiding structures typically vary slowly along the direction of the wave propagation to avoid scattering losses. For many optical structures, it can be assumed that along the wave propagation z, the dielectric constant distribution $\varepsilon\ (x, y, z)$ varies slowly, such as the electromagnetic field derivative. The transversal electromagnetic field components are decoupled from the longitudinal ones, and a Helmholtz equation for the transversal components is obtained. The medium is approximated with a cascade of sections with a constant refractive index distribution along the propagation direction. Within each section, a numerical integration is performed. The solution search involves an algorithm calculating the value of the field distribution at the updated position starting from the known field distribution at the previous position.

In addition to these numerical techniques, implemented in a number of different ways in several commercial codes, an optical engineer can employ ad hoc refined solving algorithms, based on modal solutions, anisotropic propagation, CMT, ion population rate equations describing active medium behaviour, non-linear Schrodinger's equation and so on. In the following section, a few examples are reported.

2.2 Ad hoc developed and hybrid numerical tools

2.2.1 Anisotropic waveguides and non-linear effects

As mentioned in the previous section, commercial numerical codes based on FEM and FDTD allow the investigation of the propagation in anisotropic non-homogeneous media. However, writing a home-made code based on modal solution can still be useful from the point of time consumption/computational cost, above all if the found solutions are employed as a first step of deeper investigations. The interest towards the home-made approach increases if coupled structures or propagating modes are considered. In this section, the modal solution in anisotropic waveguides is recalled [14–17]. This anisotropic propagation analysis is very intriguing when used together with the CMT, in order to understand some peculiar physical mechanisms and efficiently investigate complex devices [18–42]. CMT will be briefly recalled in the next section. The calculated electromagnetic modal refractive effective indices or propagation constants and the electromagnetic field profiles of the optical modes can be employed for the design of interferometric integrated optic sensors [18], non-linear integrated optic devices such as optical switches/couplers [20–22], WDM wavelength converters [23], wavelength shifters [26–28,30], optical transistors [25,31–36], optical temperature sensors [41,42]. As an example, the design of an optical temperature sensor will be illustrated.

2.2.1.1 Modal investigation of anisotropic waveguides/devices: guided and leaky mode propagation

Anisotropic propagation is a topic of great interest, e.g. for the design of waveguiding structures made in, or involving, lithium niobate (LN), potassium titanyl phosphate $KTiOPO_4$ (KTP) and liquid crystals. In all these cases, an efficient modelling can be based on a modal solution of the Helmholtz wave equation by taking into account the proper dielectric tensor expression. The approach is here recalled for $LiNbO_3$, but it is similar for all the above-mentioned cases. LN crystals are widely used in non-linear optical devices because of their large transparency range, high non-linearities and robustness. Integrated optics researchers and optical engineers widely use this material for the construction of many high-performance devices.

Figure 2.1 schematically illustrates an actual structure used for the construction of electrodes, e.g. for electro-optic devices as non-linear modulators. It is a slab-graded-index waveguide, obtained via titanium diffusion Ti:$LiNbO_3$ or via proton exchange PE:$LiNbO_3$ technology. It includes a metal layer embedded between the core region and the dielectric overlay allowing the microwave or the quasi-static

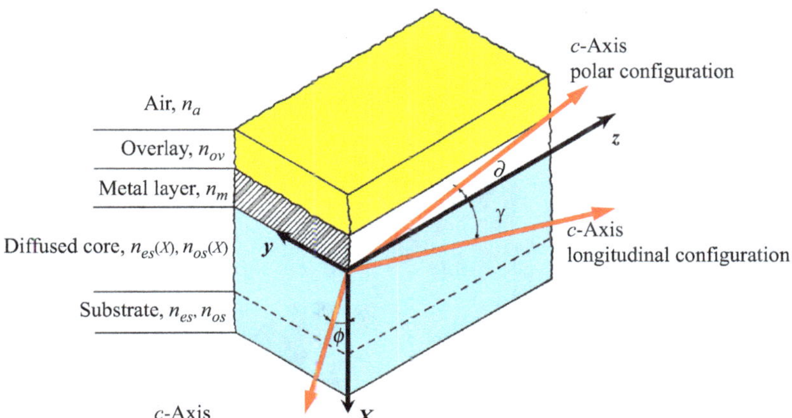

Figure 2.1 Typical slab-graded-index waveguide used in electro-optic modulators, with a metal film embedded between the core region and the dielectric overlay

electric field application. The guided and leaky mode propagation in anisotropic birefringent waveguides depends on the orientation of the crystal principal axes (a, b, c) [14–17]. The optical c-axis can be referred to a coordinate plane of the laboratory system. This case is welcome since a unique c-axis angle can identify the orientation of the crystal principal axes. The polar, the longitudinal and the equatorial configurations occur when the optical c-axis lies in the yz, xy, xz planes, respectively. The corresponding identifying angles are θ, ϕ, γ. A second rank tensor $\varepsilon(x)$ is employed to express the dielectric properties of rotated uniaxial waveguides. It has elements depending on the x direction, due to the titanium diffusion/proton exchange:

$$\varepsilon(x) = A^T \left[\varepsilon(x)^d\right] A \tag{2.1}$$

The superscript T denotes the transposition matrix operation. If the diagonal tensor $\varepsilon(x)^d$ is referred to the crystal principal axes a, b, c, a proper matrix of direction cosines allows one to evaluate the dielectric tensor components which can assume different expressions/configurations depending on the c-axis orientation. In particular, the tensor expression for polar, longitudinal and equatorial configurations is reported in Table 2.1, where the square of the ordinary/extraordinary refractive index and ordinary/extraordinary dielectric constant are linked by the relations $n_o^2 = \varepsilon_o$; $n_e^2 = \varepsilon_\varepsilon$.

To obtain an efficient modelling, the plane wave solution of the Maxwell equations can be imposed with the absence of charge and current sources. Six equations for the electromagnetic field components can be written. The well-known

Table 2.1 Tensor expression for polar, longitudinal and equatorial configurations

Configuration	c-Axis angle	Dielectric tensor expression $\varepsilon(x)=A^T\left[\varepsilon(x)^d\right]A$
Polar	θ	$\begin{bmatrix} n_o^2 & & (n_e^2-n_o^2)\sin(\vartheta)\cos(\vartheta) \\ & n_o^2\cos(\vartheta)^2+n_e^2\sin(\vartheta)^2 & \\ & (n_e^2-n_o^2)\sin(\vartheta)\cos(\vartheta) & n_e^2\cos(\vartheta)^2+n_o^2\sin(\vartheta)^2 \end{bmatrix}$
Longitudinal	ϕ	$\begin{bmatrix} n_o^2\sin(\phi)^2+n_e^2\cos(\phi)^2 & (n_e^2-n_o^2)\sin(\phi)\cos(\phi) & \\ (n_e^2-n_o^2)\sin(\phi)\cos(\phi) & n_o^2\cos(\phi)^2+n_e^2\sin(\phi)^2 & \\ & & n_o^2 \end{bmatrix}$
Equatorial	γ	$\begin{bmatrix} n_o^2\cos(\gamma)^2+n_e^2\sin(\gamma)^2 & & (n_e^2-n_o^2)\sin(\gamma)\cos(\gamma) \\ & n_o^2 & \\ (n_e^2-n_o^2)\sin(\gamma)\cos(\gamma) & & n_e^2\cos(\gamma)^2+n_o^2\sin(\gamma)^2 \end{bmatrix}$

conditions imposed for the slab's investigation are the invariance along the y-coordinate direction $\partial/\partial y$ and the exponential modal dependence $\exp(-jk_0nz + j\omega t)$. The losses due to the metal layer are expressed via a complex refractive index $n_m = n_{rm} + jn_{im}$. The lossy guided and the leaky propagation modes can be modelled via the complex refractive effective index $n = n_r + jn_i$. The field components parallel to the layer interfaces are linked by a first-order differential equation system. This system is derived from Maxwell's curl equations and can be solved analytically in the homogeneous layers [14–17]. The solution of the first-order linear differential equation is particularly simple in the isotropic layers. In fact, in this case, pure transversal electric (TE) or transversal magnetic (TM) polarizations occur. Anisotropic mode coupling between TE and TM waves can occur in uniaxial media because of the optical c-axis rotation. In anisotropic guided propagation, the electromagnetic fields can have both ordinary and extraordinary polarization components. In the core diffused region, the refractive ordinary and extraordinary indices are depending on the x direction. A closed solution of this system does not exist. A solution is obtained by a suitable numerically integration. As example, the Gear predictor–corrector method can be successfully employed, via the transformation matrix linking the electromagnetic field components at the abscissae x and x_0 as follows:

$$\tilde{E}(x) = (x, x_0)\tilde{E}(x_0) \tag{2.2}$$

where the vector $\tilde{E}(x)$ is the electromagnetic field transversal with respect to the diffusion direction x, at the generic depth x in the diffused core, while x_0 is the abscissa at the beginning of the diffusion.

In the model, the layers made of air, overlay, metal are homogeneous and isotropic. This is the case for which the differential equation system can be analytically solved. The field components in the three media can be easily expressed by means of an overlapping of exponential functions. Therefore, the eigenvalues of the matrix system, for the polar, the longitudinal and equatorial dielectric tensor configuration, respectively, are the following:

$$\begin{aligned}
\mu_{Pos} &= \pm\sqrt{n^2 - n_{os}^2} \\
\mu_{Pes} &= \pm\sqrt{\frac{n^2\left(n_{es}^2\cos(\vartheta)^2 + n_{os}^2\sin(\vartheta)^2\right)}{n_{os}^2} - n_{es}^2} \\
\mu_{Los} &= \pm\sqrt{n^2 - n_{os}^2} \\
\mu_{Les} &= \pm n_{os}\sqrt{\frac{n^2 - n_{es}^2}{n_{es}^2\cos(\phi)^2 + n_{os}^2\sin(\phi)^2}} \\
\mu_{Es} &= jn\frac{\left(n_{es}^2 - n_{os}^2\right)\sin(\gamma)\cos(\gamma)}{n_{os}^2\cos(\gamma)^2 + n_{es}^2\sin(\gamma)^2} \pm n_{os}^2\sqrt{\frac{n^2 - \left(n_{os}^2\cos(\gamma)^2 + n_{es}^2\sin(\gamma)^2\right)}{n_{os}^2\cos(\gamma)^2 + n_{es}^2\sin(\gamma)^2} n_{os}^2 n_{es}^2}
\end{aligned} \tag{2.3}$$

where μ_{Pos}, μ_{Pes} and μ_{Los}, μ_{Les} are the ordinary and extraordinary transversal eigenvalues evaluated for the polar and the longitudinal configurations, respectively. The electromagnetic field profile in the substrate is modelled via an exponential function, $\exp(\pm\mu x)$, describing a decreasing or increasing electromagnetic field amplitude along the waveguide depth x. In fact, the sign of the eigenvalue μ is considered negative or positive if waves are guided or leaky, respectively. In other words, for the guided modes, the electromagnetic field vanishes in the substrate, while for the leaky modes, it increases, describing an incoming electromagnetic power from the core towards the substrate; this rule is verified for both the ordinary and extraordinary components. For the equatorial configuration, there is a unique transversal eigenvalue μ_{Es} expression. The choice $\pm$ for the sign of the two eigenvalues is made similarly to that performed in the polar and longitudinal case [14–17]. The boundary conditions for the parallel electromagnetic components at the interfaces of the homogeneous layers, namely air-overlay, overlay-metal, metal-core and at the diffusion depth of waveguiding region lead to the determinantal/eigenvalue equation. The solutions/roots of the determinantal equation are the complex propagation constants of the electromagnetic modes. They can be calculated by using any numerical root-finding algorithm, e.g. by following the tangent plane method optimization strategy. A good convergence criterion can be based on the relative decrease of the determinantal equations value. More precisely, the comparison or ratio between the value calculated at the nth iteration and the initial value is considered a good item to identify the convergence condition. A good decreasing ratio between the two values is heuristically identified close to 10^{-8}–10^{-10} [14–17]. The imaginary part of the complex refractive index found via the tangent plane method is practically coincident with the extinction coefficient of the guided modes calculated by following the conventional perturbation method, generally applied for loss-less guided solution [14]. The described investigation allows excellent results with reduced computation cost. The accuracy and the congruency of the solutions found via a computer code based on the aforesaid theory was tested and fully demonstrated via the comparison with the experimental investigation [16,17]. As an example, in [16,17] the waveguide under test was a Ti:$LiNbO_3$ type, fabricated via the diffusion of 530 Å thick Ti layer at 1,000 °C in dry atmosphere and for a diffusion time of 10 h on Y-cut substrate. Two digital goniometers were employed, the first to measure the launching laser angle with respect to the waveguide plane, allowing the mode-matching angle evaluation; the second one to measure the direction of the optical c-axis with respect to the laboratory axes x, y, z, allowing the azimuth (polar) angle θ evaluation. A rutile prism, having ordinary and extraordinary refractive indices $n_{op} = 2.8656$ and $n_{ep} = 2.5838$, respectively, and a base angle equal to 49.45°, was employed as input optical coupler. A similar prism was employed to couple out the light beam, and its birefringence decoupled the TE and TM polarizations, allowing one to project on a screen well-separated TE and TM m-line patterns. Both the input and the output prisms were assembled on a single support. Via the single support, they were simultaneously maintained in-line with respect to the laser beam direction.

The polar configuration was investigated with this m-line characterization set-up. The azimuth angle θ was varied by rotating the waveguides with respect to the prisms. The IWKB (inverse Wentzel–Kramers–Brillouin) method was employed for waveguide optical spectroscopic characterization, by recovering the refractive index profile. The IWKB characterization was developed where TE and TM decoupling was complete and apparent, i.e. for $\theta = 90°$. The core refractive index profile was semi-Gaussian with extraordinary and ordinary depths $d_e = 3.94$ µm and $d_o = 5.24$ µm. The extraordinary and ordinary refractive index changes at the surfaces were $\Delta n_e = 0.019$ and $\Delta n_o = 0.010$. These values were considered in the simulation of the waveguide. Figure 2.2 shows the real part n_r of the refractive effective index versus the azimuth angle θ, for the guided and leaky modes. The simulated values (full curves) are compared with the experimental values (red circles) obtained by employing the two optical goniometers as previously described. The dashed curves illustrate the limiting values below which the transition from guided to leaky propagation occurs. In fact, they represent the values for which the eigenvalue is zero. When the eigenvalue sign changes from negative to positive, this indicates that the ordinary and extraordinary natures of wave change

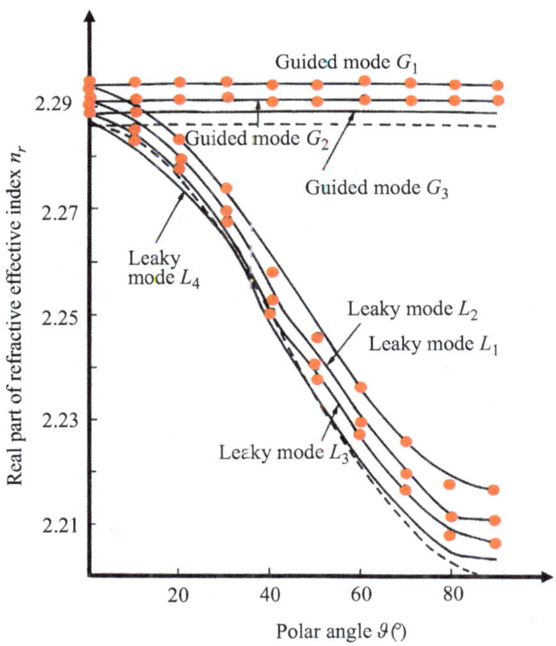

Figure 2.2 Comparison between simulation and experiment; curves of the real part n_r of the refractive effective index of guided and leaky modes versus the polar angle θ, for a Ti-diffused $LiNbO_3$ waveguide. The red circles indicate measured values via m-lines. Reproduced with adaption from [17], courtesy of the Electromagnetics Academy

from guided to leaky in the substrate. As an example of application, the transition from guided to leaky modes was proposed for the design of optical transistor in [33].

2.2.1.2 Integrated optical devices based on periodically poled lithium niobate (PPLN)

The modal investigation of LN and the application of CMT have allowed us to do the design of several devices in addition to the electro-optic modulators. A strong interest towards this strategic material occurred in the early 1990s for the potential applications predicted by using the cascade of non-linear second-order susceptibility $\chi^{(2)}:\chi^{(2)}$. For instance, in [20], it was shown by simulation that the non-linear refraction due to the cascade process can produce much larger effects than the electronic third-order non-linearity. These results promised several advantages, such as a reduced switching power of non-linear couplers. This was confirmed by other simulations reported in [21,22]. Moreover, wavelength conversion/shifting, phase modulation and all optical amplifications based on the cascaded second-order non-linearity in LN waveguides have been extensively investigated, especially in periodically poled LN (PPLN) waveguides [23–35]. Most of these LN devices were based on the interaction between the modes at the fundamental wavelength and at the second harmonic (SH).

The design of an optical thermometer based on this approach is illustrated in this section. The dependence of second harmonic generation (SHG) on the phase mismatch between the interacting waves, due to the variation of grating period of the inverted ferroelectric domains, is exploited as key principle for the sensor operation. The thermal dispersion of waveguide refractive index [37,38] causes a change of the mode propagation constants. In addition, the thermal expansion of LN [39] induces a variation of grating period of the inverted ferroelectric domains and of the efficiency of SH generation. SH generation occurs at the exact phase matching or at least at a proper/small phase-mismatch between the fundamental and SH interacting waves. This necessary condition can be obtained by following different techniques. A first technique, which is simply based on anisotropic propagation, is the birefringence phase matching (BIRPM) [20–22]. The most largely employed technique is the quasi-phase matching (QPM) technique, e.g. in PPLN waveguides [23–35]. However, other approaches were investigated, such as the use of both QPM and BIRPM [36], and the thermal dispersion [37–40].

In [41], an optical thermometer was proposed; it was based on the thermal dispersion of LN waveguides, fabricated via the annealing proton exchange–periodically poled LN (APE–PPLN) techniques. APE–PPLN waveguides allow high SHG efficiency. In these structures, the grating constitutes a periodic modulation of the second-order susceptibility elements. It allows the selection of the interacting modes having TM polarization in *Z*-cut $LiNbO_3$ substrates. For *Z*-cut $LiNbO_3$ crystal cut, the involved non-linear coefficient is d_{33}, having value larger than d_{31}, which is pertaining to the case in which both TM- and TE-polarized modes are involved. Moreover, the gratings fabricated by the poling technique are integrable with other waveguiding structures on the same substrate. The following expression is considered, linking the polarization vector with the electric field:

$$P_i = \varepsilon_0 \left(\sum_j \chi_{ij} E_j + \sum_{j,k} \chi^{(2)}_{ijk} E_j E_k + \sum_{j,k,l} \chi^{(3)}_{ijkl} E_j E_k E_l + \cdots \right) \tag{2.4}$$

where P_i is the i-th component of the polarization vector $\mathbf{P}$; E_j is the j-th component of the electric field (the additional terms E_k and E_l are considered for the cases of second-order and third-order nonlinear susceptibilities, respectively); 1, 2 and 3 are the crystal axes X, Y and Z, with $Z \equiv c$ optical axis of crystal, $i, j, k, l = 1, 2, 3$.

By considering the three electric fields at the same frequency ω, for an LN crystal, the following tensor expression can be obtained:

$$\begin{bmatrix} P_1(2\omega) \\ P_2(2\omega) \\ P_3(2\omega) \end{bmatrix} = 2 \begin{bmatrix} 0 & 0 & 0 & 0 & d_{31} & -d_{22} \\ -d_{22} & d_{22} & 0 & d_{31} & 0 & 0 \\ d_{31} & d_{31} & d_{33} & 0 & 0 & 0 \end{bmatrix} \times \begin{bmatrix} E_1^2(\omega) \\ E_2^2(\omega) \\ E_3^2(\omega) \\ 2E_2(\omega)E_3(\omega) \\ 2E_1(\omega)E_3(\omega) \\ 2E_1(\omega)E_2(\omega) \end{bmatrix} \tag{2.5}$$

For the interacting modes having TM polarization in Z-cut APE waveguides, as the axes 1, 2 and 3 correspond to the laboratory axes y, z and x, the final expressions are as follows:

$$\begin{aligned} P_y(2\omega) &= 0 \\ P_z(2\omega) &= 2d_{22}E_z^2(\omega) + 2d_{31}E_z(\omega)E_x(\omega) \\ P_x(2\omega) &= d_{31}E_z^2(\omega) + d_{33}E_x^2(\omega) \end{aligned} \tag{2.6}$$

Since the non-linear coefficient d_{33} in LN crystals is much larger than d_{31}, the second one can be neglected in the investigation. The CMT is applied to the APE–PPLN waveguide. The total electromagnetic field $\bar{E}(x,y,z,t)$, $\bar{H}(x,y,z,t)$ in the waveguide is expressed as an overlapping of the unperturbed propagation modes $\bar{E}_i(x,y,z,t)$, $\bar{H}_i(x,y,z,t)$ suitably normalized to the unit power:

$$\bar{E}(x,y,z,t) = \sum_{i=1}^{2} A_i(z)\, \bar{E}_i(x,y,z,t) \tag{2.7}$$

$$\bar{H}(x,y,z,t) = \sum_{i=1}^{2} A_i(z)\, \bar{H}_i(x,y,z,t) \tag{2.8}$$

$i = 1$ refers to the fundamental λ_f and $i = 2$ to the SH λ_{sh} wavelength. According to the CMT, the coefficient A_i [$W^{1/2}$] is the perturbed amplitudes of the propagating modes. The squared modulus of the perturbed amplitudes represents the mode power along the z propagation direction. The propagation modes can be calculated via the simple effective index method (SEIM) [14–17,23], previously illustrated, or via a numerical technique as the FEM. For non-rotated Z-cut crystal, the dielectric constant tensor is in diagonal configuration and the formalism is simple. To perform a realistic investigation, for the diffused channel waveguides, the extraordinary $n_e(x,y)$ and ordinary $n_o(x,y)$ refractive index profiles are considered semi-Gaussian along the x and Gaussian along the y coordinates:

$$n_{o,e}(x,y) = n_{os,es} + n_{o,e} \cdot e^{-\left(x/d_{o,e}\right)^2} \cdot e^{-(y/w)^2} \tag{2.9}$$

where $d_{o,e}$ is the diffusion depth and w is the channel width. The eigenvalue/determinantal equation is obtained as illustrated in the previous section by imposing the continuity of the tangential components of the electromagnetic fields by exploiting the transformation matrix formalism and via the numerical integration [14–17,23].

The grating made of inverted ferroelectric domains can be modelled by considering the evolution of the non-linear coefficient d_{33} along the z propagation direction. It can be expressed via a Fourier series. A grating profile having d_{33}, duty-cycle $D = 50\%$ and grating period Λ_g, for a first-order QPM condition between the fundamental and the SH waves, gives two coupled wave equations:

$$\begin{aligned}\frac{\partial A_1}{\partial z} &= -\frac{j\omega_f \varepsilon_0 d_{33}}{\pi} A_2 A_1^* e^{-j\beta z} \int_{-\infty}^{+\infty}\int_{-\infty}^{+\infty} E_2 E_1^* E_1^* \, dx\, dy \\ \frac{\partial A_2}{\partial z} &= -\frac{j\omega_{sh} \varepsilon_0 d_{33}}{2\pi} A_1^2 e^{j\beta z} \int_{-\infty}^{+\infty}\int_{-\infty}^{+\infty} E_1^2 E_2^* \, dx\, dy\end{aligned} \tag{2.10}$$

where ε_0 is the dielectric constant of vacuum. The mode mismatch is minimized or eliminated via the grating contribution. It has the following expression $\Delta\beta = \beta_{sh} - 2\beta_f - 2\pi/\Lambda_g$, where β_{sh} and β_f are the SH and the fundamental mode propagation constants, ω_{sh} and ω_f the SH and the fundamental radian frequency. For a different duty-cycle, (2.10) changes because different values of Fourier series coefficients are involved.

In the optical thermometer, the thermal behaviour of the material is taken into account via the thermal expansion and via refractive index dispersion [37–42]. The Sellmeier equations for the extraordinary and ordinary refractive index of the samples used in [41,42], as a function of both the wavelength and the temperature are as follows:

$$\begin{aligned} n_o^2(\lambda, T') &= A_o + \frac{B_o + C_o T'^2}{\lambda^2 - \left(D_o + E_o T'^2\right)^2} + F_o \lambda^2 \\ n_{es}^2(\lambda, T') &= A_e + B_e T'^2 + \frac{C_e + D_e T'^2}{\lambda^2 - \left(E_e + F_e T'^2\right)^2} + G_e \lambda^2 \end{aligned} \tag{2.11}$$

where λ is the wavelength in microns, T' is the temperature expressed in Kelvin and the constants, A_o, B_o, C_o, D_o, E_o, F_o, and A_e, B_e, C_e, D_e, E_e, F_e, G_e, are optimized by matching (2.11) with the measured ordinary and extraordinary refractive indices, respectively [37–39]. The Sellmeier equations (2.11) modelling the investigated

$LiNbO_3$ waveguides exhibit the following parameters: $A_o = 4.9130$, $B_o = 0.1173$, $C_o = 1.65{\cdot}10^{-8}$, $D_o = 0.212$, $E_o = 2.7{\cdot}10^{-8}$, $F_o = -0.0278$ and $A_e = 4.5567$, $B_e = 2.605{\cdot}10^{-7}$, $C_e = 0.0970$, $D_e = 2.70{\cdot}10^{-8}$, $E_e = 0.201$, $F_e = 5.4{\cdot}10^{-8}$, $G_e = -0.0224$.

The thermal expansion induces an increase of waveguide length $L(T)$ and of the grating period $\Lambda_g(T)$. It is modelled by the expansion coefficient α:

$$\alpha = \frac{1}{L(T_0)} \cdot \frac{dL(T_0)}{dT} \tag{2.12}$$

In (2.12), for a Z-cut crystal, the thermal coefficient is $\alpha = 0.75{\cdot}10^{-5}$ $[°C]^{-1}$ in the temperature range from 0 °C to 500 °C [39].

A channel waveguide can exhibit a high sensitivity to the thermal variations. An APE–PPLN waveguide in Z-cut crystal configuration, with channel width ($2w$) equal to 8.4 μm, channel depth $d = 3.62$ μm, ordinary and extraordinary refractive index change $\Delta n_o = 0.01$ and $\Delta n_e = 0.02$, waveguide length $L_g = 26$ mm, non-linear coefficient $d_{33} = 14$ pm/V, duty-cycle of PPLN grating $D = 0.6$, grating period $\Lambda = 16.3$ μm, was considered in [41,42]. The set-up used for the characterization of the PPLN waveguide sensor versus the temperature included as source is a tunable laser in the range 1500–1580 nm, coupled to an EDFA (erbium-doped fibre amplifier) with a maximum output power of 30 dBm, a Peltier cell assured a temperature control within 0.1 °C.

The simulated and measured efficiencies of SH generation versus the wavelength are in good agreement [41,42]. In particular, the efficiency is expressed as the ratio $\eta = (P_{sh}(z = L))/\left(P_f(z = 0)\right)^2$, for T=333.15K and 334.15K, $P_f(z = 0)$ is the power of the TM_{00} propagation mode at the fundamental wavelength, and at the waveguide input section, ($z = 0$) and $P_{sh}(z = L)$ are the power of the TM_{00} propagation mode at the generated SH wavelength, at the waveguide output section ($z = L$). The peak/maximum of conversion efficiency, i.e. the QPM condition $\Delta\beta = 0$, shifts towards longer wavelengths by increasing the temperatures. The change of QPM wavelength is $\Delta\lambda_{QPM} \cong 0.14$ nm, changing from 1,553.36 to 1,554.00 nm when the temperature changes from $T = 333.15$K [41,42]. A linear SGH power versus the temperature was obtained from 423.5K up to 673.15K [41].

Similar performances can be obtained via titanium in diffused ($Ti{:}LiNbO_3$) PPLN waveguides. The $Ti{:}LiNbO_3$ waveguides exhibit, as an advantage, the possibility of operating even at temperatures larger than 400 °C without any damage which, on the contrary, occurs for temperature larger than 100 °C–150 °C when using APE technology.

2.2.2 Modelling of integrated evanescent field sensors

The design of an evanescent field optical sensor can be performed via the conventional electromagnetic mode investigation used for slab, channel or ridge waveguides. As an example, for monitoring the hydrocarbon pollution of water, the sensor can be constituted by an optimized glass planar waveguide with a polymeric overlay. The detection of the contaminant can be performed in the UV wavelength region, in the range from 250 to 400 nm. In this region, several hydrocarbons exhibit high absorption peaks, and the water loss is negligible [43,44]. Another interesting wavelength region is the medium infrared. In both the cases, optimized glasses are required. If the sensor is surrounded by the water containing hydrocarbon, at the equilibrium, the hydrocarbon concentration C_p in the polymer is very high if compared to that in the water C_w. The distribution constant $K_{p/w} = C_p/C_w$ is a typical parameter of the used polymer and depends on the particular hydrocarbon. The polymer, in the absence of contaminant, is practically lossless. Its refractive index is real in the absence of hydrocarbon. The polymer infiltrated/impregnated by the hydrocarbon and the water with the distributed hydrocarbon both exhibit absorption losses and thus are modelled via a complex refractive index. In the hypothesis of small concentrations, the real part of the complex refractive index of the polymer impregnated by hydrocarbon is almost constant. On the contrary, the imaginary part almost linearly depends on the concentration of the hydrocarbon. For low contaminant concentrations, the propagation mode exhibits a power attenuation almost proportional to the hydrocarbon concentration. In addition, the polymer is water-repellent. This allows the correct sensor operation since the interference of the analyte bands with the broad bands of the water OH groups is avoided or at least can be neglected. To enhance the sensor performance, the optimal operation wavelength must be chosen where the hydrocarbon absorption peak occurs.

The evanescent field sensor principle is very simple. The light intensity, in a cell which contains a chemical species, is modelled by the Lambert–Beer law:

$$I = I_0 10^{-\varepsilon Cz} = I_0 e^{-\alpha_m Cz} \tag{2.13}$$

where I_0 is the light intensity travelling in the cell at the wavelength where the species absorption occurs; I is the light transmitted at the output of the cell having length z [cm]; C [mol/l] is the equilibrium molar concentration of the species, ε [l/(mol×cm)] is the molar extinction coefficient, α_m [l/(mol×cm)] is the molar absorption coefficient.

The imaginary part of the complex refractive index $n_c = n - jk$ of the medium in the cell is as follows:

$$\frac{4\pi}{\lambda}k = \frac{\varepsilon C}{\log_{10}(e)} = \alpha_m C \tag{2.14}$$

where λ [cm] is the peculiar line absorption wavelength of the analyte, i.e. the monitored species.

The electromagnetic study of a waveguide sensor covered by a polymeric overlay and surrounded by hydrocarbon-polluted water can be performed with a numerical method as FEM. The phase and amplitude mode dependence on the z propagation direction is modelled by $\partial/\partial z = -jk_0(n_{ef} - jk_{ef})$, where $n_{cef} = n_{ef} - jk_{ef}$ is the complex refractive effective index, $k_0 = 2\pi/\lambda$ is the wave number of vacuum and $k_0 n_{cef}$ the complex propagation constant of the mode. An alternative to FEM or other numerical approach could be the electromagnetic analysis of the channel waveguides carried out by means of the SEIM or EIM previously recalled for LN waveguides. The investigation of the 2D structures in the x, y plane is performed with the analysis of two auxiliary slab waveguides with infinite size along the y and x direction, respectively, according to the Marcatili approach. The temporary refractive effective index n_t of the first slab is calculated by imposing the invariance along y-coordinate ($\partial/\partial y = 0$). As second step, the invariance along x-coordinate ($\partial/\partial x = 0$) is imposed for the analysis of an auxiliary waveguide having a core with the refractive index previously calculated (temporary refractive effective index n_t). In this way, the solutions have hybrid polarization. The modal characteristics are calculated by imposing the continuity of the electromagnetic field components at the interfaces of the planar waveguide having step-index profile. For diffused waveguides, the method is the same described for PE- or Ti-diffused LN waveguide in Section 2.2.1.1.

The imaginary part k_{ef} (modal extinction coefficient) of the complex refractive effective index n_{cef} allows the evaluation of the power absorption coefficient of the guided propagation mode:

$$\alpha_{ef} = 2k_0 k_{ef} = k_{ef}\frac{4\pi}{\lambda} \tag{2.15}$$

The optical power in the waveguide depends on the coordinate z:

$$P(z) = P_0 e^{-\alpha_{ef} z} \tag{2.16}$$

where P_0 is the light power launched at the wavelength where the species absorption line occurs. By applying the perturbation theory, power absorption coefficient of the guided propagation mode is calculated:

$$\alpha_{ef} = 2\frac{k_0}{\eta_0}\frac{\iint\limits_{\infty} n(x,y)k(x,y)\left|\bar{E}(x,y)\right|^2 dA}{\mathcal{Re}\left\{\iint_{\infty}\left[\bar{E}(x,y)\times\bar{H}^*(x,y)\right]\cdot\bar{a}_z dA\right\}} \tag{2.17}$$

where η_0 is the characteristic impedance of the vacuum, $n(x,y)$ and $k(x,y)$ are the real and the imaginary part of the waveguide refractive index profile along the transversal directions x and y; $P_0 = 1/2\Re e\left(\iint_{\infty}\left[\bar{E}(x,y)\times\bar{H}^*(x,y)\right]\cdot\bar{a}_z dA\right)$ is the real part of the power flux, through the waveguide section, along the propagation direction z (a_z being the versor of z-axis); $\bar{E}(x,y), \bar{H}(x,y)$ are the electric and magnetic fields pertaining to the unperturbed mode. In the case of guided modes, the FEM numerical solution perfectly agrees with the calculation of (2.17). In [43,44] different kinds of waveguides were investigated for the detection of chlorobenzene and toluene.

Five waveguide configurations were investigated, via FEM, by considering both guided TE- and TM-like fundamental mode polarizations [44]. A very complex channel waveguide was also designed, with a rectangular groove on the top of the 15% G_eO_2-doped region and an air hole in the 5.5% G_eO_2-doped region [44]. It was characterized by a Polydimethylsiloxane (PDMS) thickness $t = 1.1$ µm; thickness of

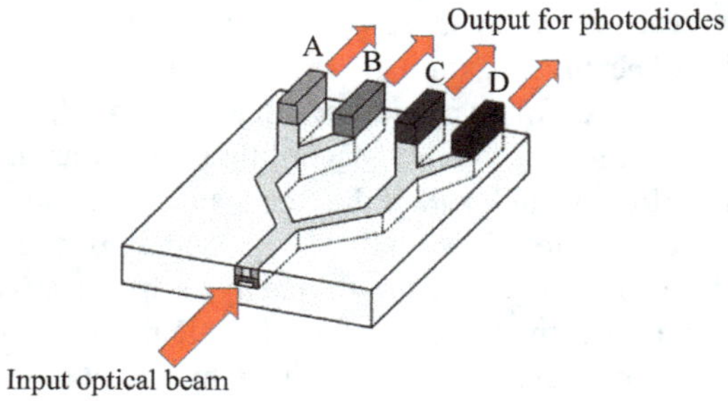

Figure 2.3 Sketch of the sensor array. Adapted from [44], reprinted by permission from Springer

the 5.5% G_eO_2-doped region $h_1 = 285$ nm; thickness of the 15% G_eO_2-doped region $h_2 = 215$ nm; groove width $w_g = 140$ nm, groove thickness $h_g = 215$ nm, air hole width $w_h = 500$ nm, air hole thickness $h_h = 70$ nm. This waveguide was identified as the optimal one and exhibited the best sensor performance. It allowed only the propagation of TE-like fundamental mode polarization. A sensor having length $L = 1.18$ mm exhibited an absorbance slope $S^A_{TE} = 2.52\times10^{-2}$ ppm^{-1}. A sensor array was designed in order to simultaneously detect the concentration of toluene, benzene, chlorobenzene and ethylbenzene. The sensor array is sketched in Figure 2.3. In this array, each sensor employs a different polymer, thus giving a different output.

2.2.3 Modelling of rare-earth-doped devices

The modelling of rare-earth-doped waveguides is generally performed by considering the light–matter interaction, via the rate equation for the ion level concentrations, and the power propagation equations. Examples of waveguide modelling are reported in [45,46]. In this section, the modelling of a rare-earth-doped micro-disk will be illustrated. Micro-disks are fundamental elements in a number of optical systems for which high-Q factor cavities are required. They are sometimes preferred to other micro-resonators, such as microsphere [47,48], since they can be designed and integrated within complex optical circuits by employing a single planar substrate. Moreover, they allow excellent mechanical stability and low-cost fabrication [49]. Among the techniques used for fabricating micro-disks, there is the well-known photolithography and reactive ion etching process. Micro-disks can exhibit very high-quality factors, close to $Q = 8\times10^6$. Rare-earth-doped micro-disks are interesting structures operating at well-selected wavelengths in order to obtain optical amplification or microlasers. These devices can be refined to operate in the mid-IR wavelength range, where promising active sensors can be based on the strong absorption of the vibrational transitions pertaining to many organic species and biomolecules.

2.2.3.1 Modelling of integrated optical micro-disks

As described in [49], a well-validated approach can be followed for the simulation of active WGM micro-resonators. In particular, the modelling of rare-earth-doped micro-disks can be divided into three main steps regarding three different physical mechanisms:

1. The properties of the bounded electromagnetic field in the passive structure. In this step, the modal analysis giving the electromagnetic field profile confined by both micro-resonator and optical waveguides is simulated by neglecting the rare earth.
2. The optical coupling of micro-disk with a waveguide or two different waveguides constituting the input/output sections of the device. For this mechanism, the simulation of the coupling between the resonator and the waveguides is performed by employing CMT.
3. The light–matter interaction. For the modelling of this process, the optical gain/lasing is evaluated via the rate equations for the rare-earth ions population densities [45–49].

The rare-earth-doped micro-disk is the part of the device providing the optical gain. It is generally coupled to two ridge waveguides, optimized by considering the different operating wavelengths, one for the signal and the other one for the pump. Sometimes a single coupling waveguide is considered. The sketch of the device is illustrated in Figure 2.4, where $A^P_{in,\ WGM}$, $A^P_{out,WGM}$ are the pump amplitudes of electromagnetic field at the input and output waveguide sections; A^P_{WGM} is the pump amplitude inside the micro-disk; $A^S_{in,WGM}$, $A^S_{out,WGM}$, A^S_{WGM} are the signal amplitudes of electromagnetic fields similarly defined. The other parameters in Figure 2.5 will be defined in the following subsections; they are pertaining to the coupling mechanism and intrinsic cavity decay rates [49].

2.2.3.2 Whispering gallery mode approximation

The exact calculation of the micro-disk electromagnetic modes requires heavy calculations. The solution of the Maxwell equation is found by employing the electromagnetic field expression via the cylindrical Bessel function. The boundary conditions at the micro-disk dielectric/air interface, i.e. at the lateral cylindrical

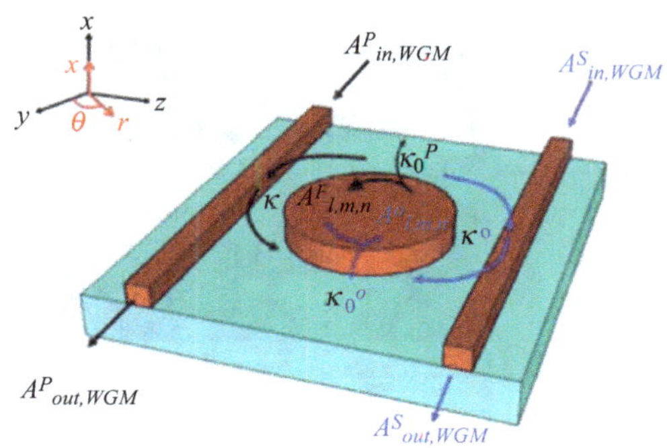

Figure 2.4 Sketch of the micro-disk coupled to two ridge waveguides

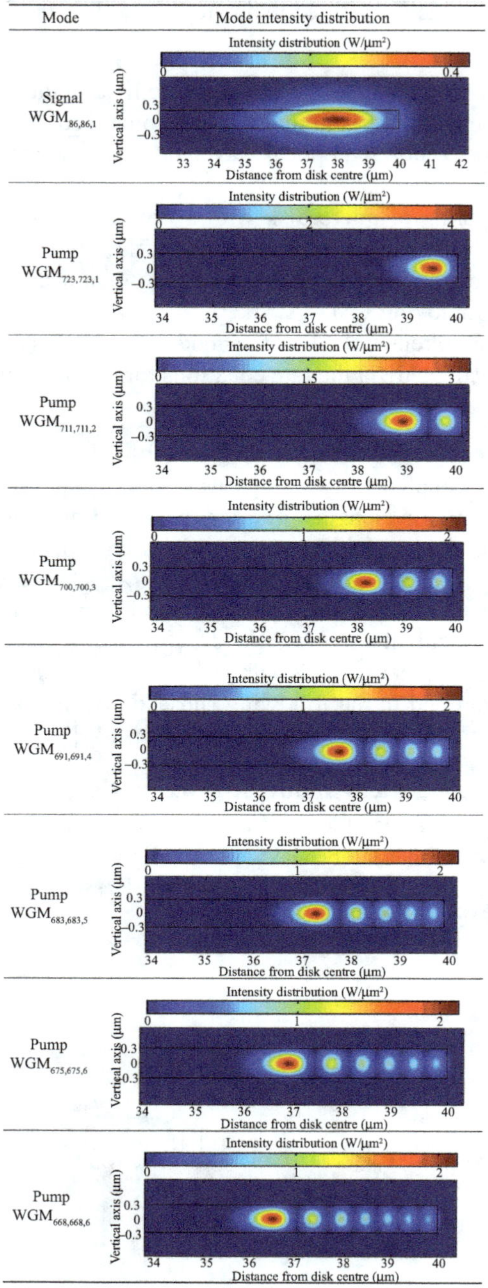

Figure 2.5 Light intensity distribution of the signal and of the seven pump modes reported in Table 2.3, with total mode power normalized to 1 W. Micro-disk radius is $R_{\mu disk} = 40\ \mu m$ and height is $h_{\mu disc} = 600\ nm$

walls and at the top and bottom cylinder basis surfaces, are imposed. An approximated approach allowing a strong reduction of calculation cost can be followed in order to calculate the electromagnetic field in a micro-disk: it is known as whispering gallery mode approximation [49]. The micro-disk is considered as a 1D slab waveguide. The effective refractive index n_{eff} of an approximated three-layer structure made of air, micro-disk region and substrate is searched via SEIM. This method can be used to obtain useful preliminary guidelines for designing the micro-disk. In order to perform a refined design, an FEM solver can be efficiently employed [49,53]. It is useful especially in the case of resonators with non-traditional geometry, such as micro-disks with edge inclination or exotic geometries.

2.2.3.3 CMT applied to rare-earth-doped micro-disk

The CMT is a very efficient approach for simulating the optical coupling between the micro-disk and the waveguides. Even with the typical approximations/assumptions made in this model, a very wide range of devices can be investigated. The model requires in principle the weak coupling and the consequent slowly varying amplitude approximation. The rate equations describing the time evolution of the amplitude of the electromagnetic field in the micro-disk are written as function of (i) the coupled pump and signal power of each WGM propagating into the micro-disk and (ii) N_i, the concentration of ions in ith level of the rare-earth element.

The time domain evolution of the amplitude A for the pump A^P_{WGM} and the signal A^s_{WGM} of the internal cavity electromagnetic field can be obtained by solving the suitable differential equations [50]. In particular, these equations link the aforesaid amplitudes with $A^P_{in,WGM}$ and $A^P_{out,WGM}$, namely the amplitudes of the pump electromagnetic field at the input and at the output ridge waveguide sections and coupled with the WGM mode of the micro-disk; N_i is the rare-earth ion population concentration of ith energy level; $\sigma_{ij}(\omega_{WGM})$ is the rare-earth cross section at frequency ω_{WGM} from level i to level j. Other parameters pertaining to the micro-resonator and coupling waveguides are $\tau = 2\pi\rho_0 n_{eff}/c$, the circulation time inside the micro-resonator, R_0, the micro-resonator radius; n_{eff}, the WGM effective refractive index of pump and signal; c, the speed of light in vacuum; ω_{WGM}, the WGM resonant frequency; $\omega = \omega_{in} - \omega_{WGM}$, the frequency detuning of the input signal with respect to the WGM resonance frequency; $\tau_0 = 1/\kappa_0^2 = Q/\omega_{WGM}$, the intrinsic lifetime; Q, the intrinsic quality factor; κ_0 is the intrinsic cavity decay rate; τ_{ext} is the coupling lifetime.

$$\begin{aligned}
\frac{dA^P_{WGM}}{dt} &= \left(-\frac{2}{\tau_{ext}} - \frac{1}{\tau_0} - g^P_{WGM} + i\omega\right)A^P_{WGM} - i\sqrt{\frac{2}{\tau_{ext}\tau}}A^P_{in,WGM} \\
\frac{dA^S_{WGM}}{dt} &= \left(-\frac{2}{\tau_{ext}} - \frac{1}{\tau_0} - g^S_{WGM} + i\omega\right)A^S_{WGM} + \frac{c}{2n_{eff}}\sum\nolimits_q N_i\sigma_{ji}(\omega_{WGM})\Omega^S_{WGM}A_0 \\
&\quad - i\sqrt{\frac{2}{\tau_{ext}\tau}}A^S_{in,WGM}
\end{aligned} \tag{2.18}$$

In particular, the light–matter interaction due to the rare earth can be modelled via g^P_{WGM} and g^s_{WGM}, which are the optical gain/absorption coefficients related to the particular WGM expressed by the following equation:

$$
\begin{aligned}
g^p_{WGM} &= \frac{c}{2n_{eff}}\left[N_i\sigma_{ji}(\omega_{WGM})\Omega^p_{WGM} - N_i\sigma_{ij}(\omega_{WGM})\Omega^p_{WGM}\right] \\
g^s_{WGM} &= \frac{c}{2n_{eff}}\left[N_i\sigma_{ji}(\omega_{WGM})\Omega^s_{WGM} - N_i\sigma_{ij}(\omega_{WGM})\Omega^s_{WGM}\right]
\end{aligned}
\tag{2.19}
$$

Ω^P_{WGM}, Ω^S_{WGM} are the overlap factors of each WGM with the rare-earth profile inside the doped region A_d for the pump and the signal, expressed by the following equation:

$$
\begin{aligned}
\Omega^P_{WGM} &= \iint_{A_d}\left|E^P_{WGM}\right|^2 ds \\
\Omega^S_{WGM} &= \iint_{A_d}\left|E^S_{WGM}\right|^2 ds
\end{aligned}
\tag{2.20}
$$

where E^P_{WGM}, E^S_{WGM} are the normalized electric field on the cross-sectional plane for pump and signal. The coupling lifetime can be calculated by the following equation:

$$
\tau_{ext} = \frac{2\tau}{\kappa^2} \tag{2.21}
$$

where κ is the cavity decay rate or coupling coefficient [50].

The gain g of the micro-disk amplifier is [49,53]

$$
g = \left|\frac{A^S_{OUT,WGM}}{A^S_{IN,WGM}}\right|^2 = \left|\sqrt{1-\frac{\tau}{\tau_{ext}}} + j\sqrt{\frac{\tau}{\tau_{ext}}}\frac{A^S_{WGM}}{A^S_{IN,WGM}}\right|^2 \tag{2.22}
$$

Suitable WGMs allowing high overlapping are those with the lowest radial modal number $n=1$ and lowest polar number $p = 1$. The electromagnetic field in the micro-resonator and in the waveguides, and the propagation constants can be obtained using an FEM solver. The overall loss depends on different loss phenomena: losses due to the bulk absorption, the scattering radiation, the tunnelling of light from micro-resonator. The intrinsic quality factor Q can be calculated as reported in [49,51–54].

2.2.3.4 Rare-earth doping of micro-disks

Rare-earth doping is used in order to activate passive crystalline and amorphous materials, thus obtaining optical gain. For the near-infrared wavelength region, there are a lot of dopants that allow efficient lasing in silica and germanate-silicate glasses, e.g. erbium, ytterbium, praseodymium, dysprosium and neodymium. Silica glass becomes absorbing beyond 2 μm. Rare-earth ions could allow useful transitions in mid-IR wavelength range between 3 and 10 μm. Therefore, in order to obtain lasing in mid-IR, other materials that are transparent at these wavelengths and have low phonon energy are required. Optimal results are obtained with tellurite, fluoride and chalcogenide glasses. Chalcogenide glasses allow the propagation at longest wavelengths, and a wide literature is available [55–65]. Novel pumping schemes in both pulsed and CW operation have been recently proposed to enhance mid-IR lasing [63,64]. A model suitable to design a chalcogenide rare-earth-doped micro-disk has been reported [49,53,66].

The interaction of the WGMs with the rare-earth-doped micro-resonator has been investigated by using the rate equation model.

With reference to the trivalent erbium ions in a chalcogenide glass: the main electronic transitions and the energy transfers are taken into account in the following points. The model includes several transitions:

- pump transition at the wavelength 0.98 μm;
- pump transition close to wavelength 0.806 μm;
- stimulated emission transition at the wavelength 4.5 μm;
- stimulated emission transition at the wavelength 2.7 μm;
- stimulated emission transition at the wavelength 1.55 μm and
- spontaneous energy transfer transitions occurring in pairs of Er^{3+} ions.

The pertaining rate equations are as follows [47]:

$$\left\{\begin{aligned}
\frac{dN_1}{dt} &= N_2\left(W_{21} + C_{up}N_2 + C_{24}N_4 + A_{21}\right) + N_3(C_3N_3 + A_{31}) + N_4A_{41} + N_5A_{51}\\
&\quad + N_6A_{61} - N_1(R_{13} + R_{14} + W_{12} + 2C_{16}N_6 + C_{14}N_4)\\
\frac{dN_2}{dt} &= N_1(W_{12} + 2C_{16}N_6 + C_{14}N_4) + N_3(W_{32} + A_{32}) + N_4(C_{14}N_1 + C_4N_4 + A_{42})\\
&\quad + N_5A_{52} + N_6A_{62} - N_2\left(W_{21} + C_{24}N_4 + 2C_{up}N_2 + A_{21}\right)\\
\frac{dN_3}{dt} &= N_1R_{13} + N_4A_{43} + N_5A_{53} + N_6A_{63} - N_3(W_{32} + 2C_3N_3 + A_{31} + A_{32})\\
\frac{dN_4}{dt} &= -N_4(A_{41} + A_{42} + A_{43} + C_{14}N_1 + 2C_4N_4 + C_{24}N_2) + 2C_{16}N_1N_6\\
&\quad + C_{up}N_2^2 + N_5A_{54} + N_6A_{64}\\
\frac{dN_5}{dt} &= N_6A_{65} - N_5(A_{51} + A_{52} + A_{53} + A_{54})\\
\frac{dN_6}{dt} &= C_3N_3^2 + C_4N_4^2 + C_{24}N_2N_4 - N_6(A_{61} + A_{62} + A_{63} + A_{64} + A_{65} + 2C_{16}N_1)
\end{aligned}\right. \tag{2.23}$$

In these equations, the spontaneous emission rate between levels i and j is expressed via the branching ratio β_{ij} between level i and j, τ_i the ion lifetime of level i.

$$A_{ij} = \frac{\beta_{ij}}{\tau_i} \tag{2.24}$$

C_{ij} takes into account the spontaneous energy transfer mechanisms, W_{ij} is the signal transition rate and R_{ij} is the pump transition rate modelling the contribution to the signal and pump rate of each WGM propagating into the micro-resonator [47].

In [47], all the spectroscopic parameters measured on the erbium-doped chalcogenide glass $Ga_5Ge_{20}Sb_{10}S_{65}$ are reported; thus, a realistic simulation can be performed. Similarly, in [49], all the praseodymium-doped micro-disk parameters and the pertaining rate equation are reported:

$$\begin{aligned}
&\frac{dN_1}{dt} = -(R_{14} + W_{12})N_1 + N_4A_{41} + N_2A_{31} + N_2\left(\frac{1}{T_2} + \frac{1}{\tau_2} + W_{21}\right) = 0\\
&\frac{dN_2}{dt} = W_{12}N_1 + N_3\left(\frac{1}{T_3} + A_{32}\right) + N_4A_{42} - N_2\left(\frac{1}{T_2} + \frac{1}{\tau_2} + W_{21}\right) = 0\\
&\frac{dN_3}{dt} = N_4\left(\frac{1}{T_4} + A_{43}\right) - N_3\left(\frac{1}{T_3} + A_{32} + A_{31}\right) = 0\\
&\frac{dN_4}{dt} = R_{14}N_1 - N_4\left(\frac{1}{T_4} + A_{43} + A_{42} + A_{41}\right) = 0\\
&N_1 + N_2 + N_3 + N_4 = N_{tot}
\end{aligned} \tag{2.25}$$

where τ_i and T_i are the ion lifetimes for pure-radiative and non-radiative transition from level i, respectively.

2.2.3.5 Design of an erbium-doped micro-disk

An erbium-doped chalcogenide micro-disk was simulated in [53–67]. The investigation was performed for the signal wavelength $\lambda_s = 4500$ nm and the pump wavelength $\lambda_p = 800$ nm; the micro-disk and the waveguides were made of gallium lanthanum sulphide glass, with erbium concentration $C_{Er} = 2.8\times10^{20}$ ions/cm^3. The waveguide height $h_w = 1.0$ μm and width $w = 2.5$ μm were identified in order to obtain the best trade-off between signal waveguiding and evanescent field coupling needs. A preliminary study was developed in order to recognize the value of gap G_P between micro-disk and waveguide pump that maximizes the gain. The gain of the amplifier is defined as the ratio of the signal power at output P_{out}^{signal} and input P_{in}^{signal} section of the signal waveguide:

$$G = 10\log_{10}\left(\frac{P_{out}^S}{P_{in}^S}\right) \tag{2.26}$$

Table 2.2 Characteristic parameters of the simulated erbium-doped micro-disk

Parameter	Value
Disc radius, $R_{\mu disk}$	40 μm
Disc thickness, $h_{\mu disk}$	600 nm
Gap between signal waveguide and disk, G_s	1.9 μm
Height of waveguides, h_w	1.0 μm
Width of waveguides, w	2.5 μm
Refractive index, n_s at $\lambda_S = 4500$ nm	2.35
Refractive index, n_p at $\lambda_P = 800$ nm	2.42
Uniform erbium concentration, C	2.8×10^{20} ions/cm^3

Table 2.3 Signal and pump WGM modes

	Resonance wavelength [nm]	$l = m$ parameter	n Parameter
Signal	4487.95	86	1
Pump	800.19	723	1
	799.63	711	2
	800.42	700	3
	800.37	691	4
	800.09	683	5
	800.41	675	6
	800.14	668	7

Table 2.2 reports the geometrical characteristics of the simulated erbium-doped micro-disk [53,67].

The micro-disk exhibited hundreds of resonating WGMs. The simulation was carried out by imposing $p = m$. WGMs with $p \neq m$ can be neglected because they exhibit low overlapping factor Ω^{y}_{WGM} between mode distribution and the uniform rare-earth-doping profile. $N = 73$ WGM resonances with n parameter varying from 1 to 8 and m parameter varying from 56 to 91 were identified. Among all these solutions, one exhibited a strong interaction with rare-earth ions. It was the fundamental one $WGM_{86,86,1}$, having resonant wavelength close to the nominal wavelength $\lambda_S = 4.5$ µm, where the $({}^4I_{9/2} \rightarrow {}^4I_{11/2})$ erbium emission occurs.

The WGMs resonating in the micro-disk within a wavelengths range of 200 nm centred on the nominal pump wavelength $\lambda_p = 800$ nm were searched. In the case of the pump, $N = 2{,}008$ different modes, with n parameter varying from 1 to 8 and m parameter varying from 560 to 760, were identified. The first seven radial order pump modes more strongly interacted with the signal via the rare earth and were considered for simulation. They are reported in Table 2.3 [67].

Figure 2.5 shows the light intensity distribution and the resonant wavelengths of the signal and pump modes of Table 2.3 with the total mode power normalized to 1 W.

The output signal strongly varies with the gap G_S between signal waveguide and micro-disk. For a gap larger than $G_S = 3$ µm, the coupling strength is low. It is worthwhile noting that for large gap values, the inversion of rare-earth population is weak. The coupling can be increased by reducing the gap, but drawbacks can occur. In fact, for a gap less than $G_S = 1.8$ µm, the coupling is too strong, and a non-negligible part of the signal leaves the micro-disk before optical amplification, thus lowering the gain level. The maximum output signal power $P^S_{out} = 10.21$ µW is simulated for a gap $G_S = 2.1$ µm when the input pump power is $P^P_{in} = 0.5$ W, and when the pump power increases, an increase of population inversion occurs, and the gain increases till saturation. Therefore, lasing at $\lambda_S = 4{,}500$ nm in the erbium-doped chalcogenide micro-disk was obtained in the simulation. These results were in good agreement with those reported in [53].

2.2.3.6 Design of a praseodymium-doped micro-disk

The structure is made of a praseodymium-doped Pr^{3+}:$Ga_5Ge_{20}Sb_{10}Se_{65}$ selenide glass micro-disk coupled with two ridge waveguides, one designed for signal propagation at $\lambda_s = 4700$ nm and the other one for the pump at $\lambda_p = 1{,}550$ nm. The geometrical parameters of the structure illustrated in Figure 2.6 are w_p and w_s, the widths of pump and signal waveguides; h_w, $h_{\mu disc}$, h_{buffer}, the thicknesses of waveguides, micro-disk and buffer; G_p and G_s, the gaps, i.e. distances between the micro-disk and the pump/signal waveguides. An Si substrate is approximated in the simulation as a semi-infinite layer. It represents a feasible mechanical basis for the device. A finite buffer thickness $h_{buffer} = 5$ μm of a slightly different chalcogenide glass allows one to prevent the interaction between light and Si substrate. The two coupling waveguides are made of the same glass as the buffer layer of $Ga_5Ge_{20}Sb_{10}S_{65}$ sulphide glass.

The geometrical parameters of the micro-disk and the coupling waveguides are chosen as a good trade-off by considering three main needs/objectives:

1. suitable micro-disk radius in order to have resonance at both the pump and signal wavelengths;
2. suitable waveguide size in order to have single-mode propagation at both the pump and signal wavelengths and to maximize the overlapping factor Ω;
3. the same thickness for waveguides and micro-disk for a feasible fabrication process (e.g. RF sputtering).

All the spectroscopic parameters for glass and rare earth are reported in [49]. In the design, the following geometrical parameters are identified: height $h_w = 1.2$ μm and width $w_s = 3.5$ μm for the signal waveguide; the micro-disk and

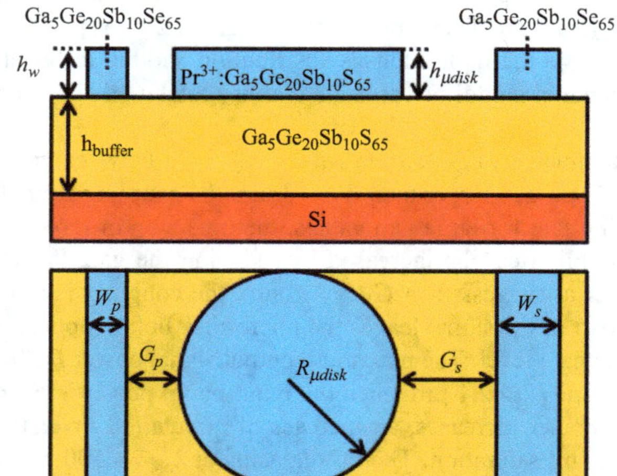

Figure 2.6 Praseodymium-doped Pr^{3+}:$Ga_5Ge_{20}Sb_{10}Se_{65}$ micro-disk. Adapted with permission from [49] © The Optical Society

coupling waveguides thickness, $h_{\mu disk} = h_w = 1.2$ μm; pump waveguide width $w_p = 0.5$ μm allowing single-mode propagation. The fundamental mode polarization is quasi-TE for the signal waveguide and quasi-TM for pump waveguide. Table 2.4 reports a few parameters of the simulated praseodymium-doped micro-disk [49].

In the simulation, the fundamental mode $WGM_{194,1,1}$ at the signal wavelength $\lambda_s = 4700$ nm is considered. It strongly interacts with the first seven WGMs at the pump wavelength $\lambda_p = 1{,}550$ nm [49]. The nominal simulation parameters are the input signal power $P^S_{in} = -30$ dBm, input pump power $P^P_{in} = 100$ mW, praseodymium concentration $C_{Pr} = 3000$ ppm, gaps $G_s = 4.45$ μm and $G_p = 200$ nm.

Figure 2.7 illustrates the maximum output signal power $P^S_{out,MAX}$ as a function of input pump power P^P_{in} and praseodymium concentration C_{Pr}; the input signal power is $P^{signal}_{in} = -30$ dBm. The maximum output signal power $P^S_{out,MAX}$ is the largest value calculated by varying the signal gap G_s in the range 0.5–5.5 μm, the input pump power P^P_{in} in the range 0.01–50 mW and the praseodymium concentration C_{Pr} in the range 100–10000 ppm. It is apparent that output signal power can reach few microwatts.

Figure 2.8 illustrates the optical gain g, as a function of the gap G_s; the input pump power is $P^P_{in} = 0.01$ mW. As expected, the input pump power is too weak to

Table 2.4 Characteristic parameters of the praseodymium-doped micro-disk

Dimension	**Value [μm]**
Micro-disk radius, $R_{\mu disk}$	65.0
Micro-disk and waveguide thicknesses, $h_{\mu disk} = h_w$	1.2
Buffer thickness, h_{Buffer}	5.0
Signal waveguide width, w_s	3.5
Pump waveguide width, w_p	0.5

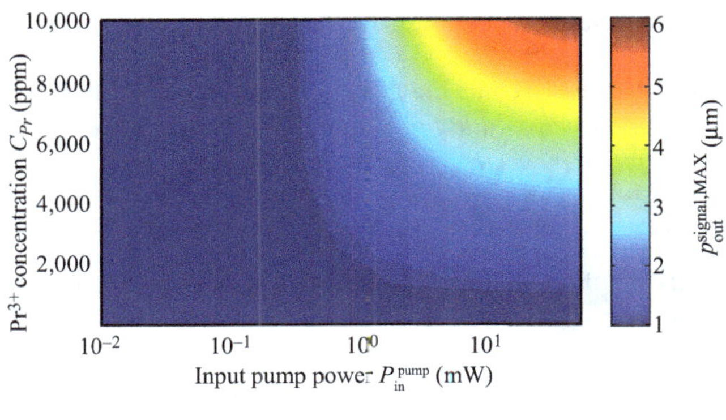

Figure 2.7 Maximum output signal power $P^S_{out,MAX}$ versus the input pump power P^P_{in} and praseodymium concentration C_{Pr}, input signal power $P^S_{in} = -30$ dBm. Adapted with permission from [49] © The Optical Society

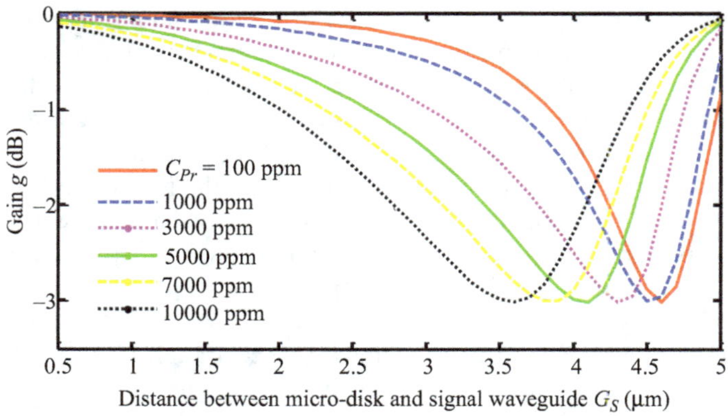

Figure 2.8 Optical gain g versus the gap, i.e. the distance G_s between micro-disk and signal waveguide for different praseodymium concentrations. The input pump power is $P_{in}^P = 0.01$ mW*; input signal power is* $P_{in}^S = -30$ dBm

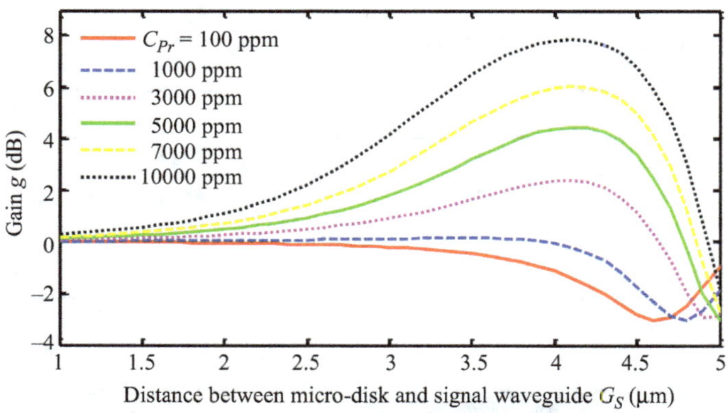

Figure 2.9 Optical gain g versus the gap, i.e. the distance G_s between micro-disk and signal waveguide for different praseodymium concentrations. Input pump power $P_{in}^P = 50$ mW*, input signal power is* $P_{in}^S = -30$ dBm

achieve the inversion of population. The optical losses in the micro-disk are higher than the gain and no amplification is obtained.

Figure 2.9 illustrates the optical gain g versus the gap G_s; the input pump power is $P_{in}^P = 50$ mW. In this case, the maximum gain is $g \approx 7.9$ dB, obtained for $C_{Pr} = 10000$ ppm and $G_s \approx 4.1$ μm.

The micro-disk has been simulated to investigate the laser behaviour: the maximum slope efficiency 8.08×10^{-4} is obtained for Pr^{3+} concentration of

10,000 ppm and input pump power in the range 2–6 mW, with a pump threshold 0.4 mW [49]. This result suggests the construction of a novel micro-laser for mid-IR applications.

2.3 Conclusion

Several commercial computer codes which strongly help the design of integrated optical devices are available on market: they are very versatile and powerful tools. However, the modelling via home-made computer can be advantageously employed in many niche investigations, to reduce the computation cost in the design of particularly complex structures. A hybrid approach is often the best trade-off solution. Here, as examples, the application of FEM/CMT methods has been illustrated with reference to SH generation and to evanescent field sensors but also to simulate rare-earth-doped micro-disks.

References

[1] Scarmozzino R., Gopinath A., Pregla R., and Helfert S. 'Numerical techniques for modeling guided-wave photonic devices'. *IEEE Journal of Selected Topics in Quantum Electronics*. 2000; **6**(1):150–162.

[2] Lavrinenko A.V., Lægsgaard J., Gregersen N., Schmidt F., and Søndergaard T. *Numerical Methods in Photonics*. Boca Raton, FL: CRC Press Taylor & Francis Group, LLC; 2015.

[3] Yee K.S. 'Numerical solution of initial boundary value problem involving Maxwell's equations in isotropic media'. *IEEE Transactions on Antennas and Propagation*. 1966; **14**(3):302–307.

[4] Taflove A. 'Application of the finite-difference time-domain method to sinusoidal steady-state electromagnetic-penetration problems'. *IEEE Transactions on Electromagnetic Compatibility*. 1980; **22**(3):191–202.

[5] Kunz K.S. and Luebbers R.J. *The Finite Difference Time Domain Method for Electromagnetics*. Boca Raton, FL: CRC Press Taylor & Francis Group, LLC; 1993.

[6] Taflove A. and Hagness S.C. *Computational Electrodynamics: The Finite-Difference Time-Domain Method* 3rd edn. Boston, MA: Artech House; 2005.

[7] Yu W., Yang X., Liu Y., Mittra R., and Muto A. *Advanced FDTD Methods. Parallelization, Acceleration, and Engineering Applications*. Boston, MA: Artech House; 2011.

[8] Sullivan D.M. *Electromagnetic Simulation Using the FDTD Method*, 2nd edn. Hoboken, NJ: Wiley-IEEE Press; 2013.

[9] Fernandez F.A. and Lu Y. 'Variational finite element analysis of dielectric waveguides with no spurious solutions'. *Electronics Letters*. 1990; **26**(25): 2125–2126.

[10] Abid Z.E., Johnson K. L., and Gopinath A. 'Analysis of dielectric guides by vector transverse magnetic fields finite elements'. *Journal of Lightwave Technology*. 1993; **11**(10):1545–1549.

[11] Rahman B.A.M. and Davies J.B. 'Finite-element analysis of optical and microwave waveguide problems'. *IEEE Transactions on Microwave Theory and Techniques*. 1983; **32**(1):20–28.

[12] Koshiba M., Hayata K., and Suzuki M. 'Improved finite-element formulation in terms of the magnetic fields vector for dielectric waveguides'. *IEEE Transactions on Microwave Theory and Techniques*. 1985; **33**(3): 227–233.

[13] Sujecki S. *Photonics Modelling and Design*. 1st edn. Boca Raton, FL: CRC Press Taylor & Francis Group, LLC; 2017.

[14] De Sario M., D'Orazio A., Petruzzelli V., and Prudenzano F. 'Leaky mode propagation in planar multi-layer inhomogeneous birefringent waveguides: polar dielectric tensor configuration'. *Journal of Physics D: Applied Physics*. 1992; **25**(8):1172–1181.

[15] D'Orazio A., De Sario M., Petruzzelli V., and Prudenzano F. 'Leaky mode propagation in planar multilayer birefringent waveguides: longitudinal dielectric tensor configuration'. *IEEE Journal of Lightwave Technology*. 1994; **12**(3):453–462.

[16] Prudenzano F., D'Orazio A., De Sario M., and Petruzzelli V. 'Comparison between the performance of Ti:$LiNbO_3$ and H:$LiNbO_3$ rotated optical axis waveguides – summary'. *Journal of Electromagnetic Waves and Applications*. 1997; **11**(4):547–559.

[17] Prudenzano F., D'Orazio A., De Sario M., and Petruzzelli V. 'Comparison between the performance of Ti:$LiNbO_3$ and H:$LiNbO_3$ rotated optical axis waveguides'. *Progress in Electromagnetic Research*. 1997; **16**:227–267.

[18] D'Orazio A., De Sario M., Ficarella G., Grando D., Petruzzelli V., and Prudenzano F. 'Design and demonstration of interferometric integrated-optic sensors in Ti:$LiNbO_3$ waveguides'. *Fiber and Integrated Optics*. 1997; **16** (4):369–386.

[19] Loridat J., Heidmann S., Thomas F., *et al*. 'All integrated lithium niobate standing wave Fourier transform electro-optic spectrometer'. *Journal of Lightwave Technology*. 2018; **36**(20):4900–4907.

[20] Schiek R. 'All-optical switching in the directional coupler caused by nonlinear refraction due to cascaded second-order nonlinearity'. *Optical and Quantum Electronics*. 1994; **26**(4):415–431.

[21] Prudenzano F., Ciminelli C., D'Orazio A., Petruzzelli V., and De Sario M. 'Exact analysis of cascaded second-order nonlinearity in rotated $LiNbO_3$ couplers'. *Optical and Quantum Electronics*. 1999; **31**(8):655–674.

[22] Prudenzano F., Ciminelli C., D'Orazio A., Petruzzelli V., M. De Sario. 'Performance enhancement of nonlinear lithium niobate couplers via double titanium and magnesium diffusion'. *Physica E – Low Dimensional System & Nanostructures*. 1999; **5**(1–2):84–97.

[23] Petruzzelli V. and Prudenzano F. 'Propagation modes in periodically poled lithium niobate waveguides exploiting cascaded second order nonlinearity'. *Fiber and Integrated Optics*. 2001; **20**(4):347–365.

[24] Busacca A.C., Cino A.C., Riva-Sanseverino S., Ravaro M., and Assanto G. 'Silica masks for improved surface poling of lithium niobate'. *Electronics Letters*. 2005; **41**(2):92–94.

[25] Assanto G., Torelli I., and Trillo S. 'All-optical phase controlled amplitude modulator'. *Electronics Letters*. 1994; **30**(9):733–735.

[26] Gallo K., Assanto G., and Stegeman G.I. 'Efficient wavelength shifting over the erbium amplifier bandwidth via cascaded second order processes in lithium niobate waveguides'. *Applied Physics Letters*. 1997; **71**(8): 1020–1022.

[27] Tehranchi A., Ahlawat M., Bostani A., and Kashyap R. 'Flexible all-optical wavelength shifters using strong focusing in a wideband engineered PPLN'. *IEEE Photonics Technology Letters*. 2016; **28**(18):1924–1927.

[28] Wang J., Sun J., Zhang X., and Huang D. 'All-optical tunable wavelength conversion with extinction ratio enhancement using periodically poled lithium niobate waveguides'. *Journal of Lightwave Technology*. 2008; **26**(17): 3137–3148.

[29] Wang J., Sun J., Zhang X., Huang D., and Fejer M.M. 'All-optical format conversions using periodically poled lithium niobate waveguides'. *IEEE Journal of Quantum Electronics*. 2009; **45**(2):195–205.

[30] Gallo K. and Assanto G. 'Analysis of lithium niobate all-optical wavelength shifters for the third spectral window'. *Journal of the Optical Society of America B*. 1999; **16**(5):741–753.

[31] Mazroa D. 'BPSK phase regeneration based on second-order nonlinearities in periodically poled lithium niobate'. *Journal of Lightwave Technology*. 2013; **31**(15):2501–2507.

[32] Bogoni A., Wu X., Nuccio S.R., and Willner A.E. '640 Gb/s all-optical regenerator based on a periodically poled lithium niobate waveguide'. *Journal of Lightwave Technology*. 2012; **30**(12):1829–1834.

[33] Lee K.J., Parmigiani F., Liu S., *et al.* 'Phase sensitive amplification based on quadratic cascading in a periodically poled lithium niobate waveguide'. *Optics Express*. 2009; **17**(22):20393–20400.

[34] Amin J., Pruneri V., Webjorn J., Russell P.St.J., Hanna D.C., and Wilkinson J.S. 'Blue light generation in a periodically poled Ti:$LiNbO_3$ channel waveguide'. *Optics Communications*. 1997; **135**(1–3):41–44.

[35] Myers L.E., Eckardt R.C., Fejer M.M., Byer R.L., Bosemberg W.R., and Pierce J.W. 'Quasi-phase-matched optical parametric oscillators in bulk periodically poled $LiNbO_3$'. *Journal of the Optical Society of America B*. 1995; **12**(11):2102–2116.

[36] D'Orazio A., De Sario M., Petruzzelli V., and Prudenzano F. 'All-optical amplification via the interaction among guided and leaky propagation modes in lithium niobate waveguides exploiting the cascaded second order non-linearity'. *Optical and Quantum Electronics*. 2003; **35**(1):47–68.

[37] Edwards G.J. and Lawrence M. 'Short communication: a temperature-dependent dispersion equation for congruently grown lithium niobate'. *Optical and Quantum electronics*. 1984; **16**(4):373–374.

[38] Jundt D.H. 'Temperature-dependent Sellmeier equation for the index of refraction, n_e, in congruent lithium niobate'. *Optics Letters*. 1997; **22**(20): 1553–1555.

[39] Kim Y.S. and Smith R.T. 'Thermal expansion of lithium tantalate and lithium niobate single crystals'. *Journal of Applied Physics*. 1969; **40**(11): 4637–4641.

[40] Dmitrev V.G., Gurzadyn G.G., and Nikogosyan D.N. *Handbook of Nonlinear Optical Crystals*. Heidelberg, DE: Springer; 1995.

[41] Del Rosso T., Margheri G., Sottini S., *et al.* 'An optical thermometer exploiting periodically poled lithium niobate for monitoring the pantographs of high-speed trains'. *IEEE Sensors Journal*. 2007; **7**(3):417–425.

[42] Margheri G., Del Rosso T., Trigari S., *et al.* 'Temperature sensing in E.M.D. environment with periodically poled lithium niobate devices'. *Proceedings of SPIE Photonics Europe;* Strasbourg, France, Apr 2006. Proceedings Volume 6183 Integrated Optics, photonics, and Photonic Integrated Circuits.

[43] D'Orazio A., De Sario M., Giasi C.I., Mescia L., Petruzzelli V., and Prudenzano F. 'Design of planar optic sensors for hydrocarbon detection'. *Optical and Quantum Electronics*. 2004; **36**(6):507–526.

[44] Prudenzano F., Mescia L., Allegretti L.A., *et al.* 'Design of an optical sensor array for hydrocarbon monitoring'. *Optical and Quantum Electronics*. 2009; **41**(1):55–68.

[45] D'Orazio A., De Sario M., Mescia L., *et al.* 'Design of Er^{3+} doped SiO_2-TiO_2 planar waveguide amplifier'. *Journal of Non-Crystalline Solids*. 2003; **322**(1–3):278–283.

[46] D'Orazio A., De Sario M., Mescia L., *et al.* 'Design of praseodymium doped optical waveguides'. *Optical Engineering*. 2003; **42**(3):765–772.

[47] Mescia L., Bia P., De Sario M., Di Tommaso A., and Prudenzano F. 'Design of mid-infrared amplifiers based on fiber taper coupling to erbium-doped microspherical resonator'. *Optics Express*. 2012; **20**(7):7616–7629.

[48] Mescia L., Bia P., Losito O., and Prudenzano F. 'Design of mid-IR Er^{3+}-doped microsphere laser'. *IEEE Photonics Journal*. 2013; **5**(4):*1501308*.

[49] Palma G., Falconi M.C., Stareki F., *et al.* 'Design of praseodymium-doped chalcogenide micro-disk emitting at 4.7 μm'. *Optics Express*. 2017; **25**(6): 7014–7030.

[50] Sergent S. and Semond F. 'Chap. 4 (Al,Ga)N micro-disk cavities' in Choi A. H.W. (ed.). *Handbook of Optical Microcavities*. Boca Raton, FL: Pan Stanford Publishing Pte. Ltd.; 2014. pp 131–156.

[51] Little B.E., Laine J.P., and Haus H.A. 'Analytic theory of coupling from tapered fibers and half-blocks into microsphere resonators'. *Journal of Lightwave Technology*. 1999; **17**(4):704–715.

[52] Little B.E., Chu S.T., Haus H.A., Foresi J., and Laine J.P. 'Microring resonator channel dropping filters'. *Journal of Lightwave Technology*. 1997; **15**(6):998–1005.

[53] Al Tal F., Dimas C., Hu J., Agarwal A., and Kimerling L.C. 'Simulation of an erbium-doped chalcogenide micro-disk mid-infrared laser source'. *Optics Express*. 2011; **19**(13):11951–11962.

[54] Borselli M., Johnson T.J., and Painter O. 'Beyond the Rayleigh scattering limit in high-*Q* silicon micro-disks: theory and experiment'. *Optics Express*. 2005; **13**(5):1515–1530.

[55] Palma G., Bia P., Mescia L., *et al.* 'Design of fiber coupled Er^{3+}: chalcogenide microsphere amplifier via particle swarm optimization algorithm'. *Optical Engineering*. 2014; **53**(7):071805 1–071805 8.

[56] Wang P., Ding M., Lee T., *et al.* 'Packaged chalcogenide microsphere resonator with high Q-factor'. *Applied Physics Letters*. 2013; **102**(13): 131110 1–131110 5.

[57] Elliot G.R., Hewak D.W., Murugan G.S., and Wilkinson J.S. 'Chalcogenide glass microspheres; their production, characterization and potential'. *Optics Express*. 2007; **15**(26):17542–17553.

[58] Wang P., Murugan G.S., Brambilla G., *et al.* 'Chalcogenide microsphere fabricated from fiber tapers using contact with a high-temperature ceramic surface'. *IEEE Photonics Technology Letters*. 2012; **24**(13):1103–1105.

[59] Charrier J., Anne M.L., Lhermite H., *et al.* 'Sulphide $Ga_xGe_{25-x}Sb_{10}S_{65\ (x=0,\ 5)}$ sputtered films: fabrication and optical characterizations of planar and rib optical waveguides'. *Journal of Applied Physics*. 2008; **104**(7):073110 1–073110 8.

[60] Palma G., Falconi M.C., Nazabal V., *et al.* 'Modeling of whispering gallery modes for rare earth spectroscopic characterization'. *IEEE Photonics Technology Letters*. 2015; **27**(17):1861–1863.

[61] Kadono K., Yazawa T., Jiang S., Porque J., Hwang B.C, and Peyghambarian N. 'Rate equation analysis and energy transfer of Er^{3+}-doped Ga_2S_3-GeS_2-La_2S_3 glasses'. *Journal of Non-Crystalline Solids*. 2003; **331**(1–3):79–90.

[62] Palma G., Falconi M.C., Starecki F., *et al.* 'Novel double step approach for optical sensing via microsphere WGM resonance'. *Optics Express*. 2016; **24**(23):26956–26971.

[63] Falconi M.C., Palma G., Starecki F., *et al.* 'Design of an efficient pumping scheme for mid-IR Dy^{3+}:$Ga_5Ge_{20}Sb_{10}S_{65}$ PCF fiber laser'. *IEEE Photonics Technology Letters*. 2016; **28**(18):1984–1987.

[64] Palma G., Falconi M.C., Starecki F., *et al.* 'Design of rare-earth doped chalcogenide microresonators for biosensing in mid-IR'. *18th International Conference on Transparent Optical Networks;* Trento, IT, Jul 2016.

[65] Anne M.-L., Keirsse J., Nazabal V., *et al.* 'Chalcogenide glass optical waveguides for infrared biosensing'. *Sensors*. 2009; **9**(9):7398–7411.

[66] Han Z., Zhang L., Kimerling L.C., and Agarwal A.M. 'Integrated mid-infrared laser based on an Er-doped chalcogenide microresonator'. *IEEE Journal of Selected Topics in Quantum Electronics*. 2015; **21**(1): 1603007.

[67] Palma G., Falconi M.C., Starecki F., *et al.* 'Design of rare-earth doped chalcogenide microresonators for biosensing in mid-IR'. *18th International Conference on Transparent Optical Networks;* Trento, IT, Jul 2016.

Chapter 3

Analytical modelling of active integrated resonators

Yann G. Boucher[1,2]

In spite of its underlying simplifying assumptions, analytical modelling remains a powerful approach to understand the processes at work in photonic structures. In this chapter, our aim is to provide the reader with some basic tools, such as linear algebra at introductory level, that enable one to access the main properties of active integrated resonators, keeping in mind the physical phenomena. We shall concentrate our investigations on steady-state behaviour in the spectral domain.

The outline is as follows: in Section 3.1, we stress the difference between the optical properties of the materials and their guided-wave modal counterpart; we then present possible models for the gain, based on classical rate equations: these take explicitly into account the fundamental mechanisms of pumping and saturation. Section 3.2 is devoted to some important tools: transfer matrix formalism (TMF), coupled-mode theory (CMT), and scattering parameters, which work for active as well as passive devices. In Section 3.3, we apply them to structures we are likely to meet, from directional couplers to micro-cavities, including distributed-feedback (DFB) amplifiers. We examine resonators of different topologies, in the rectilinear or annular configurations, with special emphasis on their spectral properties. In Section 3.4, we extend the formalism into the non-linear regime, up to the oscillation condition of a laser cavity and beyond, giving simple indications upon the diode-like behaviour of a steady-state single-mode laser emitter. Amplified spontaneous emission (ASE) is dealt with in Section 3.5, in the frame of extended (3×3) TMF that takes explicitly into account the internal sources of spontaneous emission. Extensions and perspectives are presented in Section 3.6.

3.1 Material and structural parameters

Active integrated photonics can involve different kinds of materials, from semiconductor hetero-junctions to doped glasses of various chemical compositions, each with its own specific optical properties: refractive index (n), material gain (or losses),

[1]CNRS, Institut FOTON, Lannion Cedex, France
[2]École Nationale d'Ingénieurs de Brest, Brest Cedex, France

and spectral rate of spontaneous emission (r_{sp}). According to the needs, several paths can be followed, from the simplest one, where all the parameters are assumed constant over the spectral range of interest, to much more elaborate approaches taking into account the microscopic properties, such as field–matter interactions depicted in the frame of the density matrix derivation of the atomic susceptibility [1]. The first rule for any successful analytical model is to keep it as straightforward as possible: if several descriptions are available, it is usually best to choose the simplest one… provided it remains compatible with the required degree of precision.

3.1.1 From material to modal optical properties

For instance, in a direct-gap semiconductor, absorption (or gain) depends on the joint density of states as well as on the occupation factors in the valence and conduction bands; these can be related to the quasi-Fermi levels, determined by the free carrier concentration N, itself a function of the pumping [2–4]. In bulk materials such as GaAs, a mere polynomial approximation is often preferred, with a parabolic gain spectrum for any given value of N. In rare-earth-doped glasses, the spectral behaviour of absorption and emission cross-sections (σ_a and σ_e) is sometimes empirically approximated by a set of suitably weighted Gaussian distributions [5]. Whenever closed-form expressions are within reach, it would be a shame not to use them!

Whatever the situation, it should be kept in mind that a confined field propagating in a waveguide is best described by its *modal* properties, which means its effective index (n_{eff}) as well as its modal gain (g_m), or modal losses in the case of attenuation. Even if materials were totally non-dispersive, modal dispersion would still occur. A first-order expansion around a given reference angular frequency ω_1 enables one to write

$$n_{eff}(\omega) = n_{eff}(\omega_1) + \frac{\partial n_{eff}}{\partial \omega}(\omega - \omega_1)$$

First-order dispersion is also well described by the group index n_g:

$$n_g = n_{eff} + \omega \frac{\partial n_{eff}}{\partial \omega}$$

For the sake of clarity, the waveguides considered in this chapter are single mode: for a given state of polarization, they support only a fundamental mode in their transverse plane. Time dependence is taken as $\exp(+i\omega t)$. The complex wavevector of the field confined in the waveguide can be written as

$$\beta = n_{eff}\, k_0 + i\frac{g_m}{2}$$

where $k_0 = (\omega/c)$ is the wavevector in vacuum $[\mathrm{rad} \cdot \mathrm{m}^{-1}]$. Within this convention, the positive gain constant g_m is related to the variation of optical intensity I $[\mathrm{W} \cdot \mathrm{m}^{-2}]$ propagating from the left to the right along the conventional z-axis, according to

$$\frac{1}{I}\frac{\partial I}{\partial z} = g_m > 0$$

In slab (planar) waveguides, effective index and modal gain can be retrieved exactly by solving the confinement condition. The calculation is not so straightforward in other waveguides, but, qualitatively, it is useful to keep in mind some general tendencies:

- The modal gain g_m [m^{-1}] can be roughly interpreted as

  $$g_m = \Gamma\, g_{mat} - \alpha_s$$

  where g_{mat} [m^{-1}] is the net bulk material gain, Γ is the confinement factor (the fraction of the mode in the active guide), and α_s [m^{-1}] is an extra term taking into account all other possible loss mechanisms (such as roughness-induced interface scattering).
- A first-order expansion shows that adding gain as a perturbation into a slab waveguide does not modify its effective index (at least to the first order).
- Since, in most materials of interest, the imaginary part of the complex index is always much smaller than its real part, it is often reasonable to neglect its effect. Not when studying the amplification itself, of course! But consider the reflection at an interface between an active material of complex index n_1 and a passive one of real index n_2, for instance, while the Fresnel coefficients become formally complex valued, the actual phase change is so vanishingly small that it is much simpler (and no less accurate in practice) to replace n_1 with Re[n_1] in the formula.

3.1.2 Energy balance and rate equations

Another key feature of the active medium is provided by the rate equations, expressing the energy balance between the field and the matter [6]. Their precise form depends on the actual properties of the material. Only relevant spectroscopic levels are kept in the picture. Let us examine some useful cases.

3.1.2.1 Ideal three-level system

We start with the ideal three-level system – an idealized model well suited for optically pumped Erbium transitions (Figure 3.1).

Let N_p denote the concentration [m^{-3}] of atoms in level (p), with $N_1 + N_2 + N_3 = N_{tot}$. Rate equations read

$$\frac{\partial N_1}{\partial t} = -(W_P + W)\, N_1 + (A + W)\, N_2$$

$$\frac{\partial N_2}{\partial t} = W\, N_1 - (A + W)\, N_2 - \gamma_{32}\, N_3$$

$$\frac{\partial N_3}{\partial t} = W_P\, N_1 - \gamma_{32}\, N_3$$

where W_P is the pumping rate [s^{-1}], γ_{32} is the non-radiative relaxation rate from (3) to (2) [s^{-1}], and A and W are the spontaneous and stimulated emission rates, respectively [s^{-1}]. Parameter W is a measure of the optical intensity at the (1–2)

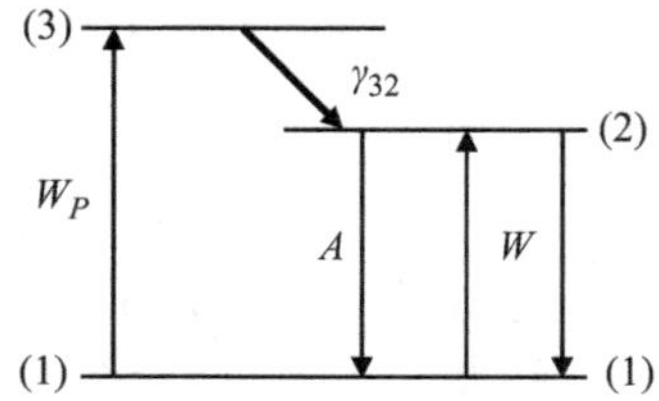

Figure 3.1 Three-level system: (1) and (2) denote the lower and upper levels of the optical transition we are interested in; A and W denote the spontaneous and stimulated emission rates, respectively. Optical pumping (at a shorter wavelength) occurs from the fundamental level (1) to an intermediary level (3). Non-radiative relaxation towards level (2) is so efficient that it can be considered instantaneous

transition. In the ideal approximation, γ_{32} is so high that non-radiative relaxation from (3) to (2) is assumed instantaneous so that $N_3 \approx 0$.

Rate equations express a temporal balance and can be very useful in a whole range of time behaviours (transient, harmonic, small-signal, and so on). However, from now on, let us concentrate on the steady-state behaviour ($\partial/\partial t = 0$). Using reduced parameters $Y_P = W_P/A$, $Y = W/A$, we are left with two linear equations connecting N_1 and N_2:

$$(Y_P + Y)\, N_1 = (1 + Y)\, N_2, \qquad N_1 + N_2 = N_{tot}$$

It is straightforward to establish the steady-state value of the populations:

$$N_1 = N_{tot} \frac{1 + Y}{1 + Y_P + 2Y}, \qquad N_2 = N_{tot} \frac{Y_P + Y}{1 + Y_P + 2Y}$$

Of paramount importance is the quantity $\Delta N = N_2 - N_1$: the inversion of population marks the boundary between attenuation and amplification.

$$\Delta N = N_{tot} \frac{Y_P - 1}{Y_P + 1 + 2Y}$$

Assuming, for the sake of simplicity, that the absorption and emission cross-sections are equal ($\sigma_a = \sigma_e = \sigma$), we are left with an excellent approximation for the material gain:

$$g_{mat} = \sigma \Delta N = g_0 \frac{Y_P - 1}{Y_P + 1 + 2Y}$$

In this expression, $g_0 = \sigma N_{tot}$ is the maximum available material gain [m^{-1}]. Several points should be emphasized:

- Material gain g_{mat} explicitly depends on the total volume concentration N_{tot} of active atoms in the system.
- Spectral dependence appears through g_0 (via σ).

- When the signal is so small that $Y \ll 1$, the modal gain is called *linear* or *unsaturated*:

$$g_{lin} = \Gamma g_0 \frac{Y_P - 1}{Y_P + 1} - \alpha_s$$

- For pumping levels $Y_P < 1$, such a medium is actually absorbing. $Y_P = 1$ corresponds to the *transparency* (for which stimulated processes of absorption and emission exactly compensate). Only for $Y_P > 1$ can one expect actual amplification.
- The maximum value g_0 is but an asymptotic limit. In an optical resonator, it should be compared with the oscillation threshold value, especially if a laser effect is expected.
- When Y can no longer be neglected, we speak of *saturation*.
- The same model without pumping ($Y_P = 0$) is excellent for describing absorption, linear or saturated.

3.1.2.2 Ideal four-level system

Let us now investigate the ideal four-level system – an idealized model well suited for optically pumped neodymium transitions (Figure 3.2).

Let N_p denote the concentration [m^{-3}] of atoms in level (p), with $N_0 + N_1 + N_2 + N_3 = N_{tot}$. Rate equations read

$$\frac{\partial N_0}{\partial t} = -W_P\, N_0 + \gamma_{10}\, N_1$$

$$\frac{\partial N_1}{\partial t} = -(\gamma_{10} + W)\, N_1 + (A + W)\, N_2$$

$$\frac{\partial N_2}{\partial t} = W\, N_1 - (A + W)\, N_2 + \gamma_{32}\, N_3$$

$$\frac{\partial N_3}{\partial t} = W_P\, N_0 - \gamma_{32}\, N_3$$

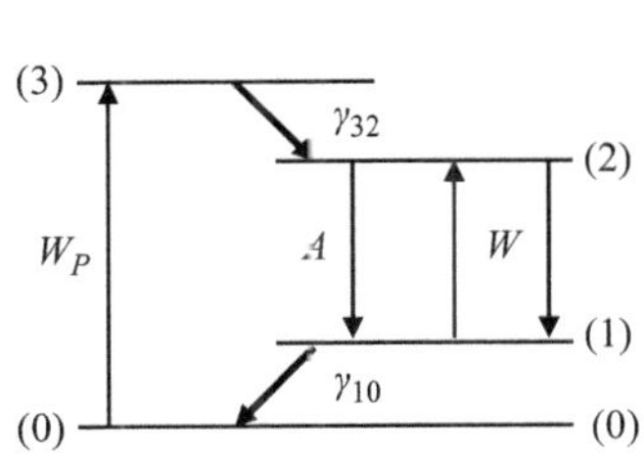

Figure 3.2 Four-level system: (1) and (2) denote the lower and upper levels of the optical transition we are interested in; A and W denote the spontaneous and stimulated emission rates, respectively. Optical pumping (at a shorter wavelength) occurs from the fundamental level (0) to an intermediary level (3). Non-radiative relaxation processes from (3) to (2) and from (1) to (0) are so efficient that they can be considered instantaneous

where W_P is the pumping rate [s^{-1}], γ_{32} is the non-radiative relaxation rate from (3) to (2) [s^{-1}], γ_{10} is the non-radiative relaxation rate from (1) to (0) [s^{-1}], and A and W are the spontaneous and stimulated emission rates, respectively [s^{-1}]. Parameter W is a measure of the optical intensity at (1–2) transition. In the ideal approximation, both γ_{10} and γ_{32} are so high that non-radiative relaxation from (3) to (2) and from (1) to (0) is assumed instantaneous so that $N_1 \approx 0$ and $N_3 \approx 0$.

In the steady-state regime ($\partial/\partial t = 0$), using reduced parameters $Y_P = W_P/A$, $Y = W/A$, we get eventually

$$Y_P\, N_0 = (1+Y)\, N_2, \qquad N_0 + N_2 = N_{tot}$$

It is straightforward to establish the steady-state value of the populations:

$$N_0 = N_{tot}\frac{1+Y}{1+Y_P+Y}, \qquad N_2 = N_{tot}\frac{Y_P}{1+Y_P+Y}$$

Since $N_1 = 0$, $\Delta N = N_2$: in such a system, optical gain is obtained as soon as pumping occurs:

$$\Delta N = N_{tot}\frac{Y_P}{Y_P+1+Y}$$

Assuming, for the sake of simplicity, that the absorption and emission cross-sections are equal ($\sigma_a = \sigma_e = \sigma$), we are left with an excellent approximation for the material gain:

$$g_{mat} = \sigma\, \Delta N = g_0 \frac{Y_P}{Y_P+1+Y}$$

In this expression, $g_0 = \sigma\, N_{tot}$ is the maximum available gain [m^{-1}]. When the signal is so small that $Y \ll 1$, the linear (unsaturated) modal gain assumes the following form:

$$g_{lin} = \Gamma g_0 \; \frac{Y_P}{Y_P+1} - \alpha_s$$

3.1.2.3 Open two-level system

It is sometimes useful to consider an open two-level system – a simplistic model where only the upper and lower levels of the optical transition are taken into account (Figure 3.3).

Let N_p denote the concentration [m^{-3}] of atoms in level (p). Rate equations read

$$\frac{\partial N_1}{\partial t} = -(\gamma_1 + W)\, N_1 + (A + W)\, N_2$$

$$\frac{\partial N_2}{\partial t} = \Lambda_P + W\, N_1 - (A + W + \gamma_2)\, N_2$$

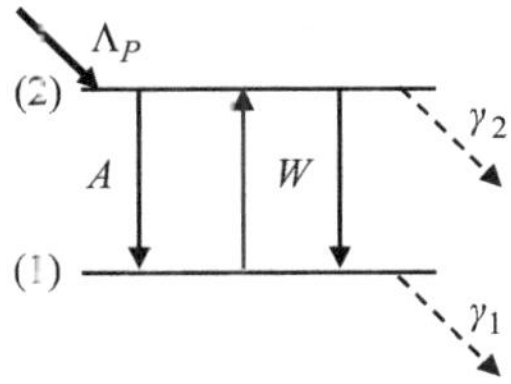

Figure 3.3 Two-level system: (1) and (2) denote the lower and upper levels of the optical transition we are interested in; A and W denote the spontaneous and stimulated emission rates, respectively. Optical pumping is described through pumping rate Λ_P. Non-radiative relaxation processes from (2) and (1) to exterior levels are described by γ_2 and γ_1

where Λ_P is the pumping rate [$m^{-3} \cdot s^{-1}$], γ_p is the non-radiative relaxation rate from level (*p*) [s^{-1}], and *A* and *W* are the spontaneous and stimulated emission rates, respectively [s^{-1}]. In this *open* scheme, the total concentration of atoms ($N_1 + N_2$) concerned with light-matter interaction is not constant but depends on the pumping level.

In the steady-state regime ($\partial/\partial t = 0$), we get eventually

$$N_1 = \Lambda_P \frac{A + W}{\gamma_1(A + \gamma_2) + W(\gamma_1 + \gamma_2)}, \quad N_2 = \Lambda_P \frac{\gamma_1 + W}{\gamma_1(A + \gamma_2) + W(\gamma_1 + \gamma_2)}$$

In such a system, optical gain is possible only if $\gamma_1 > A$:

$$\Delta N = \Lambda_P \frac{\gamma_1 - A}{\gamma_1(A + \gamma_2) + W(\gamma_1 + \gamma_2)}$$

Assuming, for the sake of simplicity, that the absorption and emission cross-sections are equal ($\sigma_a = \sigma_e = \sigma$), we are left with an excellent approximation for the material gain:

$$g_{mat} = \sigma \, \Delta N = g_0 \frac{Y_P}{1 + Y}$$

In these expressions, we have defined new reduced parameters:

$$g_0 \, Y_P = \sigma \, \Lambda_P \frac{\gamma_1 - A}{\gamma_1(A + \gamma_2)}, \quad Y = W \frac{\gamma_1 + \gamma_2}{\gamma_1(A + \gamma_2)}$$

In this expression, g_0 can be any arbitrary reference gain [m^{-1}]: only the product g_0 Y_P is defined. When the signal is so small that $Y \ll 1$, the linear (unsaturated) modal gain assumes the following form:

$$g_{lin} = \Gamma g_0 Y_P - \alpha_s$$

This is probably the simplest model of gain that takes into account both pumping and saturation. Its only drawback stems from the fact that it would seem to imply that the amplification is potentially unlimited, which is quite unphysical. Of course, as any model, this one should not be extrapolated outside of its range of validity.

3.1.2.4 The special case of semiconductors

In previous examples, atomic transitions are pumped optically. In direct-gap semiconductors, electrical pumping is preferred: in a p–n junction, an electric current is responsible for sending electrons into the conduction band, leaving holes in the valence band. The useful parameter for describing the medium is N, density of free carriers [m^{-3}], whereas the field can be described by P, an effective density of photons [m^{-3}], proportional to the optical intensity at the transition frequency (assuming a single wavelength to be involved). The rate equation reads

$$\frac{\partial N}{\partial t} = \Lambda_P - R(N) - G(N)\,P$$

where Λ_P is the pumping rate [$m^{-3}\cdot s^{-1}$], $R(N)$ is the non-stimulated recombination rate (including spontaneous emission, non-radiative relaxation, and Auger processes), and $G(N)$ is the stimulated recombination rate [s^{-1}].

3.2 Transfer matrix formalism and scattering parameters

3.2.1 Classical (2×2) transfer matrix formalism

TMF stems from the linearity of Maxwell's equations; it is a powerful tool for describing the spectral properties of multilayer or multisection structures. Several equivalent forms can be used: we follow closely Yariv's conventions [7,8]. Consider first a 1D multilayer structure, with all interfaces orthogonal to the z-axis (Figure 3.4). Let n_0 and n_S denote the index of refraction of the two semi-infinite media that surround the structure.

For a given angular frequency ω, a given angle of incidence, a given eigenstate of polarization (either TE or TM), the electromagnetic field in the structure is completely determined, at any arbitrary abscissa z, by two components only, F^+ and F^-, respectively, propagating from the left to the right and vice versa. For the TE state, it is convenient to use the electric components (E^+, E^-) of the

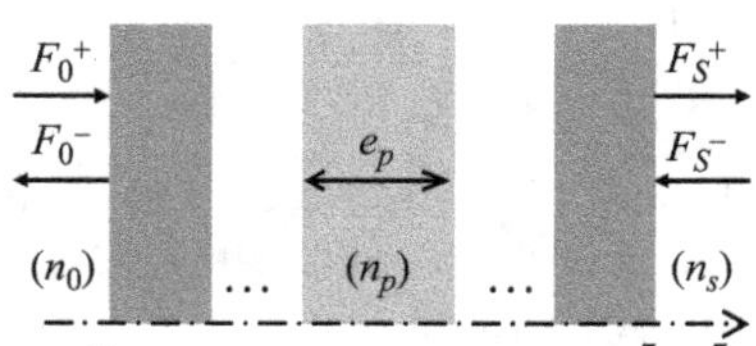

Figure 3.4 Schematic representation of a multilayer structure made of a succession of N linear, homogeneous, isotropic layers, each one characterized by refractive index (n_p) and thickness (e_p), sandwiched between two semi-infinite media (n_0) and (n_S)

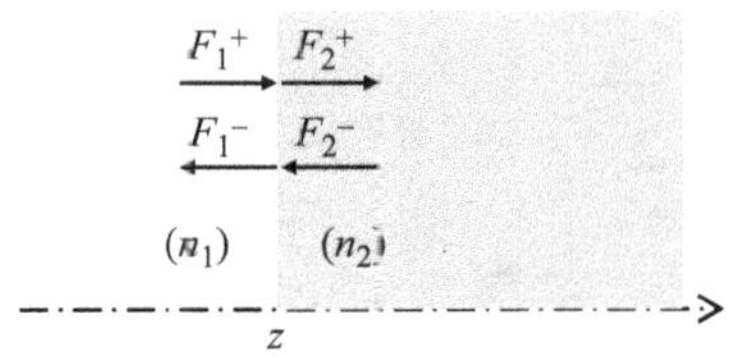

Figure 3.5 Interface between two linear, homogeneous, isotropic, non-magnetic media of refractive indices (n_1) and (n_2)

electromagnetic field, whereas their magnetic counterparts (H^+, H^-) may prove more suitable for the TM state. Let $k_0 = (\omega/c)$ denote the wavevector in vacuum and $k_{//}$ the component of the incoming wavevector parallel to the first interface – an invariant throughout the whole structure. The latter can always be decomposed into elementary transformations, corresponding either to interfaces (Fresnel coefficients) or to free-space propagation (phase change). In layer number p, let $k_{pz} = (k_0^2\, n_p^2 - k_{//}^2)^{1/2}$ denote the z-component of the wavevector.

Continuity conditions at the interface separating two media of respective indices (n_1, n_2) enable one to establish the following relationships (Figure 3.5).

For the TE state (or s state), we get eventually

$$\begin{pmatrix} E_1^+ \\ E_1^- \end{pmatrix} = \frac{1}{2}\begin{pmatrix} 1+q_s & 1-q_s \\ 1-q_s & 1+q_s \end{pmatrix}\begin{pmatrix} E_2^+ \\ E_2^- \end{pmatrix}, \quad q_s = \frac{k_{2z}}{k_{1z}}$$

A similar expression holds for the TM state (or p state):

$$\begin{pmatrix} H_1^+ \\ H_1^- \end{pmatrix} = \frac{1}{2}\begin{pmatrix} 1+q_p & 1-q_p \\ 1-q_p & 1+q_p \end{pmatrix}\begin{pmatrix} H_2^+ \\ H_2^- \end{pmatrix}, \quad q_p = \frac{n_1^2\, k_{2z}}{n_2^2\, k_{1z}}$$

Free-space propagation is responsible for a phase change (Figure 3.6).

Whatever the state of polarization, we get

$$\begin{pmatrix} F_1^+ \\ F_1^- \end{pmatrix} = \begin{pmatrix} \exp(+i\, k_z d) & 0 \\ 0 & \exp(-i\, k_z d) \end{pmatrix}\begin{pmatrix} F_2^+ \\ F_2^- \end{pmatrix}$$

Then, the transfer matrix of the multilayer can be written as the dot product, in the right order, of elementary matrices associated with basic transformations.

3.2.2 Complex reflectance and transmittance

Once the total transfer matrix $[M]$ is established, we can relate its components to experimentally measurable quantities, such as transmittance and reflectance. Whatever the boundary conditions (Figure 3.7), we get:

$$\begin{pmatrix} F_0^+ \\ F_0^- \end{pmatrix} = \begin{pmatrix} M_{11} & M_{12} \\ M_{21} & M_{22} \end{pmatrix}\begin{pmatrix} F_S^+ \\ F_S^- \end{pmatrix}$$

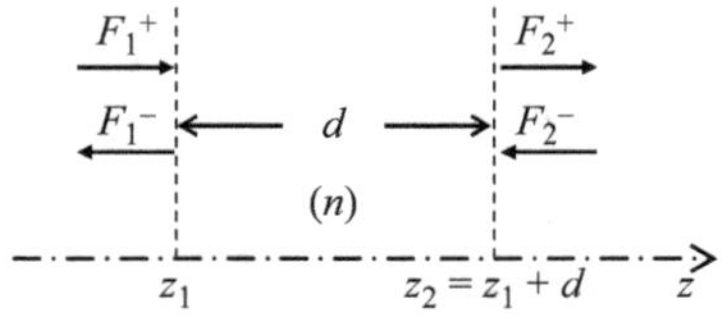

Figure 3.6 Free-space propagation over a distance d in a linear, homogeneous, isotropic medium

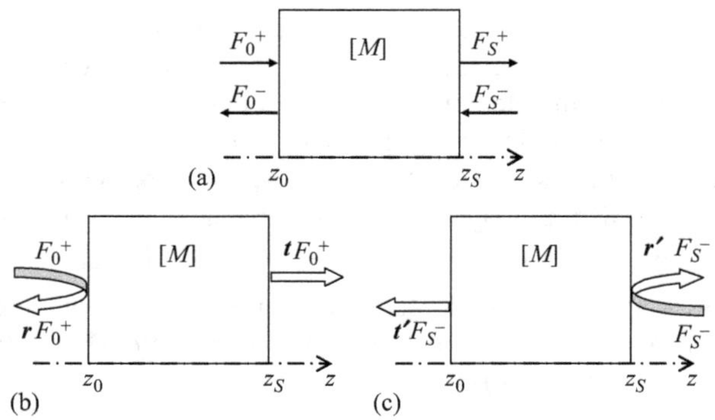

Figure 3.7 Transfer matrix between two semi-infinite media: (a) arbitrary boundary conditions. Transmittance and reflectance: (b) from left to right and (c) from right to left

For an incoming wave propagating from the left to the right ($F_S^- = 0$)

$$t = \left.\frac{F_S^+}{F_0^+}\right|_{F_S^-=0} = \frac{1}{M_{11}}, \qquad r = \left.\frac{F_0^-}{F_0^+}\right|_{F_S^-=0} = \frac{M_{21}}{M_{11}}$$

For an incoming wave propagating from the right to the left ($F_0^+ = 0$)

$$t' = \left.\frac{F_0^-}{F_S^-}\right|_{F_0^+=0} = \frac{\det(M)}{M_{11}}, \qquad r' = \left.\frac{F_S^+}{F_S^-}\right|_{F_0^+=0} = \frac{-M_{12}}{M_{11}}$$

It should be noted that if $[M]$ connects electric fields, $\det(M) = k_{Sz}/k_{0z}$ depends only on the surrounding media. If $n_0 = n_S$, then $\det(M) = 1$, no matter what the materials are. In particular, attenuation or gain inside the structure has no influence whatsoever on the determinant (another invariant). From an experimental point of view, the reflection and transmission in power are

$$T = T' = \frac{\mathrm{Re}[\det(M)]}{|M_{11}|^2}, \quad R = |r|^2 = \left|\frac{M_{21}}{M_{11}}\right|^2, \quad R' = |r'|^2 = \left|\frac{M_{12}}{M_{11}}\right|^2$$

We would like to emphasize that while reflection can be asymmetrical, transmission (in terms of power) remains unconditionally reciprocal, at least with dielectric (non-magnetic) materials in the linear regime.

In integrated structures, the formalism is basically the same, except that the z-component k_z is replaced with the modal propagation constant β. Besides, instead of using the electric E or magnetic H components of the field according to the state of polarization, waves in both waveguides can be suitably normalized in such a way that they represent directly the square root of the spectral density $I(\omega)$ of optical intensity (per unit angular frequency), expressed in $[\mathrm{W} \cdot \mathrm{m}^{-2} (\mathrm{rad} \cdot \mathrm{s}^{-1})^{-1}]$.

3.2.3 Partial matrices and internal fields

Whatever the structure, one may need to describe not only the spectral behaviour of the transfer function but also, at a given frequency, the precise distribution of energy along the propagation axis. In order to do so, it is advised to use 'partial' transfer matrices that connect any intermediary abscissa z to the boundaries, as schematically depicted in Figure 3.8.

For any internal abscissa $z \in [z_0, z_S]$, let $[L_{(z)}]$ and $[R_{(z)}]$ denote the partial matrices connecting z_0 to z and z to z_S, respectively, such as $[M] = [L_{(z)}] \cdot [R_{(z)}]$. Partial matrices are established in the same way as $[M]$. The internal fields (F^+, F^-) at abscissa z should verify

$$\begin{pmatrix} F^+ \\ F^- \end{pmatrix} = \begin{pmatrix} R_{11} & R_{12} \\ R_{21} & R_{22} \end{pmatrix} \begin{pmatrix} F_S^+ \\ F_S^- \end{pmatrix}$$

Bearing in mind that $F_S^+ = tF_0^+$, one can express the internal fields as a linear combination of input fields (F_0^+, F_S^-):

$$F^+ = R_{11}\, t\, F_0^+ + R_{12}\, F_S^-$$

$$F^- = R_{21}\, t\, F_0^+ + R_{22}\, F_S^-$$

The total field $F_{tot} = F^+ + F^-$ can now be used in order to plot the standing-wave pattern for any boundary conditions.

3.2.4 Scattering parameters

Scattering parameters are another way, very popular in microwave engineering (and very convenient in practice), of describing the linear relationship between waves coming in and out of a given structure, seen as a multi-port network [9].

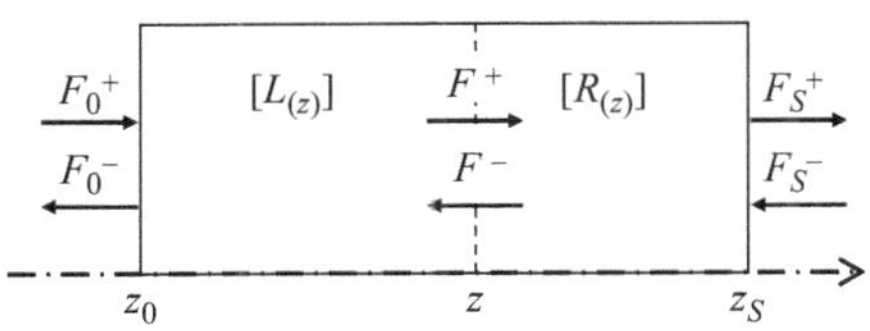

Figure 3.8 Partial transfer matrices [L] and [R]

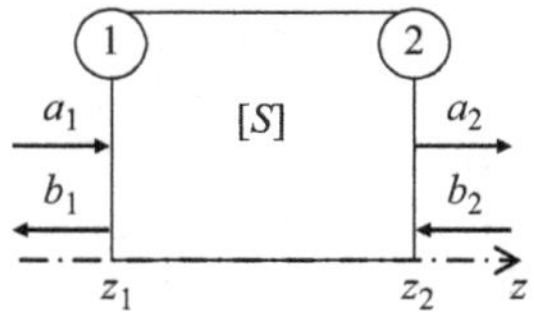

Figure 3.9 Scattering matrix of a two-port network, expressing the linear relationship between input (a_1, a_2) and output (b_1, b_2) waves at 'ports' 1 and 2

Instead of connecting two abscissas, this formalism expresses the output fields as a function of the input fields, as schematically depicted in Figure 3.9. At 'port' number p, the input and output are traditionally denoted as a_p and b_p, respectively.

Between 'ports' 1 and 2, we get

$$\begin{pmatrix} b_1 \\ b_2 \end{pmatrix} = \begin{pmatrix} S_{11} & S_{12} \\ S_{21} & S_{22} \end{pmatrix} \begin{pmatrix} a_1 \\ a_2 \end{pmatrix} = \begin{pmatrix} r & t' \\ t & r' \end{pmatrix} \begin{pmatrix} a_1 \\ a_2 \end{pmatrix}$$

The main advantage of such an approach is that the S_{pq} matrix elements, also called 'S-parameters', have a clear physical meaning: diagonal term S_{pp} is the reflectance at port p, whereas non-diagonal term S_{pq} is the transmittance from port q to port p. These quantities are all experimentally relevant (they are the data one gets directly from a network analyser, for instance). Their major drawback stems from the fact that, contrary to transfer matrices, the S-matrix of a chain structure is *not* given by the dot product of the elementary S-matrices. Another example is provided in Appendix A with the directional coupler.

Another graphical tool imported from electrical engineering is the 'signal-flow graph', representing by oriented arrows the possible connections between the ports involved.

3.3 Some classical resonators

3.3.1 Fabry–Pérot resonator

Schematically drawn in Figure 3.10, the Fabry–Pérot resonator (FPR) is made of a linear, homogeneous, isotropic (and possibly active) zone sandwiched between two reflectors, represented by their matrices $[A]$ and $[B]$ [10]. The precise nature of the mirrors (simple cleaved facet or multilayer) is not crucial. For the sake of simplicity, we assume $n_0 = n_S$.

Transfer matrix $[M]$ reads

$$[M] = [A][Z][B]$$

Of particular interest is the M_{11} component:

$$M_{11} = A_{11}\, B_{11} e^{+i\,\beta L} + A_{12}\, B_{21} e^{-i\,\beta L}$$

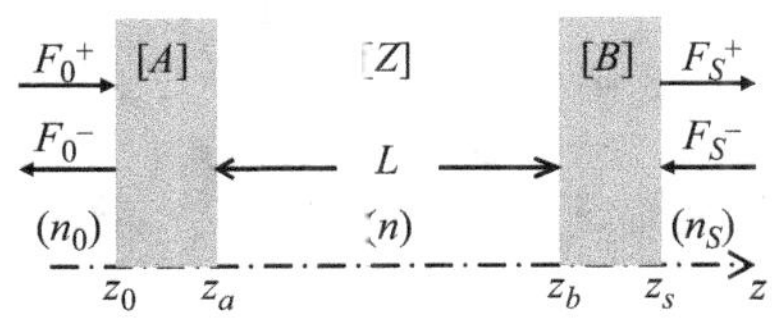

Figure 3.10 Schematic representation of a Fabry–Pérot resonator (FPR) made of a (possibly) active zone [Z] sandwiched between two reflectors [A] and [B]

Transfer function t is readily expressed as

$$t = \frac{1}{M_{11}} = \frac{t_A\, t_B\, e^{-i\,\beta L}}{1 - r'_A\, r_B\, e^{-2i\,\beta L}} = \frac{t_0}{1 - R_{eff}\, e^{-i\,\Phi}}$$

In this expression, numerator $t_0 = t_A\, t_B\, e^{-i\,\beta L}$ is the single-path transmission between z_0 and z_S (product of the complex transmittances of all the elementary components the wave must pass through), whereas denominator $D = 1 - r'_A r_B\, e^{-2i\beta L} = 1 - R_{eff} e^{-i\Phi}$ is the unmistakable signature of a feedback loop. If $r'_A = \rho_A\, e^{-i\varphi_A}$, $r_B = \rho_B\, e^{-i\varphi_B}$, then $R_{eff} = \rho_A \rho_B\, e^{2\mathrm{Im}[\beta L]}$, $\Phi = 2\mathrm{Re}[\beta L] + \varphi_A + \varphi_B$. Phase Φ and effective reflection R_{eff} represent the overall phase change and the amplitude change over a cavity round trip, respectively.

In terms of power transmission, we get the so-called Airy function:

$$T(\Phi) = |t|^2 = \frac{T_0}{1 + m_{FP}\sin^2(\Phi/2)}$$

where T_0 is the transmission at resonance, and m_{FP} is a spectral selectivity parameter:

$$m_{FP} = \frac{4R_{eff}}{(1 - R_{eff})^2}$$

For passive symmetric structures, $T_0 = 1$. We draw in Figure 3.11 the reduced Airy function $T(\Phi)/T_0$.

For small values of reflectivity ($R_{eff} \ll 1$), transmission is sine-like and can be described in terms of two-wave interference. For high enough values of R_{eff}, each resonance becomes sharply defined and can be described as a Lorentzian line:

$$T(\omega) \approx T(\omega_q)\frac{\gamma_q^2}{(\omega - \omega_q)^2 + \gamma_q^2}$$

The full width at half maximum (FWHM) is $2\,\gamma_q$. The 'quality factor' (or *Q-factor*) of a given resonance at angular frequency ω_q – such as $\Phi(\omega_q) = 2q\pi$ – is given by the ratio of the resonance frequency over the FWHM:

$$Q = \frac{\omega_q}{2\,\gamma_q}$$

The higher the Q-factor, the sharper the resonance.

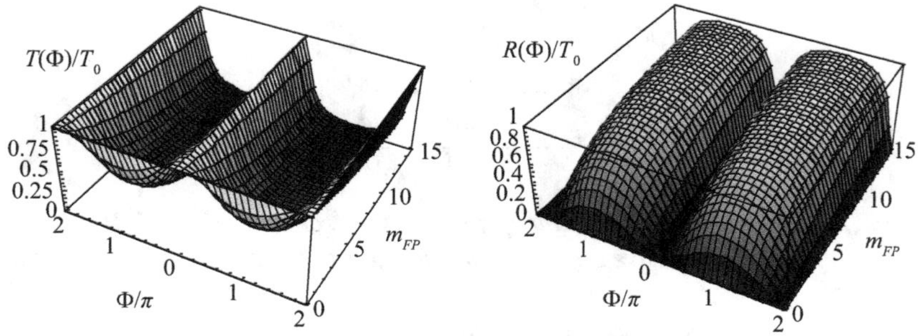

Figure 3.11 Reduced Airy function $T(\Phi)/T_0$ and complementary reflection $R(\Phi)/T_0$ as a function of m_{FP}

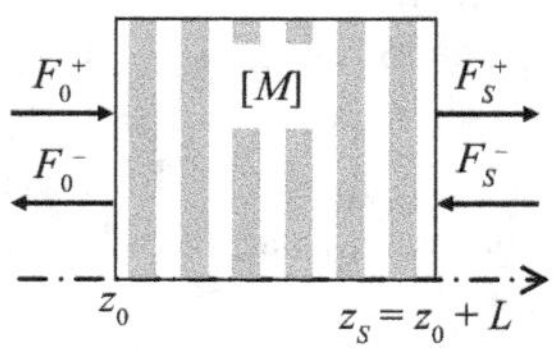

Figure 3.12 Distributed Bragg reflector of length L

3.3.2 Distributed Bragg reflector (DBR) and distributed feedback (DFB)

Apart from interface and free-space (or free waveguide) propagation, the distributed Bragg reflector (DBR) deserves a specific matrix (Figure 3.12). A periodic modulation of the (effective) index is responsible for spectrally selective reflection around the so-called Bragg wavelength. The fundamental reason is the apparition of a coherent interference mechanism, *distributed* over the whole structure (by contrast with *localized* mirrors), quite similar to what happens to electrons in crystals – hence the concurrent denominations 'photonic bandgap' or '1D photonic crystal' [11,12].

In terms of CMT, the periodic zone is easily described by a single matrix. Consider a sine-like Λ-periodic modulation of the effective permittivity:

$$n_{eff}^2(z) = n_m^2 + i\,\varepsilon_m'' + \varepsilon_{(1)}' \cos\left(2\beta_B(z - z_0) - \psi\right)$$

where $\beta_B = \pi/\Lambda$ is the reference Bragg wavevector, n_m is the average (real) effective index, ε_m'' is a constant background gain/loss term, $\varepsilon_{(1)}'$ is the amplitude of the (real-valued) modulation, and ψ is a phase of geometric origin. In the most general case, one could also consider a sine-like modulation of the imaginary part (periodic distribution of gain or losses), but in this chapter, we will limit ourselves to pure index coupling.

If the average gain/loss contribution and the modulation are treated as a perturbation ($\varepsilon''_m, \varepsilon'_{(1)} \ll n_m^2$), and neglecting any extra reflection at the boundaries, one gets eventually

$$M_{11} = \left(\cosh(\gamma L) + i\frac{\delta + i\,\alpha}{\gamma}\sinh(\gamma L)\right)\exp(+i\beta_B L)$$

$$M_{12} = \left(+i\frac{\kappa\, e^{+i\,\psi}}{\gamma}\sinh(\gamma L)\right)\exp(-i\beta_B L)$$

$$M_{21} = \left(-i\frac{\kappa\, e^{-i\,\psi}}{\gamma}\sinh(\gamma L)\right)\exp(+i\beta_B L)$$

$$M_{22} = \left(\cosh(\gamma L) - i\frac{\delta + i\,\alpha}{\gamma}\sinh(\gamma L)\right)\exp(-i\beta_B L)$$

where $\delta = n_m\, k_0 - \beta_B$, detuning with respect to the Bragg condition; $\kappa = \varepsilon'_{(1)} k_0/4n_m$, the modulus of the coupling constant, of phase ψ; $\alpha = \varepsilon''_m k_0/2n_m$, the average amplitude gain (half the modal gain); and $\gamma = \left(\kappa^2 - (\delta + i\,\alpha)^2\right)^{1/2}$.

In a lossless DBR, $\alpha = 0$, so that $\gamma = (\kappa^2 - \delta^2)^{1/2}$. Two regimes are easily distinguished:

- if $\gamma^2 > 0$ ($\delta < \kappa$), evanescence (forbidden band);
- if $\gamma^2 < 0$ ($\delta > \kappa$), propagation.

But the general expression holds whatever the level of attenuation ($\alpha < 0$) or amplification ($\alpha > 0$). It is the basis for the CMT investigation of DFB lasers [13–15].

The transmission/reflection spectra of a lossless DBR (pure index coupling) are represented in Figure 3.13 in reduced coordinates.

In terms of dimensionless parameters (κL, δL), the maximum reflection coefficient scales as

$$R_{Max} = \tanh^2(\kappa L)$$

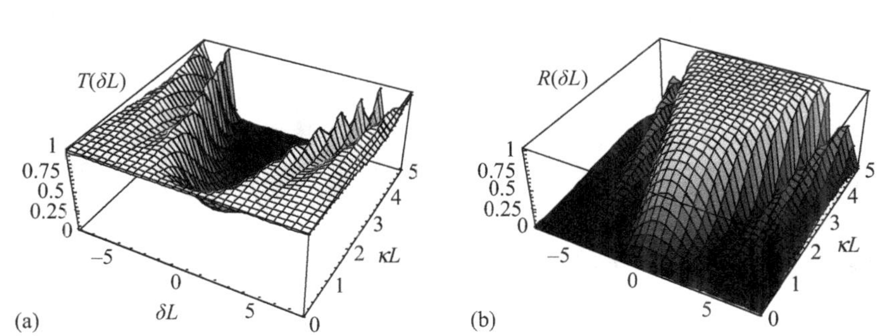

Figure 3.13 Passive DBR: (a) transmission and (b) reflection

whereas the width of the 'forbidden band', defined by $\gamma L \in \mathbb{R}$, is given by

$$\Delta(\delta L) = 2\kappa L$$

The lossless DBR can not only be seen as a mere reflector (when used in the vicinity of the Bragg wavelength) but also as a resonator in its own right: for any detuning $\delta L = \pm\delta_q L$ such as

$$\delta_q L = \left((\kappa L)^2 + q^2\pi^2\right)^{1/2}, \quad q \in \mathbb{N}^*$$

we observe a vanishing reflection ($R = 0$), a total transmission ($T = 1$), and a spectral resonance. The latter is especially pronounced at the edge of the photonic bandgap ($\delta L = \pm\delta_1 L$): the stronger the coupling κL, the sharper the resonance. It is instructive to draw the spatial distribution of the fields, including the standing-wave pattern (Figure 3.14). At resonance, the internal density of power can be significantly higher than the input or output power: the cavity acts as an 'energy condenser'.

As soon as gain enters the picture, the resonance is shifted but becomes even more pronounced. The structure is then considered as a DFB amplifier and can reach the laser regime [13], as will be seen in Section 3.4.

3.3.3 Quarter-wave-shifted DBR or DFB (QWS-DBR or QWS-DFB)

The quarter-wave-shifted (QWS)-DBR/DFB structure is made of two similar segments of passive DBR or active DFB, separated by a phase shift ensuring that the second part of the periodic modulation is in phase opposition with respect to the first (Figure 3.15). As a consequence, a spectrally narrow transmission window opens in the very centre of the forbidden band (the photonic equivalent to a 'default mode' in crystallography). The whole structure can be thought of as a specific instance of FPR, except that the cavity is reduced to a single abscissa, with all the gain (if any) lying in

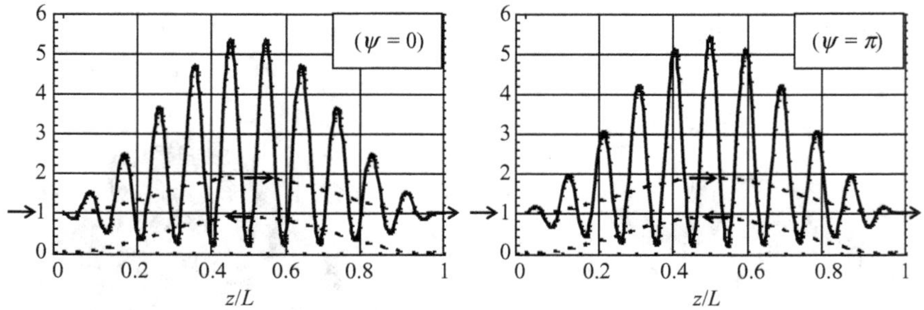

Figure 3.14 Spatial distribution of internal fields $|F^+|^2$, $|F^-|^2$, and standing-wave pattern $|F^+ + F^-|^2$ at resonance ($\delta L = \delta_1 L$) in a ten-period lossless DBR ($\kappa L = 3$, $\psi = 0$), normalized with respect to $F_0^+ = 1$, for $F_S^- = 0$. The precise position of nodes and antinodes strongly depends on phase ψ (left: $\psi = 0$; right: $\psi = \pi$)

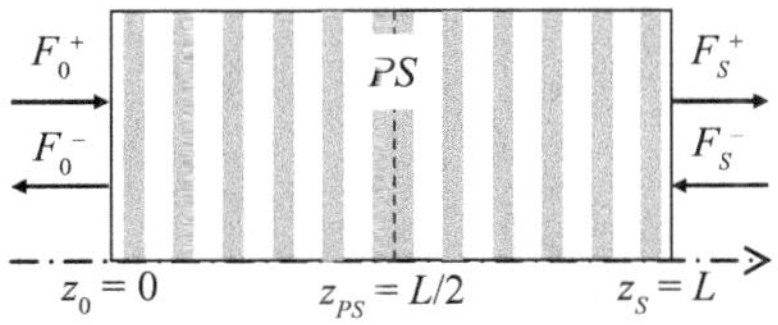

Figure 3.15 Quarter-wave-shifted DBR or DFB, with a phase shift (PS) located at the centre

the surrounding reflectors. The main interest of such a structure is its intrinsically single-mode behaviour, with a sharp transmission at the Bragg wavelength.

From a modelling point of view, the structure is made of two DFB sections that share basically the same parameters, except for the phase of their coupling constants. Let $[A]$ and $[B]$ denote the matrices associated with both halves, $[M] = [A]$ $[B]$ the complete matrix. $[A]$ is a DFB of length $(L/2)$, characterized by a coupling constant κ with phase ψ_A:

$$A_{11} = \left(\cosh\left(\frac{\gamma L}{2}\right) + i\frac{\delta + i\,\alpha}{\gamma}\sinh\left(\frac{\gamma L}{2}\right)\right)\exp\left(+i\,\frac{\beta_B L}{2}\right)$$

$$A_{12} = \left(+i\frac{\kappa e^{+i\,\psi_A}}{\gamma}\sinh\left(\frac{\gamma L}{2}\right)\right)\exp\left(-i\frac{\beta_B L}{2}\right)$$

$$A_{21} = \left(-i\frac{\kappa e^{-i\,\psi_A}}{\gamma}\sinh\left(\frac{\gamma L}{2}\right)\right)\exp\left(+i\frac{\beta_B L}{2}\right)$$

$$A_{22} = \left(\cosh\left(\frac{\gamma L}{2}\right) - i\frac{\delta + i\,\alpha}{\gamma}\sinh\left(\frac{\gamma L}{2}\right)\right)\exp\left(-i\,\frac{\beta_B L}{2}\right)$$

Matrix $[B]$ is almost the same, except for the phase of the coupling constant – in its non-diagonal terms, ψ_A is formally replaced by ψ_B, with

$$\psi_B = \psi_A - \beta_B L + \pi$$

The transmission/reflection spectra of a passive QWS-DBR are represented in Figure 3.16.

Note the 100% transmission at $\delta L = 0$.

3.3.4 Ring-like topology: the micro-ring resonator (MRR)

Up to now, we have only considered rectilinear topologies. The picture would not be complete, however, without ring-like resonators. We limit ourselves to 1D micro-rings [16], but the description can be extended to 2D (micro-discs) or 3D (microspheres). The simplest MRR is schematically described in Figure 3.17.

The coupling zone is well described in terms of co-directional CMT (Appendix A). Without loss of generality, the coupler can be thought of as lossless, localized at $z = z_c$ and well described by parameters $t_c = \cos(\chi d)$ and $k_c = -i\,\sin(\chi d)$, with $\chi d \in [0, \pi/2]$, so that $t_c^2 - k_c^2 = 1$.

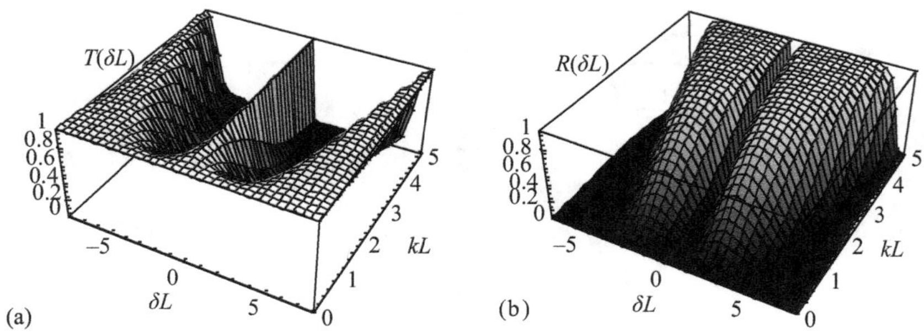

Figure 3.16 QWS-DBR: (a) transmission and (b) reflection

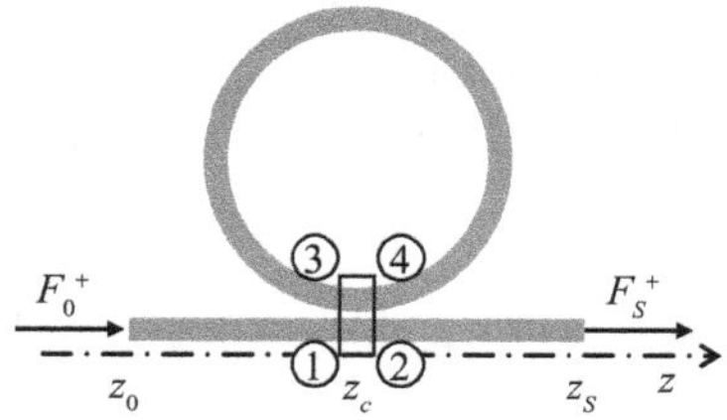

Figure 3.17 Schematic depiction of a micro-ring resonator (MRR), with the port denomination around the (supposedly) localized directional coupler

The calculation of the transfer function between ports (1) and (2) is a simple matter of algebraic manipulation. We can connect the fields around the coupler according to

$$\begin{cases} b_2 = a_1 t_c + a_3 k_c \\ b_4 = a_1 k_c + a_3 t_c \end{cases}$$

From now on, we note $t_c = \rho$ and $k_c = -i\,(1-\rho^2)^{1/2}$, with $\rho \in [0, 1]$. Parameter ρ is a measure of the fraction of field, which is fed back into the loop after each round trip and plays formally the same role in an MRR as $\rho_A\,\rho_B$ in an FPR. Besides, the ring is responsible for a phase shift $\phi = \mathrm{Re}[\beta\, L]$ where L is the perimeter and β is the ring propagation constant; and also (possibly) for an amplitude change $\sigma = \exp(\mathrm{Im}[\beta\, L])$:

$$\begin{cases} a_3 = b_4\,\sigma e^{-i\phi} \\ a_4 = b_3\,\sigma e^{-i\phi} \end{cases}$$

In the case of amplification, $\sigma > 1$; in the case of attenuation, $\sigma < 1$; and in a wholly transparent ring, $\sigma = 1$. We get eventually

$$\begin{pmatrix} b_1 \\ b_2 \end{pmatrix} = \begin{pmatrix} 0 & t_\Omega \\ t_\Omega & 0 \end{pmatrix} \begin{pmatrix} a_1 \\ a_2 \end{pmatrix}$$

$$t_\Omega = \left\{ \rho - \frac{(1-\rho^2)\,\sigma\, e^{-i\,\phi}}{1-\rho\,\sigma\, e^{-i\,\phi}} \right\}$$

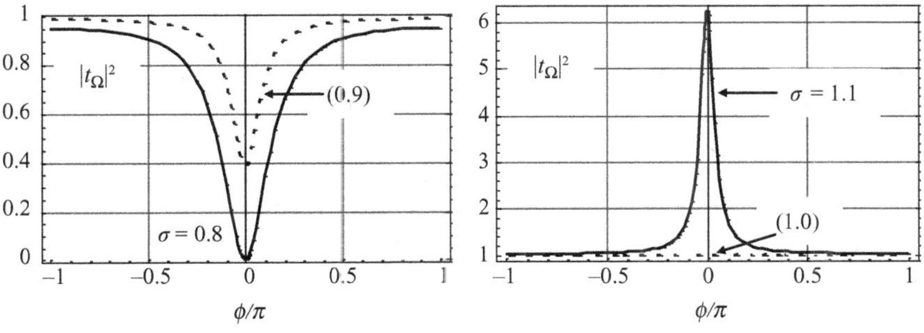

Figure 3.18 Transmission of an MRR as a function of ϕ for $\rho = 0.8$ and $\sigma = 0.8$ (critical coupling), $\sigma = 0.9$ (over-coupling), $\sigma = 1$ (transparency), and $\sigma = 1.1$ (amplification)

Under this form, the MRR transfer function t_Ω is seen to result from interference between the direct path through the coupler and the path through the ring itself. Denominator $D = 1 - \rho\sigma e^{-i\phi}$ is the unmistakable signature of a feedback loop and resembles that of an FPR, with $R_{eff} = \rho\sigma$. One can simplify further

$$t_\Omega = \frac{\rho - \sigma\, e^{-i\,\phi}}{1 - \rho\,\sigma\, e^{-i\,\phi}} = |t_\Omega|\, e^{-i\,\varphi_\Omega}$$

Note that such a structure remains unidirectional: in the ideal case, no reflection occurs.

From a spectral point of view, the behaviour is slightly different from that of an FPR, since in a completely lossless ring ($\sigma = 1$), $|t_\Omega|^2 = 1$, whatever the value of ρ and whatever the phase ϕ (i.e. whatever the frequency). The transmission in intensity is drawn in Figure 3.18 for different values of σ.

The phase φ_Ω of the transfer function (Figure 3.19) is worth a comment.

The ring is responsible for a specific structural dispersion, made manifest by the slope of the quasilinear function φ_Ω in the vicinity of a resonance. This slope can be quantified by a dispersion factor. In a transparent MRR, we get

$$F_D = \left.\frac{d\varphi_\Omega}{d\phi}\right|_{\phi=0} = \frac{1+\rho}{1-\rho}$$

More generally, as long as $\sigma > \rho$, we get

$$F_D = \left.\frac{d\varphi_\Omega}{d\phi}\right|_{\phi=0} = \frac{\sigma\;(1-\rho^2)}{(\sigma-\rho)(1-\sigma\rho)}$$

The higher the ρ, the steeper the slope; by way of corollary, the smaller the bandwidth of the linear zone.

Of particular interest, in this chapter, is the ring resonator with gain ($\sigma > 1$): from selective amplification to laser oscillation.

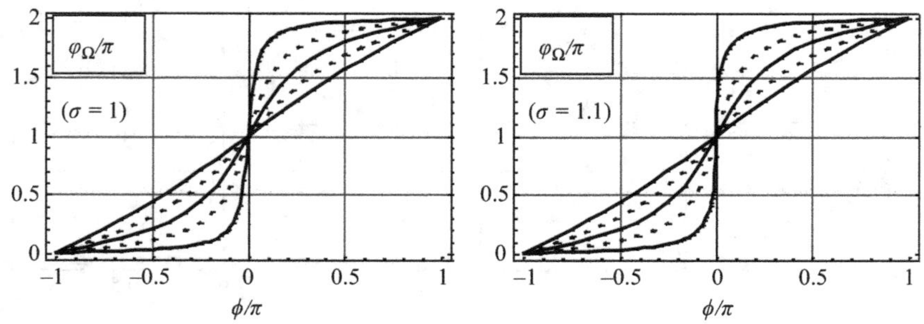

Figure 3.19 Left: phase φ_Ω of the transfer function t_Ω for a transparent ring (σ = 1), for $\rho = 0.1$, 0.3, 0.5, 0.7, or 0.9: the higher the ρ, the steeper the slope. Right: the same phase, for an amplifying ring ($\sigma = 1.1$)

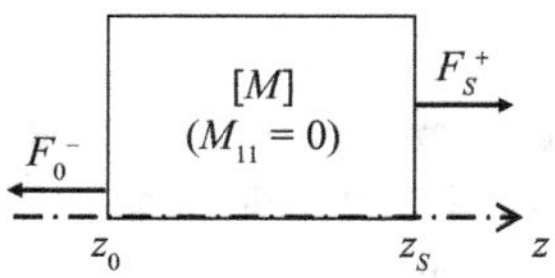

Figure 3.20 Oscillation condition: output fields must be present even without any incoming fields ($F_0^+ = F_S^- = 0$). This is only possible if $M_{11} = 0$, or $|t| \to \infty$

3.4 Oscillation condition: threshold and beyond

3.4.1 Matrix oscillation condition

Classical TMF is basically a linear tool. However, it can be extended up to give the threshold condition of a laser oscillator (Figure 3.20).

As a matter of fact, in any active resonant structure, the feedback loop can be such that output signals are emitted even when no incoming fields are present. From a mathematical point of view, we have to look for a non-vanishing solution to

$$\begin{pmatrix} 0 \\ F_0^- \end{pmatrix} = \begin{pmatrix} M_{11} & M_{12} \\ M_{21} & M_{22} \end{pmatrix} \begin{pmatrix} F_S^+ \\ 0 \end{pmatrix}$$

In order to ensure $M_{11}F_S^+ = 0$ with $F_S^+ \neq 0$, we have to guarantee

$$M_{11} = 0$$

In other words, the oscillation condition corresponds to a case when transfer function $t = 1/M_{11}$ becomes infinite.

3.4.2 FPR threshold condition

For the sake of illustration, let us consider an FPR made of an active zone sandwiched between two passive reflectors, as described earlier.

The threshold condition can also be recovered by cancelling denominator D, that is

$$R_{eff}\, e^{-i\Phi} = 1$$

In this specific case, two conditions must be simultaneously verified: $R_{eff} = \rho_A\, \rho_B e^{2\mathrm{Im}[\beta L]} = 1$ and $\Phi \equiv 0\ \ [2\pi]$, with $\beta = n_{eff} k_0 + i(g_m/2)$. The first one stipulates that linear (unsaturated) gain should compensate for the losses and gives the value of the threshold gain:

$$g_{th}L = -\ \ln(\rho_A\, \rho_B)$$

The second one indicates that only discrete frequencies satisfying a phase requirement can be made to oscillate: we recover the frequency comb of the Airy function, with a free spectral range (FSR) given by

$$\Delta\omega = \frac{\pi c}{n_g L}$$

It should be emphasized that threshold condition $M_{11} = 0$ is much more general and can be applied to other kinds of active structures.

3.4.3 DFB threshold condition

For the sake of clarity, let us restrict ourselves to the specific case of an index-coupled DFB structure, without extra reflection at both ends. In that case, the threshold condition reads

$$\cosh(\gamma L) + i\, \frac{\delta L + i\,\alpha L}{\gamma L}\, \sinh(\gamma L) = 0$$

where $\gamma = \left(\kappa^2 - (\delta + i\,\alpha)^2\right)^{1/2}$. It involves a transcendental equation that can be solved numerically to an arbitrary precision. For a given value of κL, it gives a discrete set of couples (δL, αL): the stronger the coupling, the lower the threshold. The main difference with the FPR is schematically illustrated by their respective modal cartography (Figure 3.21): the closer one gets to the DFB forbidden band, the smaller the threshold gain; besides, the FSR is not strictly constant. By way of consequence, a DFB will oscillate on one of the resonance frequencies closest to the forbidden band, whereas an FPR with a broad gain spectrum can support several longitudinal modes simultaneously. Depending on the desired regime (from single-mode emission to mode-locked pulse generation), one oscillator can prove better suited than the other.

3.4.4 QWS-DFB threshold condition

The QWS-DFB is intrinsically single mode, at the Bragg frequency. Its threshold condition can be written as

$$M_{11} = A_{11}B_{11} + A_{12}B_{21} = 0$$

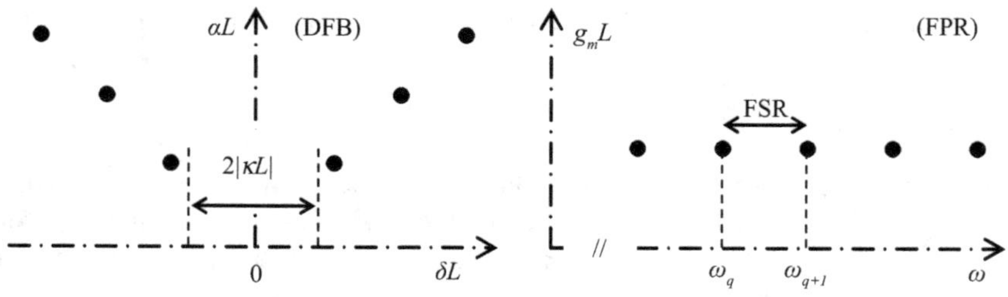

Figure 3.21 Schematic comparison between the DFB and the FPR, in terms of modal cartography

or, which amounts to the same

$$r'_A\, r_B = r_B^2 = 1$$

where $\delta = 0$ and $\gamma = (\kappa^2 + \alpha^2)^{1/2}$. Since the oscillating frequency is fixed, the only unknown is the threshold gain α_{th}. The higher κL, the lower $\alpha_{th} L$.

3.4.5 MRR threshold condition

The MRR denominator D is very similar to that of an FPR. With $\beta = n_{eff} k_0 + i\ (g_m/2)$ the modal propagation constant of the ring, the threshold condition is obtained by cancelling D:

$$g_{th} L = -2\,\ln(\rho)$$

The resonances are distributed along a frequency comb, with an FSR given by

$$\Delta\omega = \frac{2\pi c}{n_g L}$$

The only difference with that of an FPR is a factor of 2, connected to the fact that the ring remains unidirectional (length L is travelled over just once).

3.4.6 Laser emission (above threshold)

We have established the threshold condition, which means the minimal amount of pumping needed for achieving oscillation. What can we tell for higher levels of pumping? Is it even possible to extend our simple tools well into this highly non-linear regime? For the sake of illustration, let us assume a steady-state single-mode lasing regime with a gain medium well described by an ideal three-level system:

$$g_m = \Gamma g_0 \frac{Y_P - 1}{Y_P + 1 + 2Y} - \alpha_s$$

Under threshold, the light emitted at resonance remains negligible so that $Y = 0$. Once the oscillation threshold is known, one can determine the corresponding pumping level:

$$g_{th} = \Gamma g_0 \frac{Y_{Pth} - 1}{Y_{Pth} + 1} - \alpha_s \quad \Rightarrow \quad Y_{Pth} = \frac{\Gamma g_0 + g_{th} + \alpha_s}{\Gamma g_0 - g_{th} - \alpha_s}$$

Now, what happens above threshold, that is for $Y_P > Y_{Pth}$? Of course, oscillation still occurs: the saturated gain is 'clamped' to its threshold value, the excess of energy being integrally converted into a laser field ($Y > 0$). It means that condition $M_{11} = 0$ is not only valid at threshold but also can be extended into the non-linear laser regime, for describing oscillation.

$$\Gamma g_0 \frac{Y_P - 1}{Y_P + 1 + 2Y} - \alpha_s = g_{th} \quad \Rightarrow \quad \frac{Y_P - 1}{Y_P + 1 + 2Y} = \frac{Y_{Pth} - 1}{Y_{Pth} + 1}$$

As a result, we distinguish the two main regimes:

- below threshold: $Y_P < Y_{Pth}$, $g_m = g_{lin}$, $Y = 0$
- above threshold: $Y_P > Y_{Pth}$, $g_m = g_{th}$, $Y = \frac{Y_P - Y_{Pth}}{Y_{Pth} - 1}$

Even in the frame of such a simplistic model, we can recover clear tendencies (Figure 3.22): the lower the ratio (threshold gain)/(maximum available gain), the lower the threshold, and the steeper the slope of the diode-like $Y(Y_P)$ curve.

An implicit assumption lies beyond the idea that in the steady-state laser regime, saturated gain remains clamped to its threshold value: gain is supposed longitudinally uniform at the scale of the active zone and saturated by the spatially averaged intensity, neglecting any standing-wave effect. This holds true for an MRR and may still be reasonable in an FPR but can be questioned in a QWS-DFB, where the internal field is far from uniform: on the contrary, the maximum density of power is to be found in the vicinity of the phase shift. If need be, it is always possible to decompose the structure into substructures with locally constants optical parameters,

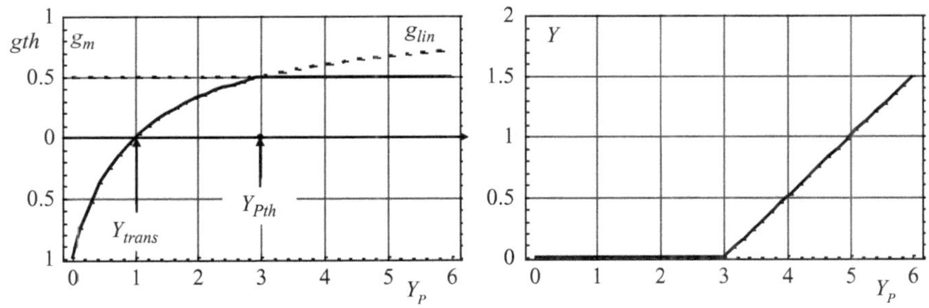

Figure 3.22 Schematic illustration of the typical behaviour of the gain (g_m) and the laser field (Y) as a function of pumping (Y_P), in an ideal three-level system with α_s=0, g_{th}=Γg_0/2. For the record, Y_P=1 corresponds to the optical transparency in the medium (the onset of stimulated amplification)

in order to better follow the saturation distribution, but at the price of a greater complexity, where the analytical approach may lose some of its natural appeal.

Similar calculations can be performed with other gain media.

3.5 Amplified spontaneous emission

Up to now, we have systematically neglected ASE that occurs as soon as spontaneous emission takes place in a gain medium. In some instances, though, this effect may deserve a closer look. In terms of modelling, the main trouble with internal sources of radiation stems from the fact that they break the proportionality between input and output signals: as a matter of fact, for high enough pumping but still below threshold, emitted light can be detected even without any incoming fields (at least at the same frequency). By way of consequence, traditional (2×2) TMF does not hold any more. However, in 1991, Weber and Wang have proposed an extended (3×3) version of TMF that works remarkably well in this case [17].

3.5.1 Extended (3×3) transfer matrix formalism

Extended (3×3) TMF works as follows [17–20]: assume a matrix relationship between both extremities (z_0, z_S) under the following form:

$$\begin{pmatrix} F_0^+ \\ F_0^- \\ 1 \end{pmatrix} = \begin{pmatrix} M_{11} & M_{12} & M_{13} \\ M_{21} & M_{22} & M_{23} \\ 0 & 0 & 1 \end{pmatrix} \begin{pmatrix} F_S^+ \\ F_S^- \\ 1 \end{pmatrix}$$

In this expression, the M_{ij} coefficients of the first two columns are the same as in the (2×2) matrix, whereas the third column contains source terms, the form of which will be shortly specified. Let us call (B_0^-, B_S^+) the 'background' fields (F_0^-, F_S^+) one gets when no input field is present ($F_0^+ = F_S^- = 0$), as schematically depicted in Figure 3.23.

A straightforward calculation shows that

$$B_S^+ = \frac{-M_{13}}{M_{11}}, \qquad B_0^- = \frac{(-M_{13}M_{21} + M_{23}M_{11})}{M_{11}}$$

The apparent dissymmetry between the two fields is only a consequence of the matrix convention that expresses the fields at abscissa z_0 as a function of the fields at abscissa $z_S > z_0$ (and not the other way round).

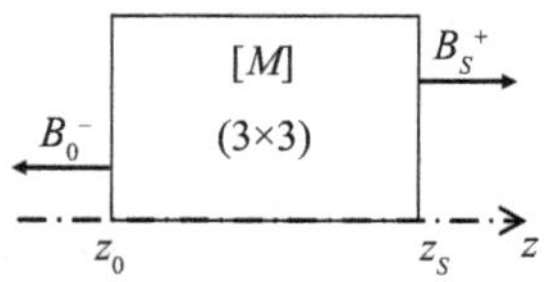

Figure 3.23 Extended (3×3) transfer matrix between two semi-infinite media: because of ASE, background fields (B_0^-, B_S^+) are emitted at both ends even though no input fields are present, even below threshold

It should be pointed out that whatever the precise content of source terms (M_{13}, M_{23}), emitted fields appear projected into the transfer function $t = 1/M_{11}$: they bear the spectral signature of the structure. As soon as 1962, Kastler had experimentally observed and theoretically explained the filtering of photoluminescence emitted by excited atoms when placed inside a resonant cavity [21].

From a practical point of view, any matrix $[M]$ can still be decomposed into the dot product of elementary transformations. Besides, the source terms (Z_{13}, Z_{23}) of transfer matrix $[Z]$ of any internal active zone can be written as

$$Z_{13} = -(u^+)\, Z_{11}, \quad Z_{13} = -(u^+)Z_{21} + (u^-)$$

where (u^+) and (u^-) represent the equivalent fields of spontaneous emission that couple into the mode at both ends of the zone under consideration; these terms were initially introduced by Choi *et al.* [22] for dealing with multisection semiconductor lasers. Since they result from the integration, over the whole active zone, of all sources of spontaneous emission, only their amplitude is experimentally relevant, whereas their phase is a random quantity that usually cannot be determined. They are best defined through their quadratic properties. For example, in an active homogeneous zone of length L,

$$\langle |(u^+)|^2 \rangle = \langle |(u^-)|^2 \rangle = D_{sp}(\omega, z)\, L \left(\frac{e^{2\mathrm{Im}[\beta L]} - 1}{2\mathrm{Im}[\beta L]} \right)$$

$$\langle (u^+)(u^-)^* \rangle = D_{sp}(\omega, z)\, L \left(\frac{\sin\left(\mathrm{Re}[\beta L]\right)}{\mathrm{Re}[\beta L]} \right) e^{\mathrm{Im}[\beta L]}$$

where $D_{sp}(\omega, z)$, expressed in $[(\mathrm{W}\cdot\mathrm{m}^{-2})\ \mathrm{m}^{-1}\ (\mathrm{rad}\cdot\mathrm{s}^{-1})^{-1}]$, is the spectral and spatial distribution of spontaneous intensity (per unit axial distance and angular frequency) so that the infinitesimal contribution dI to the spectral density of intensity that takes its source at abscissa z is $dI = D_{sp}(z)\ dz$. D_{sp} is a real-valued positive quantity that depends on the source level:

$$D_{sp}(\omega, z) = \beta_{sp} \hbar\omega\, r_{sp}(\omega)$$

In this expression, β_{sp} is the (small) fraction of spontaneous emission (otherwise emitted isotropically over 4πsr) that actually couples into the guided mode, and $r_{sp}(\omega)$ is the spectral rate of spontaneous emission (per unit time and volume), in $[\mathrm{s}^{-1}\cdot\mathrm{m}^{-3}\ (\mathrm{rad}\cdot\mathrm{s}^{-1})^{-1}]$. For instance, in a direct-gap semiconductor, it is connected to the joint density of states $\rho_J(\omega)$ through

$$r_{sp}(\omega) = K\, \rho_J(\omega)\, f_c\, (1 - f_v)$$

where (f_c, f_v) are the Fermi occupation factors in the conduction or valence bands, and K depends on the material. On the other hand, in an atomic medium such as seen in Section 3.1,

$$r_{sp}(\omega) = AN_2\, g(\omega) \qquad g(\omega) = \frac{\sigma_e(\omega)}{\int \sigma_e(\omega')\, d\omega'}$$

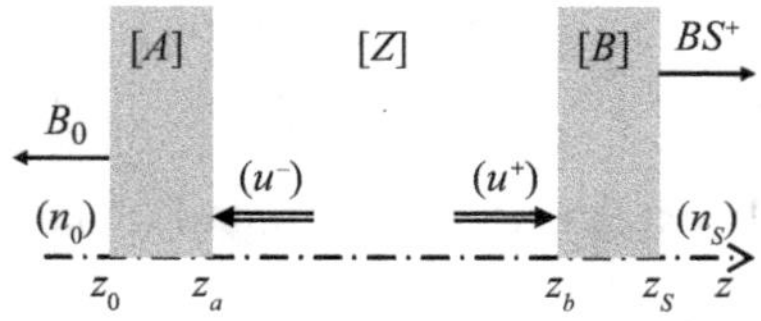

Figure 3.24 Schematic representation of an active zone [Z] between reflectors [A] and [B]. The equivalent field of spontaneous emission (u^+) or (u^-) that couples into the mode at an end (z_a or z_b) should not be confused with the actual value of the field at the same abscissa

Normalized lineshape $g(\omega)$ reproduces the spectral behaviour of the emission cross-section.

It should be remarked that in an active homogeneous zone, the 'mutual coherence term' $\langle (u^+)(u^-)^* \rangle$ tends to vanish as soon as the length L is much longer than the wavelength ($L/\lambda \gg 1$), a condition quite naturally met in most cases. The calculation can be extended to a non-homogeneous distribution of the source term [19].

3.5.2 Active FPR

For the sake of illustration, consider an FPR made of a single active, homogeneous zone between two reflectors [A] and [B], as depicted in Figure 3.24.

We get eventually

$$B_S^+ = \frac{(u^+)\, t_B + (u^-)\, r'_A\, e^{-i\beta L}\, t_B}{D}, \qquad B_0^- = \frac{(u^-)t'_A + (u^+)r_B \mathrm{e}^{-i\beta L} t'_A}{D}$$

Once again, numerator takes the form of a single-pass transfer function, whereas denominator D is responsible for spectral filtering. We would like to emphasize that field (u^+) or (u^-) should not be confused with the actual value of the field at the same abscissa: they represent the fields that would be emitted at the ends of the active zone if the latter were perfectly matched (no reflection). On the other hand, fields (F_a^+, F_a^-) at abscissa z_a^+ and (F_b^+, F_b^-) at abscissa z_b^- are given by

$$F_a^+ = \frac{(u^+)\, r_B e^{-i\beta L}\, r'_A + (u^-) r'_A}{D}, \qquad F_a^- = \frac{(u^+) r_B e^{-i\beta L} + (u^-)}{D}$$

$$F_b^+ = \frac{(u^+) + (u^-)\, r'_A e^{-i\beta L}}{D}, \qquad F_b^- = \frac{(u^+) r_B + (u^-) r'_A e^{-i\beta L} r_B}{D}$$

The same information can be expressed in terms of 'extended scattering parameters', as schematically depicted in Figure 3.25.

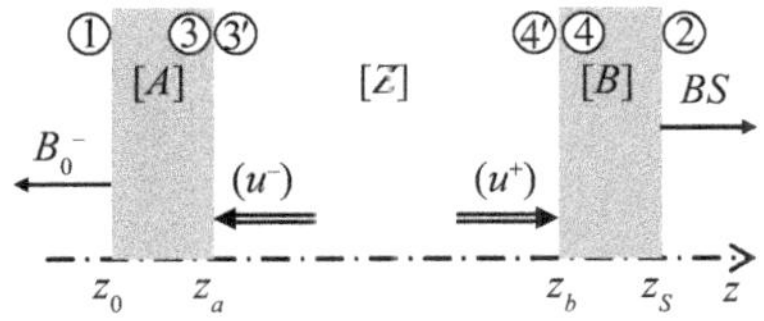

Figure 3.25 *The same active zone [Z] between reflectors [A] and [B], in terms of S-parameters. Note the convention for numbering the ports*

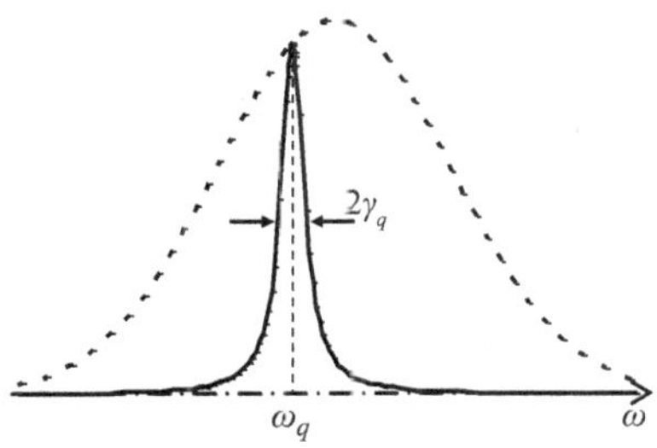

Figure 3.26 *Schematic illustration of the projection of ASE into the cavity resonance. The broader intrinsic emission spectrum [a.u.] is drawn in dashed lines, the actual emitted spectrum in straight lines [a.u.]*

Calling (3) the port at z_a and (4) the port at z_b, we get

$$\begin{cases} b_3' = a_3' e^{-i\beta L} + (u^-) \\ b_4' = a_4' e^{-i\beta L} + (u^+) \end{cases} \quad \begin{cases} b_3' = a_3 \\ b_3' = a_4 \end{cases}$$

Besides, boundary conditions ensure that

$$\begin{cases} b_1 = a_1 r_A + a_3 t_A' \\ b_3 = a_1 t_A + a_3 r_A' \end{cases} \quad \begin{cases} b_2 = a_4 t_B + a_2 r_B' \\ b_4 = a_4 r_B + a_2 t_B' \end{cases}$$

As expected, we recover eventually the same result. In terms of electrical engineering, the (u^-) and (u^+) terms can be thought of as contributions coming from accessory 'generators' inserted at both the ends of the active circuit (just before the reflectors). The projection of ASE into the cavity resonance is schematically drawn in Figure 3.26.

3.5.3 Active DFB

The situation is slightly more complex in a DFB zone, where the equivalent fields are affected by non-homogeneous spectral filtering. The overall picture remains the same, except for the precise form of quadratic quantities $\langle |(u^+)|^2 \rangle$, $\langle |(u^-)|^2 \rangle$, $\langle (u^+)(u^-)^* \rangle$: this is left as an exercise to the motivated reader (the answer is to be found in Appendix B).

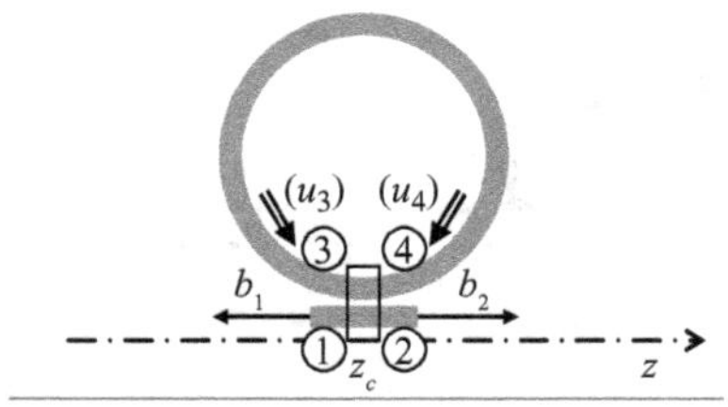

Figure 3.27 Schematic depiction of an MRR with the internal sources of spontaneous emission

If the DFB active zone $[Z]$ is subject to arbitrary boundary conditions (such as extra reflectors at both ends), then the total matrix is $[M]=[A][Z][B]$.

3.5.4 Active MRR

The ring resonator is best described in terms of scattering parameters. We take into account the internal sources through equivalent fields of spontaneous emission (u_3), (u_4) that couple into the mode at internal ports (3) and (4) (Figure 3.27).

We still assume the coupler itself to be described by

$$\begin{cases} b_2 = a_1\rho - ia_3\left(1-\rho^2\right)^{1/2} \\ b_4 = a_3\rho - ia_1\left(1-\rho^2\right)^{1/2} \end{cases}$$

The internal fields are now connected through

$$\begin{cases} a_3 = b_4\sigma e^{-i\phi} + (u_3) \\ a_4 = b_3\sigma e^{-i\phi} + (u_4) \end{cases}$$

As a result, if $a_1 = a_2 = 0$, we get eventually

$$b_1 = \frac{-i\left(1-\rho^2\right)^{1/2}(u_4)}{1-\rho\sigma e^{-i\phi}}, \quad b_2 = \frac{-i\left(1-\rho^2\right)^{1/2}(u_3)}{1-\rho\sigma e^{-i\phi}}$$

Once again, each output field is written as the ratio of a 'single-path transfer function' over a 'feedback loop' denominator.

3.5.5 Multisection structures

As far as multisection structures made of several active zones are concerned, rectilinear and annular topologies should be distinguished. In a rectilinear structure, the total (3×3) transfer matrix is obtained by the dot product of individual (3×3) matrices. Let us call $(u_p^{\pm})$ the equivalent fields of active zone number p. The total background fields (B_0^+, B_S^-) are easily obtained as linear combinations of all $(u_p^{\pm})$, with weighting coefficients that ultimately depend on their position along the chain. In order to evaluate the corresponding intensities, one must keep in mind a very simple rule: since sources distributed at different abscissas are not

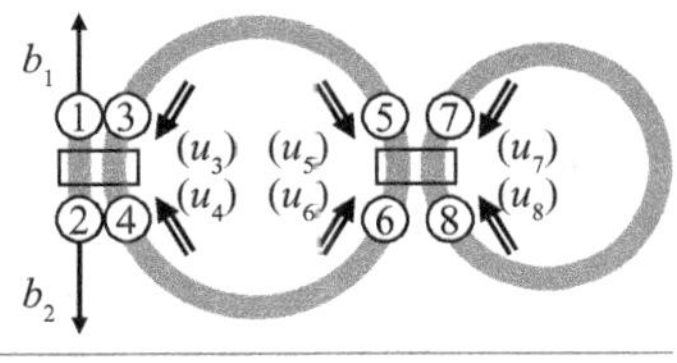

Figure 3.28 Coupled-ring MRR with internal sources of spontaneous emission

spatially coherent, any quadratic cross product that involves different active zones should vanish:

$$\text{Whenever}\, p \neq q, \left\langle \left(u_p^+\right)\left(u_q^\pm\right)^*\right\rangle = \left\langle \left(u_p^-\right)\left(u_q^\pm\right)^*\right\rangle = 0$$

This is the basis for any rectilinear coupled-cavity scheme.

In a structure combining several ring sections [23,24], scattering parameters are the obvious choice (Figure 3.28):

Consider the coupled-ring MRR of Figure 3.28; let us characterize the two couplers by parameter (ρ_1, ρ_2). Between internal ports (5), (6), (7), and (8), we get

$$\begin{cases} b_5 = a_6\rho_2 - ia_8\left(1-\rho_2^2\right)^{1/2} \\ b_6 = a_5\rho_2 - ia_7\left(1-\rho_2^2\right)^{1/2} \end{cases} \quad \begin{cases} b_7 = a_8\,\rho_2 - i\,a_6\left(1-\rho_2^2\right)^{1/2} \\ b_8 = a_7\,\rho_2 - i\,a_5\left(1-\rho_2^2\right)^{1/2} \end{cases}$$

$$\begin{cases} a_7 = b_8\sigma_2\, e^{-i\phi_2} + (u_7) \\ a_8 = b_7\sigma_2\, e^{-i\phi_2} + (u_8) \end{cases}$$

The same kind of relationships holds in the first ring (where we have to distinguish the contributions of both half-rings). The calculation (left to the patient reader) may seem cumbersome but remains straightforward; actually, it is only a matter of algebraic manipulation.

3.6 Conclusion, possible extensions, and perspectives

For describing multisection-integrated structures, we have introduced a few algebraic tools that include not only optical gain but also sources of spontaneous emission. As we have seen, the formalism is extendable into the non-linear regime, since it takes explicitly into account the saturation of the active zone by the internal field. A theoretical connection can be made between extended (3×3) TMF and the generalized transfer function of the resonators, a semiclassical approach that describes in a self-consistent way the behaviour of the emitter throughout threshold [25].

The index, hence the phase, can also be affected by saturation. In this chapter, for the sake of clarity, we have always implicitly neglected the possible shift of the resonance frequency. A detailed description of threshold crossing in a single-mode MRR, including frequency pulling, can be found in [26].

Since we work with non-linear media in cavity, bistability or multistability regimes can also be achieved [27,28].

Technical difficulties could arise if several non-linear zones must be described simultaneously, since every zone is subject to saturation through the averaged value of the total internal field: it may be necessary to solve a series of coupled non-linear equations. The problem would become *non-local*, both spatially and spectrally.

Appendix A Codirectional coupler

The codirectional coupler is another key element of integrated optics. It is conveniently described in terms of CMT [29]. Assume a coupler of length d located between abscissas z_1 and z_2, as schematically depicted in Figure A.1. Waves in both waveguides can be suitably normalized in such a way that they represent not an electric (or magnetic) field, but the square root of the optical intensity (per unit angular frequency). Let β_u and β_v denote the wavevectors of isolated waveguides u and v, and χ the codirectional coupling constant, assumed real valued and positive.

The equation evolution of the waves along the coupler is well described by

$$i\frac{\partial}{\partial z}\begin{pmatrix} F_u^+ \\ F_v^+ \end{pmatrix} = \begin{pmatrix} \beta_u & \chi \\ \chi & \beta_v \end{pmatrix}\begin{pmatrix} F_u^+ \\ F_v^+ \end{pmatrix}$$

In the absence of coupling ($\chi = 0$), each field propagates in its waveguide according to

$$F_u^+(z) = F_u^+(0)\exp(-i\beta_u z)$$
$$F_v^+(z) = F_v^+(0)\exp(-i\beta_v z)$$

As soon as coupling occurs ($\chi \neq 0$), both fields exchange energy. It is convenient to define the slowly varying envelopes (A_u, A_v), average wavevector β and detuning Δ such as

$$F_u^+(z) = A_u(z)\exp(-i\beta z), \quad F_v^+(z) = A_v(z)\exp(-i\beta z)$$
$$\beta = \frac{\beta_u + \beta_v}{2}, \quad \Delta = \frac{\beta_u - \beta_v}{2}$$

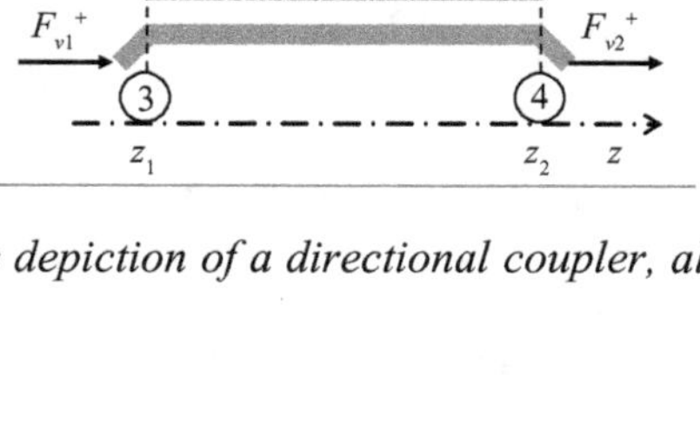

Figure A.1 Schematic depiction of a directional coupler, also seen as a four-port network

(A_u, A_v) evolve according to

$$i\frac{\partial}{\partial z}\begin{pmatrix} A_u \\ A_v \end{pmatrix} = \begin{pmatrix} +\Delta & \chi \\ \chi & -\Delta \end{pmatrix}\begin{pmatrix} A_u \\ A_v \end{pmatrix}$$

Introducing $\Gamma = \left(\chi^2 + \Delta^2\right)^{1/2}$, a straightforward calculation shows that

$$\begin{pmatrix} A_u(z) \\ A_v(z) \end{pmatrix} = \begin{pmatrix} \cos(\Gamma(z-z_1)) - i\frac{\Delta}{\Gamma}\sin(\Gamma(z-z_1)) & -i\frac{\chi}{\Gamma}\sin(\Gamma(z-z_1)) \\ -i\frac{\chi}{\Gamma}\sin(\Gamma(z-z_1)) & \cos(\Gamma(z-z_1)) + i\frac{\Delta}{\Gamma}\sin(\Gamma(z-z_1)) \end{pmatrix}\begin{pmatrix} A_u(z_1) \\ A_v(z_1) \end{pmatrix}$$

Reverting to the total fields, we get, between z_1 and $z_2 = z_1 + d$

$$\begin{pmatrix} F_{u2}^+ \\ F_{v2}^+ \end{pmatrix} = \exp(-i\beta d)\begin{pmatrix} \cos(\Gamma d) - i\frac{\Delta}{\Gamma}\sin(\Gamma d) & -i\frac{\chi}{\Gamma}\sin(\Gamma d) \\ -i\frac{\chi}{\Gamma}\sin(\Gamma d) & \cos(\Gamma d) + i\frac{\Delta}{\Gamma}\sin(\Gamma d) \end{pmatrix}\begin{pmatrix} F_{u1}^+ \\ F_{v1}^+ \end{pmatrix}$$

Formally, this matrix has exactly the same structure as the Jones matrix one uses to describe the transformation of a state of polarization by a birefringent plate. It can be rewritten as

$$\begin{pmatrix} F_{u2}^+ \\ F_{v2}^+ \end{pmatrix} = e^{-i\beta d}\begin{pmatrix} t_c & k_c \\ k_c & t_{c*} \end{pmatrix}\begin{pmatrix} F_{u1}^+ \\ F_{v1}^+ \end{pmatrix}$$

It holds whatever the boundary conditions. For instance, assuming an incoming signal on the A waveguide only ($F_{v1}^+ = 0$), we can follow the sine-like energy exchange as a function of the abscissa:

$$\left|\frac{F_u^+(z)}{F_{u1}^+}\right|^2 = \cos^2(\Gamma(z-z_1)) + \frac{\Delta^2}{\Gamma^2}\sin^2(\Gamma(z-z_1))$$

$$\left|\frac{F_v^+(z)}{F_{u1}^+}\right|^2 = \frac{\chi^2}{\Gamma^2}\sin^2(\Gamma(z-z_1))$$

One can define a characteristic interaction length $L_c = \pi/(2\Gamma)$ as the distance over which the exchange is maximal. The total amount $\eta_{\max}$ of energy transfer is limited by the detuning Δ:

$$\eta_{\max} = \frac{\chi^2}{\Gamma^2} = \frac{\chi^2}{\chi^2 + \Delta^2}$$

The evolution of intensity in both waveguides is illustrated for different values of the detuning in Figure A.2.

It is only for perfect phase matching ($\Delta = 0$) that one can expect a total transmission from one waveguide to the other. We would like to point out, however, that this condition is rarely necessary – especially for high-Q MRR structures, where only a tiny fraction of energy transfer is required. One should also note that for perfectly phase-matched couplers, direct transmission t_c and cross-coupling k_c are in quadrature (because of the imaginary 'i' factor).

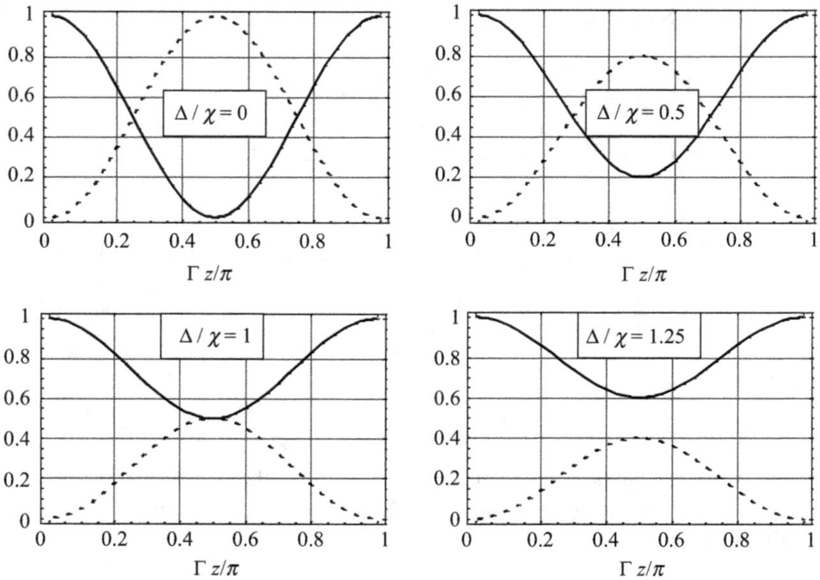

Figure A.2 Transmission in intensity as a function of distance, for different values of the (reduced) detuning. In straight lines, $|F_u^+(z)/F_{u1}^+|^2$*; in dashed lines,* $|F_v^+(z)/F_{u1}^+|^2$

In terms of scattering parameters, taking into account both possible directions of propagation, one gets eventually

$$\begin{pmatrix} b_1 \\ b_2 \\ b_3 \\ b_4 \end{pmatrix} = e^{-i\beta d} \begin{pmatrix} 0 & t_c & 0 & k_c \\ t_c & 0 & k_c & 0 \\ 0 & k_c & 0 & t_c^* \\ k_c & 0 & t_c^* & 0 \end{pmatrix} \begin{pmatrix} a_1 \\ a_2 \\ a_3 \\ a_4 \end{pmatrix}$$

Besides, by conveniently translating the reference planes, as shown in Figure A.3, it is always possible to describe the coupler as if wholly localized at any intermediary abscissa $z_c \in [z_1, z_2]$. For instance, $z_c = (z_1 + z_2)/2$.

In order to shift the reference planes, one must take into account the phase change along each waveguide, each associated with its own reference wavevector. For instance

$$a_1' = a_1 \exp(-i\beta_u d/2), \quad a_3' = a_3 \exp(-i\beta_v d/2)$$

$$b_2' = b_2 \exp(+i\beta_u d/2), \quad b_4' = b_4 \exp(+i\beta_v d/2)$$

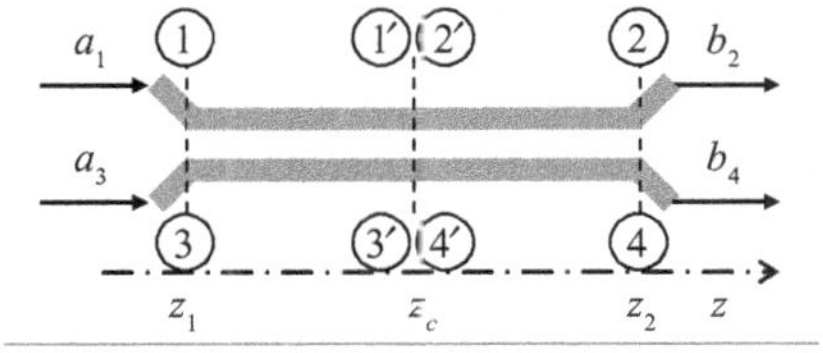

Figure A.3 *Schematic depiction of a directional coupler, also seen as a four-port network, with new ports (1′, 2′, 3′, 4′) located at $z = z_c$*

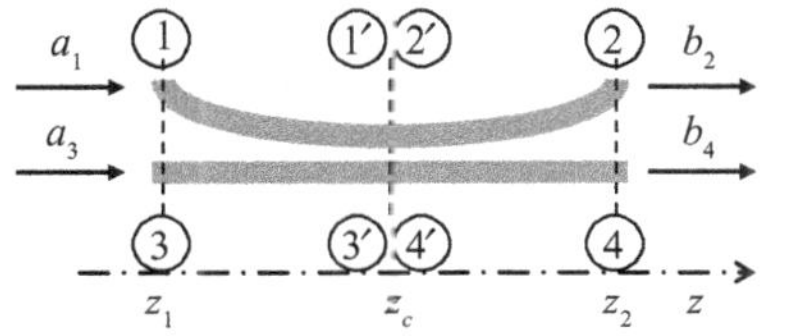

Figure A.4 *Schematic depiction of a 'weighted' directional coupler, between a straight waveguide and a curved waveguide. This four-port network, with new ports (1′, 2′, 3′, 4′) located at $z = z_c$, can be described by an effective coupling strength ($\chi_{eff}\, d$)*

After some algebra, one gets eventually

$$\begin{pmatrix} b_1' \\ b_2' \\ b_3' \\ b_4' \end{pmatrix} = \begin{pmatrix} 0 & t_c e^{+i\,\Delta d} & 0 & k_c \\ t_c e^{+i\,\Delta d} & 0 & k_c & 0 \\ 0 & k_c & 0 & t_c^* e^{-i\,\Delta d} \\ k_c & 0 & t_c^* e^{-i\,\Delta d} & 0 \end{pmatrix} \begin{pmatrix} a_1' \\ a_2' \\ a_3' \\ a_4' \end{pmatrix}$$

For a perfectly phase-matched lossless coupler, it is convenient to assume

$$\chi d \in [0, \pi/2], \;\; t_c = \cos(\chi d) \in \mathbb{R}^+, \;\; k_c = -i \, \sin(\chi d), \;\; \text{so that} \;\; t_c^2 - k_c^2 = 1.$$

For coupling a straight (rectilinear) waveguide to a ring resonator, one can resort to complex FDTD calculations, or one can treat the coupler as an instance of 'weighted' coupling, where the coupling 'constant' χ is replaced by a continuously varying coefficient $\chi(z)$ [29], as schematically depicted in Figure A.4.

For smooth enough variations, the final result remains the same as before, except that the effectively reduced coupling ($\chi_{eff}\, d$) takes the form

$$\left(\chi_{eff} d\right) = \int_{z_1}^{z_2} \chi(z)\, dz$$

In practice, it is quite legitimate not only to treat the coupling as localized but also to use the simplified expressions of (t_c, k_c) pertaining to the phase-matched case. The only difference is a slight phase change that is barely relevant from an experimental point of view.

Appendix B Partial matrices and source terms in an index-coupled DFB

Consider an active DFB zone of length L between abscissas $z_a = 0$ and $z_b = L$. For any internal abscissa z, call $[L]$ and $[R]$ the partial matrix from z_a to z and from z to z_b, respectively; $[Z] = [L]\ [R]$ is the complete matrix. $[L]$ is a DFB of length ($z - z_a$), characterized by a coupling constant κ with phase ψ.

$$L_{11} = \left(\cosh(\gamma(z - z_a)) + i\frac{\delta + i\,\alpha}{\gamma}\sinh(\gamma(z - z_a))\right)\exp(+i\beta_B(z - z_a))$$

$$L_{12} = \left(+i\frac{\kappa\, e^{+i\,\psi}}{\gamma}\sinh(\gamma(z - z_a))\right)\exp(-i\beta_B(z - z_a))$$

$$L_{21} = \left(-i\frac{\kappa\, e^{-i\,\psi}}{\gamma}\sinh(\gamma(z - z_a))\right)\exp(+i\beta_B(z - z_a))$$

$$L_{22} = \left(\cosh(\gamma(z - z_a)) - i\frac{\delta + i\,\alpha}{\gamma}\sinh(\gamma(z - z_a))\right)\exp(-i\beta_B(z - z_a))$$

The $[R]$ matrix stands for a DFB of length ($z_b - z$), with the same coupling constant κ, in terms of modulus at least. However, extra care should be taken with the *phase* of its non-diagonal terms:

$$R_{11} = \left(\cosh(\gamma(z_b - z)) + i\frac{\delta + i\,\alpha}{\gamma}\sinh(\gamma(z_b - z))\right)\exp(+i\beta_B(z_b - z))$$

$$R_{12} = \left(+i\frac{\kappa\, e^{+i\,\psi}}{\gamma}\sinh(\gamma(z_b - z))\right)\exp(+i\beta_B(z_b - z - 2L))$$

$$R_{21} = \left(-i\frac{\kappa\, e^{-i\,\psi}}{\gamma}\sinh(\gamma(z_b - z))\right)\exp(-i\beta_B(z_b - z - 2L))$$

$$R_{22} = \left(\cosh(\gamma(z_b - z)) - i\frac{\delta + i\,\alpha}{\gamma}\sinh(\gamma(z_b - z))\right)\exp(-i\beta_B(z_b - z))$$

The source distribution (per unit frequency and distance) is $D_{sp}(\omega,z)$. One can show that the three spectral densities of intensity associated with fields (u^+), (u^-) are

$$\langle|(u^+)|^2\rangle = \frac{1}{|Z_{11}|^2}\int_{z_a}^{z_b} D_{sp}(\omega, z)\ |L_{11} - L_{12}|^2 dz$$

$$\langle|(u^-)|^2\rangle = \frac{1}{|Z_{11}|^2}\int_{z_a}^{z_b} D_{sp}(\omega, z)\ |R_{11} + R_{21}|^2 dz$$

$$\langle (u^+)(u^-)^* \rangle = \frac{1}{|Z_{11}|^2} \int_{z_a}^{z_b} D_{sp}(\omega, z)\,(L_{11} - L_{12})(R_{11} + R_{21})^* dz$$

Let us also assume a spatially uniform distribution ($\partial D_{sp}/\partial z = 0$). For the special case of an index-coupled DFB structure, a rather cumbersome but otherwise straightforward calculation leads eventually to the following results; with $T_Z(\omega) = 1/|Z_{11}|^2$, $\gamma' = \mathrm{Re}[\gamma]$, and $\gamma'' = \mathrm{Im}[\gamma]$:

$$\langle |(u^+)|^2 \rangle = \langle |(u^-)|^2 \rangle = \left(\frac{D_{sp}\, L\, T_Z(\omega)}{2|\gamma|^2} \right) \times \left\{ \begin{array}{l} \left(\frac{\sinh(2\gamma' L)}{2\gamma' L} \right) \left(|\gamma|^2 + |\delta + i\,\alpha|^2 + |\kappa|^2 \right) + \\ \left(\frac{\sin(2\gamma'' L)}{2\gamma'' L} \right) \left(|\gamma|^2 - |\delta + i\,\alpha|^2 - |\kappa|^2 \right) + \\ \left(\frac{\cosh(2\gamma' L) - 1}{2\gamma' L} \right) \left(-\mathrm{Im}[(\delta + i\,\alpha)\,\gamma *] \right) + \\ \left(\frac{1 - \cos(2\gamma'' L)}{2\gamma'' L} \right) \left(-\mathrm{Re}[(\delta + i\,\alpha)\,\gamma *] \right) \end{array} \right\}$$

The mutual coherence term is slightly more complicated:

$$\langle (u^+)(u^-)^* \rangle = \left(\frac{D_{sp}\, L\, T_Z(\omega)}{2|\gamma|^2} \right) \times e^{-i\beta_B L} \times (K' - K'')$$

$$K' = \left(\frac{\sinh(\gamma' L)}{\gamma' L} \right) \times 2\,\kappa\, e^{i\,\psi} (\gamma' \sin(\gamma'' L) + \delta \cos(\gamma'' L))$$

$$K'' = \left(\frac{\sin(\gamma'' L)}{\gamma'' L} \right) \times 2\kappa e^{i\,\psi} (\gamma'' \sinh(\gamma' L) + \delta \cosh(\gamma' L))$$

A word of caution: it could be prudent to check the result one more time before using it!

Appendix C Glossary (acronyms used in the text)

ASE amplified spontaneous emission
CMT coupled-mode theory
DBR distributed Bragg reflector
DFB distributed feedback
FPR Fabry–Pérot resonator
FSR free spectral range
MRR micro-ring resonator

TMF transfer matrix formalism
QWS quarter-wave shift

References

[1] Yariv A. *Quantum Electronics*, Second edition. New York: Wiley, 1975.
[2] Agrawal G.P. and Dutta N.K. *Long Wavelength Semiconductor Lasers*. New York: Van Nostrand Reinhold, 1986.
[3] Basu P.K. *Theory of Optical Processes in Semiconductors*. Oxford: Clarendon Press, 1997.
[4] Rosencher E. and Vinter B. *Optoelectronics*. Cambridge: Cambridge University Press, 2002.
[5] Desurvire E. *Erbium-Doped Fiber Amplifiers*. Hoboken: Wiley-Interscience, 2002.
[6] Siegman A.E. *Lasers*. Mill Valley: University Science Books, 1986.
[7] Yariv A. and Yeh P. *Optical Waves in Crystals*. New York: Wiley, 1984.
[8] Yeh P. *Optical Waves in Layered Media*. New York: Wiley, 1988.
[9] Pozar D.M. *Microwave Engineering*, Fourth edition. Hoboken: Wiley & Sons, 2012.
[10] Born M. and Wolf E. *Principles of Optics*, Seventh edition (augmented). Cambridge: Cambridge University Press, 1999.
[11] Joannopoulos J.D., Johnson S.G., Winn J.N., and Meade R.D. *Photonic Crystals – Molding the Flow of Light*, Second edition. Princeton: Princeton University Press, 2008.
[12] Markoš P. and Soukoulis C.M. *Wave Propagation*. Princeton: Princeton University Press, 2008.
[13] Kogelnik H. and Shank C.V. "Coupled-Wave Theory of Distributed Feedback Lasers". *Journal of Applied Physics*. 1972, vol. 43 (5), pp. 2327–35 [doi: 10.1063/1.1661499].
[14] Ghafouri-Shiraz H. and Lo B.S.K. *Distributed Feedback Laser Diodes*. Chichester: Wiley, 1996.
[15] Morthier G. and Vankwikelberge P. *Handbook of Distributed Feedback Laser Diodes*. London: Artech House, 1997.
[16] Rabus D.G. *Integrated Ring Resonators*. Berlin: Springer-Verlag, 2007.
[17] Weber J.-P. and Wang S. "A New Method for the Calculation of the Emission Spectrum of DFB and DBR Lasers". *IEEE Journal of Quantum Electronics*. 1991, vol. 27 (10), pp. 2256–66; Weber J.-P. "Correction to 'A New Method for the Calculation of the Emission Spectrum of DFB and DBR Lasers'". *IEEE Journal of Quantum Electronics*. 1993, vol. 29 (1), p. 296 [doi: 10.1109/3.97269].
[18] Boucher Y. "Extended (3×3) Transfer Matrix Formalism: From Amplified Spontaneous Emission to Threshold-Crossing". *Research Signpost: Recent Research Developments in Optics*. 2003, vol. 3, pp. 177–204.

[19] Boucher Y.G. "Theoretical Investigation of Amplified Spontaneous Emission in an Active Structure by Extended (3×3) Transfer Matrix Formalism: The Case of a Non-Uniform Longitudinal Distribution of Emitters". *Journal of the European Optical Society-Rapid Publications*. 2006, vol. 1, (06027), pp. 1–6 [doi: 10.2971/jeos.2006.06027].
[20] Kasap S., Ruda H., and Boucher Y. *The Cambridge Handbook of Optoelectronics and Photonics*. Cambridge: Cambridge University Press, 2009 (paperback edition 2012).
[21] Kastler A. "Atomes à l'intérieur d'un interféromètre Perot-Fabry". *Applied Optics*. 1962, vol. 1 (1), pp. 17–24 [doi: 10.1364/AO.1.000017].
[22] Choi H.K., Chen K.L., and Wang S. "Analysis of Two-Section Coupled-Cavity Semiconductor Lasers". *IEEE Journal of Quantum Electronics*. 1984, vol. QE-20 (4), pp. 385–93 [doi: 10.1109/JQE.1984.1072397].
[23] Ghisa L., Dumeige Y., Thi Kim N.N., Boucher Y.G., and Féron P. "Performances of a Fully Integrated All-Optical Pulse Reshaper Based on Cascaded Coupled Nonlinear Microring Resonators". *Journal of Lightwave Technology*. 2007, vol. 25 (9), pp. 2417–26. [doi: 10.1109/JLT.2007.901529].
[24] Trebaol S., Nguyên T.K.N., Tavernier H., Ghisa L., Dumeige Y., and Féron P. "Artificial Dispersion of Active Optical Coupled Resonator Systems". *Comptes-Rendus de Physique*. 2009, vol.10 (10), pp.964–79. [doi: 10.1016/j.crhy.2009.10.014].
[25] Stéphan G.M. "A Semi-Classical Theory of the Laser Transition". *Physical Review A*. 1997, vol. 55 (2), pp.1371–84. [doi: 10.1103/PhysRevA.55.1371].
[26] Boucher Y.G. and Féron P. "Generalized Transfer Function: A Simple Model Applied to Active Single-Mode Microring Resonators". *Optics Communications*. 2009, vol. 282, pp. 3940–47. [doi:10.1016/j.optcom.2009.06.048].
[27] Boyd R.W. *Nonlinear Optics*. New York: Academic Press, 1992.
[28] Dumeige Y., Ghisa L., and Féron P. "Dispersive Multistability in Microring Resonators". *Journal of Optics A: Pure and Applied Optics*. 2006, vol. 8 (7), pp. S483–9. [doi:10.1088/1464-4258/8/7/S27].
[29] Tamir T. (ed.). *Guided-Wave Optoelectronics*. Berlin: Springer-Verlag, 1990.

Chapter 4

Modelling of nanophotonic non-linear metasurfaces

Antonino Calà Lesina[1,2,3,4,5,6,] *Pierre Berini*[1,2,3] *and Lora Ramunno*[1,2]

4.1 Introduction

The invention of the laser in 1960, followed shortly thereafter by the first experiment on second-harmonic generation (SHG), initiated the field of non-linear optics [1]. When light impinges on a material, the bound and free electrons within the material can be displaced non-linearly by the incident field if the light is intense enough. The electron dynamics produces a scattered field which, along with the altered incident field, determines the optical response of the material. This response is described by the polarization induced in the medium $P(\vec{r}, t)$, i.e. the dipole moment per unit volume:

$$P(\vec{r}, t) = P_L(\vec{r}, t) + P_{NL}(\vec{r}, t), \tag{4.1}$$

where we have separated the linear polarization

$$P_L(\vec{r}, t) = \varepsilon_0 \chi^{(1)} E(\vec{r}, t) \tag{4.2}$$

from the non-linear polarization, which in the perturbative regime takes the form

$$P_{NL}(\vec{r}, t) = P^{(2)}(\vec{r}, t) + P^{(3)}(\vec{r}, t) + \cdots = \varepsilon_0[\chi^{(2)} E^2(\vec{r}, t) + \chi^{(3)} E^3(\vec{r}, t) + \cdots], \tag{4.3}$$

[1]Department of Physics, University of Ottawa, Ottawa, Canada
[2]Centre for Research in Photonics, University of Ottawa, Ottawa, Canada
[3]School of Electrical Engineering and Computer Science, University of Ottawa, Ottawa, Canada
[4]Hannover Centre for Optical Technologies, Leibniz Universität Hannover, Hannover, Germany
[5]Cluster of Excellence PhoenixD (Photonics, Optics, and Engineering – Innovation Across Disciplines), Hannover, Germany
[6]Institut für Transport- und Automatisierungstechnik, Fakultät für Maschinenbau, Leibniz Universität Hannover, Hannover, Germany

where ε_0 is the vacuum permittivity, $\chi^{(1)}$ is the linear susceptibility, and $\chi^{(2)}$ and $\chi^{(3)}$ are the second- and third-order non-linear susceptibilities. Here all the susceptibilities are written as if dispersionless, but they can be dispersive. Further, a full tensorial treatment is generally required, and though tensor notation is not explicitly used in (4.1)–(4.3), it is understood that these are tensor equations. When the incident field $E(\vec{r},t)$ has low intensity, the term in $\chi^{(1)}$ predominates and the amplitude of the scattered field varies linearly with $E(\vec{r},t)$. By increasing the amplitude of the incident field, the $\chi^{(2)}$ and $\chi^{(3)}$ terms start playing a role, and the scattered field becomes non-linearly dependent on the amplitude of the incident field. This transition to the non-linear regime takes place when the amplitude of the incident field becomes comparable to interatomic fields, which are of the order of 10^{11} V/m [2]. In this case, the non-linear polarization acts as a source term in the wave equation

$$\vec{\nabla} \times \vec{\nabla} \times \vec{E}(\vec{r},t) + \frac{1+\chi^{(1)}}{c^2}\frac{\partial^2 \vec{E}(\vec{r},t)}{\partial t^2} = -\mu_0 \frac{\partial^2 \vec{P}_{NL}(\vec{r},t)}{\partial t^2} \tag{4.4}$$

which describes the propagation of the generated non-linear signal in a non-magnetic medium having dispersionless $\chi^{(1)}$, where c is the speed of light in vacuum.

For an incident monochromatic field of angular frequency ω, the second-order non-linear polarization $P^{(2)}(\vec{r},t)$ can lead to a field oscillating at 2ω (SHG), whereas the third-order non-linear polarization $P^{(3)}(\vec{r},t)$ a field oscillating at 3ω (third-harmonic generation, THG). If two monochromatic fields oscillating at ω_1 and ω_2 excite the medium at the same time, the second-order non-linearity can mix the fields to create, among the others, the sum-frequency generation (SFG) term at $\omega_1 + \omega_2$ and the difference-frequency generation (DFG) term at $\omega_1 - \omega_2$.

The non-linear response of natural materials is usually weak enough that large interaction volumes are needed to produce a non-linear signal which is intense enough to be used in applications. As the wave at the fundamental frequency propagates through the volume, it interacts with atoms which, in turn, can act as dipole emitters at the non-linear frequency, such as in the case of SHG and THG, for example. These non-linear emitters have to oscillate with precise phase relations in order for the non-linear radiation to constructively interfere and reach a high amplitude at the point where the non-linear signal exits the device. Such 'phase matching' can be challenging to achieve due to the dispersive nature of the materials.

An opportunity to achieve phase matching in bulk materials comes from optical metamaterials. Metamaterials are artificial materials engineered at the nanoscale, i.e. they are composed by nanostructures arranged on a 3D lattice (3D metamaterials) or a 2D lattice (2D metamaterials or metasurfaces), and they exhibit properties not found in natural materials. This new paradigm is possible due to the advances in nanofabrication techniques and the ability to manipulate matter at the nanoscale. Examples of 3D metamaterials where phase matching has been achieved are epsilon-near-zero (ENZ) materials [3], negative refractive index materials [4], and non-linear photonic crystals [5,6]. However, 3D metamaterials are generally difficult to manufacture.

Metasurfaces are ultrathin devices which hold the promise of circumventing the strict phase matching requirements of traditional bulky devices [7] and decrease the dimensions of integrated optical systems [8]. They offer new degrees of freedom to control how light propagates in integrated all-optical circuits and can be used to improve light coupling efficiency, such as in surface and vertical grating couplers, and to realize on-chip optical operations, such as polarization rotation, mode conversion, beam splitting, and spectral filtering [9]. In recent years, research on metasurfaces has advanced rapidly due to the ease of fabrication with respect to their 3D metamaterial counterpart. To understand how a metasurface works, we can think of the relation between a field distribution on an aperture (near field) and the field radiated by the aperture in the far field. The far field is obtained by applying the spatial Fourier transform to the near field, and conversely, knowing the far field, we can calculate the field distribution on an aperture required to create that far field. This near-field distribution can be artificially created by a collection of scatterers arranged on a surface with subwavelength spacing. Such an engineered surface is a metasurface, and the scatterers upon it are called meta-atoms [10–12]. Meta-atoms can be metallic, dielectric, and hybrid metallic/dielectric nanostructures, and one of their main functions is to scatter the incident light where the properties of the scattered light, such as polarization, amplitude, phase, and direction of propagation, can be controlled.

In plasmonic nanoantennas, the re-radiation of light is possible due to localized surface plasmon resonances which are the oscillation modes of the free electrons within the metallic structure excited by the incident light. These resonant modes can confine and enhance the amplitude of the incident light over subwavelength distances, as shown for different geometries in Figure 4.1. It is in these regions of the enhanced linear near field that non-linear optical processes would occur for

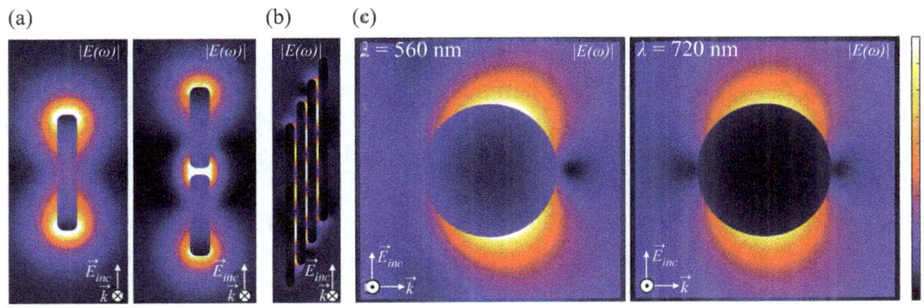

Figure 4.1 Field distribution in (a) gold monopole nanoantenna of length 174 nm (left) and gold dipole nanoantenna of length 255 nm with a gap of 10 nm (right) at their resonance, (b) four shifted nanorods of length 700 nm at their resonance, and (c) gold sphere of radius 75 nm at two wavelengths

intense enough excitation. Figure 4.1(a) shows the electric field for a monopole gold nanoantenna (left) and a gold dipole nanoantenna (right) excited at their resonance by a linearly incident electric field oriented along the axis of the nanoantennas. We observe that at resonance there is field enhancement at the extremities of the two nanostructures and in the gap of the dipole nanoantenna (bright areas). Figure 4.1(b) shows the field distribution for a set of four nanorods spatially shifted along the axis of the rods for an incident field parallel to the axis; the high density of hotspots between the nanorods and the fact that the field enhancement is present for every incident field polarization (though we only show one here, oriented along the vertical axis) make this configuration interesting for non-linear plasmonics [13]. Figure 4.1(c) shows the field enhancement surrounding a gold sphere of radius 75 nm in air for two wavelengths; at the shorter wavelength, we observe that the excited mode is bent to the right (the forward direction) due to interference effects, while for the longer wavelength, the sphere acts more like a dipolar emitter. Looking at the field distributions in Figure 4.1, we see that the enhanced near fields occur close to the metallic surface, and this makes possible to control the properties of the scattered light in a more precise way compared to what is obtainable in bulk materials [14]. Plasmonic nanostructures have been exploited in subwavelength plasmonic guiding, nanofocusing [14], nanoantennas [15], linear metasurfaces [10,11], and non-linear metasurfaces, as we will see in the next section.

Metals suffer from losses in the optical regime and dielectric nanostructures have been studied to overcome this limitation [12]. High refractive index materials, such as semiconductors, provide a mode volume which is not limited to the interfaces such as for metals. In Figure 4.2, we show the electric and magnetic modes (Mie-type resonances) of a silicon sphere of radius 300 nm. In particular, in Figure 4.2(a), we show the absorption and extinction coefficients which identify the resonance wavelengths. In Figure 4.2(b) and (c), we show the electric (top) and magnetic (bottom) modes at six resonance wavelengths highlighted in Figure 4.1(a) for TM (in-plane electric field) and TE (out-of-plane electric field) excitation, respectively. In this case, the simultaneous excitation of electric and magnetic modes (the magnetic field in the case of plasmonic nanostructures is much weaker) allows more design freedom. For example, electric and magnetic dipole modes can be engineered to create a Huygens' source [16,17] which gives unidirectional radiation in metasurfaces, or the anapole mode to have a nonzero near field and zero radiated power in the far field [18].

Both plasmonic metasurfaces and dielectric metasurfaces have been extensively studied for a wide range of applications, such as flat optical devices for focusing, beam steering and beam shaping [19–22], biosensing [23], colour production [24–28], conformal and active metasurfaces [29,30], and enhancement and structuring of the non-linear signal [31].

As the research on metasurfaces continues to progress, simulation tools need to become more sophisticated to handle, for example non-locality, non-linearity, electro-optical effects, quantum effects, and multi-physics approaches. In this chapter, we will discuss an advanced application of metasurfaces, the generation,

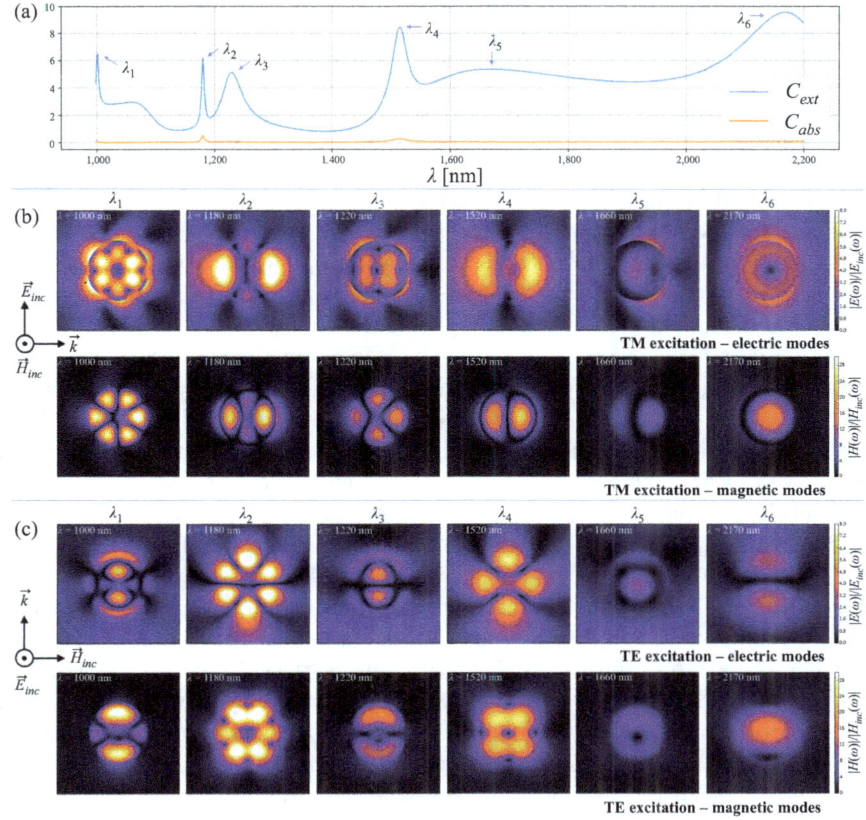

Figure 4.2 Mie-type resonances for a silicon sphere with radius 300 nm: (a) extinction and absorption coefficients, (b) TM excitation: electric modes (top) and magnetic modes (bottom), and (c) TE excitation: electric modes (top) and magnetic modes (bottom)

enhancement, and control of non-linear signal via non-linear metasurfaces. We will introduce the computational methods for nanophotonic metasurfaces, focusing in particular on the approaches for the simulation of non-linear processes. Finally, we will provide a few examples from our research.

4.2 Non-linear metasurfaces

Looking at the expressions for the non-linear polarizations in (4.3), e.g. $P^{(2)}(\vec{r}, t) = \varepsilon_0 \chi^{(2)} E^2(\vec{r}, t)$, it is clear that in order to locally enhance the non-linear generation, we need to enhance the non-linear susceptibility and/or the local electric field $E(\vec{r}, t)$. The enhancement and localization of the incident field in hotspots, as reported in plasmonic [32] and dielectric [33] nanostructures, can significantly increase non-linear processes over nanoscale volumes. This opens new opportunities for non-linear optics beyond the requirement of phase matching in bulk materials, as well as for advanced techniques to effectively increase the non-linear susceptibility via metamaterial engineering. The enhancement and control of the non-linear emission from metallic, dielectric, and hybrid nanostructures has been exploited in recent years to realize so-called non-linear metasurfaces, as extensively discussed in the following review papers [34–38].

The generation of non-linear optical signals does not occur in the same way in all materials because non-linear processes are subject to symmetry constraints. For example, the second-order non-linear susceptibility $\chi^{(2)}$ is zero in bulk materials with a centrosymmetric atomic lattice [2], and this prevents materials with centrosymmetric lattices, such as gold and silver, to exhibit second-order non-linear response in the bulk. One may design a meta-atom to break this symmetry to allow SHG, and a well-known example is the split-ring resonator [39]. However, SHG has been reported for nanoantennas with centrosymmetric geometries, and this was attributed to local symmetry breaking on the surface of the nanostructure. Third-order non-linear effects do not have these constraints.

Plasmonic metasurfaces have been applied to enhance SHG [40,41], THG [42,43], DFG [44], four-wave mixing (FWM) [45,46], and even high-harmonic generation (HHG) [47]. In contrast to plasmonics, high-index dielectric nanoparticles provide a mode volume that is not limited to interfaces and thus may lead to higher conversion efficiencies. Dielectric metasurfaces have been applied to enhance SHG [48], THG [49], many non-linear processes at the same time (SHG, THG, fourth-harmonic generation, SFG, two-photon absorption-induced photoluminescence, FWM, and six-wave mixing) [50], and HHG [51].

The field enhancement which drives the enhancement of the non-linear process can be in-plane – as in the case of a gap nanoantenna containing a non-linear material in its gap [52] – or out-of-plane – as in the case of a subwavelength thin dielectric layer sandwiched between an optically thick metal film and an array of metal nanostructures [53]. Non-planar solutions have also been proposed to maximize the non-linear emission by minimizing the destructive interference of the non-linear emitters in the far field [54]. We summarize here different methods for enhancing non-linear generation:

- exploiting the linear field enhancement in the proximity of a metallic nanostructure due to localized surface plasmon resonances (see Figure 4.1) [32];
- exploiting the linear field enhancement in the bulk of high index dielectric nanostructures via Mie-type electric and magnetic resonances, and multipolar resonances (see Figure 4.2, this is contrary to plasmonic nanostructures, where the field inside the metal is almost zero) [33,49,55–57];
- increasing the number of hotspots per unit area, for example, via shifted nanorods [13];
- increasing the field intensity of the fundamental resonant mode in the gap of a plasmonic nanoantenna which contains a non-linear material in its gap [31,52,58,59];

- increasing the non-linear susceptibility of the material, for example, by exploiting the inverse proportion between the Kerr non-linear refractive index n_2 and the linear refractive index in ENZ materials [60,61] (to observe this enhancement, the ENZ condition is needed for the real part of the linear dielectric permittivity while maintaining a relatively low imaginary part);
- coupling the metasurface on an ENZ material [62];
- using ENZ materials to provide a field enhancement over a larger volume respect to plasmonic hotspots [3];
- designing a nanostructure which supports a resonance at both fundamental and harmonic wavelengths, optimizing the spatial overlap between the fundamental and the harmonic modes, and/or increasing the quality factor of the modes involved [63–65];
- exploiting the linear field enhancement in periodic arrays of nanostructures arranged with a spacing of the order of the wavelength, exploiting plasmonic surface lattice resonances [66–68];
- exploiting a stack of semiconductor media, the multi-quantum well, with a plasmonic nanostructure on top to couple in-plane light polarization into out-of-plane polarization for optimal excitation of the multi-quantum well non-linearity [41];
- achieving strong non-linearity via metamaterial design, even at frequencies where the constituent materials show a negligible non-linear response [69];
- exploiting the enhancement associated with Fano resonances in plasmonic [70,71] and dielectric [72] metasurfaces; and
- exploiting the non-locality of the electron response [73,74].

Besides enhancing the non-linear signal, controlling its phase and polarization is important for many practical applications. Controlling the phase of the field scattered by the meta-atom allows one to shape the radiation in the far field. In passive linear metasurfaces, the phase can be controlled by using a different nanostructure for each phase or by exploiting the Pancharatnam–Berry phase or geometric phase, which allows us to use the same nanostructure and achieve phase control simply by rotating the nanostructure by a certain angle at each location in the array, obtaining a so-called geometric metasurface [22,75]. The geometric phase approach works also in the non-linear regime, but the phase of the non-linear signal depends on the order of the non-linearity [31,76].

Controlling the polarization of the non-linear signal is important for non-linear beam structuring. For example, in [77], the mode hybridization of a metallic and a dielectric nanoparticle in close proximity is exploited not only to enhance the SHG emission but also to control its polarization, while in [31], the linear polarization is used to drive the non-linear emitter with the same polarization.

In general, the non-linear emission control has been experimentally demonstrated for non-linear beam shaping [78], non-linear photonic crystals [79], non-linear holography in plasmonic [80] and all-dielectric [81] metasurfaces, beam steering, focusing, and polarization manipulation [82–84].

4.3 Non-linear simulations

As fabrication techniques become more precise, they allow the creation of ever smaller nanostructures and very small gaps. There is a corresponding need to advance the simulation tools to model the behaviour of these systems. This includes the ability to simulate non-locality, quantum effects, and non-linearity, to name a few. In this section, we will focus on simulating non-linearity. The research on the non-linear generation from nanostructures is still in its early days, and some basic topics are still under debate. For example, it is still not entirely clear from where the non-linear emission originates in hybrid nanostructures composed of a metallic nanoantenna and a non-linear material in its gap [52,58,59,85,86], or in a plasmonic nanostructure on top of a non-linear substrate [87]. This is due to the fact that experiments measure the far-field radiation, and identifying where the non-linear emission is generated is not straightforward. Studies have demonstrated that small changes in the geometry of a nanostructure can lead to dramatic changes in the non-linear far field, and this suggests that it is possible to optically characterize the nanoantenna morphology by looking at the non-linear far field [88]. Experiments on measuring the non-linear emission in the near field present a promising solution to this problem. For example, the second-harmonic near field generated from a gold monopole nanoantenna has recently been probed via an aluminium nanoantenna resonating at the second harmonic and placed in the near field of the primary nanoantenna [89]. The near field suggests that the effective second-order emitter is oriented perpendicularly to the axis of the monopole, although the excitation field is polarized along the axis of the nanoantenna, which is counterintuitive.

It is in cases such as these where simulations provide valuable support for experiments. In fact, simulations allow us to (1) estimate quantities which are not accessible in experiments; (2) switch effects on and off so that single contributions can be isolated and investigated; (3) understand the effect of changing materials, shapes, and physical response without the need to run expensive experiments; (4) optimize the design and thus minimize fabrication iterations; and (5) gain insight into the physical phenomena by powerful visualization and movies.

Many are the computational electromagnetic techniques that over the years have been applied to nanophotonics in the linear regime. These include the finite-difference time-domain (FDTD) technique [90,91], the finite-element method (FEM) [92], discontinuous Galerkin methods [93,94], volume integral equation (VIE) methods [95], the discrete dipole approximation method [96], and surface integral equation (SIE) methods [88]. A classification of simulation problems with an indication of the methods which are best suited for each is reported in [97].

For non-linear optics, there are fewer options. In the linear regime, analytical solutions are available for simple problems, such as Mie theory for the scattering of a plane wave by a sphere. So, too, in the non-linear regime, analytical solutions can be derived, for example for spheres [98–100] and nanoshells [101], but these are few. Among the computational techniques which have been used for non-linear simulations, we have FEM [102,103], the SIE method [88,104,105] (best suited for surface emission, such as in SHG from metallic nanostructures), the VIE method [106] (better suited for volume generation, such as THG emission from nanostructures), and time-domain methods such as FDTD [52,107].

In this section, we will focus on the FDTD method, which is the most widely used computational technique in nanophotonics. FDTD is relatively easy to implement, it is extremely versatile, capable of modelling different material properties, complex geometries, and non-linear optical processes can be directly incorporated within it. Due to its near-linear scalability, it is also among the most suitable methods for parallel processing and supercomputing [108]. We will discuss the differences between a non-linear simulation and a linear one and describe two of the most used strategies to conduct non-linear simulations: the two-step approach consisting two linear simulations and direct non-linear generation approach wherein the non-linear optical process is directly incorporated within the simulation.

4.3.1 General considerations for a non-linear simulation

In linear FDTD simulations where the frequency-domain response is desired, a broadband pulse is generally used to excite the system, and a discrete Fourier transform (DFT) is calculated for all the frequencies of interest for a subset of points in the simulation domain. At the end of the simulation, the DFT of the simulated field is normalized with respect to the DFT of the incident signal at the same frequency. This is equivalent to exciting the system with a monochromatic electric field of 1 V/m. Compared to linear simulations, there are some differences which have to be taken into account when a non-linear simulation is conducted:

- To obtain the frequency-domain response, we must excite with a narrowband signal which does not contain the non-linear signal of interest.
- Since the incident field does not contain the non-linear signal frequency, there is no need for normalization.
- Since the incident field is narrowband or monochromatic, this results in a long pulse and a longer simulation time.
- We need to ensure that the amplitude of the incident field is strong enough so that the non-linear signal is able to emerge from the background numerical noise.
- The space step and the time step for the FDTD simulation have to be small enough to resolve the non-linear frequency.

4.3.2 Two-step approach

This strategy solves the non-linear problem by decomposing the simulation into two linear simulations. This approach uses the undepleted pump approximation, which assumes that the non-linear process is weak enough that pump is not affected by the non-linear process, or, in other words, that the non-linear polarization is much smaller than the linear polarization. The first step computes the linear field distribution at the fundamental wavelength. The non-linear polarization is calculated from the linear field of the first simulation, and this non-linear polarization is used as the source term in the wave equation in the second step. The wave equation can be solved either analytically for those geometries with an analytical solution (such as for the non-linear emission from a surface) or numerically for more complicated geometries.

Another two-step approach is called 'non-linear scattering theory' and was recently applied to predict the non-linear properties of metamaterials [109]. This method is based on the Lorentz reciprocity theorem and combines the results of the following two simulations: the simulation to calculate the pump field distribution which drives the non-linear polarization, and the linear simulation of the field distribution at the non-linear frequency. The actual non-linear generated field is then calculated by the overlap of the non-linear polarization and the mode distribution at the non-linear frequency.

In the case of THG, the third-harmonic field generated by the nanostructure is

$$E(3\omega) \propto \iiint \sum_{i=\{x,y,z\}} P_i(3\omega)E'_i(3\omega)dV, \tag{4.5}$$

where

$$P_i(3\omega) = \sum_{jkm} \chi_{ijkm}E_j(\omega)E_k(\omega)E_m(\omega), \tag{4.6}$$

and where $E'_i(3\omega)$ and $E_i(\omega)$ are the ith components of the fields for the linear field distributions at 3ω and ω excited by sources at 3ω and ω, respectively. The prime on $E'_i(3\omega)$ serves to distinguish this field mode, obtained by linear simulations, from the actual non-linear generated signal $E(3\omega)$.

In the case of THG, the FDTD method is a good choice for calculating $E'_i(3\omega)$ and $E_i(\omega)$ because the non-linear susceptibility is nonzero over the entire volume of the nanostructure, and the fields $E'_i(3\omega)$ and $E_i(\omega)$ have to be known over the entire volume as well. For SHG, SIE methods may be a better choice for the calculation of $E'_i(2\omega)$ and $E_i(\omega)$, since the susceptibility tensor is nonzero only on the surface.

Since both two-step approaches outlined here use the frequency-domain linear field distribution to calculate the non-linear polarization, they work well for monochromatic excitation. If the excitation has a broader spectrum and we are interested in tracing the dynamics of the non-linear generation and the transient fields, this method is not ideal, and we should use the direct non-linear generation approach described in the next section.

4.3.3 Direct non-linear generation approach

The non-linear optical process may be directly integrated into the FDTD algorithm by solving a non-linear equation at each time step. This approach does not require the undepleted pump approximation. We show here how to implement SHG for a dispersionless medium ($\chi^{(1)}$ and $\chi^{(2)}$ are frequency independent). We start with Ampère–Maxwell's law

$$\frac{\partial \vec{D}(\vec{r},t)}{\partial t} = \vec{\nabla} \times \vec{H}(\vec{r},t) - \vec{J}_{free}(\vec{r},t), \tag{4.7}$$

where we can assume $\vec{J}_{free} = 0$, and the constitutive relation for the displacement field

$$\vec{D}(\vec{r},t) = \varepsilon_0 \vec{E}(\vec{r},t) + \vec{P}(\vec{r},t). \tag{4.8}$$

By substituting (4.8) into (4.7), and using (4.1)–(4.3), we obtain

$$\vec{\nabla} \times \vec{H}(\vec{r},t) = \varepsilon_0 \frac{\partial \vec{E}(\vec{r},t)}{\partial t} + \varepsilon_0 \chi^{(1)} \frac{\partial \vec{E}(\vec{r},t)}{\partial t} + \frac{\partial \vec{P}_{NL}(\vec{r},t)}{\partial t}, \tag{4.9}$$

which can be written in discretized form for numerical implementation. For one Cartesian component, this is

$$\nabla \times H^{n+1/2} = \varepsilon_0 \varepsilon_r \frac{E^{n+1} - E^n}{\Delta t} + \frac{P_{NL}^{n+1} - P_{NL}^n}{\Delta t}, \tag{4.10}$$

where the index n means that the equation is evaluated at the instant $n\Delta t$, and $\varepsilon_r = 1 + \chi^{(1)}$. By introducing the non-linear polarization for SHG, $P_{NL} = \varepsilon_0 \chi^{(2)} E^2$, we obtain

$$\frac{\Delta t}{\varepsilon_0} \nabla \times H^{n+1/2} = \varepsilon_r (E^{n+1} - E^n) - \chi^{(2)} (E^{n+1})^2 - \chi^{(2)} (E^n)^2. \tag{4.11}$$

In order to find E^{n+1}, we need to solve a second-order equation, and the solution is

$$E^{n+1} = \frac{-\varepsilon_r \pm \sqrt{\varepsilon_r^2 + 4\chi^{(2)} b}}{2\chi^{(2)}}, \tag{4.12}$$

which is valid for $\varepsilon_r^2 + 4\chi^{(2)} b \geq 0$,

where

$$b = \frac{\Delta t}{\varepsilon_0} \nabla \times H^{n+1/2} + \varepsilon_r E^n + \chi^{(2)} (E^n)^2. \tag{4.13}$$

The '−' solution in (4.12) is not physical, so from now on we will consider the '+' solution only. If $4\chi^{(2)} b \ll \varepsilon_r^2$, then we obtain

$$E^{n+1} \approx \frac{b}{\varepsilon_r}, \tag{4.14}$$

assuming ε_r is not zero. In the case of $\varepsilon_r^2 + 4\chi^{(2)} b < 0$, we need to introduce an approximation to solve (4.11). For a very small time step, which is usually the case for non-linear simulations, we can assume $E^{n+1} \approx E^n$ and rewrite (4.11) as

$$\frac{\Delta t}{\varepsilon_0} \nabla \times H^{n+1/2} = \varepsilon_r (E^{n+1} - E^n) + \chi^{(2)} E^n E^{n+1} - \chi^{(2)} (E^n)^2, \tag{4.15}$$

which now has a solution

$$E^{n+1} = \frac{b}{\varepsilon_r + \chi^{(2)} E^n}. \tag{4.16}$$

Another way to simulate non-linearity is by leveraging the auxiliary differential equation (ADE) method [110,111]. The ADE method is used to simulate linear dispersion, via Drude, Lorentz, and Drude+2CP models [108], to name a few, and introduces an additional updating equation for the polarization field in the leapfrog FDTD algorithm. To understand the ADE method, we start by substituting (4.8) into (4.7):

$$\vec{\nabla} \times \vec{H}(\vec{r},t) = \varepsilon_0 \frac{\partial \vec{E}(\vec{r},t)}{\partial t} + \frac{\partial \vec{P}(\vec{r},t)}{\partial t}, \tag{4.17}$$

and by discretizing it, for one Cartesian component, we obtain

$$\nabla \times H^{n+1/2} = \varepsilon_0 \frac{E^{n+1} - E^n}{\Delta t} + \frac{P^{n+1} - P^n}{\Delta t}, \tag{4.18}$$

from which we can derive E^{n+1} as a function of E^n, P^{n+1}, P^n, and $\nabla \times H^{n+1/2}$. The polarization P^{n+1} is unknown, and we need to solve for P^{n+1} before we can calculate E^{n+1}.

For dispersive materials, the linear polarization is

$$\vec{P}(\omega) = \varepsilon_0 \chi^{(1)}(\omega)\vec{E}(\omega), \tag{4.19}$$

and in the case of the Drude model, the frequency-dependent linear susceptibility is

$$\chi^{(1)}(\omega) = -\frac{\omega_D^2}{\omega(\omega + i\gamma)}, \tag{4.20}$$

where ω_D is the plasma frequency and γ is the damping coefficient.

The time-domain form of (4.19) is

$$\frac{\partial^2 \vec{P}(\vec{r},t)}{\partial t^2} + \gamma \frac{\partial \vec{P}(\vec{r},t)}{\partial t} = \varepsilon_0 \omega_D^2 \vec{E}(\vec{r},t), \tag{4.21}$$

which can be discretized as

$$\frac{P^{n+1} - 2P^n + P^{n-1}}{\Delta t^2} + \gamma \frac{P^{n+1} - P^{n-1}}{2\Delta t} = \varepsilon_0 \omega_D^2 E^n, \tag{4.22}$$

and from this, we can derive P^{n+1} as a function of E^n, P^n, and P^{n-1}, which is the auxiliary equation we use in the FDTD leapfrog scheme. Thus, the sequence of steps in the FDTD algorithm becomes

$$H^{n+1/2} = f(H^{n-1/2}, \nabla \times E^n), \tag{4.23}$$

$$P^{n+1} = f(E^n, P^n, P^{n-1}), \tag{4.24}$$

$$E^{n+1} = f(E^n, P^{n+1}, P^n, \nabla \times H^{n+1/2}). \tag{4.25}$$

The ADE approach is well suited for simulating advanced material properties, such as non-linearity [107] and non-locality [112]. This requires adding an extra term to the linear polarization field in the time domain as the simulation runs. In the case of non-linearity, the non-linear polarization term can have arbitrary complexity, it can be dispersionless, dispersive, and can even incorporate the hydrodynamic model which describes the current density of the electrons in the medium and includes non-linear terms as well [73,112–114]. For example, starting from the ADE method for the Drude model reported earlier, we can simulate THG in a metal by adding to P a non-linear polarization term P_{NL}, which for one Cartesian component is

$$P^n_{NL_x} = \chi^{(3)} E^n_x |E^n|^2, \tag{4.26}$$

where $|E^n|^2 = (E^n_x)^2 + (E^n_y)^2 + (E^n_z)^2$, and $\chi^{(3)}$ is constant (isotropic dispersionless non-linearity). The FDTD leapfrog scheme then becomes

$$H^{n+1/2} = f(H^{n-1/2}, \nabla \times E^n), \tag{4.27}$$

$$P^n_{NL} = f(\chi^{(3)}, E^n), \tag{4.28}$$

$$P^n = P^n + P^n_{NL}, \tag{4.29}$$

$$P^{n+1} = f(E^n, P^n, P^{n-1}), \tag{4.30}$$

$$E^{n+1} = f(E^n, P^{n+1}, P^n, \nabla \times H^{n+1/2}). \tag{4.31}$$

4.4 Examples

4.4.1 Enhancement of second-order non-linear processes

In our research, we used a two-step approach to calculate the amplitude of the DFG signal generated from a slab of a non-centrosymmetric material with a large $\chi^{(2)}$, such as GaAs, with a plasmonic metasurface embedded in the material [44]; this is sketched in Figure 4.3(a). The metasurface was defined by repeating the unit cell (that we call 'bolt' and is contained within the darker square in the middle) along x and z. This allows us to simulate the metasurface by applying periodic boundary conditions to the 'bolt'. Two monochromatic plane wave signals oscillating at ω_1 and ω_2 and polarized in the xz plane at 45° excite the device simultaneously. The x and z components of the incident fields are all enhanced in the gap of the 'bolt', and due to the fact that GaAs belong to the crystal class $\bar{4}$ 3 m, a non-linear polarization term is generated at $\omega_3 = \omega_1 - \omega_2$ in the direction perpendicular to the surface. In this research project, we conducted the linear simulation to calculate the field distributions at ω_1 and ω_2, we then used those fields to calculate the non-linear polarization $P_y(\omega_3) \propto E_x(\omega_1)E^*_z(\omega_2) + E_z(\omega_1)E^*_x(\omega_2)$, which was finally used to calculate the amplitude of the DFG signal emitted by the material, demonstrating an enhancement over bulk of almost two orders of magnitude near the surface. In [44], we used an analytical formula to compute this amplitude, but the $P(\omega_3)$ term could have been used as a source term in a new linear simulation and the field at ω_3 monitored in the far-field region.

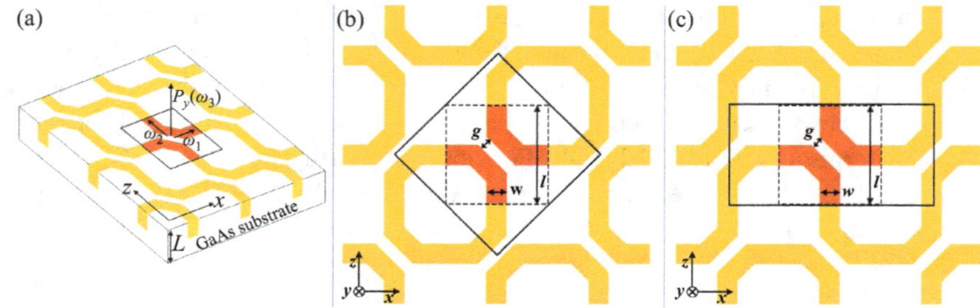

Figure 4.3 Metasurfaces to enhance second-order processes in the gap: (a) metasurface for linearly polarized light and (b and c) metasurfaces for unpolarized light

The density of the hotspots (i.e. the regions of field enhancement) can boost the non-linear generation. The metasurface in Figure 4.3(a) was designed to have maximum efficiency when the fields at ω_1 and ω_2 are linearly polarized at 45° (for polarization at 135°, we do not get any field enhancement in the gap and thus no non-linear generation enhancement). In Figure 4.3(b) and (c), we show two additional designs obtained by repeating the 'bolt' along the surface in different ways, which results in the larger unit cells highlighted by a black solid line. These metasurfaces not only enhance DFG, but they could also be used to enhance other second-order processes, such as SHG and SFG. Even though the simulation domain is now larger, the advantage of these two additional metasurfaces is that they are suitable for unpolarized incident light since the gaps are not all oriented in the same direction, meaning that at least few gaps will 'light up' for any incident linear polarization.

4.4.2 Understanding non-linear emission in hybrid nanostructures

As mentioned in the previous section, there is some debate regarding where the non-linear generation occurs in hybrid nanostructures. Simulations can help one to answer this question since they allow us to investigate the non-linear emission from each single material composing the hybrid nanostructure. Using the direct non-linear generation approach, we have performed a study on a dipole gold nanoantenna with a material with the linear permittivity of indium tin oxide (ITO) in its gap on a dielectric substrate [52]. Assuming a value for the third-order non-linear susceptibility $\chi^{(3)}$ of gold, we have identified the values for $\chi^{(3)}$ of the gap material such that the non-linear signal measured in the far field comes primarily from gold or the gap material. In Figure 4.4, we show the non-linear near field at 3ω for a monochromatic excitation at ω with incident polarization along the axis of the dipole (top) and the corresponding radiation pattern for two cut planes through the centre of the dipole nanoantenna (bottom). This is done for the case when $\chi^{(3)}$ is enabled only in the gap material (Figure 4.4(a)), only in the gold (Figure 4.4(b)), and in both gold and gap (Figure 4.4(c)). We can appreciate the differences in the

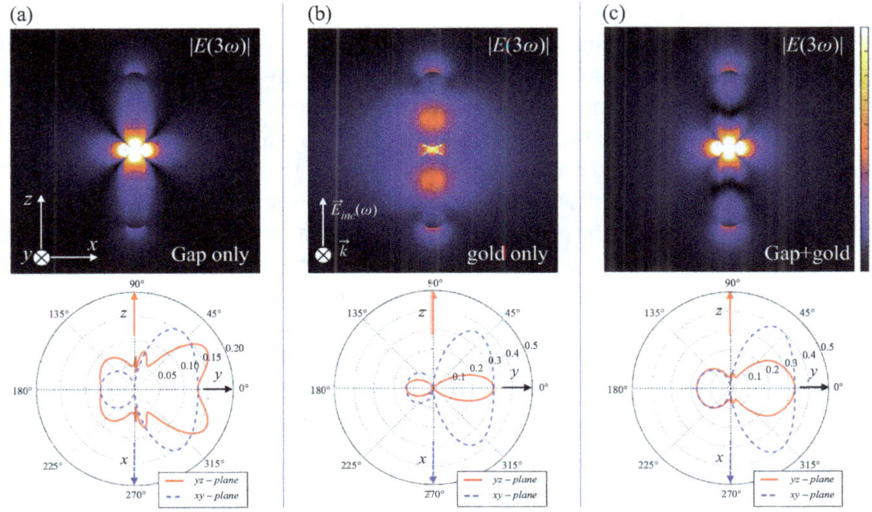

Figure 4.4 Third-harmonic near-field (top) and non-linear radiation pattern (bottom) for a hybrid dipole nanoantenna with a non-linear material in the gap when the non-linear emission is enabled in (a) gap only, (b) gold only, and (c) gap and gold

near fields and in the radiation patterns for the three different cases: a more diffracted radiation for the emission from the gap material only, a more directive emission for emission from gold, and something in between when the gap material and gold equally contribute to the THG far field. The far-field patterns prove to be very good metrics to identify the properties of a non-linear emitter, including size and shape of the emitter, as also reported in [88].

4.4.3 Vectorial control of third-order non-linear emission

In order to shape the non-linear radiation produced by a metasurface, control over phase, polarization, and amplitude of each single nanostructure is required. In [31], the linear field enhancement in the gap of a plasmonic nanoantenna was exploited to control the non-linear emission from a non-linear material placed in the gap. In this case, the control of the phase has been achieved by progressively rotating a butterfly-shaped nanoantenna in the plane of the metasurface based on the Pancharatnam–Berry or geometric phase. The linear field enhancement in the gap of the nanoantenna was used to control the polarization of the third-order non-linear emission. The control of the THG emission from each single nanostructure has allowed the demonstration of a highly structured far-field beam at the third harmonic.

Controlling the non-linear emission from nanostructures has plenty of applications in non-linear metasurfaces as described earlier. In [31], our butterfly nanoantenna was used to control the polarization and the phase of the linear field in its gap when excited at a specific working frequency ω_c. At the working frequency, the linear field in the gap shows a phase which changes linearly with the angle of the linear polarization of the incident light, or in other words, the phase changes linearly with the orientation of the nanoantenna under circular polarization illumination, which is a desired condition for geometric metasurfaces [75]. The linear field in the gap at ω_c also demonstrates a polarization that is always perpendicular to the gap axis and an amplitude that is insensitive to the polarization of the incident radiation. Thus, we have the complete polarization, amplitude, and phase control required for creating structured non-linear beams.

In Figure 4.5, we illustrate the process for designing a geometric metasurface that creates a non-linear beam carrying orbital angular momentum (OAM). Simulations are conducted using the direct non-linear generation approach for THG. In Figure 4.5(a), we start with an arrangement of dipoles progressively rotated in the plane according to parameter γ, which is the number of full rotations of the dipoles over a 2π angle. The background colour indicates the phase, which we associate with the orientation of the dipoles in the same figure. A non-linear dipole driven by the linear excitation has the same polarization direction, which is represented by the direction of the arrow, and has a phase which is n times the phase of the linear field, where n is the order of the non-linear process. In this example, we use $\gamma = 0.5$ and $n = 3$ for THG, as shown in Figure 4.5(b). Since the phase of the field within the gap of the butterfly nanoantenna changes linearly with incident polarization angle, the distribution of linear dipoles in Figure 4.5(a) can be realized by using the butterfly nanoantennas with the gap axis oriented along the arrow, as shown in Figure 4.5(c). This metasurface design, with gold butterfly antennas embedded in a non-linear material, was simulated with FDTD under left circularly polarized monochromatic excitation at ω_c. In Figure 4.5(d), we show a snapshot from the time-domain simulation (left) and the field distribution at ω_c (right). The non-linear far fields are shown in Figure 4.5(e). The simulation confirms that the metasurface is able to produce a linear field at ω_c, which reproduces the distribution of dipoles in Figure 4.5(a), and a non-linear field at $3\omega_c$ which recreates the configuration shown in Figure 4.5(b). The non-linear near field calculated within the simulation is then projected to the far field via a spatial Fourier transform and the beam that we obtain is shown in Figure 4.5(e) for all the components in cylindrical coordinates, which demonstrate that the beam is carrying an OAM state $l = 2$. The results in Figure 4.5(e) look very similar (not shown) to the far-field beam obtained by applying the same procedure to the near field described in Figure 4.5(b), thus confirming the validity of the metasurface approach to create structured non-linear beams, as we reported in [31] for an OAM state $l = 41$.

4.5 Conclusions

Ultrathin metasurfaces have the potential to disrupt the photonics industry by allowing the fabrication of all-optical integrated devices with significantly smaller

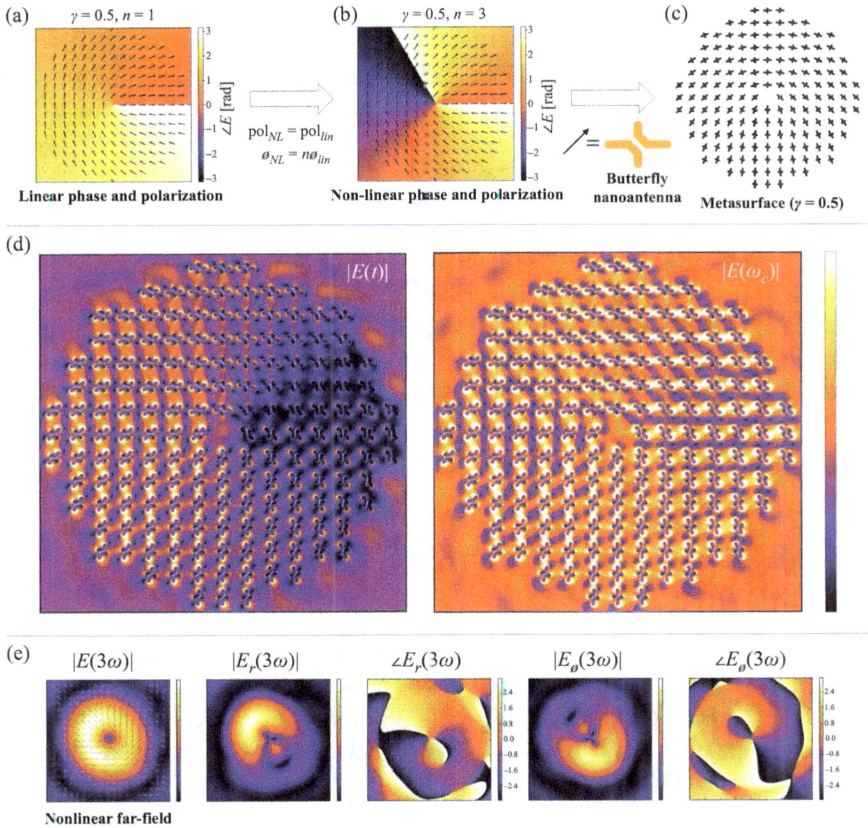

Figure 4.5 Design of a metasurface for THG structured beam carrying OAM: (a) theoretical linear near field, (b) theoretical non-linear near field, (c) metasurface design for realizing the desired linear and non-linear near fields, (d) field distribution in the time domain (left) and frequency domain (right) obtained by FDTD simulation, and (e) computed non-linear far-field beam created by the metasurface

dimensions. While the analysis of a metasurface in the linear regime is well established, the non-linear modelling is still challenging, and quantitative matching between non-linear experimental results and simulations remains unreported. Bridging this gap would provide the tools to understand and design new non-linear devices, and further research is needed to advance simulation tools. The use of artificial intelligence and deep learning is also opening exciting opportunities for the discovery of new nanostructures and metasurfaces for non-linear optics, and for the optimization of their optical properties.

Acknowledgements

Computations were performed on the SOSCIP Consortium's Blue Gene/Q computing platform using an in-house parallel 3D-FDTD software. SOSCIP is funded by the Federal Economic Development Agency of Southern Ontario, the Province of Ontario, IBM Canada Ltd, Ontario Centres of Excellence, Mitacs, and Ontario academic member institutions. We acknowledge financial support from SOSCIP, the National Sciences and Engineering Research Council of Canada, and the Canada Research Chairs program. A. C.L. acknowledges the Bundesministerium für Buldung und Furschung (German Federal Ministry of Education and Research) under the Tenure-Track Programme, and the Deutsche Forschungsgemeinschaft (DFG, German Research Foundation) under Germany's Excellence Strategy within the Cluster of Excellence PhoenixD (EXC 2122, Project ID 390833453).

References

[1] Franken PA, Hill AE, Peters CW, *et al.* Generation of optical harmonics. Physical Review Letters. 1961;7(4):118–119.
[2] Boyd RW. Nonlinear Optics, 3rd Edition. San Diego, USA: Academic Press; 2008.
[3] Liberal I and Engheta N. Near-zero refractive index photonics. Nature Photonics. 2017;11(3):149–158.
[4] Rose A, Huang D, and Smith DR. Controlling the second harmonic in a phase-matched negative-index metamaterial. Physical Review Letters. 2011;107(6):063902.
[5] Berger V. Nonlinear photonic crystals. Physical Review Letters. 1998;81:4136–4139.
[6] Xu T, Switkowski K, Chen X, *et al.* Three-dimensional nonlinear photonic crystal in ferroelectric barium calcium titanate. Nature Photonics. 2018; 12(10):591–595.
[7] Wang C, Li Z, Kim MH, *et al.* Metasurface-assisted phase-matching-free second harmonic generation in lithium niobate waveguides. Nature Communications. 2017.
[8] Shaltout AM, Kim J, Boltasseva A, *et al.* Ultrathin and multicolour optical cavities with embedded metasurfaces. Nature Communications. 2018.
[9] Cheben P, Halir R, Schmid JH, *et al.* Subwavelength integrated photonics. Nature. 2018.
[10] Yu N, Genevet P, Kats MA, *et al.* Light propagation with phase discontinuities: generalized laws of reflection and refraction. Science. 2011; 334(6054):333–337.
[11] Meinzer N, Barnes WL, and Hooper IR. Plasmonic meta-atoms and metasurfaces. Nature Photonics. 2014;8(12):889–898.
[12] Kuznetsov AI, Miroshnichenko AE, Brongersma ML, *et al.* Optically resonant dielectric nanostructures. Science. 2016;354(6314):aag2472.

[13] Calà Lesina A, Berini P, and Ramunno L. Shifted plasmonic nanorods to enhance the density of hot-spots for surface-based nonlinear optics. In: Conference on Lasers and Electro-Optics. Washington, DC: OSA; 2016. p. JW2A.109.
[14] Gramotnev DK and Bozhevolnyi SI. Plasmonics beyond the diffraction limit. Nature Photonics. 2010;4(2):83–91.
[15] Novotny L, and van Hulst N. Antennas for light. Nature Photonics. 2011; 5(2):83–90.
[16] Decker M, Staude I, Falkner M, *et al.* High-efficiency dielectric Huygens' surfaces. Advanced Optical Materials. 2015;3(6):813–820.
[17] Picardi MF, Zayats AV, and Rodríguez-Fortuño FJ. Janus and Huygens dipoles: near-field directionality beyond spin-momentum locking. Physical Review Letters. 2018;120(11):117402.
[18] Miroshnichenko AE, Evlyukhin AB, Yu YF, *et al.* Nonradiating anapole modes in dielectric nanoparticles. Nature Communications. 2015;6(1):8069.
[19] Yu N and Capasso F. Flat optics with designer metasurfaces. Nature Materials. 2014;13(2):139–150.
[20] Chen HT, Taylor AJ, and Yu N. A review of metasurfaces: physics and applications. Reports on Progress in Physics. 2016;79(7):076401.
[21] Hsiao HH, Chu CH, and Tsai DP. Fundamentals and applications of metasurfaces. Small Methods. 2017;1(4):1600064.
[22] Karimi E, Schulz SA, De Leon I, *et al.* Generating optical orbital angular momentum at visible wavelengths using a plasmonic metasurface. Light: Science & Applications. 2014;3(5):e167.
[23] Wu C, Khanikaev AB, Adato R, *et al.* Fano-resonant asymmetric metamaterials for ultrasensitive spectroscopy and identification of molecular monolayers. Nature Materials. 2012;11(1):69–75.
[24] Guay JM, Calà Lesina A, Côté G, *et al.* Laser-induced plasmonic colours on metals. Nature Communications. 2017;8:16095.
[25] Guay JM, Calà Lesina A, Baxter J, *et al.* Topography tuning for plasmonic color enhancement via picosecond laser bursts. Advanced Optical Materials. 2018;6(17):1800189.
[26] Zhao Y, Zhao Y, Hu S, *et al.* Artificial structural color pixels: a review. Materials. 2017;10(8):944.
[27] Kristensen A, Yang JKW, Bozhevolnyi SI, *et al.* Plasmonic colour generation. Nature Reviews Materials. 2017;2(1):16088.
[28] Lee T, Jang J, Jeong H, *et al.* Plasmonic- and dielectric-based structural coloring: from fundamentals to practical applications. Nano Convergence. 2018;5(1):1.
[29] Forouzmand A, Salary MM, Inampudi S, *et al.* A tunable multigate indium-tin-oxide-assisted all-dielectric metasurface. Advanced Optical Materials. 2018;6(7):1701275.
[30] Kamali SM, Arbabi E, Arbabi A, *et al.* A review of dielectric optical metasurfaces for wavefront control. Nanophotonics. 2018;7(6):1041–1068.

[31] Calà Lesina A, Berini P, and Ramunno L. Vectorial control of nonlinear emission via chiral butterfly nanoantennas: generation of pure high order nonlinear vortex beams. Optics Express. 2017;25(3):2569.
[32] Kauranen M and Zayats AV. Nonlinear plasmonics. Nature Photonics. 2012;6(11):737–748.
[33] Smirnova D and Kivshar YS. Multipolar nonlinear nanophotonics. Optica. 2016;3(11):1241–1255.
[34] Minovich AE, Miroshnichenko AE, Bykov AY, *et al.* Functional and nonlinear optical metasurfaces. Laser & Photonics Reviews. 2015;9(2):195–213.
[35] Li G, Zhang S, and Zentgraf T. Nonlinear photonic metasurfaces. Nature Reviews Materials. 2017;2(5):17010.
[36] Krasnok A, Tymchenko M, and Alù A. Nonlinear metasurfaces: a paradigm shift in nonlinear optics. Materials Today. 2018;21(1):8–21.
[37] Panoiu NC, Sha WEI, Lei DY, *et al.* Nonlinear optics in plasmonic nanostructures. Journal of Optics. 2018;20(8):083001.
[38] Keren-Zur S, Michaeli L, Suchowski H, *et al.* Shaping light with nonlinear metasurfaces. Advances in Optics and Photonics. 2018;10(1):309.
[39] Ciracì C, Poutrina E, Scalora M, *et al.* Origin of second-harmonic generation enhancement in optical split-ring resonators. Physical Review B. 2012; 85(20):201403.
[40] Chen PY, Argyropoulos C, D'Aguanno G, *et al.* Enhanced second-harmonic generation by metasurface nanomixer and nanocavity. ACS Photonics. 2015;2(8):1000–1006.
[41] Lee J, Tymchenko M, Argyropoulos C, *et al.* Giant nonlinear response from plasmonic metasurfaces coupled to intersubband transitions. Nature. 2014;511(7507):65–69.
[42] Khurgin JB and Sun G. Plasmonic enhancement of the third order nonlinear optical phenomena: figures of merit. Optics Express. 2013;21(22):27460.
[43] Lassiter JB, Chen X, Liu X, *et al.* Third-harmonic generation enhancement by film-coupled plasmonic stripe resonators. ACS Photonics. 2014;1(11): 1212–1217.
[44] Calà Lesina A, Ramunno L, and Berini P. Dual-polarization plasmonic metasurface for nonlinear optics. Optics Letters. 2015;40(12):2874.
[45] Genevet P, Tetienne JP, Gatzogiannis E, *et al.* Large enhancement of nonlinear optical phenomena by plasmonic nanocavity gratings. Nano Letters. 2010;10(12):4880–4883.
[46] Jin B and Argyropoulos C. Enhanced four-wave mixing with nonlinear plasmonic metasurfaces. Scientific Reports. 2016;6(1):28746.
[47] Vampa G, Ghamsari BG, Siadat Mousavi S, *et al.* Plasmon-enhanced high-harmonic generation from silicon. Nature Physics. 2017;13(7):659–662.
[48] Smirnova D, Smirnov AI, and Kivshar YS. Multipolar second-harmonic generation by Mie-resonant dielectric nanoparticles. Physical Review A. 2018;97(1):013807.

[49] Shorokhov AS, Melik-Gaykazyan EV, Smirnova DA, *et al.* Multifold enhancement of third-harmonic generation in dielectric nanoparticles driven by magnetic Fano resonances. Nano Letters. 2016;16(8):4857–4861.

[50] Liu S, Vabishchevich PP, Vaskin A, *et al.* An all-dielectric metasurface as a broadband optical frequency mixer. Nature Communications. 2018;9(1):2507.

[51] Liu H, Guo C, Vampa G, *et al.* Enhanced high-harmonic generation from an all-dielectric metasurface. Nature Physics. 2018;14(10):1006–1010.

[52] Calà Lesina A, Berini P, Ramunno L, *et al.* Origin of third harmonic generation in plasmonic nanoantennas. Optical Materials Express. 2017;7(5):1575.

[53] Ding F, Yang Y, Deshpande RA, *et al.* A review of gap-surface plasmon metasurfaces: fundamentals and applications. Nanophotonics. 2018;7(6): 1129–1156.

[54] Wang F, Martinson ABF, and Harutyunyan H. Efficient nonlinear metasurface based on nonplanar plasmonic nanocavities. ACS Photonics. 2017;4(5): 1188–1194.

[55] Shcherbakov MR, Shorokhov AS, Neshev DN, *et al.* Nonlinear interference and tailorable third-harmonic generation from dielectric oligomers. ACS Photonics. 2015.

[56] Wang L, Kruk S, Xu L, *et al.* Shaping the third-harmonic radiation from silicon nanodimers. Nanoscale. 2017;9(6):2201–2206.

[57] Staude I and Schilling J. Metamaterial-inspired silicon nanophotonics. Nature Photonics. 2017;11(5):274–284.

[58] Aouani H, Rahmani M, Navarro-Cía M, *et al.* Third-harmonic-upconversion enhancement from a single semiconductor nanoparticle coupled to a plasmonic antenna. Nature Nanotechnology. 2014;9(4):290–294.

[59] Metzger B, Hentschel M, Schumacher T, *et al.* Doubling the efficiency of third harmonic generation by positioning ITO nanocrystals into the hot-spot of plasmonic gap-antennas. Nano Letters. 2014;14(5):2867–2872.

[60] Alam MZ, De Leon I, and Boyd RW. Large optical nonlinearity of indium tin oxide in its epsilon-near-zero region. Science. 2016;352(6287):795–797.

[61] Caspani L, Kaipurath RPM, Clerici M, *et al.* Enhanced nonlinear refractive index in ε-near-zero materials. Physical Review Letters. 2016;116(23): 233901.

[62] Alam MZ, Schulz SA, Upham J, *et al.* Large optical nonlinearity of nanoantennas coupled to an epsilon-near-zero material. Nature Photonics. 2018;12(2):79–83.

[63] Harutyunyan H, Volpe G, Quidant R, *et al.* Enhancing the nonlinear optical response using multifrequency gold-nanowire antennas. Physical Review Letters. 2012;108(21):1–4.

[64] Yang KY, Butet J, Yan C, *et al.* Enhancement mechanisms of the second harmonic generation from double resonant aluminum nanostructures. ACS Photonics. 2017;4(6):1522–1530.

[65] Celebrano M, Wu X, Baselli M, *et al.* Mode matching in multiresonant plasmonic nanoantennas for enhanced second harmonic generation. Nature Nanotechnology. 2015;10(5):412–417.

[66] Huttunen MJ, Rasekh P, Boyd RW, *et al.* Using surface lattice resonances to engineer nonlinear optical processes in metal nanoparticle arrays. Physical Review A. 2018;97(5):053817.
[67] Czaplicki R, Kiviniemi A, Huttunen MJ, *et al.* Less is more: enhancement of second-harmonic generation from metasurfaces by reduced nanoparticle density. Nano Letters. 2018;18(12):7709–7714.
[68] Kravets VG, Kabashin AV, Barnes WL, *et al.* Plasmonic surface lattice resonances: a review of properties and applications. Chemical Reviews. 2018;118(12):5912–5951.
[69] Neira AD, Olivier N, Nasir ME, *et al.* Eliminating material constraints for nonlinearity with plasmonic metamaterials. Nature Communications. 2015; 6(1):7757.
[70] Thyagarajan K, Butet J, and Martin OJF. Augmenting second harmonic generation using Fano resonances in plasmonic systems. Nano Letters. 2013;13(4):1847–1851.
[71] Butet J and Martin OJF. Fano resonances in the nonlinear optical response of coupled plasmonic nanostructures. Optics Express. 2014;22(24):29693.
[72] Yang Y, Wang W, Boulesbaa A, *et al.* Nonlinear Fano-resonant dielectric metasurfaces. Nano Letters. 2015;15(11):7388–7393.
[73] Krasavin AV, Ginzburg P, Wurtz GA, *et al.* Nonlocality-driven supercontinuum white light generation in plasmonic nanostructures. Nature Communications. 2016;7(1):11497.
[74] Hu H, Zhang J, Maier SA, *et al.* Enhancing third-harmonic generation with spatial nonlocality. ACS Photonics. 2018;5(2):592–598.
[75] Wen D, Yue F, Liu W, *et al.* Geometric metasurfaces for ultrathin optical devices. Advanced Optical Materials. 2018;6(17):1800348.
[76] Li G, Chen S, Pholchai N, *et al.* Continuous control of the nonlinearity phase for harmonic generations. Nature Materials. 2015;14(6):607–612.
[77] Renaut C, Lang L, Frizyuk K, *et al.* Reshaping the second-order polar response of hybrid metal-dielectric nanodimers. Nano Letters. 2019.
[78] Keren-Zur S, Avayu O, Michaeli L, *et al.* Nonlinear beam shaping with plasmonic metasurfaces. ACS Photonics. 2016;3(1):117–123.
[79] Segal N, Keren-Zur S, Hendler N, *et al.* Controlling light with metamaterial-based nonlinear photonic crystals. Nature Photonics. 2015;9(3):180–184.
[80] Almeida E, Bitton O, and Prior Y. Nonlinear metamaterials for holography. Nature Communications. 2016;7(1):12533.
[81] Gao Y, Fan Y, Wang Y, *et al.* Nonlinear holographic all-dielectric metasurfaces. Nano Letters. 2018;18(12):8054–8061.
[82] Wolf O, Campione S, Benz A, *et al.* Phased-array sources based on nonlinear metamaterial nanocavities. Nature Communications. 2015;6(1):7667.
[83] Tymchenko M, Gomez-Diaz JS, Lee J, *et al.* Gradient nonlinear Pancharatnam-Berry metasurfaces. Physical Review Letters. 2015;115(20): 207403.
[84] Wang L, Kruk S, Koshelev K, *et al.* Nonlinear wavefront control with all-dielectric metasurfaces. Nano Letters. 2018;18(6):3978–3984.

[85] Linnenbank H, Grynko Y, Förstner J, *et al.* Second harmonic generation spectroscopy on hybrid plasmonic/dielectric nanoantennas. Light: Science & Applications. 2016;5(1):e16013.

[86] de Ceglia D, Vincenti MA, and Scalora M. On the origin of third harmonic light from hybrid metal-dielectric nanoantennas. Journal of Optics. 2016; 18(11):115002.

[87] Liu X, Larouche S, Bowen P, *et al.* Clarifying the origin of third-harmonic generation from film-coupled nanostripes. Optics Express. 2015;23(15): 19565.

[88] Butet J, Thyagarajan K, and Martin OJF. Ultrasensitive optical shape characterization of gold nanoantennas using second harmonic generation. Nano Letters. 2013;13(4):1787–1792.

[89] Metzger B, Hentschel M, and Giessen H. Probing the near-field of second-harmonic light around plasmonic nanoantennas. Nano Letters. 2017;17(3): 1931–1937.

[90] Taflove A and Hagness SC. Computational Electrodynamics: The Finite-Difference Time-Domain Method. Norwood, USA: Artech House; 2005.

[91] Taflove A, Oskooi A, and Johnson SG. Advances in FDTD Computational Electrodynamics: Photonics and Nanotechnology. Norwood, USA: Artech House; 2013.

[92] Rahman BMA and Agrawal A. Finite Element Modeling Methods for Photonics. Boston, MA: Artech House; 2013.

[93] Busch K, Knig M, and Niegemann J. Discontinuous Galerkin methods in nanophotonics. Laser & Photonics Reviews. 2011;5(6):773–809.

[94] Hille A, Moeferdt M, Wolff C, *et al.* Second harmonic generation from metal nano-particle resonators: numerical analysis on the basis of the hydrodynamic Drude model. The Journal of Physical Chemistry C. 2016; 120(2):1163–1169.

[95] Martin OJF and Piller NB. Electromagnetic scattering in polarizable backgrounds. Physical Review E. 1998;58(3):3909–3915.

[96] Draine BT and Flatau PJ. Discrete-dipole approximation for scattering calculations. Journal of the Optical Society of America A. 1994;11(4):1491.

[97] Gallinet B, Butet J, and Martin OJF. Numerical methods for nanophotonics: standard problems and future challenges. Laser & Photonics Reviews. 2015;9(6):577–603.

[98] Dewitz JP, Hübner W, and Bennemann KH. Theory for nonlinear Mie-scattering from spherical metal clusters. Zeitschrift fur Physik D Atoms, Molecules and Clusters. 1996;37(1):75–84.

[99] Pavlyukh Y and Hübner W. Nonlinear Mie scattering from spherical particles. Physical Review B. 2004;70(24):245434.

[100] de Beer AGF and Roke S. Nonlinear Mie theory for second-harmonic and sum-frequency scattering. Physical Review B. 2009;79(15):155420.

[101] Butet J, Russier-Antoine I, Jonin C, *et al.* Nonlinear Mie theory for the second harmonic generation in metallic nanoshells. Journal of the Optical Society of America B. 2012;29(8):2213.

[102] Bachelier G, Russier-Antoine I, Benichou E, *et al.* Multipolar second-harmonic generation in noble metal nanoparticles. Journal of the Optical Society of America B. 2008;25(6):955.

[103] Bachelier G, Butet J, Russier-Antoine I, *et al.* Origin of optical second-harmonic generation in spherical gold nanoparticles: local surface and nonlocal bulk contributions. Physical Review B. 2010;82(23):235403.

[104] Butet J, Gallinet B, Thyagarajan K, *et al.* Second-harmonic generation from periodic arrays of arbitrary shape plasmonic nanostructures: a surface integral approach. Journal of the Optical Society of America B. 2013; 30(11):2970.

[105] Bernasconi GD, Butet J and Martin OJF. Mode analysis of second-harmonic generation in plasmonic nanostructures. Journal of the Optical Society of America B. 2016;33(4):768.

[106] Hentschel M, Utikal T, Giessen H, *et al.* Quantitative modeling of the third harmonic emission spectrum of plasmonic nanoantennas. Nano Letters. 2012;12(7):3778–3782.

[107] Varin C, Emms R, Bart G, *et al.* Explicit formulation of second and third order optical nonlinearity in the FDTD framework. Computer Physics Communications. 2018;222:70–83.

[108] Calà Lesina A, Vaccari A, Berini P, *et al.* On the convergence and accuracy of the FDTD method for nanoplasmonics. Optics Express. 2015;23(8): 10481.

[109] O'Brien K, Suchowski H, Rho J, *et al.* Predicting nonlinear properties of metamaterials from the linear response. Nature Materials. 2015;14(4): 379–383.

[110] Alsunaidi MA and Al-Jabr AA. A general ADE-FDTD algorithm for the simulation of dispersive structures. IEEE Photonics Technology Letters. 2009;21(12):817–819.

[111] Prokopidis KP and Zografopoulos DC. A unified FDTD/PML scheme based on critical points for accurate studies of plasmonic structures. Journal of Lightwave Technology. 2013;31(15):2467–2476.

[112] Calà Lesina A, Baxter J, Berini P, *et al.* Simulations in nanophotonics. In: Di Bartolo B, Silvestri L, Cesaria M, *et al.*, editors. Quantum Nano-Photonics. Dordrecht, Netherlands: Springer Netherlands; 2018. p. 117–131.

[113] Ciracì C, Poutrina E, Scalora M, *et al.* Second-harmonic generation in metallic nanoparticles: clarification of the role of the surface. Physical Review B. 2012;86(11):115451.

[114] Scalora M, Vincenti MA, de Ceglia D, *et al.* Harmonic generation from metal-oxide and metal-metal boundaries. Physical Review A. 2018; 98:023837.

Part II

Material platforms and fabrication techniques

Chapter 5

Rare-earth-doped glasses and glass ceramics for integrated optics

Thi Ngoc Lam Tran[1,2,3], Lidia Zur[1], Alessandro Chiasera[1], Andrea Chiappini[1], Wilfried Blanc[4], Monica Bollani[2], Anna Lukowiak[5], Giancarlo C. Righini[6] and Maurizio Ferrari[1]

Glass photonics is the brick corner for a huge number of integrated optics applications [1] covering, for instance, solid-state laser sources [2], superluminescent light sources [3], optical sensors [4], phosphors and luminescent devices [5], photovoltaic systems [6], communication technologies [7], quantum microcavities [8], and phase change random access memories [9]. From a general point of view, the integration is performed on a planar configuration, i.e. in planar waveguide systems [1,10], although interesting approaches with microcavities, microresonators [11], and photonic crystals [12] have been developed and continuously investigated. In this chapter, we highlight some basic aspects and applications of doped glasses in integrated optics. First, we will briefly summarize the role of rare-earth (RE) ions in integrated optics, reporting some consolidated results presented in the literature as well as some new results concerning the photon management exploited for the luminescence-enhancement mechanisms. Few lines are also assigned to recall the use of the Judd–Ofelt theory in the determination of the spectroscopic parameters, crucial for the design of more novel and efficient photonic devices based on rare-earth-doped (RED) glasses. The second section is devoted to the novel and exciting topic of transparent glass ceramics. First, we introduce this kind of materials and we focus the attention of the reader on their important optical, structural, and spectroscopic properties. Of particular interest is the role of tin-dioxide-based glass

[1]CSMFO Lab. and FBK Photonics Unit, Institute of Photonics and Nanotechnologies (IFN-CNR), Trento, Italy

[2]Institute of Photonics and Nanotechnologies (IFN-CNR), Milano, Italy

[3]Department of Materials Technology, Ho Chi Minh City University of Technology and Education, Ho Chi Minh City, Vietnam

[4]Institut de Physique de Nice, Université Côte d'Azur, CNRS UMR7010, Nice Cedex 2, France

[5]Institute of Low Temperature and Structure Research, PAS, Wrocław, Poland

[6]MiP Lab, 'Nello Carrara' Institute of Applied Physics (IFAC-CNR), Sesto Fiorentino, Italy

ceramics as effective sensitizers of the RE ions luminescence. Novel spectroscopic results regarding the Er^{3+}-activated SiO_2–SnO_2 system are also presented with a discussion about the different possible structures of integrated optics. The different kinds of nanoceramics operating in luminescence quantum yield enhancement are shortly discussed. The differences between the top-down and bottom-up fabrication techniques are presented, as well as some well-consolidated results about the transparent glass ceramics doped with RE ions. In this regard, it is worthy to note the results obtained in a hybrid glass-ceramic fabricated employing an amorphous matrix SiO_2–HfO_2 loaded by nanoparticles of lithium lanthanum tetraphosphates doped with different concentration of Eu^{3+} ions. Finally, we conclude the chapter with the perspectives of glass materials for integrated optics.

5.1 Glasses activated by rare-earth ions

Silica-based glasses have demonstrated their remarkable importance in several optical applications, especially since the invention of optical fibres [13]. The full exploitation of erbium-doped fibre amplifiers is now evident in the entertainment, sensing, and telecommunication markets. The awesome technical and scientific outcomes obtained by the team at Southampton lead by D.N. Payne paved the way to the integrated optoelectronic devices [14]. The research for developing integrated photonic systems mainly based on planar platforms is largely based on glasses, and silicate-based glasses are the most commonly employed [15]. RED silicate glasses have been developed for a large spectrum of applications using different fabrication techniques based on melting, wet chemistry, and both chemical and physical deposition techniques. One of the most critical issues in RED glasses for integrated photonics is the so-called physical clustering, i.e. the interaction among RE ions, which has an immediate consequence on the drastic reduction of the luminescence quantum yield, which in turn hampers the photonic device efficiency [16–19]. Several approaches have been attempted for avoiding this effect, including new host compositions and, as we will see in the next section, the use of transparent glass ceramics [20,21]. Among the others, the sol–gel-derived confined structures give an efficient solution to succeed in this goal. Much research activity in this field is aimed to develop optical amplifiers in planar format to provide lossless devices for communication but also integrated lasers for medical application or environmental sensing [19,20,22].

A further development in the control of luminescence properties is given by RE-activated microcavities that represent a particular class of photonic crystals [23]. Due to the possibility to obtain high-quality optical films with thickness of a fraction of a wavelength and fabricated by multi-deposition techniques, the control of the spontaneous emission properties of luminescent centres in microcavities is possible. We will discuss these systems at the end of the chapter.

Although silica-based photonic glasses and structures still remain the most commonly used, the interest to move to the near infrared (NIR) and in particular to the medium infrared (MIR) regions promoted a very strong research towards tellurite, fluoride, and chalcogenide glasses. RED tellurite glasses are very attractive materials for photonic applications, not only as optical amplifiers in the second and third telecommunications windows but also as frequency up-converters [24]. Telluride glasses have a wide transmission region (0.35–5 μm), good glass stability and corrosion resistance, the lowest vibrational energy (~780 cm^{-1}) among oxide glass formers, low process temperature, and a high refractive index, which increases the local field correction at the RE ion site and leads to an enhancement of the radiative transition rates [7,24–26]. Telluride glasses also have high non-linear refractive index and can find application in harmonic generation [27].

Fluoride glasses are also very attractive for integrated optics. Fluorides are often chosen as host materials instead of oxides because their phonon energy is lower, reducing the probability of non-radiative relaxation processes as well as increasing the transparency in the infrared spectral region. Moreover, the RE ion solubility in fluoride glasses is considerable higher than in silica [28]. In the context of integrated optics, several techniques have been used successfully for the fabrication of RED fluoride glass planar waveguides, such as physical vapour deposition (PVD) [29], F^-/Cl^- ionic exchange [30], and pulsed laser deposition [31]. Planar waveguides with excellent optical properties have been obtained by PVD and ion exchange, working in the infrared [32] and visible region, in direct or up-conversion [29] and downconversion [33] configurations.

Last but not the least, chalcogenide glasses are reaching an important position in the field of low phonon energy photonic glasses [34]. As remarked by Seddon *et al.*, chalcogenide-based glasses are the new frontier for MIR photonics, i.e. in the wavelength range from 3.0 to 50 μm, where there are the atmospheric windows of 3.0–5.0 and 8.0–12 μm and the related gas absorption fingerprints [34]. Moreover, the possibility for developing integrated structures based on these glasses would open a broad spectrum of applications covering environmental, structural, and industrial monitoring, molecular spectroscopy, free-space communications, and medical diagnostic [35,36]. Chryssou *et al.* demonstrated that the tellurite glass host material can offer advantages for the fabrication of high-gain integrated optical amplifiers [37]. The main advantages evidenced by the authors are that Er^{3+}-activated tellurite waveguides exhibit higher signal gain than Er^{3+}-doped silica waveguides and that the broad bandwidth of Er^{3+} in tellurites and its higher emission cross section can allow a flat-gain optical amplifier [37].

Among the different materials used for the preparation of optically active glasses, the silica–hafnia binary system has shown excellent optical and structural properties. It has been shown that SiO_2–HfO_2 planar waveguides activated by Er^{3+} ions can be fabricated by the sol–gel or radio frequency sputtering techniques [38,39]. There are two important points that should be considered in the fabrication of Er^{3+}-activated systems for integrated optics: (a) the low absorption cross section of the Er^{3+} ions at 980 nm and (b) the modification of the $^4I_{11/2}$ depopulation rate with the aim to populate the metastable state $^4I_{13/2}$. The first drawback is usually resolved by co-doping with Yb^{3+} ions. The second issue is more complex because some problems can be induced on the $^4I_{13/2}$ quantum yield. In the following, we will discuss two typical examples of the above-mentioned topics.

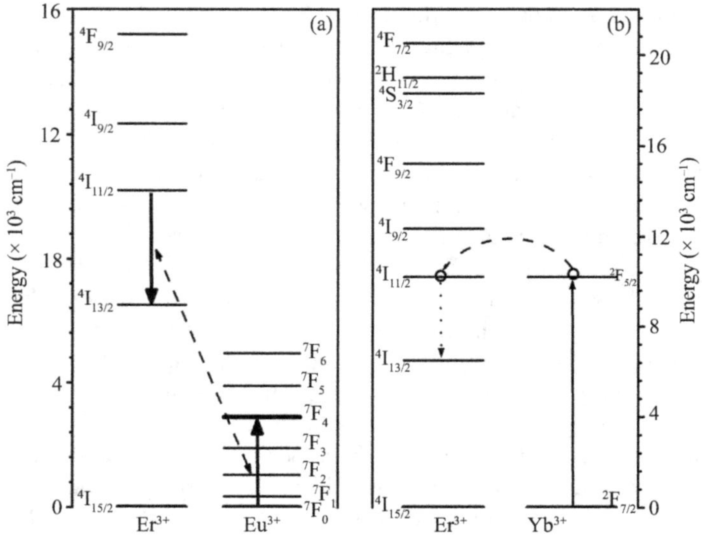

Figure 5.1 *Er^{3+}, Yb^{3+} (a) and Er^{3+}, Eu^{3+} (b) energy levels involved in the two different processes that can enhance the luminescence efficiency of the Er^{3+} ion at 1.5 μm*

Several researchers have focused their attention on the preparation of erbium-doped glasses co-doped with different kinds of RE ions. It has been showed that the introduction of ytterbium ions enhances the absorption at 980 nm, making the pumping mechanism more efficient [40–46]. The introduction of other RE ions, such as Eu^{3+}, Tb^{3+}, or Ce^{3+}, has been demonstrated to increase the population of the $^4I_{13/2}$ by the modification of the $^4I_{11/2}$ relaxation rate [47–51]. The effect of the introduction of Eu^{3+} and Ce^{3+} has been studied in several systems, including multicomponent glasses, with very-low-to-high phonon energy and thus with different non-radiative transition rates. Figure 5.1 shows two examples of the Er^{3+}, Yb^{3+} (Figure 5.1(a)) and Er^{3+}, Eu^{3+} (Figure 5.1(b)) energy levels involved in the two different processes that can enhance the luminescence efficiency of the Er^{3+} ion at 1.5 μm.

The two mechanisms have been investigated in detail in the silica–hafnia system. The waveguides have been realized by the deposition of a sol on cleaned vitreous SiO_2 substrates by dip-coating technique, using a dipping rate of 40 mm/min. The starting solution, constituted by tetraethylorthosilicate (TEOS), ethanol, deionized water, and hydrochloric acid as a catalyst, was pre-hydrolyzed for 1 h at 65°C. The molar ratio TEOS:HCl:EtOH:H_2O was 1:0.01:37.9:2. Ethanolic colloidal suspensions were prepared using $HfOCl_2$ as precursor and then added to the TEOS solutions, with an Si/Hf molar ratio of 80/20. Erbium was added to the final solutions as $Er(NO_3)_3{\cdot}5H_2O$ with an Er/(Si+Hf) concentration of 0.3 mol% [52]. Ytterbium and europium were added as $Yb(NO_3)_3{\cdot}5H_2O$ and $Eu(NO_3)_3{\cdot}5H_2O$ with different concentrations, ranging between 0.1 and 0.6 mol%. The final mixtures were left at room temperature under stirring for 16 h. The obtained sols were used for the deposition process. After every single deposition, the film was annealed in the air for 1 min placing the samples in a tubular oven at 900°C. After 10 dipping cycles, the film was annealed for 2 min at 900°C. Eventually, the films were stabilized for 5 h in air at 900°C. As a result of the procedure, transparent and crack-free waveguides were obtained. Four samples with the same Si/Hf molar ratio and with the same erbium concentration (0.3 mol%), but different molar percentage of Yb^{3+} (sample WY) or Eu^{3+} (sample WE), were prepared. A waveguide doped with 0.3 mol% of Er^{3+} (W0) was produced as a reference sample. The nominal compositional concentration for the fabricated waveguides is reported in Table 5.1.

Figure 5.2 shows the room temperature photoluminescence spectra of the $^4I_{13/2}\rightarrow{}^4I_{15/2}$ transition of Er^{3+} ions obtained upon excitation at 980 nm for the samples W0, WY01, and WY06. All the spectra present the same shape, characterized by the main emission peak at 1.53 μm, with a shoulder at ~1.55 μm and a spectral bandwidth of ~51 nm. The intensity of the emission increases with the increasing Yb^{3+} ions concentration. This is a direct evidence that the energy transfer between ytterbium and erbium, which increases the population of the $^4I_{13/2}$ erbium level, is effective. The decay curves for the waveguides W0, WY01, and WY06 exhibit the same lifetime of 6.3 ± 0.2 ms. Room temperature photoluminescence spectra of the $^4I_{13/2}\rightarrow{}^4I_{15/2}$ transition of Er^{3+} ions obtained upon excitation at 980 nm have been measured also for the WE03 and WE06 samples. As expected, the co-doping with Eu^{3+} ions does not modify the shape of the emission spectra with respect to those reported in Figure 5.2, so we do not show them; the spectral width of the emission band, measured at 3 dB from the maximum of the intensity, is again 51 nm for all the samples. The intensity of the emission does not change significantly with the Eu^{3+} ions concentration. Figure 5.3 shows the normalized decay curves of the $^4I_{13/2}$ metastable state obtained upon 980 nm excitation for the Er^{3+}/Eu^{3+}-doped planar waveguides. A clear difference is observed in respect to the decay curve measured for the reference waveguide W0. The measured lifetimes are W0: 6.3 ± 0.2 ms, WE03: 4.1 ± 0.2 ms, and WE06: 3.5 ± 0.2 ms.

The co-doping with Eu^{3+} ions introduces in the system a series of energetic levels, with values ranging between 400 ($^7F_1$) and 5,000 cm^{-1} ($^7F_6$). Consequently, the energy transfer mechanisms between Er^{3+} and Eu^{3+} ions are promoted,

Table 5.1 Compositional parameters of the 80 SiO_2–20 HfO_2 planar waveguides activated by 0.3 mol% of Er^{3+} ions

Composition (mol%)	W0	WY01	WY06	WE03	WE06
Er^{3+}	0.3	0.3	0.3	0.3	0.3
Yb^{3+}	0	0.1	0.6	0	0
Eu^{3+}	0	0	0	0.3	0.6

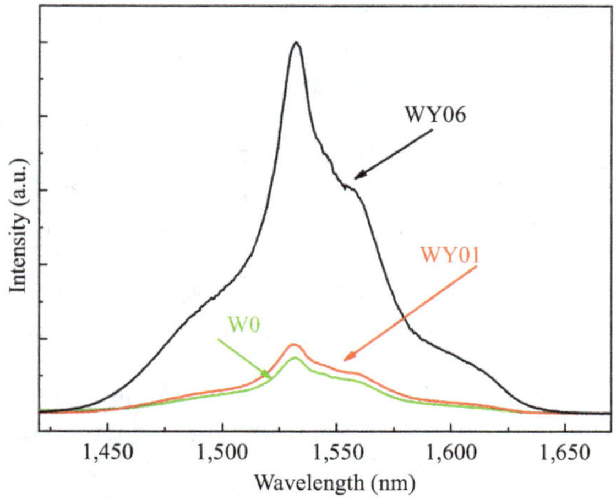

Figure 5.2 Photoluminescence spectra of $^4I_{13/2}\rightarrow^4I_{15/2}$ transition of Er^{3+} ions obtained upon excitation at 980 nm for the samples W0 (0.3% Er^{3+}), WY01 (0.3% Er^{3+}, 0.1% Yb^{3+}), and WY06 (0.3% Er^{3+}, 0.6% Eu^{3+})

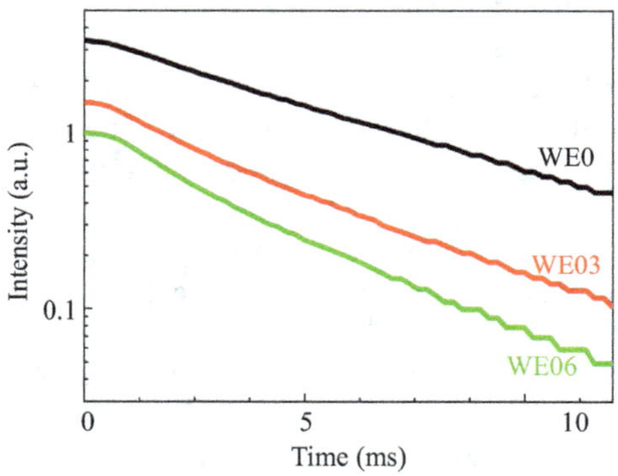

Figure 5.3 Decay curves of the Er^{3+} $^4I_{13/2}$ metastable level obtained upon excitation at 980 nm for the samples W0 (0.3% Er^{3+}), WE03 (0.3% Er^{3+}, 0.3% Eu^{3+}), and WE06 (0.3% Er^{3+}, 0.6% Eu^{3+})

lowering the lifetimes of both the $^4I_{11/2}$ and $^4I_{13/2}$ energy levels [47]. The lowering of the $^4I_{11/2}$ lifetime indicates that Eu^{3+} co-doping, in Er^{3+}-activated silica–hafnia system, efficiently promotes the population of the $^4I_{13/2}$ state. However, the lowering of the $^4I_{13/2}$ metastable state is strongly detrimental for efficient amplification at 1.5 μm affecting the emission quantum yield.

Among the most common glasses used in integrated optics, silicates and phosphates take a significant place [53,54]. As an example concerning silicate glasses, an integrated waveguide laser has been developed for biomedical sensing and produced by the sol–gel technology on silica–hafnia platform [55]. In fact,

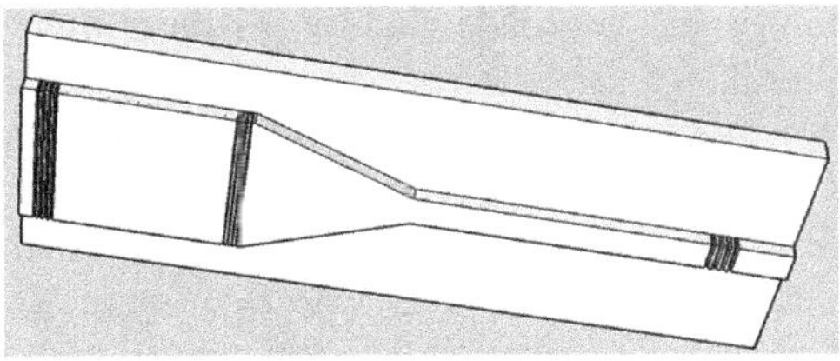

Figure 5.4 Monolithic Nd^{3+}-doped silica–hafnia sol–gel-tapered rib waveguide laser. The grating on the left is the input–output coupler, the intermediate is a partial reflecting grating for the lasing wavelength, and the last one, at the end of the taper, is a total reflecting grating to close the cavity

there is a general effort to achieve life science functionalities on a single chip for enabling fast, consistent, and low-time- and low-cost-consuming tests and diagnostic analysis. A well-recognized approach is based on optical biochemical sensors, constituted by integrated optical devices with embedded specific biochemical receptors for binding the target species to be sensed. There are several constraints in the development of such a device, which require channel waveguides and the coupling with the external light source to operate correctly. This is an important point, and the use of these platforms is often limited by the costs related to the coupling of a laser source and detectors to the chip. End-facet cutting and polishing is one of the delicate and time-consuming processes that can be overcome using on-chip grating couplers. However, the reliability of the sensor strongly depends on the spectral and intensity fluctuation of the external laser source, and the optimal solution is integrating the light source into the chip. An interesting integrated system is shown in Figure 5.4. The structure is based on a tapered rib waveguide allowing gain enhancement. The input/output grating allows coupling the pump in the waveguide and coupling out the laser. The intermediate grating is a partial reflection grating in tailoring the lasing wavelength. At the end of the taper, a total refection grating closes the cavity. This structure is an exemplary system to demonstrate as glasses are crucial for the fabrication of very effective devices for a strategic application. Peled *et al.* adopted this very innovative approach to fabricate a monolithic Nd^{3+}-doped silica–hafnia sol–gel-tapered rib waveguide laser, employing photolithography and reactive ion etching [55].

The pump at 806 nm was coupled in by a grating, which also allows the lasing output at 1,065.4 nm. A reflection of the Bragg mirror is used for the feedback. A lasing threshold of 20 mW and an output power of 2.45 mW were obtained. With a guiding layer thickness of only 504 nm, which is ideal for evanescent field sensing, this laser source enables fully monolithic optical-based lab-on-chip devices.

Phosphate glasses are of particular interest, and they are intensively investigated because of their high RE ions solubility and consequent physical and chemical clustering reduction [56] RED phosphate glasses generally exhibit a robust luminescence quantum yield, a broad window of transparency, and a nonlinear refractive index smaller than that of silicate glasses [57]. Different kinds of network modifiers can be employed to tailor the composition in order to enhance the chemical, mechanical, and thermal stability of phosphate glasses as well as the RE solubility and the RE local crystal field [58,59]. Thanks to these properties, phosphate glasses have demonstrated to be very effective for Er waveguide amplifiers fabrication [60,61] and for Nd compact lasers and amplifiers [62]. Due to the high number of possible compositions, the main objective of an important part of the current research is dedicated to the optimization of multicomponent phosphate glasses doped with RE ions, and to their assessment. The structure and the spectroscopic properties of the RED glasses are studied to develop a suitable glass composition and to choose the most promising RE concentration for optical amplifier and laser applications. As an example, we can analyse a recent paper discussing the optical properties of Nd^{3+}-doped phosphate glasses [62]. In order to determine the most effective composition, the authors use the Judd–Ofelt method to assess the important spectroscopic parameters [63,64]. This is a common method that one can use when absorption cross section, refractive index, and density of the glass are known. The experimental oscillator strengths (f_{exp}) for all the absorption bands of the RE ions can be determined from the absorption spectra by the following equation:

$$f_{exp}\left(\Psi J \rightarrow \Psi' J'\right) = \frac{mc^2}{\pi e^2}\int \varepsilon(v)dv = 4.318 \times 10^{-9}\int \varepsilon(v)dv \tag{5.1}$$

where m and e are the mass and the charge of an electron, respectively, c is the speed of light, v is the wavenumber (in cm^{-1}), and ε is the molar absorptivity. Following the Judd–Ofelt theory, the total oscillator strength is given by the following equation:

$$f_{calc}\left(\Psi J \rightarrow \Psi' J'\right) = \frac{8\pi^2 m v_0}{3h(2j+1)n^2}\left[\chi_{ed}S_{ed} + \chi_{md}S_{md}\right] \tag{5.2}$$

where S_{ed} and S_{md} are the electric and magnetic dipole line strengths; χ_{ed} and χ_{md} are the correction for the local field in a medium of refractive index n.

$$S_{ed}\left(\Psi J \rightarrow \Psi' J'\right) = \frac{8\pi^2 m v_0 \chi_{ed}}{3h(2j+1)n^2}\left[\sum_{t=2,4,6}\Omega_t|\langle i||U^t||j\rangle|^2\right] \tag{5.3}$$

$$S_{md}\left(\Psi J \rightarrow \Psi' J'\right) = \frac{e^2h^2\chi_{md}}{16\pi^2 m^2 c^2}|\langle i||L+2S||j\rangle|^2 \tag{5.4}$$

where Ω_t ($t = 2$, 4, and 6) are the Judd–Ofelt intensity parameters, m is the electron mass, v_0 is the wavenumber at the absorption maximum, h is the Plank constant, $2J + 1$ is the degeneracy of the initial level of the transition, and $\langle i||U^t||j\rangle$ are the reduced matrix elements which can be calculated using well established and easily available numerical software [65–67].

Using the Judd–Ofelt intensity parameters, it is possible to estimate the radiative transition probabilities:

$$A_r\left(\Psi J \to \Psi' J'\right) = \frac{64\pi^4 \nu_0^2 e^2}{3h(2j+1)}\left[\frac{n(n^2+1)^2}{9}S_{ed} + n^3 S_{md}\right] \tag{5.5}$$

where S_{ed} and S_{md} are the electric and magnetic dipole line strengths:

$$S_{ed}\left(\Psi J \to \Psi' J'\right) = \sum_{t=2,4,6} \Omega_t \left|\langle J||U^t||J'\rangle\right|^2 \tag{5.6}$$

$$S_{md}\left(\Psi J \to \Psi' J'\right) = \frac{h^2}{16\pi^2 m^2 c^2}\left|\langle J||L+2S||J'\rangle\right|^2 \tag{5.7}$$

The total radiative transition probability A_t for an excited level is given by the sum of all the possible relaxation paths to the terminal levels. The radiative lifetime τ_r of an electronic level is given by

$$\tau_r = \frac{1}{A_t} \tag{5.8}$$

This value should be compared with the measured lifetime $\tau_{\exp}$ in order to estimate the quantum efficiency η of the investigated systems as well as to design the pumping schema of an integrated amplifier:

$$\eta = \frac{\tau_{\exp}}{\tau_r} \tag{5.9}$$

The other important parameter for the assessment of the optimal photonic glass is the so-called fluorescence branching ratio B_r

$$B_r = \frac{A_r}{A_t} \tag{5.10}$$

that gives the relative emission intensity among the different transitions observed in an RE-activated glass.

All these parameters, when compared with the experimental data, allow the optimization of the glass composition for the development of active photonic systems.

There is another strategy to enhance the RE luminescence properties, consisting of modifying the nanostructure of the photonic system. A very powerful tool for success is the development of transparent glass ceramics.

5.2 Transparent glass ceramics activated by rare-earth ions

Glass ceramics are nanocomposite materials offering unique properties of strategic relevance in photonics [1,68]. Glass ceramics are two-phase materials consisting of nanocrystals embedded in a glass matrix; the respective volume fractions of crystalline and amorphous phase determine the properties of the glass-ceramic. When activated by RE ions, the nanocrystals allow fluorescence enhancement by strongly reducing both chemical and physical clustering through the control of the RE partition and by increasing absorption and emission cross sections. Moreover, in general, we have an effective reduction of the non-radiative relaxation processes, thanks to the lower phonon cut-off energy. Among these properties, transparency is crucial, in particular for systems such as fibres, integrated waveguides, and microresonators, where scattering losses are highly detrimental in operating conditions. There is a large quantity of appealing researches dealing with this topic.

More than 20 years ago, in an interesting paper, Champagnon *et al.* investigated the relation between glass structure and light scattering in silica glasses at low OH content [69]. In this paper, the concept of Boson peak was employed to exhaustively describe the glass structure. The authors performed Rayleigh, Raman, and Brillouin light scattering experiments, taking into account elastic and inelastic mechanisms. The conclusion of the paper was that the standard Rayleigh scattering is due to large domains of density fluctuation, and the Boson peak is related to smaller domains associated with elastic fluctuations. It is the latter phenomenon that can play a crucial role in nanostructured waveguides and fibres. Before this paper, the extrinsic scattering losses were mainly considered typically associated with the crystallization phenomena [70].

RE-activated glass-ceramic constitutes a perfect system to analyse the mechanisms related to light scattering, light propagation, and luminescence enhancement. The paper published by Tick in 1998 discussed about the possibility to fabricate low-loss glass-ceramic optical waveguides [71]. He explained that the system is compatible in principle with the physical and chemical constraints and suggested that the minimum transmission loss limit of the investigated effective medium glass ceramics is in the order of tens of decibels per kilometre, once all of the impurities can be eliminated. It is interesting to note that the general criteria for light propagation given in the paper concern nanocrystals size, narrow particle-size distribution, inter-particle spacing, and clustering. As far as scattering attenuation is concerned, it is important to note that the transparency of glass ceramics is higher than that expected from the theory of Rayleigh scattering. In this context, the density fluctuation of small domains, mentioned in the paper published 2 years after the Tick's paper, could help to explain the unexpected transparency of glass-ceramic waveguides [69,71]. On the same thought line is the experimental result demonstrating that the physical mechanism producing high transparency in glass ceramics is the low-density fluctuation in the number of scattering centres [72]. The important question put by Tick in the 1990s has had a positive answer, as demonstrated by the numerous papers reporting transparent glass-ceramic waveguides for integrated optics applications [18,20].

There are two important items in RED glass-derived waveguides when we look at integrated optics application: (a) the minimization of ion–ion interactions and non-radiative relaxation processes and (b) the minimization of fabrication steps necessary to realize the system. The first point concerns propagation losses and luminescence quantum yield and the second one fabrication costs, time consumption, and reproducibility. Glass ceramics have demonstrated to overcome the mentioned problems. In fact, glass ceramics combine the mechanical and optical properties of the glass with the characteristics of a crystal-like environment for the RE ions. This means that, due to the presence of RE ions in a crystalline matrix, the physical clustering is strongly reduced, and the chemical clustering is completely absent [16,19].

As mentioned before, the population of the RE ions is a crucial point in integrated optics. Moreover, the low absorption cross section of the electronic states of the RE ions constitutes a significant feebleness for RED photonic glasses. This is an intrinsic property of the RE ions, due to the *f*–*f* transitions forbidden by parity [74]. As a consequence, a sensitizer is mandatory. Recently, the strategic role of SnO_2 nanocrystals as extremely effective sensitizer for the RE ions has been demonstrated [75–78]. SnO_2 is a wide-band gap semiconductor with a low phonon energy cut-off and a broad transparent window [79].

Figure 5.5 shows the transmission electron microscopy (TEM) micro-image of a $70SiO_2–30SnO_2:0.5Er^{3+}$ glass ceramics planar waveguide. The black dots are SnO_2 nanocrystals and the light grey part is the SiO_2 matrix. The SnO_2 nanocrystals appear homogenously distributed with a narrow size distribution of less than 10 nm. The research on SnO_2-based glass ceramics activated by RE ions has been performed because of the requirement to develop reliable fabrication processes and to explain particular optical, structural, and spectroscopic features characterizing this system [75–80].

Figure 5.6 shows the rutile crystal structure of SnO_2 activated by Er^{3+} ions. The central atom has been substituted by an erbium atom (green), and Sn atoms are shown in grey and O in red. The figure is obtained by using XCrySDen program [81]. It is interesting to note that the RE ion is substitutional of an Sn atom and not interstitial in the SnO_2 crystals, as experimentally demonstrated in [82,83]. This result is not obvious and is crucial to figure out exactly where and how the ions are situated inside the network and the nature of their neighbours. Photoluminescence

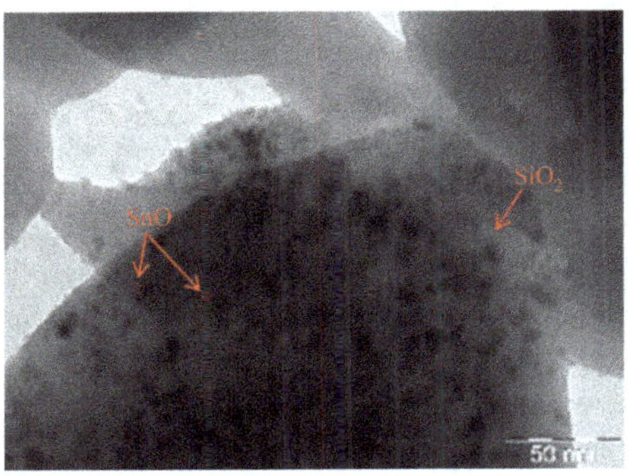

Figure 5.5 TEM micro-image of a $70SiO_2–30SnO_2:0.5Er^{3+}$ glass ceramics planar waveguide. The black dots are SnO_2 nanocrystals, and the light grey part is the SiO_2 matrix. Adapted from [73]

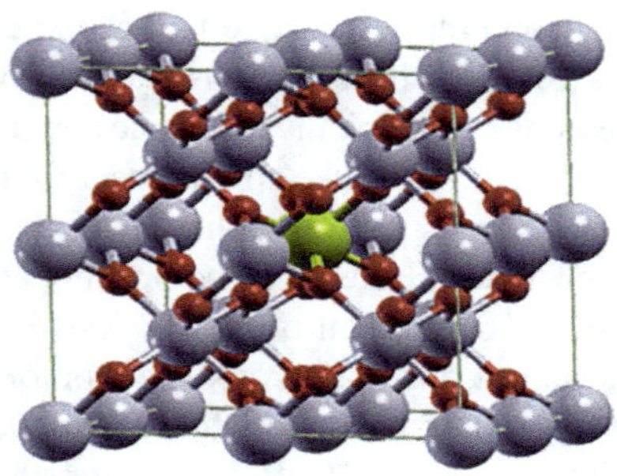

Figure 5.6 Rutile crystal structure of SnO_2 nanocrystal doped with the Er^{3+} ion. The central atom has been substituted by an Er atom (green), Sn atoms are in grey, and O in red. The figure is obtained using by XCrySDen program [81]

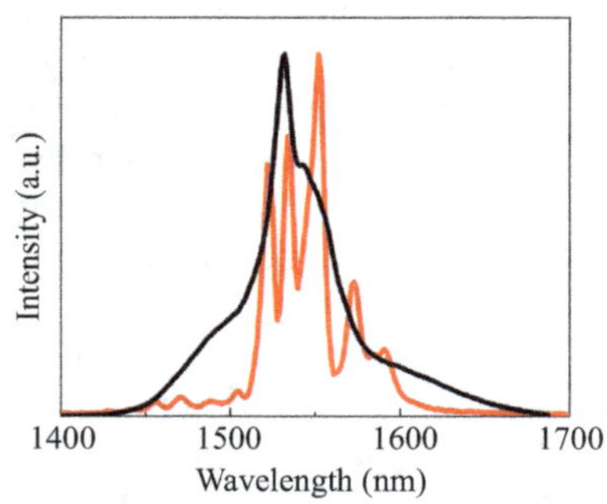

Figure 5.7 Room temperature luminescence spectrum of the $^4I_{13/2}\rightarrow{}^4I_{15/2}$ transition of Er^{3+} ions in amorphous (black spectrum) and crystalline (red spectrum) environment

spectra show narrow lines, and this feature is the fingerprint of a local crystal environment for the RE ion. The situation is clearly shown in Figure 5.7, where the $^4I_{13/2}\rightarrow{}^4I_{15/2}$ emission of Er^{3+} ions at 1,530 nm in amorphous and crystalline environment is reported. However, photoluminescence spectra do not give enough information to assess the local symmetry of the RE ion when incorporated in the SnO_2 nanocrystal. Two local different sites are allowed for the RE ion which are identified as one with symmetry C_{2h} and the other with symmetry D_{2h}.

In order to determine the position of the RE ion in the SnO_2 nanocrystals, the extended X-ray absorption fine structure (EXAFS) data reported in [82] have been numerically simulated obtaining the bond lengths Er–O and Er–Sn in the first and second coordination spheres for the two symmetries [83]. Table 5.2 shows the calculated and the EXAFS measured values.

Another useful characteristic of the SiO_2–SnO_2 system is its photorefractivity. It has been shown that the UV irradiation induces refractive index change allowing the direct writing of both channel waveguides and Bragg gratings [19,78,79]. As an

Table 5.2 Calculated and EXAFS obtained bond lengths values for the first and second coordination shell of erbium in an SnO_2 cell

	1° shell distance (Å)	**2° shell distance (Å)**
C_{2h}	1.6	1.7
D_{4h}	2.1	3.7
EXAFS^{+}	2.2	2.7

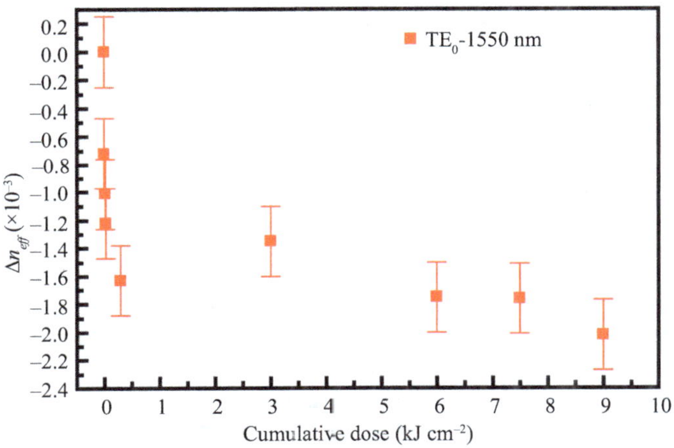

Figure 5.8 Refractive index variation of the TE_0 mode supported by an $80SiO_2$–$20SnO_2$:$0.5Er^{3+}$ planar waveguide at the wavelength of 1,550 nm as a function of cumulative dose of UV irradiation at 248 nm by a KrF excimer laser. Adapted from [73]

example, Figure 5.8 shows the negative effective index change of the fundamental mode TE_0 at the wavelength of 1,550 nm observed in an $80SiO_2$–$20SnO_2$:$0.5Er^{3+}$ planar waveguide fabricated by the sol–gel route [73]. Clearly, the demonstration of this effect paves the way to the fabrication of photonic structures such as gratings and channel waveguides just using UV irradiation.

Concerning the important role of RE sensitizer played by the SnO_2 nanocrystals, a demonstration of the process efficiency is given by the excitation spectrum reported in Figure 5.9, which shows the excitation spectrum of a $70SiO_2$–$30SnO_2$:$1.5Er^{3+}$ monolithic glass-ceramic obtained by detecting the $^4I_{13/2} \rightarrow {}^4I_{15/2}$ emission of Er^{3+} ions at 1,530 nm. The broad excitation band centred at 350 nm corresponds to the band gap of the SnO_2 nanocrystal. A comparison between the emission intensities obtained under direct and indirect excitation clearly indicates that the more efficient pumping schema is to excite in the bandgap of the SnO_2 nanocrystal, resulting in the effective sensitizing of the Er^{3+} luminescence from the metastable state $^4I_{13/2}$.

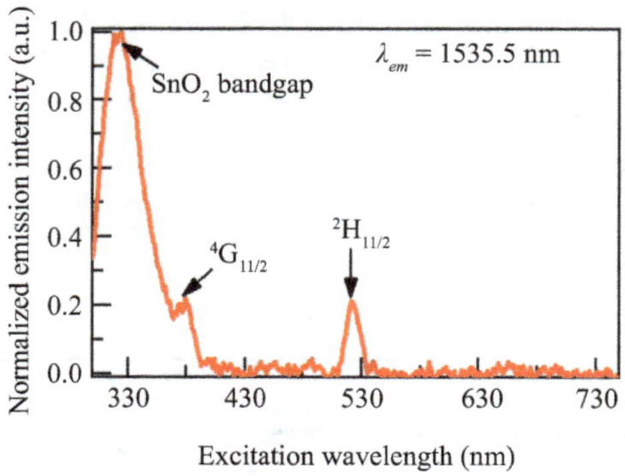

Figure 5.9 Room temperature excitation spectrum obtained by detecting the 1,535.5 nm emission of $70SiO_2$–$30SnO_2$:$1.5Er^{3+}$ monolithic glass-ceramic. The SnO_2 nanocrystals absorption band and the electronic transitions from the ground state $^4I_{15/2}$ of the Er^{3+} ion are indicated. Adapted from [73]

Other sensitizers have been recently developed always in the area of transparent glass ceramics. It has been demonstrated that the Tb^{3+} ion is an efficient donor for the Yb^{3+} ion in silica–hafnia glass ceramics with applications not only as downconverters but also as effective solar concentrators in the NIR spectral range [84,85]. Metallic nanoparticles have been also investigated as luminescence sensitizer. The aim of these researches was to study the possibility to increase the solar cells efficiency by combining the properties of RE co-doped materials with the optical sensitizing effects provided by silver doping [86–88].

Although the results are interesting from a scientific point of view, the doping with silver must take in account the presence of direct emission of different emitting species such as single Ag^+ ions, dimer, trimers, or small multimers of Ag^+. These emissions can act as noise in a photonic device and demand for a careful control of the fabrication protocols.

The peculiar properties of the SiO_2–SnO_2 glass ceramics suggest using this system to develop integrated structures such as lasers and amplifiers in planar configuration [89]. We have investigated the potential of an erbium-doped single-mode slab waveguide made of silica–tin-dioxide glass ceramics on silica as a possible candidate for achieving an optically pumped laser source emitting around 1.55 μm. Thanks to the high photorefractivity of the system, it is possible to write a Bragg grating, as it was shown in [89], such that a laser oscillation can be easily achieved for realistic values of material and structural parameters. With a suitable pumping schema, the active SiO_2–SnO_2 glass ceramics waveguide can operate as an effective optical amplifier. The feedback is given by a Fabry–Pérot resonant cavity.

Figure 5.10 presents various resonant structures that can be considered for the development of integrated active photonic structures. Considering the overall

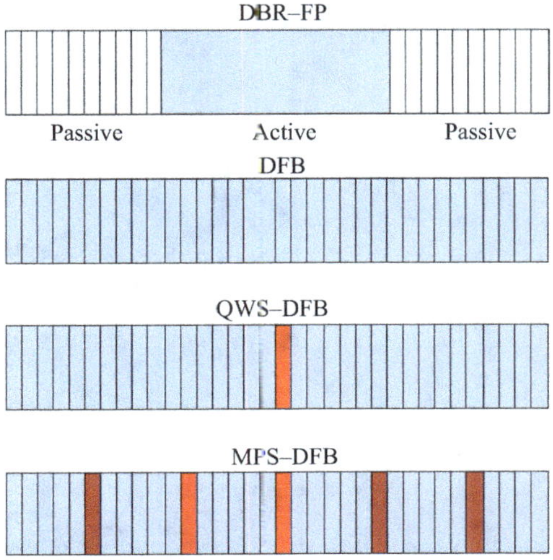

Figure 5.10 Some designed photonic structures based on erbium-activated SiO_2–SnO_2 glass-ceramic waveguides and photo-inscribed Bragg gratings: active Fabry–Pérot cavity with passive distributed Bragg reflectors (DBR–FP); active distributed-feedback structure (DFB); DFB with quarter-wave phase shift (QWS–DFB); DFB with multiple phase shifts distributed along the cavity (MPS–DFB). The red colour indicates the active medium

efficiency, the distributed-feedback (DFB) structure is more effective than the active Fabry–Pérot cavity with passive distributed Bragg reflectors (DBR–FP), where lengths devoted to amplification or feedback are more limited. Looking at the threshold gain, the DFB with quarter-wave phase shift (QWS–DFB) seems to be more appealing than the DFB one. In fact, the modelling discussed in [89] shows that the required threshold gain is two orders of magnitude lower, and the obtained laser emission is single frequency at the Bragg wavelength. This appears evident comparing the simulated transmission of DFB and QWS–DFB structures in the pump-off condition in Figures 5.11 and 5.12, respectively [90].

Moreover, the internal field presents a much higher peak in the close vicinity of the phase shift, ~40 dB higher than the emitted field at both ends. A DFB with multiple phase shifts (MPS–DFB structure) distributed along the cavity allows a better field 'flatness' but at the price of greater structural complexity. In the case of a reasonable total erbium concentration equal to 1.1859×10^{26} m^{-3}, and a stimulated emission cross section of $\sigma \approx 2 \times 10^{-25}$ m^2, the highest attainable material gain is $\sigma N_{tot} \approx 23$ m^{-1}, with $\alpha_{\max} \approx 9$ m^{-1} as modal gain. Over a typical length $L = 10$ mm, we get $(\alpha_{\max} L) \approx 0.09$: just enough to reach oscillation in the DFB structure, but comfortably higher, by two orders of magnitude, than the QWS-DFB threshold. In [89], it was demonstrated that erbium-activated SiO_2–SnO_2 glass-ceramic waveguides can achieve threshold pumping level just ~2% higher than

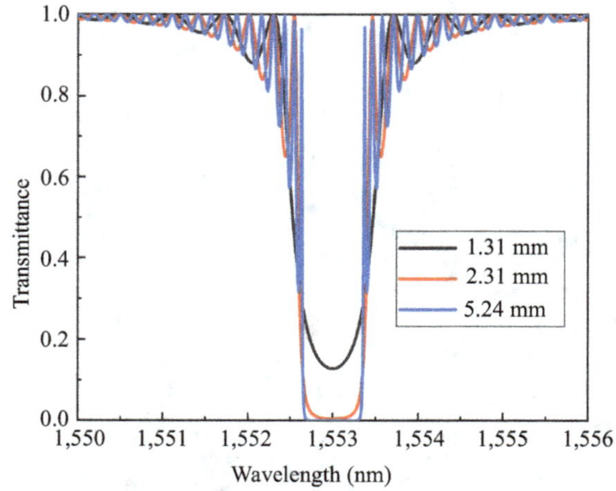

Figure 5.11 Transmission spectra of a DFB structure in the pump-off condition. The spectra are obtained with different lengths of the structure and thus different numbers of periods. The photorefractivity of $20SnO_2$–$80SiO_2$ planar waveguide was used: $\Delta n_{eff} = 1.1 \times 10^{-3}$ [90]

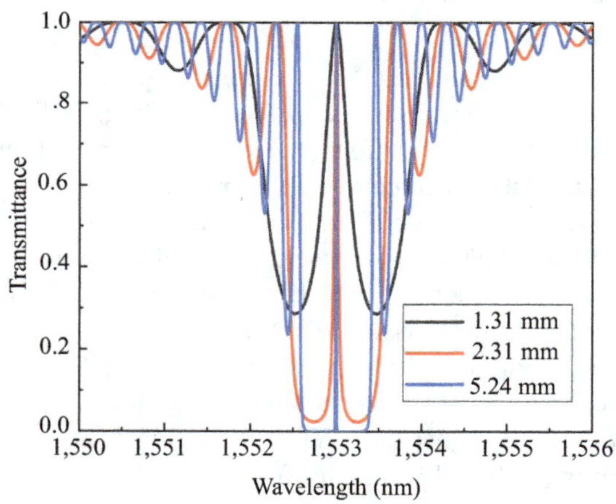

Figure 5.12 Transmission spectra of a QWS–DFB structure in the pump-off condition. The spectra are obtained with different lengths of the structure and thus different numbers of periods. The photorefractivity of $20SnO_2$–$80SiO_2$ planar waveguide was used: $\Delta n_{eff} = 1.1 \times 10^{-3}$. It must be noted that the central peak, thanks to the defect layer located in the middle of the structure, becomes more pronounced when the length of the structure increases [90]

transparency, which means low power consumption, steep slope, and good overall efficiency.

As discussed earlier, these techniques have driven the fabrication of low loss glass-ceramic planar-waveguides of various compositions, such as SiO_2–SnO_2 and SiO_2–HfO_2. The top-down approach uses an optimized thermal annealing protocol to nucleate nanocrystals in an amorphous matrix; it is of particular interest in the case of systems prepared by the sol–gel route. All the systems previously described were fabricated by the top-down technique.

The bottom-up approach is in principle more appealing but not so easy to apply if we look at the final result, i.e. transparency of the fabricated nanoceramic. The idea is to first synthesize the nanocrystals and then load them into the matrix. The evident advantage is that it is possible to start with a narrow size distribution of nanocrystals, characterized by the same structure and local symmetry for the incorporated RE ions. It is easy to imagine the advantages, in term of RE concentration, effective absorption and emission cross sections, and luminescence quantum yield, when the RE ion is substitutional and not interstitial in the nanocrystal. Unfortunately, the coalescence of the nanopowders and the difficulties related to the appropriate solution preparation for the dip-coating process, but also for the incorporation of the nanoparticles in polymeric matrices, make the bottom-up approach not very common. However, some interesting results have been obtained using this approach; for instance, it has been successfully employed to strongly reduce the propagation losses in silica–hafnia waveguides [91]. The HfO_2 nanoparticles were obtained by reflux technique and prepared in a colloidal suspension. A solution of TEOS with a suitable molar ratio of Si/Hf and doped by Er^{3+} ions using $ErNO_3{\cdot}5H_2O$ was also prepared. Finally, after separation from the colloidal suspension, the nanoparticles were added to the dipping solution. The final result is a nanoceramic planar waveguide with a homogeneous distribution of Er^{3+}-activated HfO_2 nanocrystals and propagation loss as low as 0.3 dB cm^{-1} at 1.54 μm. Figure 5.13 shows the high-resolution TEM micro-image of a silica–hafnia glass-ceramic fabricated by the bottom-up approach.

Another interesting system, also prepared by the bottom-up approach, on the basis of tetraphosphate nanocrystals embedded in silica–hafnia matrix, has been reported [92]. The precipitation method was employed for the preparation of nanoparticles of lithium lanthanum tetraphosphates doped with different concentration of Eu^{3+} ions. The Eu^{3+}:$LiLa(PO_3)_4$ nanocrystals were loaded in a sol–gel-derived silica–hafnia amorphous film, resulting in transparent glass ceramics after annealing at 900°C [92]. The film transparency, higher than 90% in the whole visible range, allowed to perform photoluminescence measurements. The typical Eu^{3+} emission bands were observed, but the interesting point is that in the glass-ceramic system, the intensity of the $^5D_0 \rightarrow {}^7F_2$ transition was strongly enhanced. The distortion parameter was 3.5 compared with the 1.2 value measured for the Eu^{3+}:$LiLa(PO_3)_4$ nanocrystals as prepared. This behaviour indicates that the local crystal field of the RE ion is much more distorted in the glass-ceramic environment. This result is important because we can exploit the nanostructure to tailor the luminescence properties.

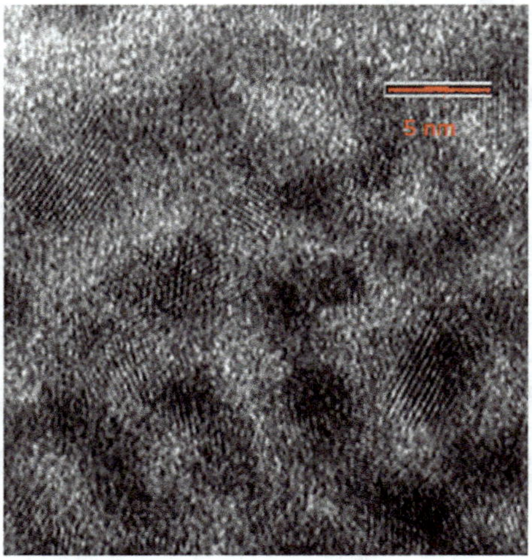

Figure 5.13 High-resolution TEM image of a silica–hafnia glass-ceramic waveguide fabricated by bottom-up approach. The HfO_2 nanocrystals with a size of ~3.5 nm are homogeneously dispersed in the amorphous matrix, and their crystalline planes are clearly visible

5.3 Summary

In this chapter, we have discussed fabrication techniques and optical, spectroscopic, morphological, and structural properties of RE-activated glasses and glass ceramics, having in mind their application in integrated optics. Concerning photonic glasses, novel materials have been presented, and attention has been paid to the characterization techniques that allow the determination of the parameters crucial for developing effective pumping schemas. Moreover, the mechanisms related to the photoluminescence enhancement have been addressed, and some mitigation strategies presented. The second part of the chapter has been focused on transparent glass ceramics and on the important role of nanocrystals in the enhancement of the energy transfer mechanism, in the *RE* solubility, as well as in the reduction of non-radiative relaxation process. The important physical problem of transparency has been also mentioned. On the basis of the recent results and on the current literature which shows a significant technical effort, with continuous advances in materials research and fabrication protocols, novel applications can be easily predicted for the photonic glass ceramics.

The novel system of silica–tin-dioxide has been presented, its photorefractivity demonstrated, and the role of SnO_2 nanocrystals as efficient *RE* luminescence sensitizers clearly proved. Looking at the state of art in glass photonics for integrated optics, it is evident that, although wet chemical and physical technologies allow to fabricate efficient materials, there is an important work to do in optimizing the fabrication protocols, in particular for a controlled embedding of the *RE* ions in the nanocrystals. The development of more efficient glass and glass-ceramic host matrices is an exciting challenge. Last but not the least, an important point is the feasibility of easy and low-cost integrated optics devices incorporating the highest number of functionalities as well the mechanical flexibility, while maintaining the optical features of the rigid counterpart [93–95].

Acknowledgements

The research activity was performed in the framework of the projects CNR-PAS 'Flexible Photonics' (2020–2021), PRIN 2019 'NOMEN' (2020–2022). WB and MF acknowledge the support of CNR-STM – Short Term Mobility program 2019–2020.

References

[1] Ferrari M. and Righini G.C. 'Glass-ceramic materials for guided-wave optics'. *Int. J. Appl. Glass Sci.* 2015; **6**: 240–248, doi: 10.1111/ijag.12129.

[2] Koechner W. *Solid-State Laser Engineering* (Berlin, Springer, 1988), pp. 28–78.

[3] Zia N., Viheriälä J., Koskinen R., Koskinen M., Suomalainen S., and Guina M. 'Fabrication and characterization of broadband superluminescent diodes for 2 μm wavelength'. *Proc. SPIE* 2016; **9768**: 97680Q, doi: 10.1117/12.2209720.

[4] Minati L., Chiappini A., Armellini C., *et al.* 'Gold nanoparticles 1D array as mechanochromic strain sensor'. *Mat. Chem. Phys.* 2017; **192**: 94–99, doi: 10.1016/j.matchemphys.2017.01.073.

[5] Righini G.C., Caldiño U., Falcony C., Ferrari M., and Pelli S. 'Rare-earth phosphors for the control of WLED's color output: State of the art'. *Proc. ICTON 2014*, Graz, Austria, July 2014. IEEE; 2014, We.A5.1, doi: 10.1109/ICTON.2014.6876519.

[6] Bouajaj A., Belmokhtar S., Britel M. R., *et al.* 'Tb^{3+}/Yb^{3+} codoped silica-hafnia glass and glass-ceramic waveguides to improve the efficiency of photovoltaic solar cells'. *Opt. Mater.* 2016; **52**: 62–68, doi: 10.1016/j.optmat.2015.12.013.

[7] Lousteau J., Boetti N., Chiasera A., *et al.* 'Er^{3+} and Ce^{3+} co-doped tellurite optical fiber for lasers and amplifiers in the near infrared wavelength region: Fabrication, optical characterization and prospects'. *IEEE Photonics J.* 2012; **4**: 194–204, doi: 10.1109/JPHOT.2011.2181974.

[8] Chiasera A., Meroni C., Scotognella F., *et al.* 'Coherent emission from fully Er^{3+} doped monolithic 1-D dielectric microcavity fabricated by rf-sputtering'. *Opt. Mater.* 2019; **87**:107–111, doi: 10.1016/j.optmat.2018.04.057.

[9] Ghazi Sarwat S. 'Materials science and engineering of phase change random access memory'. *Mat. Sci. Technol.* 2016; **33**: 1890–1906, doi: 10.1080/02670836.2017.1341723.

[10] Vázquez M.R., Sotillo B., Rampini S., *et al.* 'Femtosecond laser inscription of nonlinear photonic circuits in gallium lanthanum sulphide glass'. *J. Phys. Photonics* 2019; **1**: 015006.

[11] Righini G.C., Arnaud C., Berneschi S., *et al.* 'Integrated optical amplifiers and microspherical lasers based on erbium-doped oxide glasses'. *Opt. Mater.* 2005; **27**: 1711–1717, doi: 10.1016/j.optmat.2004.11.041.

[12] Quandt A., Ferrari M., and Righini G.C. 'Towards integrated nanoelectronic and photonic devices' in Baleanu D., Guvenc Z.B., and Tenreiro Machado J.A. (eds.). *New Trends in Nanotechnology and Fractional Calculus Applications* (Dordrecht, Springer, 2010), pp. 25–41, doi: 10.1007/978-90-481-3293-5.

[13] Ainslie B. 'A review of the fabrication and properties of erbium-doped fibers for optical amplifiers'. *J. Lightwave Technol.* 1991; **9**: 220–227, doi: 10.1109/50.65880.

[14] Clayton R. *Integrated Opto-Electronics*. 2001. Available from https://csmantech.org/OldSite/Digests/2001/PDF/6_1_Clayton_V2.pdf [Accessed 3 March 2020].

[15] DeCorby R. and Irannejad M. *Glasses for Photonic Integration* in Kasap S., Capper P. (eds.). (Cham, Springer, 2017).

[16] Auzel F. and Goldner P. 'Towards rare-earth clustering control in doped glasses' *Opt. Mater.* 2001; **16**: 93–103, doi: 10.1016/S0925-3467(00)00064-1.

[17] Monteil A., Chaussedent S., Alombert-Goget G., *et al.* 'Clustering of rare earth in glasses, aluminium effect: Experiments and modelling'. *J. Non-Cryst. Solids* 2004; **348**: 44–50, doi: 10.1016/j.jnoncrysol.2004.08.124.

[18] Ferrari M. and Righini G.C. 'Rare-earth-doped glasses for integrated optical amplifiers' in Hussain N.S. and Da Silva Santos J.D. (eds.). *Physics and Chemistry of Rare-Earth Ions Doped Glasses* (Bach, Switzerland, Trans Tech Publ., 2008), pp. 71–120.

[19] Lukowiak A., Chiasera A., Chiappini A., Righini G.C., and Ferrari M. 'Active sol-gel materials, fluorescence spectra and lifetimes' in Klein L., Aparicio M., and Jitianu A. (eds.). *Handbook of Sol-Gel Science and Technology* (Berlin, Springer, 2017), pp. 1–43.

[20] Quandt A., Ferrari M., and Righini G.C. 'Advancement of glass-ceramic materials for photonic applications' in Mishra A.K. (ed.). *Sol-Gel Based Nanoceramic Materials: Preparation, Properties and Applications* (Berlin, Springer, 2017), pp. 133–155.

[21] Bhaktha S.N.B., Beclin F., Bouazaoui M., *et al.* 'Enhanced fluorescence from Eu^{3+} in low-loss silica glass-ceramic waveguides with high SnO_2 content'. *App. Phys. Lett.* 2008; **93**: 211904–1/3, doi: 10.1063/1.3037224.

[22] Chiappini A., Pasquardini L., Nodehi S., *et al.* 'Fluorescent aptamer immobilization on inverse colloidal crystals'. *Sensors* 2018; **18**: 432611, doi: 10.3390/s18124326.

[23] Sayginer O., Chiasera A., Zur L., *et al.* 'Fabrication, modelling and assessment of hybrid 1-D elastic Fabry Perot microcavity for mechanical sensing applications'. *Ceramics Intern.* 2019; **45**: 7785–7788, doi: 10.1016/j.ceramint.2019.01.083.

[24] Wang J.S., Vogel E.M, and Snitzer E. 'Tellurite glass: A new candidate for fiber devices'. *Opt. Mater.* 1994; **3**: 187–203, doi: 10.1016/0925-3467(94)90004-3.

[25] Rolli R., Montagna M., Chaussedent S., Monteil A., Tikhomirov V.K., and Ferrari M. 'Erbium-doped tellurite glasses with high quantum efficiency and broadband stimulated emission cross section at 1.5 μm'. *Opt. Mater.* 2003; **21**: 743–748, doi: 10.1016/S0925-3467(02)00092-7.

[26] Albalawi A., Brilliant C., Chiasera A., *et al.* 'Analytical modelling of Tm-doped tellurite glass including cross relaxation process'. *Opt. Mater.* 2019; **87**: 29–34, doi: 10.1016/j.optmat.2018.06.037.

[27] Narazaki A., Tanaka K., Hirao K., and Soga N. 'Induction and relaxation of optical second-order nonlinearity in tellurite glasses' *J. Appl. Phys.* 1999; **85**: 2046–2051, doi: 10.1063/1.369500.

[28] Adam J.L. 'Lanthanides in non-oxide glasses'. *Chem. Rev.* 2002; **102**: 2461–2476, doi: 10.1021/cr010305b.

[29] Boulard B., Dieudonné B., Gao Y., Chiasera A., and Ferrari M. 'Up-conversion visible emission in rare-earth doped fluoride glass waveguides'. *Opt. Eng.* 2014; **53**: 071814, doi:10.1117/1.OE.53.7.071814.

[30] Haquin H., Fonteneau G., and Adam J-L. 'Recent developments in ion-exchanged fluoride glass planar waveguides'. *J. Non-Cryst. Solids* 2003; **326–327**: 460–463, doi: 10.1016/S0022-3093(03)00453-8.

[31] Ganser D., Gottmann J , Mackens U., and Weichmann U. 'Pulsed laser deposition of fluoride glass thin films'. *Appl. Surf. Sci.* 2010; **257**: 954–959, doi: 10.1016/j.apsusc.2010.07.100.

[32] Vasilief I., Guy S., Jacquier B., *et al.* 'Propagation losses and gain measurements in erbium doped fluoride glass channel waveguides by use of a double-pass technique'. *Appl. Opt.* 2005; **44**: 4678–4683, doi: 10.1364/AO.44.004678.

[33] Maalej O., Boulard B., Dieudonné B., Ferrari M., Dammak M., and Dammak M. 'Downconversion in Pr^{3+}-Yb^{3+} co-doped ZBLA fluoride glass'. *J. Lumin.* 2015; **161**: 198–201, doi:10.1016/j.jlumin.2015.01.018.

[34] Seddon A.B., Tang Z., Furniss D., Sujecki S., and Benson T.M. 'Progress in rare-earth-doped mid-infrared fiber lasers'. *Opt. Express* 2010; **18**: 26704–26719, doi: 10.1364/OE.18.026704.

[35] Seddon A.B. 'A prospective for new mid-infrared medical endoscopy using chalcogenide glasses'. *Inter. J. Appl. Glass Sci.* 2011; **2**: 177–191, doi: 10.1111/j.2041-1294.2011.00059.x.

[36] Seddon A.B. 'Mid-infrared (IR) – A hot topic: The potential for using mid-IR light for non-invasive early detection of skin cancer in vivo'. *Phys. Status Solidi B* 2013; **250**: 1020–1027, doi: 10.1002/pssb.201248524.

[37] Chryssou C.E., Di Pasquale F., and Pitt C.W. 'Er^{3+}-doped channel optical waveguide amplifiers for WDM systems: A comparison of tellurite, alumina and Al/P silicate materials'. *IEEE J. Sel. Top. Quantum Electron.* 2000; **6**: 114–120.

[38] Gonçalves R.R., Carturan G., Zampedri L., *et al.* 'Infrared-to-visible CW frequency upconversion in erbium activated silica–hafnia waveguides

prepared by sol–gel route'. *J. Non-Cryst. Solids* 2003; **322**: 306–310, doi: 10.1016/S0022-3093(03)00220-5.

[39] Chiasera A., Armellini C., Bhaktha S.N.B., *et al.* 'Er^{3+}/Yb^{3+}-activated silica–hafnia planar waveguides for photonics fabricated by rf-sputtering'. *J. Non-Cryst. Solids* 2009; **355**: 1176–1179, doi: 10.1016/j.jnoncrysol.2008.11.039.

[40] Hehlen M.P., Cockroft N.J., Gosnell T.R., and Bruce A.J., 'Spectroscopic properties of Er^{3+}- and Yb^{3+}-doped soda-lime silicate and aluminosilicate glasses' *Phys. Rev. B* 1997; **56**: 9302–9318, doi: 10.1103/PhysRevB.56.9302.

[41] Chiasera A., Tosello C., Moser E., *et al.* 'Er^{3+}/Yb^{3+}-activated silica-titania planar waveguides for EDPWAs fabricated by rf sputtering'. *J. Non-Cryst. Solids* 2003; **322**: 289, doi: 10.1016/S0022-3093(03)00217-5.

[42] Peters P.M., Funk D.S., Peskin A.P., *et al.* 'Ion-exchanged waveguide lasers in Er^{3+}/Yb^{3+} codoped silicate glass'. *Appl. Opt.* 1999; **38**: 6879–6886, doi: 10.1364/AO.38.006879.

[43] Tanabe S., Kouda T., and Hanada T. 'Excited energy migration and fluorescence decay in Yb-doped and Yb/Pr codoped tellurite glasses'. *Opt. Mater.* 1999; **12**: 35–40, doi: 10.1016/S0925-3467(98)00056-1.

[44] Philipps J.F., Töpfer T., Ebendorff-Heidepriem H., Ehrt D., and Sauerbrey R. 'Spectroscopic and lasing properties of Er^{3+}:Yb^{3+}-doped fluoride phosphate glasses'. *Appl. Phys. B* 2001; **72**: 399–405, doi: 10.1007/s003400100515.

[45] Lukowiak A., Stefanski M., Ferrari M., and Strek W. 'Nanocrystalline lanthanide tetraphosphates: Energy transfer processes in samples co-doped with Pr^{3+}/Yb^{3+} and Tm^{3+}/Yb^{3+}'. *Opt. Mater.* 2017; **74**: 159–165, doi: 10.1016/j.optmat.2017.03.025.

[46] Zur L., Armellini C., Belmokhtar S., *et al.* 'Comparison between glass and glass-ceramic silica-hafnia matrices on the down-conversion efficiency of Tb^{3+}/Yb^{3+} rare earth ions'. *Opt. Mater.* 2019; **87**: 102–106, doi: 10.1016/j.optmat.2018.05.008.

[47] Strohhöfer C., Kik P.G., and Polman A. 'Selective modification of the Er^{3+} $^4I_{11/2}$ branching ratio by energy transfer to Eu^{3+}'. *J. Appl. Phys.* 2000; **88**: 4486–4490, doi: 10.1063/1.1311824.

[48] Strohhöfer C. and Polman A. 'Enhancement of Er^{3+} $^4I_{13/2}$ population in Y_2O_3 by energy transfer to Ce^{3+}'. *Opt. Mater.* 2001; **17**: 445–451, doi: 10.1016/S0925-3467(01)00103-3.

[49] Mattarelli M., Montagna M., Zampedri L., *et al.* 'Effect of Eu^{3+} and Ce^{3+} codoping on the relaxation of Er^{3+} in silica-hafnia and tellurite glasses'. *Phys. Status Solidi C* 2007; **4**: 793–796, doi: 10.1002/pssc.200673774.

[50] Boulard B., Alombert G., Savelii I., *et al.* '$Er^{3+}/Yb^{3+}/Ce^{3+}$ co-doped fluoride glass ceramics waveguides for application in the 1.5 μm telecommunication window'. *Adv. Sci. Technol.* 2010; **71**: 16–21, doi: 10.4028/www.scientific.net/AST.71.16.

[51] Soltys M., Górny A., Ferrari M., *et al.* 'Near-infrared emission in barium gallo-germanate glasses co-doped with Ce^{3+} and Pr^{3+} for broadband optical amplifiers'. *Proc. SPIE* 2018; **10683**: 106832N-1/5, doi: 10.1117/12.2306373.

[52] Berneschi S., Soria S., Righini G.C., *et al.* 'Rare-earth-activated glass ceramic waveguides'. *Opt. Mater.* 2010; **32**: 1644–1647, doi:10.1016/j.optmat.2010.04.035.

[53] Gan F. and Xu L. (eds.). *Photonic Glasses* (Singapore, World Scientific, 2006).

[54] Martellucci S., Chester A.N., and Bertolotti M. (eds.). *Advances in Integrated Optics* (New York, Springer, 1994).

[55] Peled A., Chiasera A., Nathan M., Ferrari M., and Ruschin S. 'Monolithic rare-earth doped sol-gel tapered rib waveguide laser'. *Appl. Phys. Lett.* 2008; **92**: 221104, doi: 10.1063/1.2936961.

[56] Yamane M. and Asahara Y. (eds.). *Glasses for Photonics*. (Cambridge, Cambridge University Press, 2000)

[57] Azkargorta J., Iparraguirre I., Balda R., and Fernández J. 'On the origin of bichromatic laser emission in Nd^{3+} doped fluoride glasses'. *Opt. Express* 2008; **16**: 11894–11906, doi: 10.1364/OE.16.011894.

[58] Campbell J.H. and Suratwala T.I. 'Nd-doped phosphate glasses for high-energy/high peak-power lasers'. *J. Non-Cryst. Solids* 2000; **263&264**: 318–341, doi: 10.1016/S0022-3093(99)00645-6.

[59] Neelima G., Krishnaiah K.V., Ravi N., Suresh K., Tyagarajan K., and Prasad T.J. 'Investigation of optical and spectroscopic properties of neodymium doped oxyfluoro-titania-phosphate glasses for laser applications'. *Scr. Mater.* 2019; **162**: 246–250, doi: 10.1016/j.scriptamat.2018.11.018.

[60] Yan Y.C., Faber A.J., de Waal H., Kik P.G., and Polman A. 'Erbium-doped phosphate glass waveguide on silicon with 4.1 dB/cm gain at 1.535 μm'. *Appl. Phys. Lett.* 1997; **71**: 2922–2924, doi: 10.1063/1.120216.

[61] Li G.S., Zhang C.M., Zhu P.F., Jiang C., Song P., and Zhu K. 'Broadband near-infrared emission in Pr^{3+}-Er^{3+} codoped phosphate glasses for optical amplifiers'. *Ceram. Intern.* 2016; **42**: 5558–5561, doi: 10.1016/j.ceramint.2015.12.026.

[62] Ismail M.M, Batisha I.K., Zur L., Chiasera A., Ferrari M., and Lukowiak A. 'Optical properties of Nd^{3+}-doped phosphate glasses'. *Opt. Mater.* 2020; **99**: 109591, doi: 10.1016/j.optmat.2019.109591.

[63] Judd B.R. 'Optical absorption intensities of rare earth ions'. *Phys. Rev.* 1962; **127**: 750–761, doi: 10.1103/PhysRev.127.750.

[64] Ofelt G.S. 'Intensities of crystal spectra of rare-earth ions'. *Chem. Phys.* 1962; **37**: 511–520, doi: 10.1063/1.1701366.

[65] Mohan S., Thind K.S., and Sharma G. 'Effect of Nd^{3+} concentration on the physical and absorption properties of sodium-lead-borate glasses'. *Braz. J. Phys.* 2007; **37**: 1306–1313, doi: 10.1590/S0103-97332007000800019.

[66] *RELIC Software Package*. Available from https://www.lanl.gov/projects/feynman-center/deploying-innovation/intellectual-property/software-tools/relic/index.php [Accessed 3 March 2020].

[67] Hehlen M.P., Brik M.G., and Krämer K.W. '50th anniversary of the Judd–Ofelt theory: An experimentalist's view of the formalism and its application'. *J. Lumin.* 2013; **136**: 221–239, doi: 10.1016/j.jlumin.2012.10.035.

[68] de Pablos-Martin A., Ferrari M., Pascual M.J., and Righini G.C. 'Glass-ceramics: A class of nanostructured materials for photonics'. *Riv Nuovo Cimento* 2015; **38**: 311–369, doi: 10.1393/ncr/i2015-10114-0.

[69] Champagnon B., Chemarin C., Duval E., and Le Parc R. 'Glass structure and light scattering'. *J. Non-Cryst. Solids* 2000; **274**: 81–86, doi: 10.1016/S0022-3093(00)00207-6.

[70] Sagaguchi S. 'Evaluation of extrinsic scattering loss due to crystallization in fluoride optical fibers'. *J. Lightwave Technol.* 1993; **11**: 187–191, doi: 10.1109/50.212525.

[71] Tick P.A. 'Are low-loss glass–ceramic optical waveguides possible?'. *Opt. Lett.* 1998; **23**: 1904–1905, doi: 10.1364/OL.23.001904.

[72] Mattarelli M., Montagna M., and Verrocchio P. 'Ultratransparent glass ceramics: The structure factor and the quenching of the Rayleigh scattering'. *Appl. Phys. Lett.* 2007; **91**: 061911, doi: 10.1063/1.2768642.

[73] Tran T.N.L. *Tin dioxide based photonic glass-ceramics*. PhD thesis, University of Trento (Trento, 2019).

[74] Righini G.C. and Ferrari M. 'Photoluminescence of rare-earth-doped glasses'. *Riv Nuovo Cimento* 2005; **28**: 1–53, doi: 10.1393/ncr/i2006-10010-8.

[75] Zur L., Tran T.N.L., Meneghetti M., *et al.* 'Tin-dioxide nanocrystals as Er^{3+} luminescence sensitizers: Formation of glass-ceramics thin films and their characterization'. *Opt. Mater.* 2017; **63**: 95–100, doi: 10.1016/j.optmat.2016.08.041.

[76] Lukowiak A., Zur L., Tran T.N.L., *et al.* 'Sol–gel-derived glass-ceramic photorefractive films for photonic structures'. *Crystals* 2017; **7**: 61, doi: 10.3390/cryst7020061.

[77] Tran T.N.L., Massella D., Zur L., *et al.* 'SiO_2-SnO_2:Er^{3+} glass-ceramic monoliths'. *Appl. Sci.* 2018; **8**: 1335, doi: 10.3390/app8081335.

[78] Zur L., Tran T.N.L., Massella D., *et al.* 'SiO_2-SnO_2 transparent glass ceramics activated by rare earth ions'. *Proc. SPIE* 2019; **10914**: 1091411, doi: 10.1117/12.2507214.

[79] Zur L., Tran T.N.L., Meneghetti M., and Ferrari M. 'Sol-gel derived SnO_2-based photonic systems' in Klein L., Aparicio M., and Jitianu A. (eds.). *Handbook of Sol-Gel Science and Technology* (Basel, Springer, 2017), pp. 1–19.

[80] Tran L.T.N., Zur L., Massella D., *et al.* 'SiO_2-SnO_2:Er^{3+} transparent glass-ceramics: Fabrication and photonic assessment'. *Proc. SPIE* 2018; **10683**: 106832C.

[81] Kokalj A. 'Computer graphics and graphical user interfaces as tools in simulations of matter at the atomic scale'. *Comp. Mat. Sci.* 2003; **28**: 155–168, doi: 10.1016/S0927-0256(03)00104-6.

[82] Van T.T.T., Turrell S., Capoen B., *et al.* 'Erbium-doped tin-silicate sol-gel-derived glass-ceramic thin films: Effect of environment segregation on the Er^{3+} emission'. *Sci. Adv. Mater.* 2015; **7**: 301–308, doi: 10.1166/sam.2015.2022.

[83] Meneghetti M. *Tin dioxide nanocrystals as Er^{3+} luminescence sensitizers: sol-gel fabrication of transparent glass-ceramics and their characterization*. MSc thesis, University of Trento (Trento, 2017).

[84] Bouajaj A., Belmokhtar S., Britel M.R., *et al.* 'Tb^{3+}/Yb^{3+} codoped silica-hafnia glass and glass-ceramic waveguides to improve the efficiency of photovoltaic solar cells'. *Opt. Mater.* 2016; **52**: 62–68, doi: 10.1016/j.optmat.2015.12.013.

[85] Enrichi F., Armellini C., Belmokhtar S., *et al.* 'Visible to NIR down-conversion process in Tb^{3+}-Yb^{3+} codoped silica-hafnia glass and glass-ceramic sol-gel waveguides for solar cells'. *J. Lumin.* 2018; **193**: 44–50, doi: 10.1016/j.jlumin.2017.08.027.

[86] Enrichi F., Armellini C., Battaglin G., *et al.* 'Silver doping of silica-hafnia waveguides containing Tb^{3+}/Yb^{3+} rare earths for downconversion in PV solar cells'. *Opt. Mater.* 2016; **60**: 264–269, doi: 10.1016/j.optmat.2016.07.048.

[87] Enrichi F., Cattaruzza E., Ferrari M., *et al.* 'Ag-sensitized Yb^{3+} emission in glass-ceramics'. *Micromachines* 2018; **9**: 380, doi: 10.3390/mi9080380.

[88] Mattarelli M., Montagna M., Vishnubathla K., Chiasera A., Ferrari M., and Righini G.C. 'Mechanisms of silver to erbium energy transfer in silicate glasses'. *Phys. Rev. B* 2007; **75**: 125102.

[89] Boucher, Y.G., Zur, L., and Ferrari, M. 'Modal properties of an erbium-doped asymmetric single-mode slab waveguide in the glass-ceramics SnO_2–SiO_2 system'. *Opt. Mater.* 2019; **87**: 90–93, doi: 10.1016/j.optmat.2018.05.032.

[90] Massella D. *SiO_2-SnO_2:Er^{3+} planar waveguide: Computational and Optical assessment*. MSc thesis, Department of Physics, University of Trento (Trento, 2018).

[91] Jestin Y., Armellini C., Chiasera A., *et al.* 'Low-loss optical Er^{3+}-activated glass-ceramics planar waveguides fabricated by bottom-up approach'. *Appl. Phys. Lett.* 2007; **91**: 071909, doi: 10.1063/1.2771537.

[92] Lukowiak A., Wiglusz R.J., Chiappini A., *et al.* 'Structural and spectroscopic properties of Eu^{3+}-activated nanocrystalline tetraphosphates loaded in silica–hafnia thin film'. *J. Non-Cryst. Solids* 2014; **401**: 32–35, doi: 10.1016/j.jnoncrysol.2013.12.019.

[93] Li L., Lin H., Qiao S., *et al.* 'Monolithically integrated stretchable photonics'. *Light: Sci. Appl.* 2018; **7**: 17138.

[94] Chiasera A., Sayginer O., Iacob E., *et al.* 'Flexible photonics: RF-sputtering fabrication of glass-based systems operating under mechanical deformation conditions'. *Accepted for Presentation at SPIE Photonics Europe, Fiber Lasers and Glass Photonics: Materials through Applications*; Strasbourg, France, 2020; paper 11357-3.

[95] Szczurek A., Chiasera A., Ferrari M., Krzak J., and Lukowiak A. 'Flexible sol-gel coatings on polymeric materials'. *Accepted for Presentation at SPIE Photonics Europe, Fiber Lasers and Glass Photonics: Materials through Applications*; Strasbourg, France, 2020; paper 11357-4.

Chapter 6
Lithium niobate integrated optics

Cinzia Sada[1]

The main driving force behind lithium niobate ($LiNbO_3$, LN)-integrated optics (IO) was surely coming from the domain of the optical communications (OCs) and the pushing demand of greater bandwidth as well as the ability to route signals without complicated electronic interfaces. OC applications were therefore the first to exploit $LiNbO_3$ [1–3], leading later to great advantages also in other fields such as sensors development. As a matter of fact, lithium niobate benefits from excellent properties such as high optical transparency in a wide spectral range (0.33–4.6 μm), high refractive index and birefringence ($n_o \approx 2.28$, $n_e \approx 2.20$ at 0.63 μm), high electro-optic coefficients ($r_{33} \approx 30.8 \times 10^{-12}$ m/V at $\lambda = 0.63$ μm) and piezo-electric ones, chemical resistance and even biocompatibility. Furthermore, lithium niobate has a high second-order optical nonlinearity ($d_{33} \approx -33$ pm/V at $\lambda = 1.064$ μm) enabling parametric wavelength conversion and optical signal generation. Its properties can be tailored by suitable doping, either locally or in the bulk, to increase its resistance to the optical damage or its photorefractivity, respectively, to host active ions or to produce optical waveguides as well.

These features allowed one to integrate in this material complex IO circuits, including fibre pigtailed light sources, beam splitters, optical modulation systems, polarizers and polarizer combiners, if needed, as well as signal detectors, such as a photodiode or laser-fibre detector system, dependently on the application. Although widely investigated since the last 20 years, $LiNbO_3$ has still continued to attract a great attention in developing optical components for IO systems, surely driven also by the demand of high-speed optical modulators [4], but not limited to that. Since optical waveguides are the building blocks in IO, great efforts have been focused on the following four main areas:

1. to develop new, efficient, reproducible and low-cost methods for waveguides fabrication to increase the performances of these building blocks and overcome the actual constraints to achieve higher speed, higher optical damage resistance;
2. to investigate co-doping to inherently improve the material properties on which realizing optical waveguides;

[1]Physics and Astronomy Department, University of Padova, Padova, Italy

3. to combine different materials on lithium niobate and achieve hybrid configurations where $LiNbO_3$ limits could be therefore bypassed with smart solutions and
4. to investigate new fields of application of optical waveguides in which IO can be combined with other technologies. This is the typical case of opto-microfluidics, opto-biophysics and integrated optical sensors, just to cite the most recent ones.

In this chapter, a summary is presented of the recent advances in the exploitation of lithium-niobate-IO, with a highlight on the progress made in optical waveguides and on some main applications where $LiNbO_3$ played a key role and is going to open new perspectives.

6.1 Integrated optical waveguides

Local doping by thermal diffusion and proton exchange (PE) has been widely investigated for the fabrication of IO devices in LN, and up to now, they are still keeping the track of the most widely exploited techniques in industries thanks to their scalability and batch processing compatibility. However, the constant need of better performances has continued to push the research of new solutions that could provide a real advance with respect to the current state of the art. This is especially demanded for those aspects that are a drawback, such as the light-induced charge transport phenomena and the poor compatibility of the waveguide fabrication technology with ferroelectric domain engineering. Recently, the combination of several techniques has paved the way of great improvements in the waveguiding properties especially for what concerns the optical confinement and mode overlap.

In this section, the most recent advances in the optical waveguides performances and relative application will be briefly summarized.

6.1.1 Ti-in-diffused optical waveguides and co-doping

Ti-in-diffusion is one of the most widespread LN technologies to prepare optical waveguides because it guarantees good light confinement on both ordinary and extraordinary refractive indices, allowing both TE and TM propagation modes and preserving to a good extent the electro-optical properties of the material. Ti-in-diffusion is compatible with standard photolithographic techniques to obtain channel waveguides and more complicated optical circuit configurations such as optical switches, Mach–Zehnder interferometers and optical modulators, couplers with high coupling efficiencies and large bandwidth around 1.55 μm [2]. The Ti-in-diffusion process has reached a mature stage, well controlled and widely optimized from the technological point of view. Some selected measured data of the maximum refractive index jump from different published works were reviewed in [3] and references therein quoted.

For a conventional Ti:LN waveguide, which is usually fabricated by the diffusion of 6–10-μm wide, ~100-nm thick Ti strip at 1,030°C–1,060°C for 9 h, the

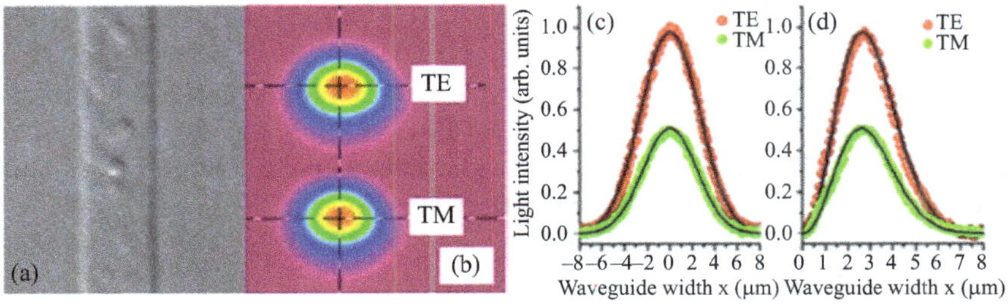

Figure 6.1 (a) Morphology image of 8-μm wide Ti:Zr:LN strip waveguide. (b) Near-field patterns of TE and TM modes guided in the 8-μm wide Ti: Zr:LN waveguide at the 1,547-nm wavelength. (c) and (d) TE-mode and TM-mode light intensity profiles along width x and depth y of 8-μm wide Ti:Zr:LN strip waveguide. © 2015 The Optical Society. Reprinted, with permission, from Reference [5]

surface Ti^{4+} concentration is around 12 × 10^{20} ions/cm^3 and Δn_o (Δn_e) is ~0.006 (0.012) at the 1.55 μm wavelength, respectively. The data, however, show a certain scattering due to the different experimental conditions and suffer from the non-linear dependence of the refractive index on the Ti concentration. In general, Δn_o lies in the range of 0.003–0.01 for *Y*-cut and 0.001–0.01 in *Z*-cut crystals, while Δn_e lies in the range of 0.002–0.04 for *Y*-cut and 0.002–0.02 for *Z*-cut. Its drawback relies mainly on the optical damage induced by waveguiding high power densities, crosstalk in switches and drift in interferometers, especially in visible range (that starts to be a limiting factor even below 1 μW at 0.633 μm). For this reason, Ti-in-diffusion has been widely exploited in the near infrared (NIR) region (wavelengths longer than 1 μm, with optimization in the telecom spectral range).

Recent works on the topics, however, have been more focused on the study of co-diffusion with other dopants to remove or, at least, smooth the optical damage. Since the 1980s, Ti-diffusion in MgO-doped lithium niobate has been widely investigated, and co-doping with at least 4.9 mol.% of MgO turned to be effective in suppressing the optical damage effect. Other dopants that have similar effect are Zn^{2+}, Sc^{3+}, In^{3+}, Tm^{3+}, Hf^{4+}, Zr^{4+} and Sn^{4+}. Among them, Sc^{3+} and Zr^{4+} showed the lower concentration thresholds of photorefractive damage, being close to 2 mol.% only. In particular, very recent advances have also been achieved in the Ti:Zr:$LiNbO_3$ and Ti:Sc:$LiNbO_3$ co-doped systems. Strip waveguides have been demonstrated by Zr-diffusion-doping followed by diffusion of 8-μm wide, 100-nm thick Ti strips on a *Z*-cut congruent substrate. Although possible, Ti-diffusion into bulk-doped Zr crystals has been less investigated due to the difficulties in growing high-quality and reproducible Zr-bulk-doped crystals at low cost. The Ti:Zr-co-diffused waveguide demonstrated to support both TE and TM in single mode at the 1.5-μm wavelength with losses $\leq$ 1.3/1.5 dB/cm for the TE/TM mode, respectively [5], as reported in Figure 6.1. The refractive index at 1.553 μm was measured to be close to the Ti:LN-doped sample, i.e. the presence of Zr slightly affects it with a contribute close to 10^{-4} for both n_o and n_e.

The compositional analysis showed that the Zr profile has a concentration greater than the threshold of photorefractive damage within a thickness that covers the 60 per cent (70 per cent) ordinary (extraordinary) index profile in the waveguide. Similar studies were performed by the same group on the Ti:Sc:$LiNbO_3$ strip waveguides fabricated by Ti-diffusion following Sc-diffusion-doping in a *Z*-cut congruent substrate [6]. In this case, the waveguides were single mode at 1.5 μm and with losses of 1.4 and 1.8 dB/cm for the TE and TM modes, respectively. The Sc compositional in-depth profile covered 80 per cent of the ordinary refractive index profile and almost 100 per cent of the extraordinary refractive index profile so that the $1/e$ level of the Sc concentration resulted to be above the threshold of photorefractive effect.

6.1.2 Proton-exchanged optical waveguides

PE consists in the immersion of a lithium niobate specimen into a liquid source of hydrogen at high temperature. Under these conditions, the replacement of lithium ions (Li^+) by hydrogen ions, or protons (H^+), causes an increase in the extraordinary refractive index (Δn_e close to 0.1) and, normally, a reduction in the non-linear coefficients. The waveguide, therefore, supports modes only on the extraordinary polarization. Despite this limitation, PE has been preferred for every application where propagation is in the visible spectrum. The most common sources of H^+ are the benzoic and the toluic acids, heated at temperatures ranging from 150°C to 400°C [3]. PE is often followed by annealing in a controlled atmosphere, leading to the so-called annealed PE to increase the high polarization rejection, to promote a better resistance to the optical damage and to recover the non-linear coefficients. For non-linear optical applications, soft PE and reverse PE (RPE) waveguides are used when the refractive index jump Δn_e can lay between 0.01 and 0.03, and better optical confinement is requested. Vapour-phase PE was also investigated to combine a strong confinement ($\Delta n_e = 0.09$) and preserved non-linearities [7], but difficulties in controlling the fabrication conditions limited its industrial application. As a valid alternative, high-vacuum PE (HiVacPE) has been recently proposed: it consists in vacuuming the exchange ampoule to decrease the water content of the acid bath to the greatest extent [8]. In this case, a refractive index contrast in *X*-cut crystals as high as 0.04 was demonstrated with low propagation losses close to 0.16 dB/cm and preserved optical non-linearity. Moreover, the HiVacPE process on *Z*-cut LN produced high-index contrast up to 0.11 without degrading the crystal non-linearity; combined with an appropriate etching process, it allowed one to achieve non-linear photonic wires.

As already cited in [3], applications to polarization control were exploiting the fact that only extraordinary refractive index variations can be achieved in PE waveguides as well as different devices already demonstrated in the years 1990–2005, such as modulators, switches, multiplexers and de-multiplexers [9] and Y branches [10]. Moreover, PE techniques in lithium niobate have achieved a significant maturity, especially in the standard Y-branch geometry, but the parasitic spectral selectivity continues to be a serious drawback. Y-branch power division as well as an oscillating power exchange between the output arms were observed together with an excess of refractive index jump within Y-branching area, all being strongly temperature dependent. To solve this issue, Skryabin *et al.* [11] used femtosecond (fs)-laser writing (see next section) on PE-exchanged Y branches by imposing extra tracks in the region of bifurcation and significantly reducing the temperature sensitivity of the Y-splitter so that the splitting ratio remained practically unchanged over the entire range of 0°C–60°C.

Finally, the feasibility was recently reported of very high-quality channel waveguides with propagation losses as low as 0.086 dB/cm and a coupling efficiency to optical fibres of 90 per cent at 1.550 μm; the splitting ratio of a set of directional couplers could be predicted with an accuracy of ±0.06 [12].

The combination of PE to other techniques was explored to get new configurations as well. In fact, it was either combined to 70-MeV argon ion irradiation or performed on photorefractive-resistant substrates such as $MgO:LiNbO_3$ and even in $Zr:LiNbO_3$-bulk-doped substrate [13], respectively, or even realized on Ti-in-diffused LN substrate as well as achieved in periodically poled structures [3 and references therein].

6.1.3 *Ion-implanted optical waveguides*

Ion implantation is a well-known technique to prepare optical waveguides in integrated photonics and modern communication systems. It consists in bombarding a surface with ions at an impact energy E_o so that they implant within the material after releasing their energy. In this process, the energy of the bombarding ion beam is transferred to the crystal lattice by way of two main processes dependent on the ions' velocity: energy loss due to electronic excitations, and nuclear collision losses. Nuclear collisions induce displacement damage, resulting in the formation of point and/or extended defects responsible for the refractive index change (see Figure 6.2). Moreover, both the electronic and nuclear processes tend to create colour centres (optical absorption sites). Consequently, an annealing treatment is commonly necessary for all the ion-implanted waveguides in crystals in order to recover the optical properties of the material.

When combined with photolithographic patterning of the surface, this procedure allows for selective implantation in 2D structures: high resolved patterns can be achieved by focusing the ion beam down to about 100 nm (focused ion beam, FIB), as already demonstrated also in $LiNbO_3$.

The first works were published on the He^+ implantation in $LiNbO_3$ at high fluences (10^{16} ions/cm^2), reporting that the ordinary refractive index had a typical barrier-type profile, while the extraordinary refractive index showed more complex changes [15]. Following the implantation of other ions at low fluences, such as B, C, O, F, Si, and P, n_e was found to be sufficient to realize high-quality guided regions [14–17], with good confinement and low losses, around 1 dB/cm. Ion implantation of Si was exploited as well, realizing monomode optical waveguides

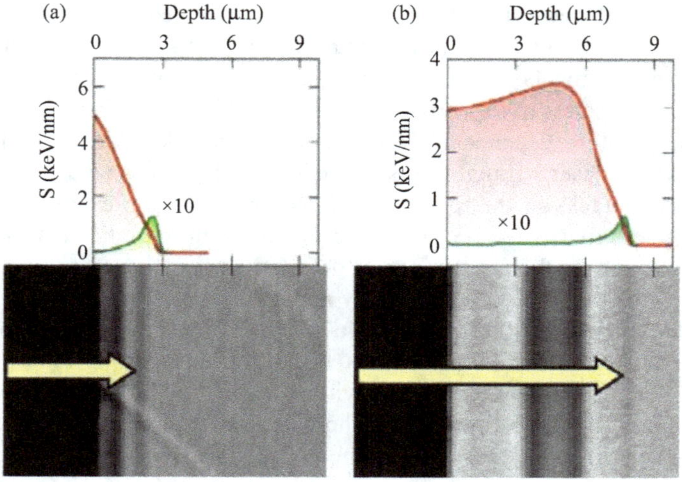

Figure 6.2 Micrographs of $LiNbO_3$ samples irradiated at a fluence of 2×10^{14} cm^{-2} with (a) Si ions at 7.5 MeV, and (b) F ions at 22 MeV. Nuclear Sn (green curve) and electronic Se (red curve) stopping powers as a function of depth are drawn for comparison. © 2009 John Wiley and Sons. Reprinted, with permission, from Reference [14]

and indicating the potential generality of the method. Waveguides in co-doped Nd (0.3 at.%):MgO (5 mol.%) *X*-cut $LiNbO_3$ crystals were also obtained using O ions [18] as well with irradiation in the regime of high electronic stopping power (~5–6 keV/nm) and ultra-low fluence (2×10^{12} cm^{-2}) [17], because of the short irradiation time required. Recently, the use of swift heavy ion (Ar) irradiation to modify refractive indices of optical materials was reported for the fabrication of single and multimode waveguides at 1.064 μm [19]. Those authors obtained surface and cladding-like optical waveguides with thicknesses of ~13, ~36 and ~23 μm at Ar implantation energies of ~120, ~240 and ~(120 + 240) MeV, respectively. The modal confinement and the optical losses were measured together with the second-harmonic generation (SHG) at 532 nm, with efficiency close to 6 per cent at room temperature.

In the case of H^+ implantation, the values of the electro-optic coefficients (r_{13} and r_{33}) were well preserved in the waveguides, independently of the implanted-proton fluence in the investigated range (1–10 × 10^{16} ions/cm^2). Also, the non-linear d_{33} coefficients had almost the same value of virgin $LiNbO_3$. Olivares *et al.* studied the SHG of 20 and 22 MeV F-implanted $LiNbO_3$ waveguides, and high SHG coefficients were obtained (50–80 per cent of a bulk $LiNbO_3$) [15].

Recent advances were reported [20] on Zr-doped $LiNbO_3$ crystals (2 and 4 mol.% of ZrO_2), where planar and channel waveguides were fabricated in *Z*-cut wafers by the double-energy proton-beam implantation at 470 keV (at the fluence of 1×10^{16} ions/cm^2) and 500 keV (at the fluence of 2×10^{16} ions/cm^2). The propagation losses of the channel waveguide were measured close to 3.6 dB/cm. The proton-implanted waveguides in Zr:$LiNbO_3$ crystals showed a relatively high optical damage threshold at the wavelength of 532 nm, which was one order of magnitude higher than that of the previously reported ion-implanted waveguides in undoped $LiNbO_3$. It was also found that the optical damage threshold (9.0×10^4 W/cm^2) of Zr: $LiNbO_3$ channel waveguides was much higher than that of the corresponding planar waveguide (2.0×10^3 W/cm^2).

6.1.4 LN waveguides by laser writing

It is well known that LN optical waveguides can be realized by scanning with a suitable focused laser beam. Depending on the physical phenomena induced in the light–matter interaction (wavelength, power and duration of the laser pulses), different types of optical waveguides can be achieved, permanently or not permanently written inside or at the surface of the crystal. Waveguides can be achieved also by exploiting the photorefractive properties of the material, as already reviewed in [3], with a typical positive refractive index jump of 10^{-3} for an extraordinarily polarized guided beam and a final transmission of up to 40 per cent. Great advances were instead reported on the exploitation of fs-laser writing: depending on the wavelength and on the impinging light power, it is possible either to ablate the surface (high power and high absorption) or damage its crystal structure, or promote photorefractive processes: typical writing speed ranges between 20 and 500 μm/s, and typical laser powers of few μJ are exploited at a wavelength of Ti sapphire lasers (795 nm, KHz rate, pulse duration of hundreds of fs) [21]. Various waveguide geometries have been demonstrated in lithium niobate crystals, such as waveguides with positive refractive index change (type I), stress-based double-line waveguides (type II) and depressed cladding waveguides (type III) or ablated ones (type IV), as sketched in Figure 6.3.

In general, when the average single-pulse energy is below the damaged threshold for $LiNbO_3$, high-quality waveguides are obtained, while slower writing velocities seem to promote larger refractive index change jumps. Different configurations can be achieved:

1. Type I: In this case, the fs-laser induces a positive refractive index change ($\Delta n > 0$) in the irradiated focal volume, which plays the role of waveguide core. Direct writing of 3D waveguiding structures in materials, and of passive devices such as Y-junctions, directional couplers or waveguide arrays/grating with feasible geometries are therefore achievable. In $LiNbO_3$ crystals, however, this type of waveguides can be generated only along the ordinary axis and has been found unstable upon temperature treatment at even 150° and to degrade the optical properties of the buk material in the irradiated region. The typical refractive index is of the order of magnitude of $\Delta n_e \sim 6 \times 10^{-4}$ (while $\Delta n_o < 0$) for light polarized along the *Z*-direction of the crystal [22]. The realization of Y branches (splitting ratio 1.1:1 and branch angle 0.5°; distance of the two branches at the exit face of the sample 60 μm, splitting loss 0.8 dB) was also successfully reported. It is worth mentioning that the final performance

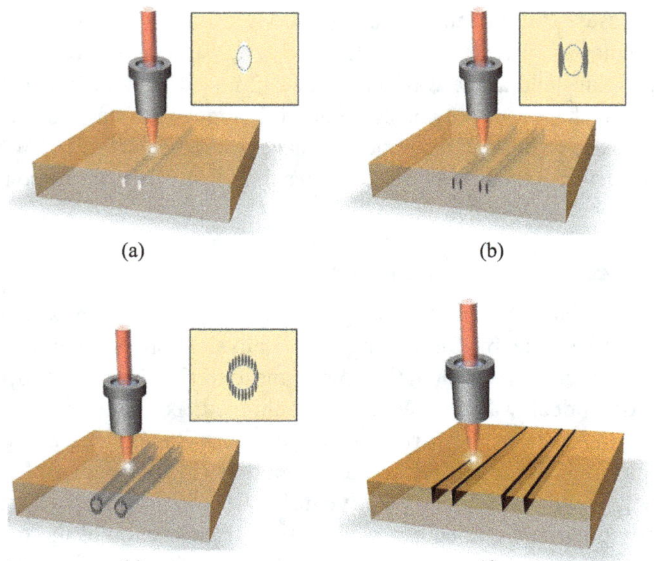

Figure 6.3 Fabrication procedure of fs-laser micromachined waveguides in crystals: (a) type I, (b) type II, (c) type III and (d) type IV configurations. The shadows represent the fs-laser-induced tracks and the dashed lines indicate the spatial locations of the waveguide cores. © 2014 John Wiley and Sons. Reprinted, with permission, from Reference [21]

strongly depends on the fabrication conditions, including average single-pulse energy and writing velocity. As already reported in [23], optical propagation losses were measured as high as 1.0 dB/cm at 0.633 μm. In periodically poled $LiNbO_3$ (PPLN) substrates, instead, lower losses were reported (0.6 dB/cm), with a typical index variation (n_e in this case) lower than 1×10^{-3}.

2. Type II: In this case, the fs-laser damages the structure of the irradiation region, inducing an expansion of the lattices in the focal volume and, consequently, a negative refractive index change ($\Delta n < 0$). If the induced stress is significantly high in the nearby region, these surrounding regions may possess a relatively high index through the stress-induced effects. The best performance was achieved by writing parallel laser tracks (with lengths of 15–30 μm or more) with suitable separation (typically 10–20 μm) so that the waveguide core is located in the region between these two tracks: the core preserves the $LiNbO_3$ properties, while the tracks play the role of barriers. This configuration is called 'double-line': in this case, the order of magnitude of propagation losses lies in the range of 2.7–4.2 dB/cm, coupling loss is estimated to be ~1.7 dB for all waveguides and the optical damage thresholds are in the range of 1.0–1.4 × 10^5 W/cm^2. Recent results of type II waveguides have been

reported also in MgO (6.5 wt.%)-doped *Z*-cut crystals [24]: Δn_e are in the range of $2.1–4 \times 10^{-3}$ for the TM modes of $MgO:LiNbO_3$ double-line waveguides. It was reported that waveguides fabricated with the same energy and diameter but different scanning velocities ($v_1 < v_2 < v_3$) exhibit different refractive index alterations ($\Delta n_1 > \Delta n_2 > \Delta n_3$). These results are promising with respect to those regarding the optical damage thresholds of the 4 mol.% Zr-doped $LiNbO_3$ planar waveguides and 5-μm width ridge waveguides produced by proton implantation, close to 2.8×10^3 and 9.2×10^4 W/cm^2, respectively [20]. The measured optical damage threshold of the $MgO:LiNbO_3$ (doped by 6.5 at.% Mg ions) 15-μm wide ridge waveguide fabricated by swift O ion irradiation is 5.6×10^3 W/cm^2.

3. Type III, or depressed cladding, waveguides consist of a core surrounded by a number of low-index tracks. A cladding region of reduced refractive index can be achieved by the irradiation of fs-laser pulses. In the transverse writing scheme, the claddings can be formed by overlapping a large number of tracks written by the focused fs-laser pulses along the boundaries of the waveguiding areas. They present some limits in downsizing of the mode field and constraints in achieving more complex configurations such as beam splitters and Mach–Zehnder interferometers. However, some recent improvements were reported by transversely writing the depressed cladding waveguides consisting only four tracks formed with slit-shaped fs-laser beams: single-mode light guiding can be achieved for both TE and TM modes at a wavelength of 1.55 μm. A reduced mode-field dimension, down to that compatible with direct coupling to the standard single-mode optical fibres, was in fact obtained. In the work by Lv *et al.* [24], type III waveguides were obtained by both dual-line and cladding configurations in *Z*-cut $MgO:LiNbO_3$ crystal by fs-laser inscription: the propagation loss of depressed cladding waveguide was demonstrated to be as low as 0.94 dB/cm at 632.8 nm wavelength. It was found that the optical damage threshold (close to 10^5 W/cm^2) of the dual-line waveguide was one order of magnitude higher than that of the cladding waveguide (close to 10^4 W/cm^2).

6.2 Ridge LN waveguide

High lateral optical confinement with respect to the previously reported cases can be achieved by using the so-called ridge waveguide configuration: in this case, a significant refractive index contrast is obtained, and the modal overlap is strongly optimized thanks to the removal of proper regions to define the waveguides borders. Ridge waveguides have been usually prepared by etching, either wet or dry [25], or by plasma-activated processes starting from a previously fabricated planar waveguide, made by conventional processes as those previously described. Most often, a mixture of HF and HNO_3 is used on the $Ti:LiNbO_3$ patterned surface by way of chromium (Cr) stripes. Low-loss monomode ridge guides with a height of up to 8–10 μm and a width between 4.5 and 7.0 μm were demonstrated by etching

Ti-planar waveguides, with propagation losses close to 0.3 dB/cm for transverse-electric and 0.9 dB/cm for transverse-magnetic polarization, respectively, at 1.55 μm wavelength [26]. Ridge waveguides demonstrated lower bending losses for small curvature radii, thus improving the level of integration, and also to reduce the modal sizes, enhancing the efficiency of non-liner devices, or lowering the threshold in integrated lasers [27]. Since imperfections at the edges could strongly affect the propagation losses [27 and references therein], the quality of the ridging process is determinant; however, the propagation losses were significantly decreased to 0.05 dB/cm by exploiting the proper etching protocol with respect to the Ti-in-diffusion step.

New advances in the state of the art were recently proven by the development of high precision machining: by employing a diamond blade (optical grade dicing) [28], high-aspect-ratio ridge waveguides were demonstrated, with depth greater than 10 μm [29], and used to successfully realize electro-optic modulators. It is worth mentioning that an optimized process for the fabrication of Ti:$LiNbO_3$ ridge waveguides with propagation losses lower than 0.2 dB/cm was demonstrated, exploiting waveguide sidewall smoothening during high-temperature Ti-diffusion [30].

The fs-laser writing technique was used as an alternative approach to fabricate ridge optical structures in dielectrics, so-called type IV direct written waveguides. It removes selected parts of the planar waveguide surface, thus constructing the ridges by laser ablation. The same technique was successfully exploited to create also microfluidic circuits engraved in $LiNbO_3$ substrates [31,32]. Recently, fs-laser writing technique was used to define ridge waveguide lasers in *Z*-cut Nd-doped $LiNbO_3$ substrates [33] as well as Y-branch waveguiding structures (1 × 2 beam splitters) which reach splitting ratios of ~1:1 at 4 μm in $LiNbO_3$ [34]. Laser ablation was also combined to get ridged waveguide in ion-implanted planar waveguides [35] for efficient beam splitters configurations, as depicted in Figure 6.4.

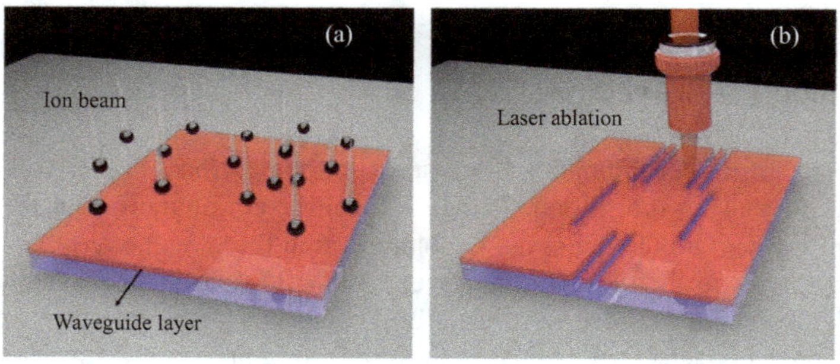

Figure 6.4 Laser ablation fabrication process on implanted $LiNbO_3$ substrate: (a) ion implantation to form planar waveguide layer and (b) femtosecond-laser ablation to define the splitter configuration. © 2018 The Optical Society. Reprinted, with permission, from Reference [35]

6.2.1 LN films for integrated optics: smart-cut, suspended waveguides and lithium niobate on insulator (LNOI)

The possibility of achieving thin lithium niobate films has attracted great attention, and several techniques have started to be investigated and optimized, including the deposition of ferroelectric thin films by molecular-beam epitaxy [33], radio frequency (RF) sputtering [36], sol–gel [37] and chemical vapour deposition [38,39]. The goal was to obtain lithium niobate waveguides integrated on a different substrate and to open the way to hybrid configurations. Despite the promising results, no disruptive advances were reported in the last decade. Difficulties in achieving high crystalline quality, the cost and long-time of deposition or the level of contaminants or the final surface quality were not at the mature level requested for batch processing, especially when embedded cladding was requested.

Another common scheme to produce thin $LiNbO_3$ thin films/membranes is to implement the crystal-ion-slicing technique. In this case, $LiNbO_3$ is bonded to a dielectric medium less refringent than $LiNbO_3$, like benzocyclobutene (BCB) or SiO_2 [40], the adhesion being optimized by Cr thin layers (around 50 nm) that play the role of electrode as well. The $LiNbO_3$/dielectric/$LiNbO_3$ sandwich undergoes ion-beam irradiation (such as H, He, F and O) and selective etching so that the $LiNbO_3$ layer between the surface and the implanted region is exfoliated, leaving a thin $LiNbO_3$ film bonded to the dielectric medium. All the process is sketched in Figure 6.5 [41]. These 'lithium niobate on insulator' (LNOI) structures present high refractive index, which can be as high as $\Delta n \sim 0.7$ thanks to the $LiNbO_3$/dielectric interface [42]. Also known as 'smart cut' in the industrial silicon-on-insulator wafer fabrication, films produced by this technique preserve basically the same crystalline properties of the starting material and its high refractive index. On the same wave, lithium-niobate-suspended structures were proposed by combining ion-implantation-induced amorphization with the etching by other microfabrication techniques such as FIB milling. This exfoliation technique was used to get $LiNbO_3$ membranes with thickness lower than 1 μm. High confinement ridge waveguides can then be achieved using smart-cut slab waveguides and lateral ion milling techniques with a lithographically defined mask. The thickness of the smart-cut waveguide is normally in the range of 0.6–0.75 μm, and the lateral width was reduced down to 1 μm, with mode field area close to 0.3–0.4 μm^2 at 1.55 μm wavelength, i.e. about one order of magnitude smaller than conventional Ti-diffused waveguides. The optical losses were reported to be lower than 10 dB/cm for both polarizations. This is of great interest for non-linear optical application, as it allows obtaining high power densities even at low input powers [3]. It was already reported that channel waveguides [43] embedded in BCB ($n = 1.55$) and SiO_2 ($n = 1.45$) can be narrower than 1 μm in order to achieve single-mode waveguiding at $\lambda = 1.55$ μm.

LNOI structures were also combined with electron-beam lithography (EBL) or Argon milling (ICP etching) to design photonic wires and more complicated structures [41–44]. The recent advances on this topic come mainly from the exploitation of optimized techniques and protocols to achieve a better resolution, an

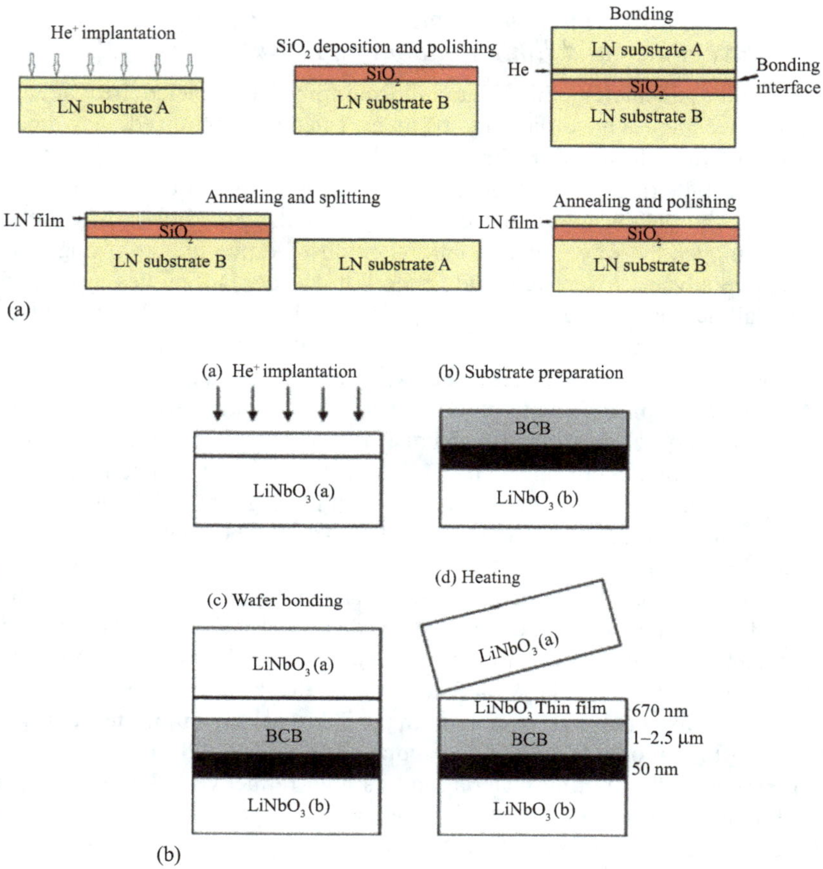

Figure 6.5 $LiNbO_3$ thin-film fabrication method: (a) a 'smart-cut' single-crystalline $LiNbO_3$ (LN) film of sub-micrometre thickness is crystal-bonded to an $SiO_2/LiNbO_3$ substrate. (b) (A) ion implantation of $LiNbO_3$ donor wafer (implantation of He^+, $E = 195$ keV, fluence 4.5×10^{16} cm^{-2}); (B) preparation of $LiNbO_3$ receptor wafer (substrate); deposition of Cr electrode and spin coating of an adhesive polymer film (BCB); (C) wafer bonding; (D) slow-ramp thermal treatment: strong bonding, $LiNbO_3$ film split-off and annealing. © 2012 John Wiley and Sons. Reprinted, with permission, from Reference [41]

improved surface and cut edge quality. fs-laser patterning, followed by the chemomechanical polishing, was used for structuring the LNOI into waveguides and micro-ring resonators (Figure 6.6). In this case, a significant optimization in decreasing the propagation losses was achieved, with losses close to 0.3 dB/cm at 1.55 μm [44], and even one order of magnitude lower (0.027 dB/cm) in the work by

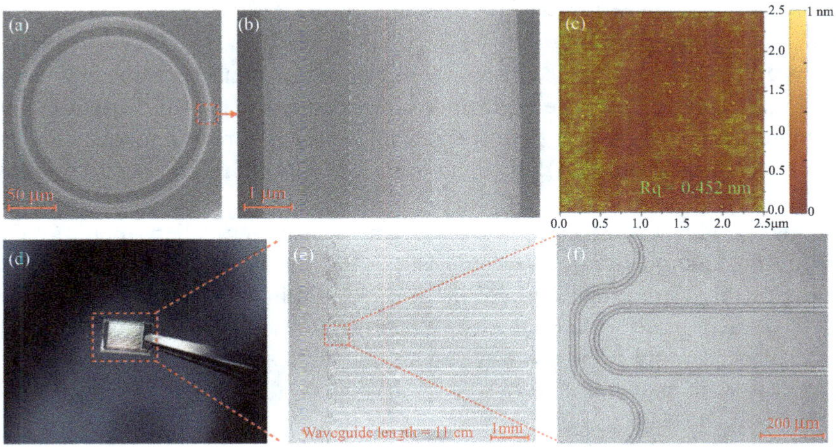

Figure 6.6 (a) Top-view scanning electron microscope (SEM) image of an LN micro-ring resonator. (b) Zoomed view of the ridge of the micro-ring resonator in (a). (c) Atomic force microscope (AFM) image of the ridge. (d) Picture of a chip consisting an 11-cm long waveguide captured by digital camera. (e) and (f) Zoomed images of the waveguides on the chip captured by an optical microscope. © 2018 Reprinted from Reference [45] under a Creative Commons Attribution License (CC BY 4.0)

Guarino *et al.* [43]. A process combining indirect wafer bonding and dry etching to produce patterns with controllable size, shape and orientation was also proposed: the areal shape of the ion-sliced $LiNbO_3$ was defined by a Cr mask patterned by EBL and plasma etching [45–46]. A hybrid silicon and lithium niobate racetrack resonator with coplanar electrodes was fabricated, with a measured tunability of 5.2 pm/V.

Moreover, ultra-thin *X*-cut $LiNbO_3$ self-suspended membranes were realized by ion-beam-enhanced etching technique or by FIB milling. Suspended membranes provide high refractive index contrast and can host high-quality waveguides with very low losses and almost the same linear and non-linear properties of bulk material [3]. Being compatible with periodic poling and doping, they can be combined in hybrid semiconductor – LN devices, and this is why 3-in. wafers of LNOI are already commercially available. Recent achievements were reported [47,48] by using a fast micro-structuring and nanostructuring process for LNOI, that is based on only two single steps, FIB milling and SiO_2 etching, which allow the realization of microstructures in thin self-suspended *Z*-cut $LiNbO_3$ membranes. LNOI wafers consisted a 0.75-μm thick *Z*-cut $LiNbO_3$ membrane bonded on a 1,200-nm thick SiO_2 layer deposited by plasma-enhanced chemical vapour deposition (PECVD) on an $LiNbO_3$ substrate. The LNOI pieces were further modified by thinning down the top $LiNbO_3$ layer by using the argon ion-beam etching (etching rate of $LiNbO_3$ 10.7 nm/min for an Ar^+ beam with an energy of 400 eV and a current density of 0.3 mA/cm^2).

Recently, an electro-optic enhancement was demonstrated in a photonic crystal (PhC) lattice milled in a ridge waveguide structured in a 1-μm thick *X*-cut $LiNbO_3$ membrane bonded to BCB [49]. Suspended micro-discs were also fabricated in thin $LiNbO_3$ on insulator (LNOI) either by argon plasma etching [50] or by fs-laser ablation [51].

Finally, advances in dielectric crystalline whispering-gallery-mode (WGM) micro-resonators in $LiNbO_3$ were reported: WGM micro-resonators suspended on silica pedestals were produced by fs-laser direct writing followed by FIB milling [51]. The micrometre-scale (diameter ~82 μm) LN resonator possesses a Q factor of ~2.5×10^5 around 1,550 nm wavelength; the fabrication process and images of the resonator are shown in Figure 6.7.

6.3 Active LN waveguides

Rare-earth doping of LN crystal has been extensively investigated to develop optical amplifiers and lasers to be fully integrated in the LN-based optical circuits: Er and Tm doping were the most studied. Although several attempts were made to use other techniques such as ion implantation and ion exchange [3], the most attractive doping technique for incorporating rare-earth ions was thermal diffusion at a temperature close to the Curie point of the crystal, thanks to its scalability and compatibility with industrial batch processing. Less appealing is the exploitation of bulk-doped substrates where the channel waveguides are realized, mainly due to higher costs of the process scalability at an industrial level. Best optimized for erbium, previous studies showed that Er^{3+} solubility in LN is limited and the diffusivity is rather low: it results in very long diffusion times (>100 h). Since the rare-earth doping does not allow to increase the refractive index of the material and therefore to obtain optical waveguiding, single-mode channel waveguides were defined by the standard in-diffusion technique of Ti stripes on rare-earth-doped substrates [52]. In this configuration, by optical pumping at $\lambda = 1.48$ μm, a wavelength-dependent gain of up to 2 dB/cm was achieved (1.53 μm $< \lambda <$ 1.61 μm): single-frequency emission at several wavelengths within the Er-gain band around $\lambda \sim 1.55$ μm and an output power of up to 8 mW were reported.

Although many attempts to improve the amplifiers/lasing properties of Ti:Er:$LiNbO_3$ systems were carried out in the past, no significant advances have been achieved in the last years. However, highly efficient erbium-doped titanium-in-diffused ridge waveguide optical amplifiers and lasers in *X*-cut congruent $LiNbO_3$ pumped at 1.486 μm were demonstrated [27]: a total single-pass small-signal internal gain of 14 dB at 1.531 μm for TM modes was achieved in 4.6-cm-long *y*-propagating waveguides for a coupled pump power of 200 mW. The laser operating at 1.561 μm had a slope efficiency of 33 per cent and a slope efficiency of 19 per cent at 1.531 μm, exceeding the best literature values for Er:Ti:$LiNbO_3$ waveguide lasers. The future perspectives are surely related to the application of the same

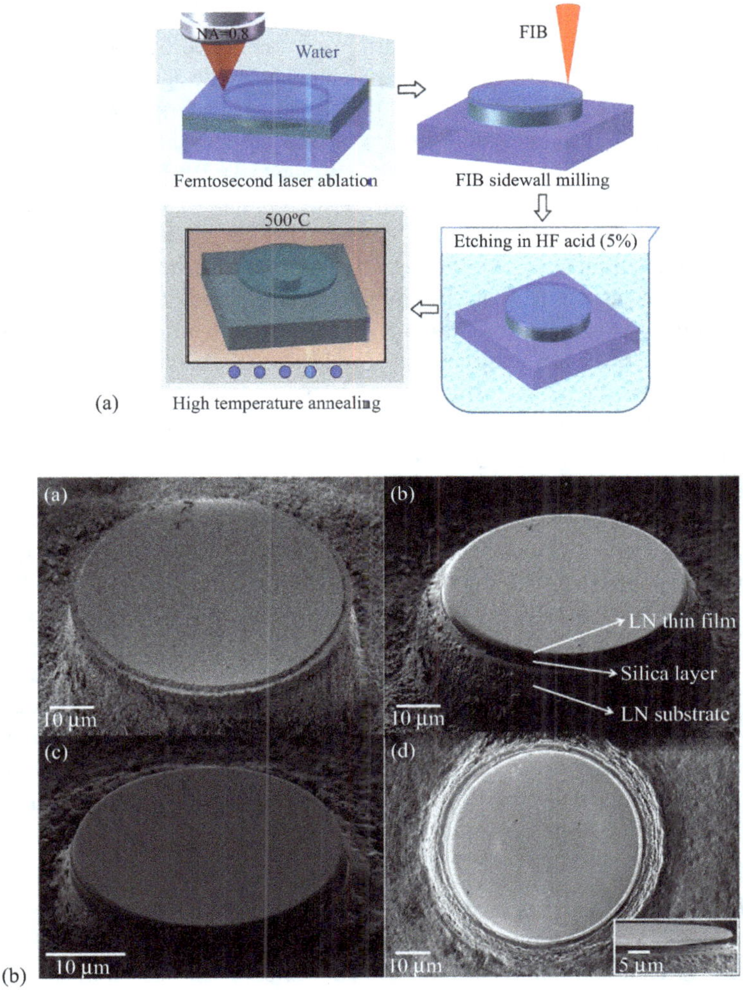

Figure 6.7 (a) Procedures of fabrication of an LN micro-resonator by water-assisted femtosecond-laser ablation, followed by FIB milling, selective chemical etching and annealing. (b) (A) SEM image of a cylindrical post formed after fs-laser ablation; (B) and (C) SEM images of two cylindrical posts with diameters of 55 and 33 μm, respectively, after FIB milling; (D) SEM image (top view) of the 55-μm diameter micro-resonator after the chemical etching and annealing. Inset in (D): side view of the micro-resonator, showing the freestanding edge of the LN micro-resonator. © 2015 Reprinted from Reference [51] under a Creative Commons Attribution License

optimized fabrication technique to optimize the amplifier/laser performance by means of erbium/ytterbium co-doping, as already modelled by Cantelar *et al.* [53].

Similar to Er, also Tm-doped LN waveguides were designed for optical amplifiers and lasers in the spectral range 1.7–1.9 μm. In particular, a simple 30-mm long Fabry–Pérot-type Ti:Tm:LN waveguide laser of very low threshold and 1,890 nm emission wavelength was proposed [54]. Longer Ti:Tm:LN waveguides were reported as optical amplifiers, and simulation results suggested gain exceeding 30 dB in 90 mm long structures. With a laser threshold at 1.89 μm at 4 mW coupled pump power, a slope efficiency was found $\approx$ 13.3per cent. Better performances were achieved in bulk co-doped Tm,Mg:$LiNbO_3$ crystal slab with high Tm^{3+} doping concentration (Tm = 2 wt.% and Mg = 5 wt.%), where an efficient continuous-wave laser emission at 1.856 μm was reported. The laser had a maximum output power of 2.62 W, realized with a slope efficiency of 19.6 per cent and a beam quality factor M^2 of 1.7 at room temperature. This result opens the way of a watt-level laser operation in Tm,Mg:$LiNbO_3$ crystal, and the output power is four orders of magnitude higher than that reported previously in Tm-doped $LiNbO_3$ [55].

Thulium-doped Ti:LN single-mode channel waveguides were also proposed for application as photon echo-based quantum memories: the storage and retrieval of photons (λ = 0.795 μm) from entangled photon pairs was in fact demonstrated, thereby temporarily creating entanglement between a photon and a collective atomic excitation [56].

As a further advance, ridge waveguide lasers were fabricated in Nd^{3+}-doped $LiNbO_3$ crystals by using the fs-laser writing technique to define ridge structures on a gradient-index planar waveguide fabricated by Zn-diffusion. This planar waveguide was formed in a *Z*-cut $LiNbO_3$ substrate homogeneously doped with a 0.23 per cent of Nd^{3+} ions [57]. The waveguide laser showed a threshold of 31 mW, with a 7 per cent of slope efficiency, respectively.

6.4 Integrated optics applications of lithium niobate

Thanks to its remarkable linear and non-linear optical properties, and chemical and mechanical stability, lithium niobate has been widely exploited for applications in IO and photonic crystal devices. Since $LiNbO_3$ has high electro-optical coefficient and high optical transparency, it has been widely used for applications in microwave telecommunications, memory units, electro-optics, modulators, optical switches, waveguides, beam deflectors, SHG, surface acoustic waves devices, parametric optical converters and data transmission, just to cite a few. In the following, some advances in IO applications of $LiNbO_3$ will be presented with a special attention to those fields in which it surely emerged as an excellent alternative to other materials.

6.4.1 Optical modulation

It is well known that lithium niobate has been widely exploited in optical modulators, benefiting from a very high intrinsic modulation bandwidth as well as switching speed. Light modulation is made by way of its electro-optic effect, to get a refractive index change by practically imposing on a voltage modulation [58]. Since the achievable index change is relatively small (typically ≈0.01–0.02 depending on the crystal cut, see Section 6.1 for more details), either large voltages or centimetres long electrode lengths are needed to gain enough modulation [2]. In this sense, these modulators are considered bulky. The demanding need of faster optical interconnects has strongly pushed for further improvements in optical modulation solutions, both digital and analogue. In this scenario, new configurations were proposed, starting from heterogeneous integration of thin-film LN on silicon substrates [59] while continuing to enhance the existing modulators performances.

Several parameters have been constantly under investigation to boost technological advances:

- decrease the optical losses;
- optimize the optical extinction ratio (OER), the half-wave voltage and tunability and
- increase/optimize modulation bandwidth and frequency response.

Great efforts have been, therefore, devoted to investigate new configurations and fabrication processes. Some of them have been successfully implemented in the last years, and others are still under investigation and are opening new perspectives for the future.

The most known $LiNbO_3$ modulators are in the Mach–Zehnder configuration, that, depending on the crystal cut, can provide different performances and applications [2,58]. The basic working principle is illustrated in Figure 6.8: an input optical field $E_i = 2E$ is equally split between two waveguide branches, and in each branch, the optical field undergoes a phase shift $\Delta\varphi$, opposite in phase for the two branches, depending on the voltage applied by the electrodes. After the recombination, the optical signal has an electric field $E(e^{j\Delta\varphi} + e^{-j\Delta\varphi}) = 2E \cos \Delta\varphi$ which results in the optical power modulation, driven by the voltage modulation V. The difference between the Z-cut and the X-cut type is the use of optical TM or TE mode for the coupling with the RF signal, so that the largest electro-optic tensor component (r_{zz}) is always exploited.

The optical waveguides are integrated with suitable patterned electrodes: X-cut, Z-cut with single drivers, or Z-cut dual drivers were presented in [4]. A silica-based buffer layer is sandwiched between the electrode and the lithium niobate crystal to guarantee either low optical losses by keeping the evanescent optical field low in the lossy metal electrodes or the velocity matching between the travelling optical and microwave fields. The difference between the single-drive and dual-drive configuration relies on the possibility of driving with different voltages the two arms of the Mach–Zehnder. Other improvements were achieved by using ridge configurations; thin plate ones and domain inverted were also revised in [4].

The switching voltage $V_\pi = \lambda G/2Lr\Gamma n_{opt}^3$ corresponds to a phase change of π and produces a theoretical zero optical output; it depends on the length L of the modulator arms, the spatial gap G between the electrodes, the overlap integral

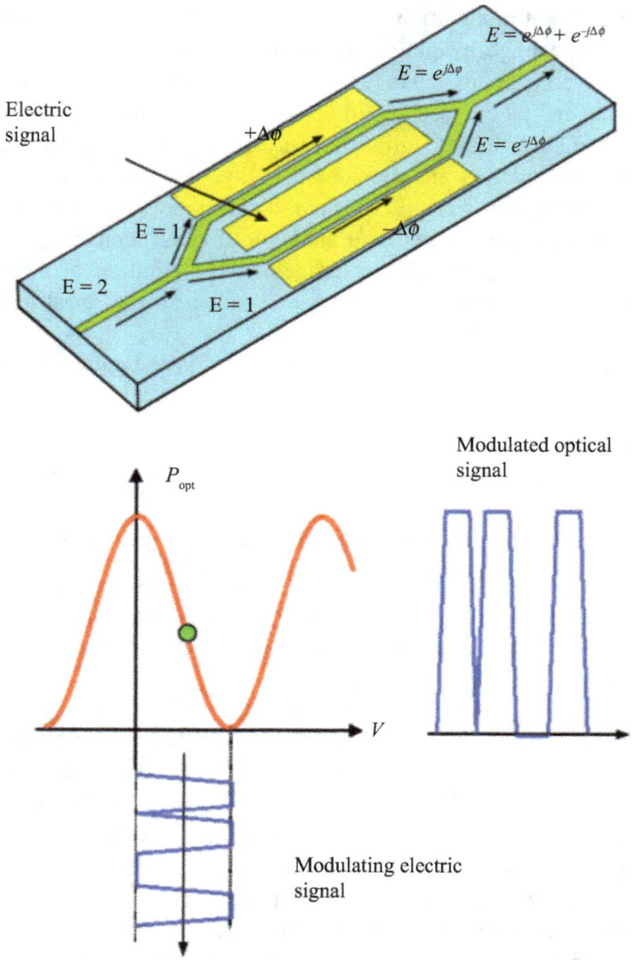

Figure 6.8 Modulator schematics and electrical-to-optical transfer curve. © 2009 John Wiley and Sons. Reprinted, with permission, from Reference [4]

Γ between the optical and microwave electric fields, the electro-optic coefficient r and the refractive index n_{opt} at the optical signal wavelength λ. Since the bandwidth BW (in GHz) is dependent on $1/L$, a trade-off exists between V_π and BW, and the ratio BW/V_π is used as a figure of merit: the best efficiency corresponds to high values of BW/V_π. Commercially available electro-optic modulators include free-space amplitude and phase modulators (of the typical length less than 10 cm) as well as fibre-coupled intensity and phase modulators. The market offers phase and intensity modulators working at 10 GHz up to 40 GHz, which are based on titanium-in-diffused $LiNbO_3$, with costs that span from 2 KEuro (10 GHz) to 5 KEuro (40 GHz) if relatively lower powers are needed, but they normally are limited in the spectral range (1.525–1.605 μm) or subintervals within. In order to support higher powers, MgO co-doped $LiNbO_3$ was exploited, as already presented in the previous sections of this chapter. The proper configuration is application dependent: in some cases, in fact, a phase/frequency shift (chirp) in the output in addition to the intensity modulation is needed. Chirp is, in fact, an important factor in high-data-rate, long-distance telecommunication systems: its optimum value (typically 0 or ~0.6) depends significantly upon the overall system architecture.

Fixed-chirp modulators are based on *Z*-cut substrates because this material creates an inequality in the push–pull phase shift between the two arms of the Mach–Zehnder interferometer. *X*-cuts are used instead when both arms of the Mach–Zehnder interferometer have to be symmetrically modulated. This symmetry ensures that the modulated output of the intensity modulator is not also shifted in phase/frequency (chirped). Zero-chirp intensity modulators, in particular, are ideal for use in metro and long-haul DWDM applications requiring less than a 2 dB power penalty for ±1,200 ps/nm dispersion. Due to chirping, the pulse increases its spectral width: the chirped pulse can, therefore, be broader than an unchirped pulse travelling through the same optical fibre. A comparison of $LiNbO_3$-based optical modulators was reported in [58].

Although these optical modulators are quite performing, the refractive index contrast $\Delta n < \sim 0.02$ between core and cladding results in large optical mode areas (even close to 10–50 μm^2 in some cases) and high bending radii. For this reason, thin-film $LiNbO_3$ on insulator has recently emerged as a promising alternative to achieve lower mode areas boosting the electro-optic efficiency. In this frame of reference, monolithic $LiNbO_3$ nanophotonic circuits have been addressed as the new challenge of high-speed modern telecommunication and in hybrid platforms where an easy-to-etch material (e.g. Si and SiN) is used as a device layer bonded to non-etched thin LN films. These heterogeneous platforms, although very promising, are still under optimization in order to get the best compromise between electro-optic bandwidth, modulation efficiency, optical confinement and operating wavelengths [4]. Recently, nanophotonic $LiNbO_3$ modulators were proposed with enhanced performances [60]; commercially available *X*-cut LN-on-insulator (LNOI) substrates were made of LN thin film (700 nm thick) bonded on top of silica (2 μm thick). The index contrast between the LN core and the SiO_2 cladding thus obtained was close to $\Delta n = 0.67$, over an order of magnitude higher than ion-diffused $LiNbO_3$ waveguides.

As far as the modulator performance is concerned [59], waveguide propagation losses in optical modulator span over a wide range, normally 0.1–11 dB/cm depending on the technologies they are based on. Different performances can be achieved: although the data have been limited to demonstrations of modulators, it emerges that the rib-loading platform offers the lowest propagation loss of $\sim$1 dB/cm among modulators, even if it is still higher than the $\sim$0.2 dB/cm values in commercial on-the-shelf devices [59]. Another parameter to compare the performances of an optical modulator is the OER, [OER] = [dB], i.e. the optical power extinction between the off/on state of a modulator operating at low frequencies (DC to a few MHz) [59]. In the case of a symmetric MZ, OER is expected to be infinite in the ideal case but, due to imperfections (including deviations from 50:50 power splitting and combining, and loss-imbalance between the arms of the MZ modulator), OER assumes finite values. In particular, OER >20 dB is required for long-haul communications, OER >30 dB is desirable in certain switching applications, while short-reach transmission can tolerate a lower OER [59]. OER (low frequency) ranges from 3 up to 25 dB, the best performances being provided by $LiNbO_3$ on SiO_2-based optical modulators. A trend of V_π and $V_\pi L$ values among compact $LiNbO_3$ MZ modulators was revised in [59]: lower values of V_π are demanded. The ion-milled and the rib-loaded variations of the thin-film LN on SiO_2 lower cladding approach (either on LN or Si substrates) offer the lowest $V_\pi L$ to date on any LN-based technology.

Finally, the acoustic properties of lithium niobate, especially close to its resonances, complicate the frequency response of the material and give different results depending on the clamping or unclamping configuration of the LN structure. Ripples can, therefore, be detected, even larger than 1 dB at frequencies as large as a few GHz. Whether it is possible to simultaneously achieve a low on/off switching voltage, an ultra-high bandwidth and a low optical loss in LN modulators have remained an outstanding question. Among the best performances, monolithically integrated LN electro-optic modulators seem to overcome such trade-offs, featuring a switching voltage of 1.4 V while supporting very high bandwidths which allow the modulator to be directly driven by a CMOS circuit but with a high optical power extinction ratio of about 30 dB [61].

6.4.2 Light generation

Lithium niobate presents a large optical non-linearity that, combined with domain engineering processes, allows for the fabrication of non-linear optical components based on the quasi-phase matching (QPM) scheme for efficient wavelength conversion. Periodical modulation in one dimension of the susceptibility tensor χ^2 was achieved by periodical poling of the crystal and was used to compensate for the wave vector mismatch caused by the different group velocities of the waves involved in a given non-linear process. The typical period of PPLN crystals is ≈10 μm (6.7 μm) for SHG at 1.5 μm (1.064 μm), respectively. SHG frequency converters, frequency tripling, optical parametric oscillators (OPOs) and difference frequency generation devices are now commercially available: by the proper combination of PPLN with common lasers, it is possible to access a wide range of wavelengths in the visible and infrared, in subintervals in the range 0.488–5.09 μm. QPM was also exploited in multiple spatial dimensions gratings giving rise to non-linear photonic crystals: experimental realization of 2D non-linear periodic structure with hexagonal symmetry in lithium niobate (HeXLN) was therefore successfully presented, with internal conversion efficiencies of 80 per cent [62].

To further improve PPLN light generation performances and allow for optical integration, waveguides on PPLN substrates were realized by Ti-in-diffusion, PE, ion implantation, fs-laser writing and smart-cut processes: the full process is mature enough to have reached a commercial diffusion. Also, Ti-diffused ridge PPLN waveguides were proposed to generate light in the 3.4–3.8 μm range: efficiencies of more than 10 per cent/W, resulting in several mW of converted radiation, were demonstrated [63]. For comparison, fs-laser writing allowed efficiencies close to 35 per cent (0.626 W SHG peak power generation in pulsed regime) in type II waveguides with QPM to generate SHG at 774 nm [64], while multiscan type I waveguides provided 18 per cent/W for SHG at 0.7835 μm [65].

As already reviewed in [3], integrated OPOs with built-in (intracavity) components like electro-optical and acousto-optical tunable filters, phase and amplitude modulators, Bragg reflectors as wavelength selective mirrors, wavelength-dependent couplers to get separate cavities for signal and idler, (tunable) pump lasers in the same structure, were already demonstrated, as well as the use of PPLN waveguides as logical elements in the realization of all-optical signal processing systems. More recently, efforts were focused on the SHG obtained in periodically poled thin films of lithium niobate bonded on silicon and rib-loaded with silicon nitride channels with ridge waveguide with cross sections of ~2 μm^2 (Figure 6.9): a non-linear conversion of 8 per cent is obtained with a pulsed input in 4 mm long waveguides, with propagation losses lower than 1 dB/cm for both the pump and SHG wavelengths [66].

6.4.3 *Integrated optical sources for quantum optics and communication*

IO have been addressed as an enabling technology for quantum optics thanks to optical confinement that enhances the efficiency of non-linear interactions by maintaining high optical power densities over distances far exceeding that

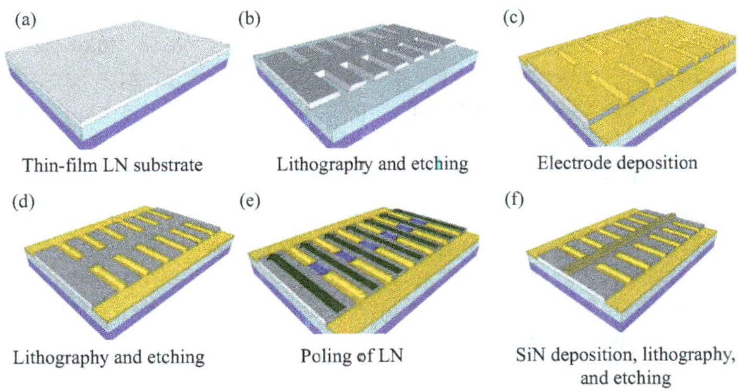

Figure 6.9 Major fabrication steps: (a) Y-cut LN on Si substrate; (b) first lithography and etching of LN; (c) metal electrode deposition; (d) lithography and etching to completely define the periodic electrodes; (e) poling of LN on Si with periodic domain reversal; (f) SiN rib definition by PECVD, lithography and etching to form the ridge waveguide. Not shown in this figure is the final deposition of a SiO_2 top cladding by PECVD. © 2016 The Optical Society. Reprinted, with permission, from Reference [66]

permitted by the diffraction limit [67]. The limiting factor still relies on the coupling of the IO devices to beams propagating in fibres or free space, and this is one of the reasons why IO is under the focus of a continuous improvement. Increasing interest was then promoted by the advent of upconversion detectors, based on non-linear interactions in IO waveguides, as well as by the realization of entangled photon-pair sources that exploit photon pairs produced by spontaneous parametric downconversion in $\chi^{(2)}$ non-linear crystals.

One of the key elements in quantum-photonic systems is a quantum memory, i.e. an interface between light and matter that allows for the storage and retrieval of photonic quantum information and must be capable of supporting the large bandwidths involved with the potential of storing, and retrieving, data at GHz bandwidths at the single-photon level. Quantum memories based on rare-earth-doped crystals were addressed as promising candidates due to their large storage bandwidths and multimode capacity: based on Ti:Er:$LiNbO_3$ [58], they were used to demonstrate high-efficiency storage, long coherence times, multimode storage, on-demand read-out at the single-photon level, storage of photonic entanglement and heralded entanglement between two crystals [67 and references therein].

A current extensive studied approach to implement a quantum memory is based on photon echoes [68,69]: Tm:Ti:$LiNbO_3$ optical waveguides were found to be excellent candidate thanks to the high performance of Ti:$LiNbO_3$ waveguide and thulium featuring a transition at 0.795 μm. At this wavelength, entangled photons are conveniently generated; high-efficiency and easy-to-operate single-photon detectors are commercially available and transmission in air presents very low absorption. Based on a quantum memory, it is possible to realize deterministic single-photon sources and provide scalable schemes for quantum computation, quantum repeaters and quantum networks [70]. Moreover, quantum networks require interference between photons produced by independent sources: the maximum achievable distance for quantum communication links is mainly limited by the inherent dark counts in the employed single-photon detectors. Single and entangled photons can be both produced via spontaneous non-linear parametric processes, and lithium niobate played a key role in the development of efficient parametric IO generators consisting of waveguide structures fabricated in PPLN. Most of these first IO source experiments were initially carried out using proton-exchanged waveguides on *Z*-cut lithium niobate: 3–4 orders of magnitude of improvement with respect to bulk sources was reported, resulting in a conversion efficiency (probability of pair creation per pump photon) of about 10^{-6} at telecom wavelengths (1.31 or 155 μm). Type 0 PE:PPLN waveguides were also demonstrated, with photons at 0.854 μm (two-photon interference raw visibility of 80 per cent when testing energy-time entanglement) and photons in the telecom range 1.31 μm ([67 and references therein]). More recently, a slightly modified version of this source, emitting paired photons at 1.31 and 1.55 μm wavelengths, led to a 98 per cent net visibility two-photon interference for maximally energy-time entangled states. The polarization entanglement source was made using two PE: PPLN waveguides mounted in a Mach–Zehnder interferometer configuration; pairs with net visibilities of nearly 100 per cent were also produced [67 and references therein], indicating the promise of such IO devices as sources for quantum communication protocols, not only in terms of high generation efficiencies but also in terms of QPM engineering.

RPE guides were also used to realize a device having a two-mode section in the PPLN followed by an adiabatic Y-junction serving as an integrated modal beam splitter, resulting in a buried and more cylindrically symmetric guide for improved guide-fibre coupling. The device was operated at 10 GHz and had a coincidence to accidental pair counting ratio of over 4,000, but no measurements of entanglement or efficiency were reported. Very recently, PPLN crystals were used to realize a broadband source of polarization-entangled photons: two periodically poled nonlinear waveguides were made by 6 cm-long titanium-in-diffused waveguide in lithium niobate (PPLN, required periods as short as 6.5 μm) [71] with experiments in teleportation. By coating a PPLN waveguide with mirror-like facets, a monolithic OPO-based source of energy-time entangled photons was also demonstrated.

Finally, direct generation and detection of high-purity photon pairs were achieved at room temperature with 3.2 μm wavelength spacing, one at 0.780 μm to match the rubidium D2 line and the other at 3.95 μm that falls in a transparent, low-scattering optical window for free-space applications [72]. The pairs were created via spontaneous parametric downconversion in a magnesium-doped periodically poled lithium niobate (MgO:PPLN) waveguide with specially designed geometry and periodic poling to support overlapping fundamental transverse modes for all light waves, and its optical domains are periodically poled to offset the large phase mismatch. Quantum correlation measurements yield a high coincidence-to-accidental ratio close to 50, which indicates the strong correlation with the extremely non-degenerate photon pairs (Figure 6.10).

6.4.4 Integrated optics and microfluidics: opto-microfluidics

In the last years, $LiNbO_3$ has been demonstrated also to be an interesting material for microfluidic applications, opening the way to integrated opto-microfluidics, i.e. IO merged with microfluidics circuitries on the same substrate, so constituting an optimal candidate for lab-on-a-chip applications. $LiNbO_3$ has been already exploited as active substrate for acoustic wave generation and particle trapping in liquids, as well as light-controlled switcher of a liquid crystal [73]. Furthermore, only recent the feasibility of microchannels fabrication in lithium niobate by means of dicing [33], laser ablation [32], micromilling [74] and even for droplet generation circuits was proven. The first demonstration of a self-aligned opto-microfluidic platform completely integrated in lithium niobate for optical detection of dispersed phases in fluids, such as droplets, was demonstrated by Bettella *et al.* [73]: by coupling a cross-shape junction for droplets generation and optical channel waveguides array obtained by Ti-in-diffusion, a fully integrated opto-microfluidics platform was realized. In this case, the same technique used to achieve ridged waveguides by high precision saws was instead exploited to engrave microfluidic channels crossing orthogonally the array of optical waveguides: the same optical

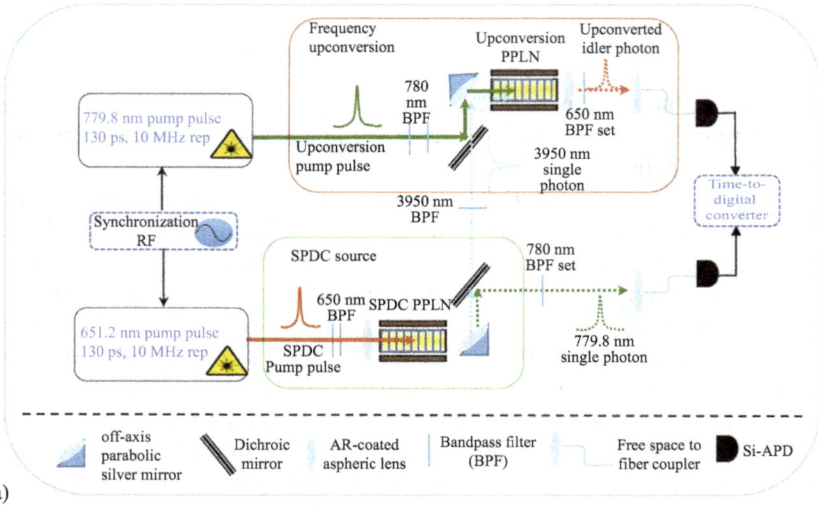

(a)

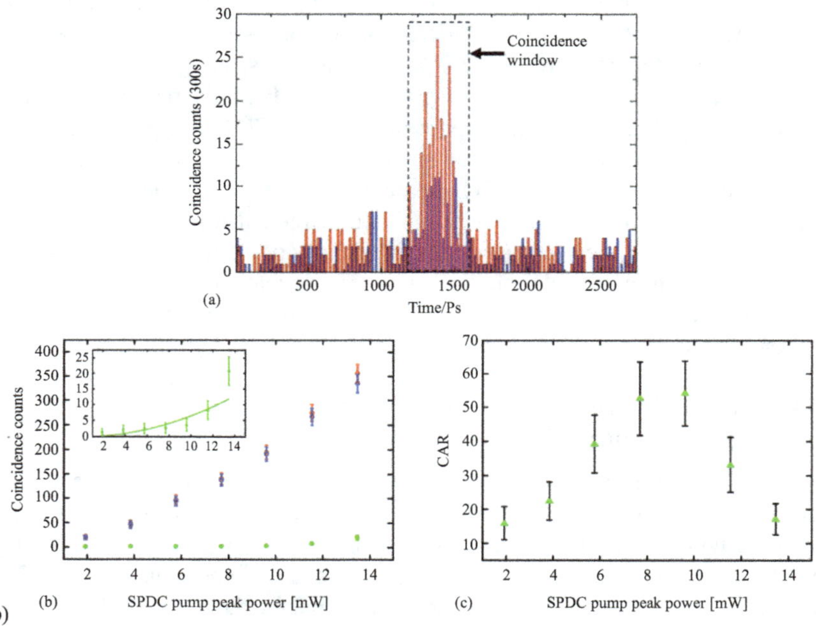

(b)

Figure 6.10 *(a) Experimental setup: A 651.2 nm pulse train with 10 MHz repetition rate and 130 ps pulse width is coupled into a 2.2-cm long PPLN waveguide through an anti-refection-coated aspheric lens. (b) (A) Histogram of the raw coincidence counts obtained with (red) and without (blue) input of 3,950 nm idler photons. Contribution from the upconversion noise photons and detector dark counts are not subtracted. (B) Measured coincidence (red square), accidental coincidence (green dot), true coincidence (blue dot), plotted as a function of pump peak power. Inset: accidental coincidence (green dot) plotted on a smaller scale with quadratic fitting (green solid line). (C) True coincidence-to-accidental coincidence ratio versus pump peak power. Each data point is acquired over an integration time of 300 s. © 2017 Reprinted from Reference [72] under Creative Commons Attribution 4.0 License*

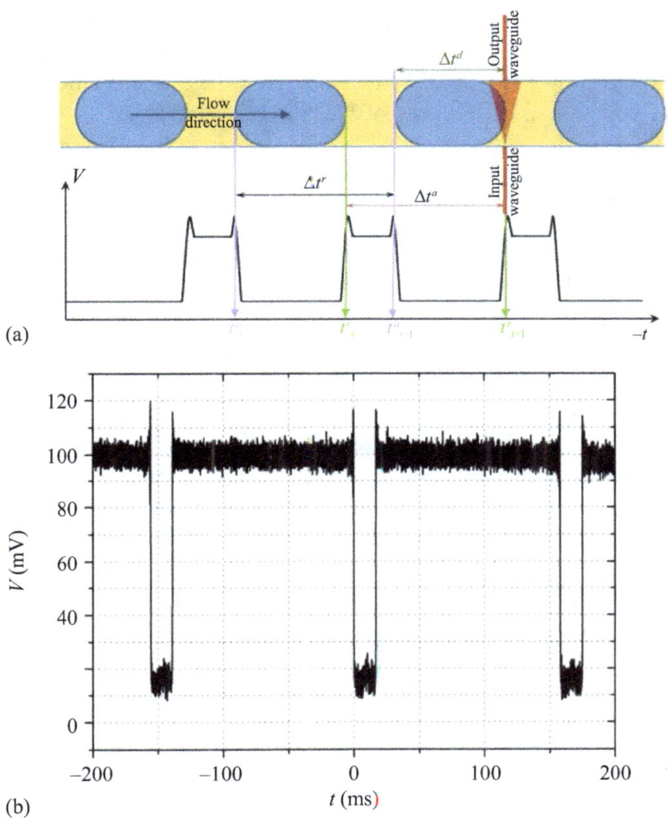

Figure 6.11 The working mechanism of the signal: (a) scheme of the time interval measured with the integrated optical trigger system during the passage of droplets, tai (violet) and tri (green) refer, respectively, to the passage of the receding and advancing menisci. The $\Delta t_d = t^a - t^r$ represents the passage time of a droplet and it is proportional to the droplet length. The $\Delta t_r = t^r{}_{i+1} - t^r{}_i$ (black), $\Delta t_a = t^a{}_{i+1} - t^a{}_i$ (red) are the duration between the passage of two consecutive menisci, receding and advancing, respectively. (b) Example of the voltage signal at the photodiode at the passage of three subsequent droplets. © 2019 Elsevier. Reprinted, with permission, from Reference [73]

features of the sidewall were in fact mandatory as in ridge waveguides. In this way, the waveguides can be exploited to illuminate fluids and objects present in the channels and, at the same time, to detect the transmitted or emitted light (see Figure 6.11).

This optical sensor is sensitive to the passage of each single droplet so that each droplet's position, size, geometry and optical properties were monitored in real time. Although the device was tested with droplets, it can be used with flowing objects of different nature, such as biological samples or particles. As a matter of fact, the high confinement of light provided by the integrated optical waveguides system not only allows the detection of objects at the microscale, but it is also highly sensitive to small differences of the refractive index.

6.5 Conclusions

IO aims is to create miniature optical circuits with potentialities similar to silicon chips that have revolutionized the electronics industry. The advantage of the optical approach, however, is that data can be processed at much higher speeds, and benefits from light–matter interactions still are mostly to be discovered. In this scenario, lithium niobate allows for integrating multiple optical stages on the same substrate and offers fascinating properties: it can be widely used to generate, confine and propagate as well as eventually modulate the light intensity that can be recorded and even used to probe and sensing applications. From this point of view, it is a resource but also a challenge to exploit its properties at best and fully integrate them together. Finally, hybrid configurations that bond lithium niobate to other materials are opening the way to new solutions and better performances, overcoming the lithium niobate limits whenever necessary, and to integrate it with a material such as silicon, the advances of which brought a disruptive positive change even in our daily life.

References

[1] Lawrence M. 'Lithium niobate integrated optics'. *Rep. Prog. Phys.* 1993, vol. 56, pp. 363–429.

[2] Wooten L., Kissa K.M., Yi-Yan A., *et al.*, 'A review of lithium niobate modulators for fiber-optic communications systems'. *IEEE J. Sel. Top. Quantum Electron.* 2000, vol. 6(1), pp. 69–82.

[3] Bazzan M. and Sada C. 'Optical waveguides in lithium niobate: recent developments and applications'. *Appl. Phys. Rev.* 2015, vol. 2. pp. 040603–23.

[4] Janner D., Tulli D., García-Granda M., Belmonte M., and Pruneri V. 'Micro-structured integrated electro-optic $LiNbO_3$ modulators'. *Laser Photonics Rev.* 2009, vol. 3(3), pp. 301–13.

[5] Zhang D.L., Kang J., Zhang Q., Du W.J., Wong W.H., and Pun E.Y.B. 'Zirconium-diffusion-doped Ti:$LiNbO_3$ strip waveguide for integrated optics'. *Opt. Mater. Express* 2015, vol. 5(8), pp. 1715–21.

[6] Zhang D.L., Cong-Xian Q., Wing-Han W., Dao-Yin D., and Pun E.Y. 'Optical-damage-resistant Ti-diffused $LiNbO_3$ strip waveguide doped with scandium'. *IEEE Photonics Technol. Lett.* 2014, vol. 26(17), pp. 1770–3.

[7] Tsou D.H., Chou M.H., Santhanaraghavan P., Chen Y.H., and Huang Y.C. 'Structural and optical characterization for vapor-phase proton exchanged lithium niobate waveguides'. *Mater. Chem. Phys.* 2002, vol. 78, pp. 474–9.

[8] Rambu A., Apetrei A.M., Doutre F., Tronche H., De Micheli M., and Tascu S. 'Analysis of high-index contrast lithium niobate waveguides fabricated by high vacuum proton exchange'. *J. Lightwave Technol.* 2018, vol. 36(13), pp. 2675–84.

[9] Wang T., Ma Z., and Hung W.K. 'Integrated-optic wavelength demultiplexer on lithium niobate by double proton exchange'. *Opt. Eng.* 2007, vol. 46(2), pp. 024601–5.

[10] Castaldini D., Bassi P., Tascu S., Aschieri P., De Micheli M.P., and Baldi P. 'Soft-proton-exchange tapers for low insertion-loss $LiNbO_3$ devices'. *J. Lightwave Technol.* 2007, vol. 25(6), pp. 1588–92.

[11] Skryabin N.N., Bukharin M.A., Kostritskii S.M., Korkishko Yu.N., Fedorov V.A., and Khudyakov D.V. 'Correction of Y-Branches on Proton-exchanged Waveguides in Lithium Niobate By Femtosecond Writing Technology' in VII International Conference on Photonics and Information Optics, 2018. KnE Energy & Physics. pp. 103–8.

[12] Lenzini F., Kasture S., Haylock B., and Lobino M. 'Anisotropic model for the fabrication of annealed and reverse proton exchanged waveguides in congruent lithium niobate'. *Opt. Express* 2015, vol. 23, pp. 1748–56.

[13] Langrock C., Roussev V.R., Nava G., *et al.* 'Nonlinear diffusion model for annealed proton- exchanged waveguides in zirconium-doped lithium niobate'. *Appl. Opt.* 2016, vol. 55(24), pp. 6559–63.

[14] Rivera A., Olivares J., Garcia G., Cabrera J.M., Agullò-Rueda F., and Agullo-Lopez F. 'Giant enhancement of material damage associated to electronic excitation during ion irradiation: the case of $LiNbO_3$'. *Phys. Status Solidi A.* 2009, vol. 206(6), pp. 1109–16.

[15] Pena-Rodrìguez O., Olivares J., Carrascosa M., Garcia-Cabanes A., Rivera A., and Agullò-López F. 'Chapter 12 Optical waveguides fabricated by ion implantation/irradiation: a review' in Goorsky M. (ed) *Ion Implantation* (London, UK, InTech, 2012).

[16] Bentini C.G., Bianconi M., Correra L., *et al.* 'Effect of low dose high energy O^{3+} implantation on refractive index and linear electro-optic properties in X cut $LiNbO_3$ planar optical waveguide formation and characterization'. *J. Appl. Phys.* 2004, vol. 92, pp. 6477.

[17] Olivares J., Garcia-Navarro A., Garcia G., *et al.* 'Buried amorphous layers by electronic excitation in ion-beam irradiated lithium niobate: structure and kinetics'. *J. Appl. Phys.* 2007, vol. 101(3), pp. 033512-1–8.

[18] Chen F., Tan Y., Wang L., Lu Q.M., and Ma H.J. 'Oxygen ion-implanted optical channel waveguides in $Nd:MgO:LiNbO_3$: fabrication, characterization and simulation'. *J. Phys. D: Appl. Phys.* 2007, vol. 40(19), pp. 5824–7.

[19] Chen C., Pang L., Lu Q., *et al.* 'Refractive index engineering through swift heavy ion irradiation of $LiNbO_3$ crystal towards improved light guidance'. *Sci. Rep.* 2017, vol. 7 pp. 10805–12.

[20] Zhang C., Yang J., Chen F., and Kong Y. 'Optical damage of $Zr:LiNbO_3$ waveguides produced by proton implantation'. *Nucl. Instrum. Methods Phys. Res. B* 2012, vol. 286, pp. 209–12.

[21] Chen F. and Vazquez de Aldana J.R. 'Optical waveguides in crystalline dielectric materials produced by femtosecond-laser micromachining'. *Laser Photonics Rev.* 2014, vol. 8(2), pp. 251–75.

[22] Bhardwaj S., Mittholiya K., Bhatnagar A., *et al.* 'Inscription of type I and depressed cladding waveguides in lithium niobate using a femtosecond laser'. *Appl. Opt.* 2017, vol. 56, pp. 5692–7.

[23] Gui L., Xu B., and Chong T.C. 'Microstructure in lithium niobate by use of focused femtosecond laser pulses'. *IEEE Photonics Technol. Lett.* 2004, vol. 16(5), pp. 1337–9.

[24] Lv J., Cheng Y., Lu Q., Vazquez de Aldana J., Hao X., and Chen F. 'Femtosecond laser written optical waveguides in z-cut MgO:$LiNbO_3$ crystal: fabrication and optical damage investigation'. *Opt. Mater.* 2016, vol. 57, pp. 169–73.

[25] Hu H., Ricken R., Sohler W., and Wehrspohn R.B. 'Lithium niobate ridge waveguides fabricated by wet etching'. *IEEE Photonics Technol. Lett.* 2007, vol. 19(6), pp. 417–9.

[26] Hu H., Ricken R., and Sohler W. 'Low-loss ridge waveguides on lithium niobate fabricated by local diffusion doping with titanium'. *Appl. Phys. B* 2010, vol. 98(4), pp. 677–9.

[27] Brüske D., Suntsov S., Rüter C.E., and Kip D. 'Efficient ridge waveguide amplifiers and lasers in Er-doped lithium niobate by optical grade dicing and three-side Er and Ti in-diffusion'. *Opt. Express* 2017, vol. 25(23), pp. 29374–9.

[28] Courjal N., Guichardaz B., Ulliac G., *et al.* 'High aspect ratio lithium niobate ridge waveguides fabricated by optical grade dicing'. *J. Phys. D: Appl. Phys.* 2011, vol. 44(30), pp. 305101–7.

[29] Courjal N., Bernal M.P., Caspar A., *et al.* 'Chapter 8 Lithium niobate optical waveguides and microwaveguides' in You K.Y. (ed). *Emerging Wavelength Technology* (Switzerland, IntechOpen, 2016).

[30] Gerthoffer A., Guyot C., Qui W., Ndao A., Bernal M.P., and Courjal N. 'Strong reduction of propagation losses in $LiNbO_3$ ridge waveguides'. *Opt. Mater.* 2014, vol. 38(4), pp. 37–41.

[31] Pozza G., Kroesen S., Bettella G., *et al.* 'T-junction droplet generator realised in lithium niobate crystals by laser ablation'. *Optofluid. Microfluid. Nanofluid.* 2014, vol. 1, pp. 34–42.

[32] Bettella G., Pozza G., Kroesen S., *et al.* 'Lithium niobate micromachining for the fabrication of microfluidic droplet generators'. *Micromachines* 2017, vol. 8(6), pp. 185.

[33] Betts R.A. and Pitt C.W. 'Growth of thin-film lithium niobate by molecular beam epitaxy' *Electron. Lett.* 1985, vol. 21(2), pp. 960–2.

[34] Li X. 'All-laser-micromachining of ridge waveguides in $LiNbO_3$ crystal for mid-infrared band applications'. *Sci. Rep.* 2107, vol. 7(1), pp. 1–7.

[35] Yao Y., Wang W., and Zhang B. 'Designing MMI structured beam-splitter in $LiNbO_3$ crystal based on a combination of ion implantation and femtosecond laser ablation'. *Opt. Express* 2018, vol. 26(15), pp. 19648–56.

[36] Lansiaux X., Dogheche E., Remiens D., Guillouxviry M., Perrin A., and Ruterana P. 'LiNbO$_3$ thick films grown on sapphire by using a multistep sputtering process'. *J. Appl. Phys* 2001, vol. 90, pp. 5274–7.
[37] Yoon J.G. and Kim K. 'Growth of highly textured LiNbO3 thin film on Si with MgO buffer layer through the sol-gel process'. *Appl. Phys. Lett.* 1996, vol. 68, pp. 2523–5.
[38] Sakashita Y. and Segawa H.J. 'Preparation and characterization of LiNbO$_3$ thin films produced by chemical-vapor deposition'. *J. Appl. Phys.* 1995, vol. 77, pp. 5995–9.
[39] Saulys D., Joshkin V., Khoudiakov M., *et al.* 'An examination of the surface decomposition chemistry of lithium niobate precursors under high vacuum conditions'. *J. Cryst. Growth* 2000, vol. 217, pp. 287–301.
[40] Poberaj G., Koechlin M., Sulser F., Guarino A., Hajfler J., and Günter P. 'Ion-sliced lithium niobate thin films for active photonic devices'. *Opt. Mater.* 2009, vol. 31, pp. 1054–8.
[41] Poberaj G., Hu H., Sohler W., and Günte P. 'Lithium niobate on insulator (LNOI) for micro-photonic devices'. *Laser Photonics Rev.* 2012, vol. 6(4), pp. 488–503.
[42] Boes A., Corocoran B., Chang L., Bowers J., and Mitchell A. 'Status and potential of lithium niobate on insulator (LNOI) for photonic integrated circuits'. *Laser Photonics Rev.* 2018, vol. 12, 1700256.
[43] Guarino A., Pobetaj G. Rezzonico D. Degli'Innocenti R., and Gunter P. 'Electro–optically tunable microring resonators in lithium niobate'. *Nat. Photonics* 2007, vol. 1, pp. 407–10.
[44] Siew S.Y., Cheung E.J.H., Liang H., *et al.* 'Ultra-low loss ridge waveguides on lithium niobate via argon ion milling and gas clustered ion beam smoothening'. *Opt. Express* 2018, vol. 26(4), pp. 4421–30.
[45] Wu R., Wang M., Xu J., *et al.* 'Low-loss lithium niobate on insulator (LNOI) waveguides of a 10 cm-length and a subnanometer surface roughness'. *Nanomaterials* 2018, vol. 8(11), pp. 910–918.
[46] Chen L., Nagy J., and Reano R.M. 'Patterned ion-sliced lithium niobate for hybrid photonic integration on silicon'. *Opt. Mater. Express* 2016, vol. 6(7), pp. 2460–7.
[47] Cai L., Han H., Zhang S., Hu H., and Wang K. 'Photonic crystal slab fabricated on the platform of lithium niobate-on-insulator'. *Opt. Lett.* 2014, vol. 39, pp. 2094–6.
[48] Diziain S., Geiss R., Steinert M., *et al.* 'Self-suspended micro-resonators patterned in Z-cut lithium niobate membranes'. *Opt. Mater. Express* 2015, vol. 5(9), pp. 2081–9.
[49] Lu H., Sadani B., Courjal N., *et al.* 'Enhanced electro-optical lithium niobate photonic crystal wire waveguide on a smart-cut thin film'. *Opt. Express* 2012, vol. 20, pp. 2974–81.
[50] Wang C., Burek M.J., Lin Z., *et al.* 'Integrated high quality factor lithium niobate microdisk resonators'. *Opt. Express* 2014, vol. 22, pp. 30924–33.

[51] Lin J., Xu Y., Fang Z., *et al.* 'Fabrication of high-Q lithium niobate microresonators using femtosecond laser micromachining'. *Sci. Rep.* 2015, vol. 5, pp. 8072-1–4.

[52] Sohler W., Hu H., Ricken R., and Quiring V. 'Integrated optical devices in $LiNbO_3$', *Opt. Photonics News* 2008, vol. 19(1), pp. 24–31.

[53] Cantelar E., Nevado R., Lifante G., and Cussò F. 'Modelling of optical amplification in Er/Yb co-doped $LiNbO_3$ waveguides', *Opt. Quantum Electron.* 2000, vol. 32(6–8), pp. 819–27.

[54] George M., Ricken R., Quiring V., and Sohler W. 'In-band pumped Ti:Tm: $LiNbO_3$ waveguide amplifier and low threshold laser'. *Laser Photonics Rev.* 2012, vol. 7(1), pp. 121–33.

[55] Zhang R., Li H., Zhang P., Hang Y., and Xu J. 'Efficient 1856 nm emission from Tm,Mg:$LiNbO_3$ laser'. *Opt. Express* 2013, vol. 21(18), pp. 20990–8.

[56] Saglamyurek E., Sinclair N., Jin J., *et al.* 'Broadband waveguide quantum memory for entangled photons'. *Nature* 2011, vol. 469(7331), pp. 512–5.

[57] de Mendìvil J.M., del Hoyo J., and Lifante G. 'Ridge waveguide laser in Nd: $LiNbO_3$ by Zn-diffusion and femtosecond-laser structuring'. *Opt. Mater.* 2016, vol. 62, pp. 353–6.

[58] Janner D., Tulli D., Belmonte M., and Pruneri V.J. 'Waveguide electro-optic modulation in micro-engineered $LiNbO_3$'. *J. Opt. A: Pure Appl. Opt.* 2008, vol. 10, pp. 104003–9.

[59] Rao A. and Fathpour S. 'Compact lithium niobate electrooptic modulators'. *IEEE J. Sel. Top. Quantum Electron.* 2018, vol. 24(4), pp. 3400114–30.

[60] Wang C., Zhang M., Stern B., Lipson M., and Lončar M. 'Nanophotonic lithium niobate electro-optic modulators'. *Opt. Express* 2018, vol. 26(2), pp. 1547–55.

[61] Wang C., Zhang M., Chen X., *et al.* 'Integrated lithium niobate electro-optic modulators operating at CMOS-compatible voltages'. *Nature* 2018, vol. 562, pp. 101–4.

[62] Broderick N.G.R., Ross G.W., Offerhaus H.L., Richardson D.J., and Hanna D.C. 'Hexagonally poled lithium niobate: a two-dimensional nonlinear photonic crystal'. *Phys. Rev. Lett.* 2000, vol. 84(19), pp. 4345–8.

[63] Langrock C., Diamanti E., Roussev R.V., Yamamoto Y., Fejer M.M., and Takesue H. 'Highly efficient single photon detection at communication wavelengths by use of up conversion in reverse proton exchanged periodically poled $LiNbO_3$ waveguides'. *Opt. Lett.* 2005, vol. 30, pp. 1725–7.

[64] Zhang S., Yao J., Shi Q., *et al.* 'Fabrication and characterization of periodically poled lithium niobate waveguide using femtosecond laser pulses'. *Appl. Phys. Lett.* 2008, vol. 92, pp. 231106–9.

[65] Osellame R., Lobino M., Chiodo N., Marangoni M., and Cerullo G. 'Femtosecond laser writing of waveguides in periodically poled lithium niobate preserving the nonlinear coefficient'. *Appl. Phys. Lett.* 2007, vol. 90, pp. 241107.

[66] Rao A., Malinowski M., Honardoost A., *et al.* 'Second-harmonic generation in periodically poled thin film lithium niobate wafer-bonded on silicon'. *Opt. Express* 2016, vol. 24(26), pp. 29941–8.

[67] Tanzilli S., Martin A., Kaiser F., DeMicheli M., Alibart O., and Ostrowsky D.B. 'On the genesis and evolution of integrated quantum optics'. *Laser Photonics Rev.* 2012, vol. 6(1), pp. 115–43.

[68] Tittel W., Afzelius M., Chanelière T., *et al.* 'Photon-echo quantum memory in solid state systems'. *Laser Photonics Rev.* 2008, vol. 4(2), pp. 244–67.

[69] De Riedmatten H., Afzelius M., Staudt M.U., Simon C., and Gisin N. 'A solid-state light-matter interface at the single-photon level'. *Nature* 2008, vol. 456, pp. 773–7.

[70] Caspani L., Xiong C., Eggleton B.J., *et al.* 'Integrated sources of photon quantum states based on nonlinear optics'. *Light: Sci. Appl.* 2017, vol. 6, pp. 17100–10.

[71] Bussieres F., Clausen C., Tiranov A., *et al.* 'Quantum teleportation from a telecom-wavelength photon to a solid-state quantum memory'. *Nat. Photonics* 2014, vol. 8, pp. 775–8.

[72] Sua Y.M., Fan H., Shahverdi A., Chen J.Y., and Huang Y.P. 'Direct generation and detection of quantum correlated photons with 3.2 μm wavelength spacing'. *Sci. Rep.* 2017, vol. 7, 17494.

[73] Bettella G., Zamboni R., Pozza G., *et al.* '$LiNbO_3$ integrated system for opto-microfluidic sensing'. *Sens. Actuators B* 2019, vol. 282, pp. 391–8.

[74] Huo D., Choong Z.J., Shi Y., Hedley J., and Zhao Y. 'Diamond micro-milling of lithium niobate for sensing applications'. *J. Micromech. Microeng.* 2016, vol. 26, pp. 095005–15.

Chapter 7

Thin-film deposition: physical techniques

Alessandro Chiasera[1]

7.1 Introduction

The aim of this chapter is to give an overview of the main physical techniques that are used for the fabrication of thin films. The technology and science required a more and more interdisciplinary approach, and – to cover all the requirements – different materials must be considered. This means that also different fabrication protocols must be developed to allow the processing of the various materials and geometries.

Among the various fabrication protocols available, the physical vapour deposition (PVD) technology plays a fundamental role in the manufacture of dielectric, metal or semiconductor films. In this technique, a condensable vapour by physical means is realized, and subsequently, the deposition of the film is achieved from this vapour. To obtain this vapour in a physical way, the most obvious idea is to heat the gas source by a hot filament; more sophisticated evaporation tools, such as a crucible, an electron beam and even laser sources, can also be employed. Another possibility is to extract atoms from a target by bombarding its surface with energetic ions, as in the sputtering deposition techniques.

A possible summary of thin-film deposition processes is reported in Figure 7.1.

The quality and the characteristics of the fabricated films are governed in PVD by numerous parameters, such as the deposition rate, substrate temperature, substrate materials and deposition atmosphere [1]. All these parameters must be monitored and controlled during the fabrication process. Each particular technique, and even each different deposition apparatus, presents specific particularities that must be tailored and defines the final result of the deposition. In this chapter, we will first give a quick hint on the main PVD techniques and, then, a more detailed vision of the sputtering process without losing the generality of the presentation.

7.2 Thermal processes

Thermal processes are particular PVD techniques in which the solid materials that must be deposited are simply heated in order to produce a vapour. Various

[1]Institute of Photonics and Nanotechnologies (IFN – CNR) CSMFO Lab. and FBK CMM, Trento, Italy

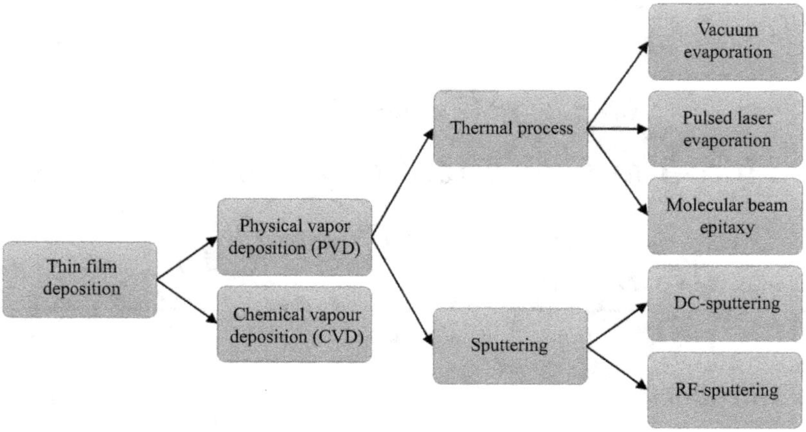

Figure 7.1 Thin-film deposition processes

procedures could be used to produce this vapour, and therefore, different kinds of thermal processes can be identified.

7.2.1 Vacuum evaporation

An evaporation method is a basic and widely used PVD procedure. The aim of this procedure is to transfer, in a controlled way, atoms from a source to a substrate placed away. The source emits atoms because is heated, and all the system is kept at low pressure in order to permit the emitted atoms from the source to reach the substrates. The schematic of a vacuum evaporation apparatus is shown in Figure 7.2.

To heat the source, the first possibility is to employ an electrical heater. Clearly, such a heater has to reach the temperature needed for the material's evaporation and, at the same time, must present a negligible evaporation pressure. Ideally, it should not contaminate the final thin film; these requirements impose to use resistance-heated evaporation sources made of inert oxide or ceramic compounds crucibles.

The electron beam heating technique attempts to overtake the problems related to the contaminations and maximum applied possible power without the evaporation of the crucibles and is becoming the preferred vacuum evaporation technique [2]. The schematic of an electron beam evaporation apparatus is shown in Figure 7.3. The material to be evaporated is placed in a water-cooled copper container, called 'hearth'. The purity of the film is guaranteed by the small number of charge melts or sublimes, so that the effective crucible is the unmelted skull material next to the cooled hearth. Moreover, multiple source units are available for either sequential or parallel deposition of more than one material. The projectile electrons are thermionically emitted from heated filaments that are shielded from the direct line of sight of both the evaporator charge and the substrate. The crucibles are maintained at a potential from 4 to 20 kV in respect to the filament, and a

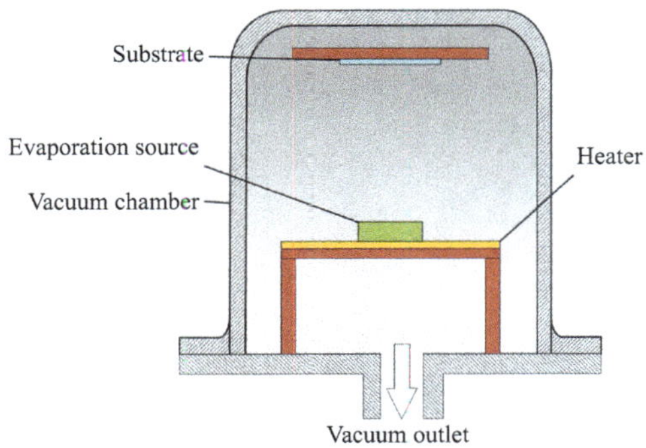

Figure 7.2 Schematic of a vacuum evaporation apparatus

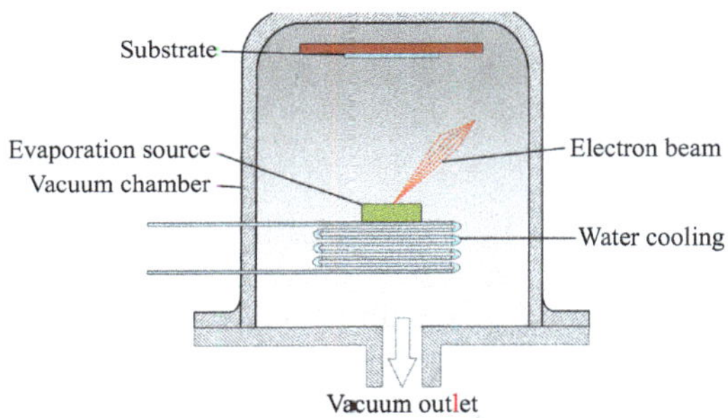

Figure 7.3 Schematic of an electron beam evaporation apparatus

transverse magnetic field is applied to deflect the electron beam and to focus it on the hearth and evaporator charge, which are kept at ground potential. It is important that the electron beam arrives at the centre of the source material, so to avoid hitting the hearth; however, to improve the deposition uniformity, a scanning of the electron beam around the surface may be necessary.

There are numerous adaptations of the principle of the electron beam evaporator, but the greater advantage of this technique is that it may also be used for depositing refractory materials that require higher temperatures than the usual crucibles for thermal deposition, still maintaining the purity of the films.

7.2.2 Pulsed laser deposition

Pulsed laser deposition (PLD) is an improved thermal deposition process where a high-power pulsed laser beam is irradiated onto a target material placed inside a vacuum chamber. In this technique, a condensable vapour is produced when a powerful laser beam strikes a target and vaporizes the surface region [3]. The working mechanism of this deposition approach can be explained using the thermal evaporation model [1,3].

Figure 7.4 Schematic of a pulsed laser deposition apparatus

A schematic example of a PLD apparatus is shown in Figure 7.4. This technique allows fabricating thin films starting from a compound material target with controlled composition. This key feature, shared with the sputtering deposition technique, can be successfully used for the fabrication of optical materials of superconductive films [4,5], where the deposition of materials with complex composition, using also a particular atmosphere, is mandatory.

The deposition is typically achieved by selecting an ultraviolet laser wavelength and a nanosecond pulse length that is strongly absorbed by a small volume of the target material. The laser absorption by the ejected material creates a plasma plume [6]. In practice, the laser beam enters in the deposition chamber through a quartz window and is focalized on the target material that can be moved or changed. The substrates could be usually kept faced to the substrate to optimize the deposition rate, but also other geometries can be used to avoid the deposition of micro-sized ejections. For this purpose, also spinning apertures synchronized with the laser pulses can be employed, to filter the macroparticles just taking advantage of the different velocity between particles and atoms [5].

This deposition technique exhibits a lot of advantages and in particular requires a relatively easy geometry and can be used with targets of different nature (glasses, crystals, sintered pellets, etc.); the final results, however, have a limited area of uniformity, and the deposition of microparticles could be an issue. To overtake this problem, more complicated geometries must be introduced.

7.2.3 Molecular beam epitaxy

In the fabrication of thin films, the substrates and the adhesion on it are, of course, fundamentals. In the epitaxial deposition protocol, this bond with the substrate is even tighter, also because the crystallographic order of the fabricated film is strongly influenced by that of the substrate. This technique is probably the queen of the fabrication techniques for semiconductor thin-film devices, also from the industrial point of view in the manufacturing processes of electronic and opto-electronic devices.

The main features that distinguish the molecular beam epitaxy (MBE) from other growth techniques are the precise reproducibility of all parameters involved during the epitaxial process, the growth conditions far from thermodynamic equilibrium and the possibility of controlling the kinetic evolution of the outermost layers of the epitaxial film. Moreover, this technique could also be used to produce multilayer structures consisting of many alternate thin layers with a single thickness on the angstrom thickness range. Dopants are evaporated onto the growing film through separate sources. In this way, the doping profile may be varied and controlled with a spatial resolution not easily achieved by other techniques.

An MBE system can be considered basically a vacuum evaporation apparatus. The substrate holder is located a few centimetres away from the openings of the effusion cells, along with the centreline of the system. The temperature of the substrate can be set during the deposition (from room temperature up to about 1,400°C), at a suitable temperature depending on the epitaxial process. It can also be heated before deposition, primarily for cleaning and/or surface reconstruction, and afterwards for various heat treatments. In the growth of compound semiconductors, the materials contained in the crucible could be the compounds themselves, their components, or different elements to be used as doping impurities. The crucible is usually made of pyrolytic boron nitride, which can stand temperatures up to about 1,400°C without harmful material dissociation on the grown layers. Standard effusion cells are limited to operate at a temperature lower than 1,200°C, which is only just within the range of that required for Si, Ge, Al, Ga evaporation. The shape can be either cylindrical or conical, with different tapering angles, depending on the material to be evaporated, as well as its diameter.

The nature and the preparation of the substrates are of crucial importance because the interactions between the evaporated atoms and the surface of the substrate are on the basis of this deposition method. In fact, during the epitaxial growth, the involved mechanism can be summarized as (i) adsorption of the evaporated atoms or molecules impinging on the surface, (ii) surface diffusion and dissociation of the absorbed atoms, (iii) atoms arrangement into the crystal lattice and (iv) thermal desorption of the species not arranged into the lattice [7]. The interaction of the atoms with the substrate surface, which of course must first be cleaned, is fundamental, and the other factors affecting the deposition are the substrate temperature, the arrival rates and the affinity with the substrate.

7.3 Sputtering

A rough definition of sputtering is provided by Sigmund [8] as 'Erosion of material surfaced by particle impact'. By this, it is meant that a sputter event is identified as the emission of atoms or molecules from a material surface produced by

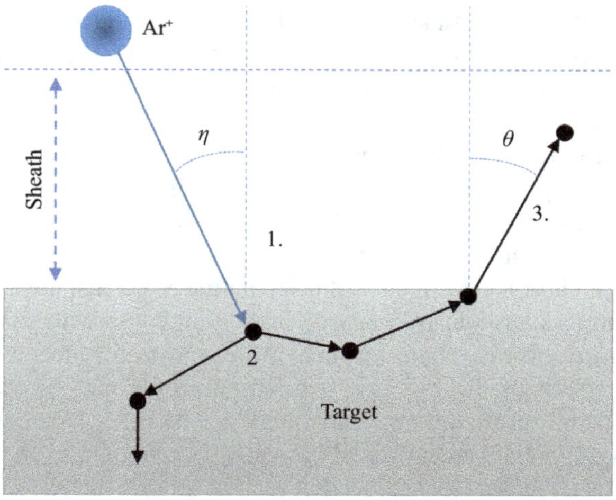

Figure 7.5 A sputtering event consists of accelerations of an ion across the cathode sheath (1), atomic collision cascade within the target (2), ejection of a target atom (3); η is the incident angle and θ the emission angle

bombarding particles. This phenomenon could be therefore successfully employed to etch the surface of the samples or to extract atoms from one or more target that will be deposited on a substrate [9].

This technique is widely used in different fields of research and development and in industrial manufacturing. Milling, etching, thinning and polishing of microstructures are possible application; sputter cleaning procedures make this process very important for the surface physics. The possibility of using this phenomenon to fabricate thin layers also makes it very appealing to produce thin films for photonic components [10]. Different approaches, however, are used for the deposition of different materials and the various possible applications. For the fabrication of optical materials, particular attention has to be taken to control the various parameters; adhesion and homogeneity of the films are crucial for the fabrication of multilayer structures. For this reason, it is important to deeply understand the sputtering mechanism, to identify which parameters of the fabrication protocol have to be optimized.

The sputtering process, when utilizing the ions from a gaseous discharge, portrayed schematically in Figure 7.5, consists of the following:

1. acceleration of an ion across the cathode sheath (the region surrounding the target);
2. penetration of the target, resulting in a series of atomic collisions and
3. backward ejection of one or more recoiling target atoms.

Considering the energy of the projectiles, the collision cascade event could be divided into three regimes:

1. Low-energy regime, or single-knock on regime: Ion–surface collisions set target atoms in motion and may give rise to separate knock-on events. If enough energy is transferred to target atoms, they overcome binding forces and sputter. Threshold energy E_{th} (between 5 and 40 eV) is the minimum energy required to do this and depends on the nature of the incident ion, the mass and the atomic number of the target atoms.
2. Linear collision cascade regime: The density of recoils is sufficiently low so that most collisions involve one moving and one stationary particle, resulting in the ejection of target atoms (sputtering). Sputtering yield in this regime is found to be proportional to the first power of the deposited energy density at the surface [8], that is $Y = \Lambda\ F_D(E)$, where Λ is a constant depending on material properties (such as the binding energy), while F_D accounts for the energy deposited at the surface and depends on type, energy and incident angle of the ion as well as target parameters.
3. Spike regime: At a higher density of recoils, the majority of the atoms within the cascade volume are simultaneously in motion. These atoms resemble a vapour, and the ejection mechanism is similar to vaporization. For this regime, theory predicts, and experiments confirm, that the sputter yield is highly non-linear with the deposited energy. However, if the aim is to deposit thin films, the homogeneity and the quality are decreased; thus, usually, in a sputtering deposition, this regime is not used, while linear cascade regime is the preferential choice.

It is clear that, to use this sputtering process to fabricate thin films, particular attention has to be devoted to the generation and the manipulation of the projectile ions, and in particular to the processing of the support plasma that is the vector that allows this technique. Plasma is a partially ionized gas, populated by singly charged positive ions, electrons and neutral gas particles. The concentration of electrons and ions is such that plasma is globally neutral, but, in the presence of electromagnetic fields, plasma exhibits a collective behaviour. Depending on electron kinetic energy and concentration, there are different regimes of plasma, but practical plasma for sputtering is the glow-discharge (kinetic energy of electrons (E_k) between 100 and 101 eV, and concentration 10^{10} cm^{-3}). There is not a fixed value of the pressure of support gas (that usually can be Ne, Ar, Kr or Xe) in the chamber; one should find a balance between the number of suitable ions that can be involved in the etching process and the free mean path of the eroded target atoms, which must be enough to keep a suitable deposition rate. The final deposition rate, in fact, depends on numerous parameters that must be controlled.

One important factor is the sputtering yield: in the linear cascade regime, it is again a function of various parameters and is important because is related not only to the duration of the total deposition process but also becomes crucial in particular cases such as the deposition of different materials together. The most important factors that influence the sputtering yield are the projectile energy dependence, the surface binding energy of the target, the projectile target and identity, the projectile angle of incidence, but the experimental yield varies also with the amount of target conditioning by the previous sputtering procedures that strongly modify the surface of the target. At present, empirical curves are the trust estimates of sputter yield in a thin-film deposition regime. With respect to the incident ion's kinetic energy, there is a sharp yield threshold (20–40 eV), followed by a rapid increase in yield with energy. After a maximum, at around 10 keV, there is a decline, stronger at very high energy (~100 keV). The threshold exists because there is a minimum energy that must be provided for a particle to escape from the target (surface binding energy of the target). The decline is due to a decrease in nuclear energy loss probability, together with the onset of electronic energy loss (projectile energy goes into stripping electrons from target atoms instead of dislodging the atoms from their lattice sites). In the literature, there are many proposals of empirical/semiempirical formulas for energy dependence of yield. Bohdansky [11] observed that yield curves for many ion target combinations are of nearly the same shape from near-threshold up to projectile energies of few keV. The empirical expression is as follows:

$$Y = \left(6.4 \times 10^{-0}\right) m_r \gamma^{5/3} E^{0.25} \left(1 - \frac{E_{\text{th}}}{E}\right)^{3.5}$$

where γ is the energy-transfer mass factor, m_r is the recoil mass in amu, E is the initial energy of the projectile in eV, and E_{th} is the threshold energy in eV. This is generally the best empirical curve that fits experimental data of different experiments. For some systems, the normalization factor can be changed [3]. Of course, also the nature and, in particular, the surface binding energy of the materials that have to be processed changes drastically the sputtering yield. When the projectile and its energy are held constant, the yield increases with the increasing volatility of the target material. A powerful example is carbon, having in the form of graphite the lowest vapour pressure and the lowest sputter yield of all elements [3]. Moreover, the sputtering yield also depends on the choice of the support gas and the projectile used to etch the material. For heavy targets, the experimental yield increases with a projectile in the order Ne→Ar→Kr→Xe. However, for light targets, the yield exhibits a maximum at an intermediate projectile mass, usually Kr. This trend is essentially due to the variations in a projected range (spatial distance between the point where projectile enters the target and when projectile is absorbed, projected on incidence direction) of the projectile [3]: a small projected range leads to the deposition of more of the projectile's energy near the surface, and thus to a higher sputter yield. Anyway, argon is the most common sputtering gas.

For technological applications, the fabrication of uniform thin films is of crucial importance; the angle effect is therefore important. The yield increases for off-normal incidence, up to a certain point, and then decreases. It is found experimentally that the near-normal dependence varies with $1/\cos(\eta)$, where η is the angle of incidence measured from the surface normal (Figure 7.5). This trend can be understood to be due to an increase in the probability of escape of the recoils for off-normal incidence. Thus, a trend Y is proportional to 1/depth, and depth can be calculated as $R_p\cos(\eta)$ (R_p is the projected range of the projectile within the target).

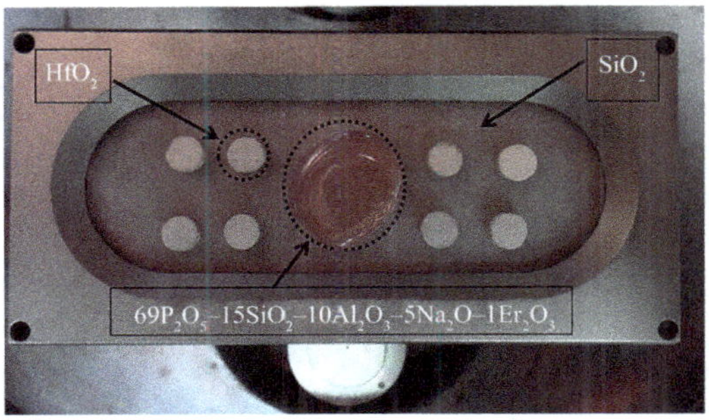

Figure 7.6 *Target for sputtering deposition made of a 15 × 5 cm² silica target on which the $69P_2O_5–15SiO_2–10Al_2O_3–5Na_2O–1Er_2O_3$ glass and four disks of HfO_2 (diameter 5 mm) were placed. More details in [12]*

Finally, the real target physical condition and its history influence the sputtering rate; it is affected by variations on the target's surface, such as (i) possible existence of oxides on the surfaces; (ii) evolution of surface morphology during sputtering and (iii) accumulation of implanted ions. Because of these different causes, the yield has been observed to both increase and decrease with dose, making difficult to model the yield for a practical application, such as thin-film deposition. Experimental measures like the one presented earlier are always performed with the target brought to a stationary state by previous conditioning.

All these considerations become fundamental when sputtering is used on targets made up of various components. In specific cases, for instance when it is required to modify strongly and often the compositions of the films or it is complicated to acquire a target with specific characteristics, it may be more convenient to use a target made of different components, as shown in Figure 7.6. This specific target was employed to fabricate phosphate-based planar waveguides starting from a particular bulk glass [12,13]. The drawbacks of this approach are the requirement of a compositional analysis to define the final composition of the films, which depends on the geometry of the target and also on the usage history of the single components of the target.

7.3.1 DC sputtering

The scheme of a DC sputtering apparatus is presented in Figure 7.7. Sputtering can be performed with different equipment, depending on the way exploited by the experimenter to switch on the plasma. DC sputtering exploits a DC high-voltage power supply to create the plasma and thus perform the deposition.

The sputtering target is the cathode of the discharge, while the anode is the substrate and/or vacuum chamber walls. Cathode–anode separation is typically a few centimetres. Argon is the most common sputtering gas, at a pressure on the order of 1.33 mbar. The plasma is created and sustained by the DC source via a mechanism that pertains to the abnormal glow regime (secondary electron emission at the cathode and impact ionization of neutral gas atoms) [3]. The ion current density to the cathode is around 1 mA/cm^2.

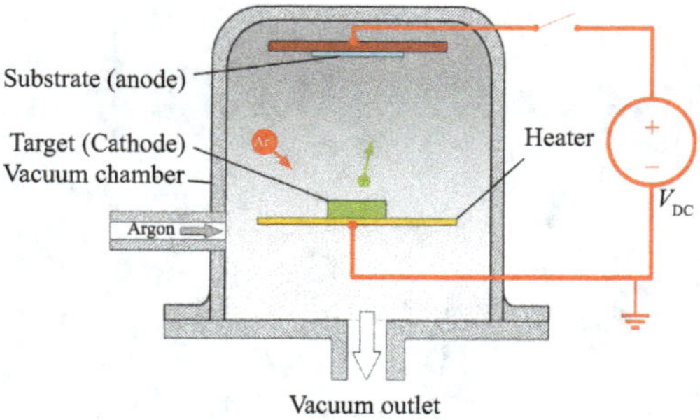

Figure 7.7 Schematic of a DC-sputtering apparatus

7.3.1.1 Cathode and anode sheaths

The interaction between the plasma and the target surface leads to the formation of regions, just near the target surface, that are positive electrically charged. These regions are called sheaths. Phenomenologically, what happens is that, due to the aleatory motion of particles in the plasma, particles move in the chamber with collisions and diffusion. Thus, there are two kinetic effects. Consider the mobility μ of charged species in the plasma, namely ions and electrons. Due to the difference in mass of ions and electrons, $\mu_e \gg \mu_{\text{ions}}$. The diffusion phenomenon is governed by Fick's law:

$$J = D\frac{dC}{dx}$$

where J is the mass flux and dC/dx is the concentration gradient. When migrating species move under the simultaneous influence of two driving forces (diffusion in a concentration gradient and drift in the applied electric field, the one used for generating the plasma), particles fluxes are as follows:

$$J_e = -n_e\mu_e E - D_e\frac{dn_e}{dx}$$

$$J_i = n_i\mu_i E - D_i\frac{dn_i}{dx}$$

where n is the volumetric density of particles (for both ions and electrons). In order to maintain the global charge neutrality of the plasma, $J_e = J_i = J$, $n_e = n_i = n$, thus

$$E = \frac{(D_i - D_e)}{n(\mu_e + \mu_i)} \cdot \frac{dn}{dx}$$

Therefore, the difference in electron and ion diffusivities produces a separation of charges, and thus, an electric field develops. Thus, physically, more electrons than ions tend to leave the plasma, establishing an electric field that hinders further electron loss but at the same time enhances ion motion. Subsequently, additional electrons are repelled, while positive ions are attracted. Therefore, the surface is

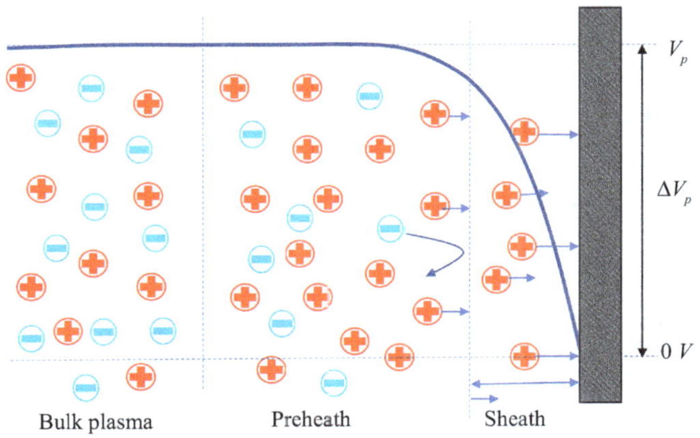

Figure 7.8 Charge distribution and voltage profile near a surface abutting glow-discharge plasma

negatively charged and continues to charge negatively at a decreasing rate until the electron flux equals the ion flux, and there is no net current (steady-state condition). Observing the region just near the target, in steady-state condition, there is a positive charge region, the sheath (Figure 7.8). Experimentally, the electric field in DC is generated with an anode–cathode couple.

Obviously, the mechanism depicted earlier is influenced by the presence of the cathode at the target and of the substrate at the anode, but the net effect is observed as well, enhancing the ion attraction at the cathode. The body of the plasma is thus the object at higher potential. The profile of the potential in DC apparatus is evident in Figure 7.9(b).

7.3.1.2 DC discharge model

The two electrodes of the experimental apparatus (Figures 7.7 and 7.9(a)) can be portrayed as diodes in an electric circuit schematic (Figure 7.10). The cathode is the electrode which attracts cations (positive ions) from the plasma, while the anode attracts anions (electrons). The body of the plasma is assumed to be equipotential.

The Kirchhoff's voltage law applied to the entire DC discharge unit requires that the potential drop across the cathode sheath is $\Delta V_{\text{cathode}} = V_p + V_{\text{DC}}$, where V_p is plasma potential and V_{DC} the applied voltage between anode and cathode. Thus, this is the maximum kinetic energy that ions striking the target can have, and it is greater than V_{DC}. The sputtering power density to the target is given by

$$p_{\text{ion}} = j_{\text{ion}}\left(V_p + V_{\text{DC}}\right).$$

It is worth noticing that most of the space between the electrodes is electric-field free. Two high-field regions (sheaths) separate the plasma body from anode and cathode. The apparent bombardment of ions to the anode (thus to the substrate) gives, in reality, no appreciable effects, due to their low kinetic energy.

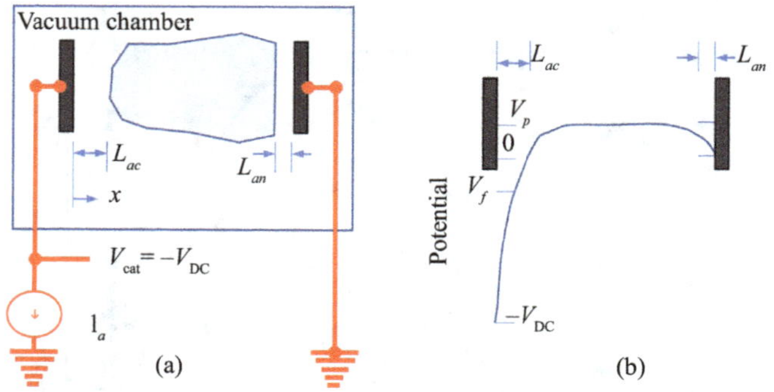

Figure 7.9 (a) Schematic diagram of DC-sputtering discharge and (b) potential profile (not to scale)

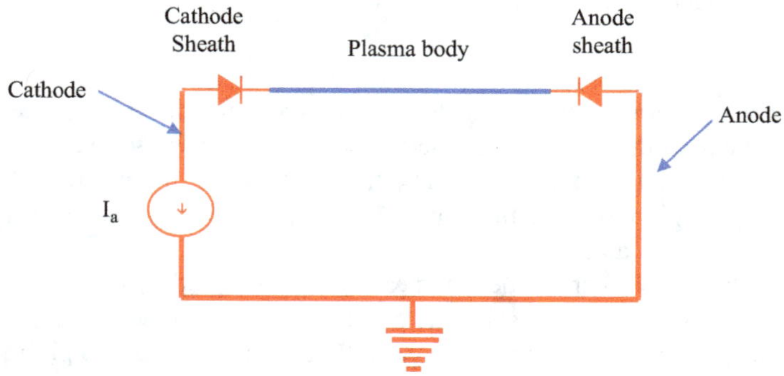

Figure 7.10 Circuit model for DC-sputtering discharge

The plasma of the discharge is sustained with free electrons generated from secondary electron emission when ions strike the surface. These are accelerated across the cathode sheath into the body of the plasma. Here, they strike neutral gas atoms, creating an ion–electron pair, that sustain the plasma.

All stated earlier are true if the target is a conductor. To place an insulator onto the cathode of DC discharge would make the discharge inoperable because of no current flow through the insulator. This is due to the potential profile (Figure 7.9 (b)) in the chamber and to the polarization effect of an insulator. In fact, the potential profile makes the dielectric material polarized, and thus, positive ions, instead of sputtering, stuck on the surface of the target, repelling other positive ions and stopping sputtering. A method to perform sputtering with an insulator target is to replace the DC power supply with radio frequency (RF) power supply, i.e. using RF sputtering.

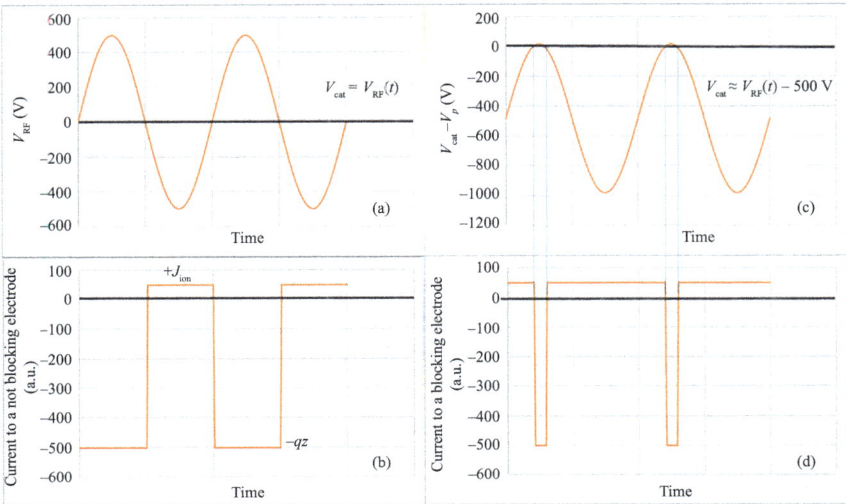

Figure 7.11 *DC self-bias. Potentials and currents for RF excitation of (a and b) nonblocking and (c and d) blocking electrodes*

7.3.2 RF sputtering

7.3.2.1 RF discharge model

Consider an input waveform $v_{RF} = V_{RF}\sin(\omega t)$ (Figure 7.11(a) and (b)):

- v_{RF} mediated in one period is 0;
- electrons are lighter than ions; thus, they react faster to the change of sign of v_{RF} and
- net current averaged in one period is not zero, thus no sputtering.

In order to reach the steady-state condition of zero net currents (obtained in DC case), v_{RF} should be summed to a negative potential DC. Moreover, it is impracticable to place DC voltage supply: plasma potential is not fixed but changes continuously in time, depending on the conditions of chamber, gas and target. This would lead the experimenter to change continuously the value of the DC voltage supply during the deposition. An efficient way to solve the problem is represented by the blocking capacitor.

The RF-sputtering apparatus is represented again as a circuit (Figure 7.12), where sheaths are represented no more with a diode only, but a diode with a capacitor placed in parallel. These capacitors represent the oscillation behaviour of sheaths and were omitted in the DC model because they were not relevant. As the first approximation, the drop potential across the anode sheaths is neglected (very small if compared to the cathode), so as the displacement current due to cathode sheath capacitance, so that all the current that flows in the circuit is due to ions and electrons moving across the cathode-sheath conduction current. The capacitor C placed on the right of the RF voltage supply is

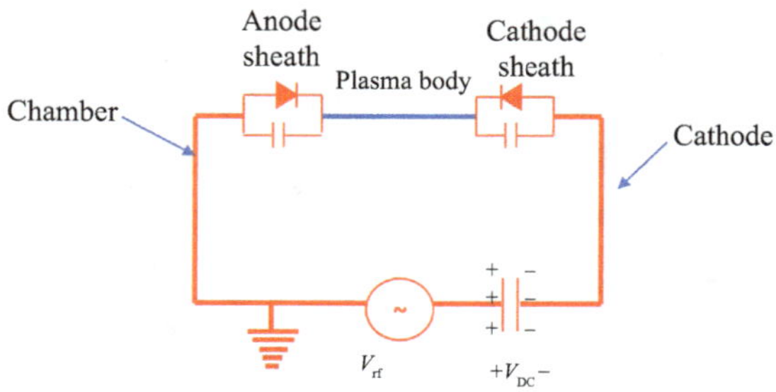

Figure 7.12 Circuit model for capacitive RF-sputtering discharge

the blocking capacitor, and, applying Kirchhoff's law to the circuit, the cathode fall is

$$v_{cat}(t) = v_{RF} - V_{DC}$$

which means that, most of the time, the potential of the target is below that of the plasma. This phenomenon is called DC self-bias and develops because negative charge accumulates on the right-hand plate. Obviously, the capacitor must be dimensioned correctly:

- charging time during electron bombardment must be not much greater than the RF period (self-bias develops quickly);
- discharging time during ion bombardment must be greater than RF period (prevents neutralization of significant amounts of negative charge on the cathode);
- most of the applied RF potential must be dropped across the cathode sheath rather than the blocking capacitor (the value of cathode fall could never become high) and
- typical values are 10–1,000 pF.

If it is true, the blocking capacitor allows reaching the zero net current condition needed to have sputtering. This condition, in the presence of the blocking capacitor, is shown in Figure 7.11(c) and (d): the area under (d) must be zero for one complete period in the steady state. The plasma sustainment mechanism is the same as in DC sputtering: the source of free electrons needed for electron impact ionization of neutrals is secondary electron emission by ions, which strike the self-biased electrode.

The sheaths of a capacitive RF discharge are formed in a manner entirely different from those of the DC discharge. The governing principle in RF is the oscillatory free-electron motion, which creates displacement current within the sheaths [3]. Electrons in the body of the plasma oscillate between the two electrodes, while ions are virtually stationary [3]. The width of the two electrodes in the body of the plasma is *d* (electrode separation, Figure 7.13), while the width of the electron cloud is smaller because any electron that comes in contact with an electrode is collected. Since the local ion and electron densities are equal within the bounds of the electron cloud, while outside the electron cloud are no electrons, the net result is a positive charge for the plasma body as making it the most positive body in the discharge.

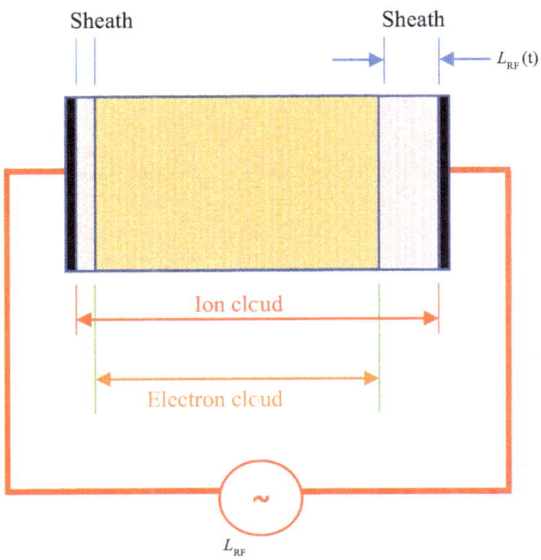

Figure 7.13 Oscillation of the free-electron cloud in an RF discharge creates sheaths

Let us assume a total current density j_{RF} sin(ωt) for the discharge. In a symmetric, homogeneous model of the RF discharge [14], each of the sheath thicknesses is of the form

$$L_{RF}(t) = L_{RF} - L_{RF} \cos(\omega t)$$

where there will be a 180° phase difference between right and left-hand sheaths. Each sheath thickness oscillates with the period of the RF current and goes to zero during each cycle: the sheath must collapse to zero briefly in order to transfer electrons to the electrode, which neutralizes the positive charge that accumulates by continuous ion bombardment. Thus, the width of the electron cloud is $d - L_{RF}$, while the amplitude of the sheath thickness is

$$L_{RF} = \frac{j_{RF}}{qn + \omega}$$

This estimation as a function of time is qualitatively correct with this simple model, but not very much accurate [3].

7.3.2.2 Matching network

In the practical use, an RF power supply (Figure 7.12) is not ideal but has an internal resistance of 50 Ω, much lower than the typical plasma resistance, around 400 Ω for plasma excited by RFs at 13.56 MHz [5]. This impedance mismatch, combined with the displacement current generated by the capacitance of the plasma, gives rise to reflected power into the power supply, overloading it and anyway giving an inefficient delivery of power. The problem can be resolved by inserting an *L–C* matching network, such as the one sketched in Figure 7.14.

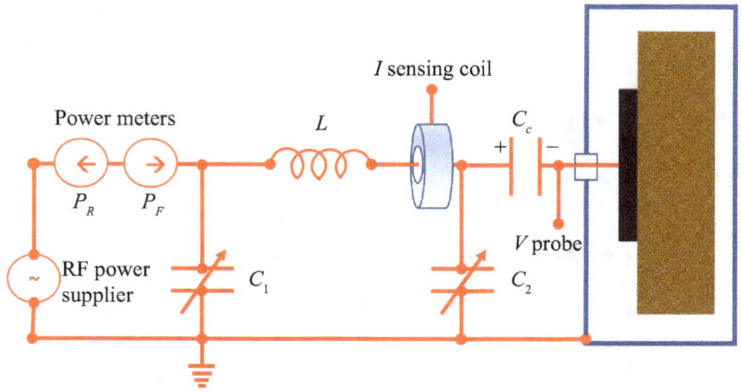

Figure 7.14 L–C matching network for high-frequency RF plasma coupling

In a crude analysis, the variable capacitor C_1 is tuned to the 50-Ω source impedance (load capacitor), about 200 pF, and the variable capacitor C_2 to the 400-Ω impedance (shunt capacitor), about 25 pF. In the inductor L, V leads I by 90°, just the opposite of a capacitor; so, when $X_L = X_C$ (impedance equal), the I–V phase is restored to zero. Then, the system network+plasma looks like a load from both the ends, so that the reflected power goes to zero. The value of L needed to match $C_1 + C_2$ at 13.56 MHz is about 6.3 μH [5].

Finally, a working apparatus for RF sputtering can be represented as in Figure 7.15, where the matching network and the blocking capacitor are in the same box because it is better to have the electronic components altogether.

7.3.2.3 Magnetron

A magnetron configuration can be implemented in a sputtering apparatus in order to achieve higher deposition rates. A permanent magnet is added under the target, in order to create a line of magnetic flux that is perpendicular to the electric field used for the ignition and sustainment of the plasma, and thus parallel to the surface of the target (Figure 7.16). The magnetic field concentrates and intensifies the plasma in the space immediately above the target, as a result of trapping of electrons near the target surface. Trapping occurs as a result of the drift of the electrons in the $-\vec{\mathbf{E}} \times \vec{\mathbf{B}}$ direction superposed to cycloidal motion (Figure 7.16(c)). The gyroradius of their orbits is given by

$$\rho = \frac{mv_\perp}{qB}$$

where m is the electron mass, q is the electron charge, B is the strength of the magnetic field, $v_\perp$ is the component of the electron velocity that is perpendicular to the flux lines. In sputtering systems, this radius is typically of the order of few millimetres; thus, the confinement near the target surface is quite effective. An electron encircles the line flux until it is scattered by another particle.

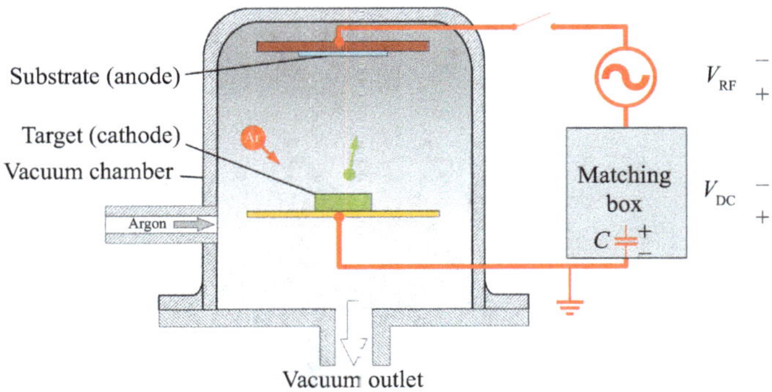

Figure 7.15 Sketch of a single-target RF-sputtering apparatus

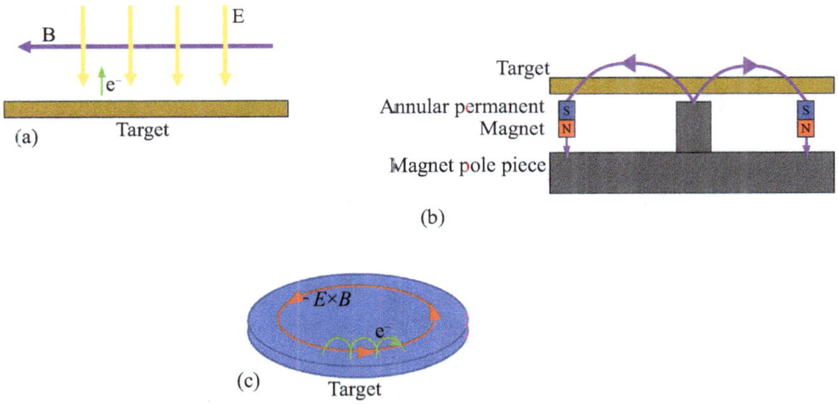

Figure 7.16 (a) Planar magnetron-sputtering arrangement: a static magnetic field is created parallel to the surface of the sputtering target to retain secondary electrons in that region; (b) annular design and (c) electrons drift in the $-\vec{\mathbf{E}} \times \vec{\mathbf{B}}$ *direction, actually executing a cycloidal path*

7.3.2.4 Control system of the thickness

As an example, let us consider the fabrication of a 1D photonic crystal: it is of crucial importance to monitor the evolution of the thickness of the deposited layers during the fabrication. Various approaches can be employed, such as in situ ellipsometry, or a conductive method where the thickness variations are monitored by measuring the electrical resistance change of the film, but here we refer to the approach that uses a vibrating quartz crystal. We need, however, to keep in account that this monitoring method is strongly influenced by the presence of RF interference generated by the plasma. A careful calibration and positioning of the sensors is therefore necessary, and these requirements could give a limitation on the resolution of the measurements. The

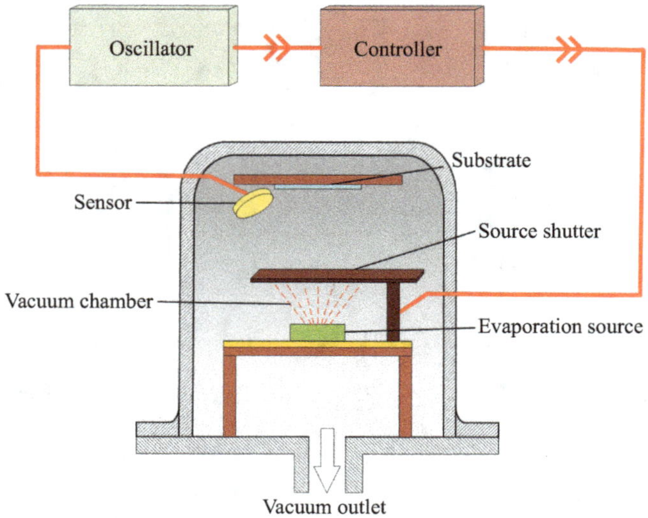

Figure 7.17 General set-up for a quartz crystal deposition monitor

vibrating quartz crystal mass-deposition monitor, or quartz crystal microbalance, is one of the most powerful and widely used diagnostic instruments in thin-film technology [5].

A possible configuration that could be used for the placing of these sensors in a deposition apparatus is reported in Figure 7.17 [15]. The physical working principle exploits the piezoelectric sensitivity of a quartz crystal with a mass added; its mass sensitivity is used to control the deposition rate and final thickness of the deposition [16]. When a voltage is applied across the faces of a properly shaped piezoelectric crystal, the crystal is distorted and changes shape in proportion to the applied voltage. At certain discrete frequencies of applied voltage, a condition of very sharp electro-mechanical resonance is encountered. When the mass is added to the face of a resonating quartz crystal, the frequency of these resonances is reduced. This change in frequency is very repeatable and is precisely understood for specific oscillating modes of quartz [15]. Electrical coupling is done with thin-film metal electrodes deposited on opposite faces of a thin quartz wafer having the proper crystallographic orientation (Figure 7.18). For deposition monitoring, one electrode is exposed to the vapour flux and proceeds to accumulate a mass of deposit.

Quantitatively, the change of mass is related to the change of resonance frequency by the following relation [15]:

$$\frac{M_f}{M_q} = \frac{\Delta F}{F_q}$$

where M_q is the mass of the uncoated quartz, F_q is its resonance frequency and ΔF the change in frequency after the coating. This monitoring method allows reaching a sub-angstrom resolution in the fabrication of thin glass-based films. However, the presence of electrical interferences induced by the plasma discharge imposes to

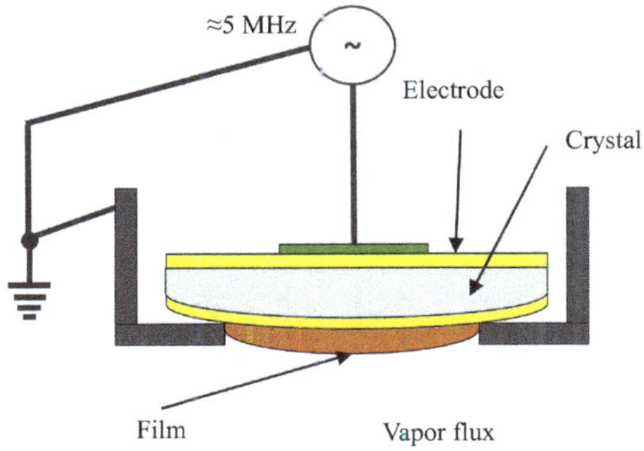

Figure 7.18 Quartz crystal deposition monitor cross section

Figure 7.19 Cross-sectional image of a magnetron RF-sputtering fabricated multilayer sample, obtained with the scanning electron microscopy. The bright and the dark areas are TiO_2 and SiO_2 layers, respectively. The substrate is located at the bottom of the image

place the balance sensor far away from the substrates. The calibration gives anyway the possibility to monitor the evolution of the thickness but with limited resolution.

Other experimental techniques such optical monitoring system for simultaneous reflectance and transmittance measurements during deposition [17] have also been demonstrated to be suitable for the fabrication of glass-based 1D photonic crystals; the most suitable monitoring system can be adopted in respect to the required resolution, material to be deposited and total thickness required to be checked. All these monitor procedures become even more important when the PVD technique is employed for the fabrication of optical multilayer structures, such as the one reported in Figure 7.19, where the reproducibility of each layer is mandatory to achieve good optical quality [16].

Not only the thickness and the refractive index of the thin film are important parameter, but in the case of multicomponent films, also the composition is of crucial importance. For that reason, quadrupole mass spectrometer sensors are commonly used in the MBE apparatus to monitor during the process the gas composition in the deposition chamber [7].

7.4 Conclusions

This chapter gives a fast overview of PVD techniques, with more details on the RF-sputtering process and apparatus. Integrated optics requires fabrication protocols for thin films that allow not only high reproducibility but also the capability to operate with different materials and substrates of disparate nature to cover all the functions required by the multidisciplinary that characterizes integrated optics. These factors make a very precise subdivision of the different deposition techniques almost useless; moreover, all the PVD processes share many characteristics. It is not possible, therefore, to identify only one protocol that fulfils all the requirements, but the combination of different techniques will be the key to integrate on the same device various materials and functionalities.

The need for new devices with different functionalities, to be used, for example, in surveying, telecommunication, lighting and energy storage applications, makes it necessary to develop manufacturing protocols suitable for large mass industrial productions, compatible with protocols and technology manufacturing already established, ability to manage different materials and allow low production and maintenance costs. PVD techniques, with the possibility to be combined together, operate with high reproducibility on different materials and are compatible with the conventional complementary metal–oxide–semiconductor (CMOS) fabrication protocols that can satisfy these requests and will surely play a fundamental role in the development of future new devices.

Acknowledgements

I would like to thank all the colleagues who have been closely collaborating with me on this topic.

References

[1] Wasa K., Kitabatake M. and Adachi H. *Thin Film Materials Technology*. (Norwich, William Andrew Publishing and Springer Verlag, 2004).

[2] Ohring M. *Materials Science of Thin Films: Deposition & Structure*. (San Diego, CA, Academic Press, 2001).

[3] Mahan J.E. *Physical Vapor Deposition of Thin Films*. (New York, NY, John Wiley & Sons, 2000).

[4] Haindl S., Hanzawa K., Sato H., Hiramatsu H. and Hosono H. 'In-situ growth of superconducting $SmO_{1-x}F_xFeAs$ thin films by pulsed laser deposition'. *Scientific Reports* 2016; 6: 35797.

[5] Smith D.L. *Thin-Film Deposition: Principles & Practice*. (New York, NY, McGraw Hill, 1995).

[6] Eason R. (ed.). *Pulsed Laser Deposition of Thin Films Applications-Led Growth of Functional Materials*. (Hoboken, New Jersey, John Wiley & Sons, 2007).

[7] Herman M.A. and Sitter H. *Molecular Beam Epitaxy*. (Berlin, Springer-Verlag, 1989).

[8] Sigmund P. 'Recollections of fifty years with sputtering'. *Thin Solid Films* 2012; 520: 6031–6049.

[9] Meroni C. *Low-Threshold Coherent Emission from Fully Er^{3+} Doped Monolithic in-Fiber 1D Dielectric Microcavity Fabricated by RF-Sputtering*. MSc Thesis, Department of Physics, University of Trento (Trento, Italy, 2018).

[10] Miller J.W., Khatami Z., Wojcik J., Bradley J.D.B. and Mascher P. 'Integrated ECR-PECVD and magnetron sputtering system for rare-earth-doped Si-based materials'. *Surface and Coatings Technology* 2018; 336: 99–105.

[11] Bohdansky J., Roth J. and Bay H.L. 'An analytical formula and important parameters for low-energy ion sputtering'. *Journal of Applied Physics* 1984; 51(5): 2861–2865.

[12] Chiasera A., Vasilchenko I., Dorosz D., *et al.* 'SiO_2-P_2O_5-HfO_2-Al_2O_3-Na_2O glasses activated by Er^{3+} ions: From bulk sample to planar waveguide fabricated by RF-sputtering'. *Optical Materials* 2017; 63: 153–157.

[13] Dorosz D., Kochanowicz M., Zmojda J., *et al.* 'Rare-Earth Doped Materials for Optical Waveguides'. *Proceedings of ICTON 2015*; Budapest, Hungary, July 2015 (IEEE), paper We.A5.3.

[14] Lieberman, M.A. and Lichtenberg, A.J. *Principles of Plasma Discharges and Material Processing*. (New York, NY, John Wiley & Sons, 2005).

[15] INFICON. *SQM-160 Multi-Film Rate/Thickness Monitor – Operating Manual*. (Bad Ragaz, Switzerland, INFICON, 2012). https://products.inficon.com/getattachment.axd/?attaName=52c9fbd2-c674-4faf-8922-93f621b63847.

[16] Chiasera A., Meroni C., Varas S., *et al.* 'Photonic band edge assisted spontaneous emission enhancement from all Er^{3+} 1-D photonic band gap structure'. *Optical Materials* 2018; 80: 106–109.

[17] Wu C.Y., Zou Y.H., Timofeev I., *et al.* 'Tunable bi-functional photonic device based on one-dimensional photonic crystal infiltrated with a bistable liquid-crystal layer'. *Optics Express* 2011; 19: 7349–7355.

Chapter 8
Thin-film deposition: chemical techniques

Anna Lukowiak[1] *and Beata Borak*[2]

8.1 Introduction

Integrated optics (IO) technology is mainly based on the guiding of electromagnetic energy at optical frequencies by thin films. This technology applies thin film to form optical circuits and components such as optical filters, modulators, amplifiers, light sources, lasers, and photodetectors, which are combined into a single device to fulfil some complex function, to finally achieve a better and more economical optical system [1]. The ultimate aim of the IO is to create miniature optical circuits similar to the silicon chips that have revolutionized the electronics industry. The advantage of the optical approach is that data can be processed at much higher speeds.

In IO, all optical devices are connected to form one complete optical circuit. These devices consist of many thin-film optical components deposited on a common substrate and interconnected by thin-film waveguides. In an IO circuit, light wave propagates along the surface in the two-dimensional plane of the film. All system's components strongly connected with the substrate minimize vibration problems and alignment [1]. Integrated optical devices are important in optical networks. They are used in the optical path of an entirely planar optical circuit or in combination with other optical elements such as fibre waveguides and free-space optics. All these devices require the deposition of thin films with very uniform properties.

A thin film is a microscopically thin layer of material, ranging from fractions of nanometres (monolayer) up to several micrometres in thickness. A stack of thin films is called a multilayer. A thin film could be deposited onto the surface of various substrates, such as metal, ceramic, semiconductor, polymer, or onto a previously deposited layer. Most of the deposition techniques allow one to control layer thickness within a few tens of nanometres.

Several methods have been reported for the processing of materials into thin films, which allow for integration into various types of devices of IO. The

[1]Institute of Low Temperature and Structure Research, PAS, Wroclaw, Poland
[2]Department of Mechanics, Materials and Biomedical Engineering, Wroclaw University of Science and Technology, Wroclaw, Poland

properties of material significantly differ when studied in the form of thin films and can be to some extent controlled by the thickness of the layer. On the other hand, layers' parameters can be also controlled by selecting the proper technique of film deposition.

Thin-film deposition techniques allow one to apply a thin film to a surface by chemical or physical processes. One can speak about chemical techniques when a volatile fluid precursor undergoes a chemical change on a solid surface (usually a chemical reaction with the substrate) to produce a desired solid layer. Chemical deposition is further categorized by the phase of the precursor: gas or liquid. Liquid precursors are utilized in (electro)plating, sol–gel, and Langmuir–Blodgett methods [2]. Precursors in gaseous form are used in chemical vapour deposition (CVD), flame hydrolysis deposition (FHD), and atomic layer deposition (ALD). Chemical methods of thin-film deposition could be also divided into other two classes: methods which need electrical sources of ions (electroplating or anodization) and methods in which chemical reactions are demanded [2]. Many of those techniques found application in fabrication of the IO components.

A crucial aspect in IO materials is the transparency in a desired wavelength range and proper refractive index. For a visible and near-infrared (NIR) range, silica-based matrices work quite well; however, in mid-infrared (and for some purposes also in NIR), other materials must be used (e.g. fluoride or chalcogenide glasses, SiN_x, and Ge/Si systems) [3–5]. But good transparency is not enough to get, for instance, an optical waveguide. To fulfil waveguiding conditions for a structure deposited on silica with refractive index around 1.45 (at $\lambda = 632.8$ nm), the guiding layer must have higher refractive index. In this case, non-oxide materials can be used (e.g. germanium-based systems) or other oxides (TiO_2, GeO_2, ZrO_2, and others), including silica-based hybrid systems. Moreover, to avoid signal losses, the film should be smooth, dense, and homogenous both in thickness and composition.

Silicon oxynitride (SiO_xN_y and SiON) shows low absorption losses in a wide spectral region. In comparison to silica materials, silicon oxynitride thin films offer much larger refractive index for the core of optical waveguide. The index of refraction for an SiON thin layer could be adjusted over the wide range between 1.45 (SiO_2) and 2.0 (Si_3N_4) [6]. By changing the nitrogen/oxygen ratio, it would be possible to obtain waveguides with high refractive index for a broad wavelength range, from visible to IR. Thin films based on this material have been considered as an interconnect thin layer in the nano-optical chip technology. Silicon oxynitride shows compatibility with the standard silicon technology, high temperature stability and can be easily grown or deposited on the substrate; it allows the fabrication of waveguides with desired characteristics and compactness [6].

Another kind of materials where refractive index is adjustable is organic–inorganic hybrids. The refractive indices of popular organic materials are low; their values are in the range from 1.4 to 1.5 and are connected with limitations of their chemical structures [7]. There are some examples of polymers having refractive indices greater than 1.7, but they show high optical dispersion and strong adsorption in the visible region, which limit their practical applications in optical devices. But it was observed that the incorporation of inorganic materials having high refractive indices into polymer structures allows one to overcome these disadvantages and to obtain systems with high refractive indices. Hybrid polyimide-TiO_2 films may reach a refractive index as high as 1.94 [8,9]. Some fabrication techniques for waveguides also achieve the high-index core by impurity diffusion or implantation [10–12].

IO can be fabricated on the surface of some crystalline materials such as silicon, silica-on-silicon (SOS), or lithium niobate (LN) and connected with waveguides [13]. There is a particular interest in IO devices fabricated in thin-film dielectrics on silicon substrates. Silicon shows good surface quality, mechanical integrity and can be prepared in the form of large-area wafers. IO technology on silicon substrates is usually carried out in silica. An SiO_2 film on a silicon substrate is a promising material to fabricate arrayed waveguide gratings, but there are mainly two problems for the fabrication cf SOS materials. One is the deposition of SiO_2 films over 20-μm in thickness without cracks, the other is the further consolidation of the deposited films [14]. Cracks cannot be tolerated because they cause large optical losses.

Another material used as a substrate is $LiNbO_3$. LN is an excellent electro-optic material with high optical transmission in the visible and NIR, with relatively large refractive index (2.15–2.2) [13]. LN can be easily doped with laser-active (rare earth) ions. Based on this, a whole family of Er-doped waveguide lasers of excellent quality have been developed, emitting in the wavelength range 1,530 nm $< \lambda <$ 1,603 nm [11].

Most useful IO devices require waveguide confinement in a stripe or channel geometry. Among all-dielectric waveguide systems, the strongest confinement is found in silicon-on-insulator (SOI) waveguides which have a silicon core (refractive index $n_{Si} = 3.48$) and silica ($n_{SiO_2} = 1.45$), or even air, cladding. The SOI waveguides so far seem to be an ideal technical solution for planar photonic integrated circuits [15].

Taking into account all above-mentioned requirements, it is obvious that to obtain layers for specific purposes, different deposition techniques will be needed because they influence the film structure and properties. This is the case of Al_2O_3, where the optical and chemo-mechanical properties of deposited films are markedly dependent on the deposition conditions, so that the refractive indices may range from as low as 1.54 up to 1.70 [16]. The choice of a specific technique of film fabrication depends not only on the desired application but also on the available facilities.

8.2 Sol–gel

Sol–gel processing (also called chemical bath deposition or chemical solution deposition) offers a low-temperature route and powerful tool for making almost all kinds of transparent materials (including amorphous bulk glasses, thin films, and fibres) with interesting optical and photonic properties, potentially suitable for the production of functional IO devices for the new generation of optical networks. In this method, precursor solutions are highly controlled for the deposition of the films, where homogeneity at molecular level due to well-mixed liquid phase is easily obtained.

The sol–gel process, as its name implies, involves the manufacture of inorganic matrices through the formation of a colloidal suspension (sol) and gelation of the sol to form a wet gel, which after drying forms the 'dry gel' state (xerogel). The chemical reactions include two main steps that occur simultaneously in the reaction medium. The first one is hydrolysis, in which a precursor (usually a metal alkoxide, e.g. tetraethoxysilane, TEOS) reacts with water:

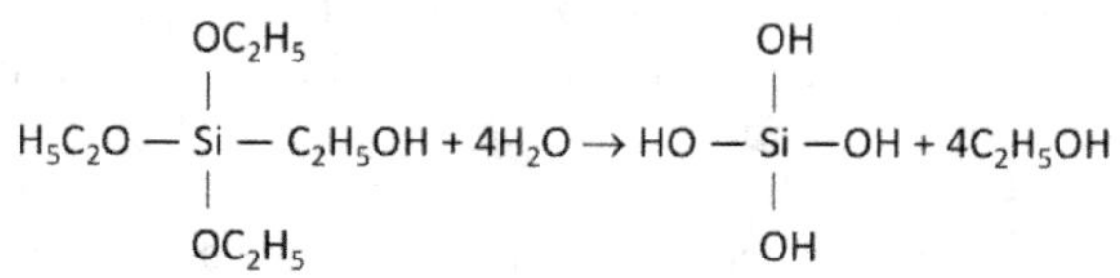

The next is a condensation step, in which the initial chemical solution turns into a sol (and then gel) where the macromolecular oxide network is obtained.

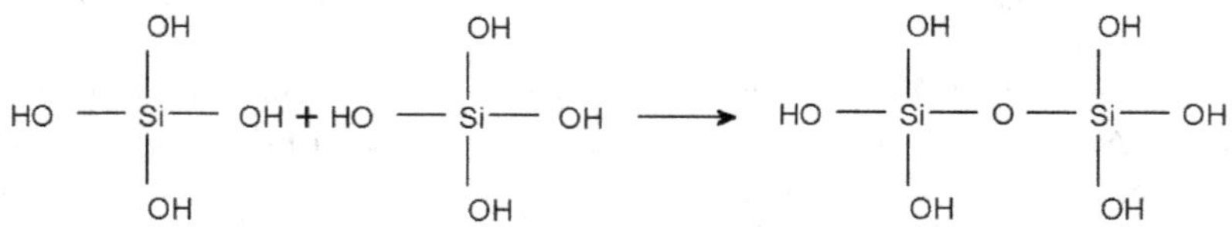

In time, a viscous liquid and, finally, a gel is formed.

The control of the hydrolysis and condensation processes by adjusting such parameters as an amount of solvents and water, catalysts, time, and temperature allows one to obtain a sol suitable for thin-film preparation. Sol can be deposited onto different substrates by various deposition techniques (spraying, spinning, dip-coating, and others), and after solvent evaporation, a solid thin film is formed. To obtain desired properties or structure, the material usually needs further processing, for instance annealing. Thermal treatment is also necessary to increase mechanical strength and to decrease porosity of the layer for better light propagation properties.

Tetraethoxysilane is probably the most widely used alkoxide in the sol–gel processes. TEOS and other organofunctional silanes are very useful to obtain amorphous silica-based networks. However, many ceramic materials can be fabricated by this route as well. Among them, TiO_2, ZnO, ZrO_2, GeO_2, V_2O_5, and hybrid systems attracted interest as materials for IO. In the latter case, titanium, aluminium, germanium, zirconium, hafnium, tin, niobium, or other oxides are used to increase the refractive index of pure silica and assure some particular optical properties [17–23].

Those structures can act as passive components in IO like optical waveguides, but the sol–gel chemistry is a unique method for producing homogenous rare-earth-doped materials for active components like lasers and amplifiers [17,24]. Increasing the rare earth concentration is a key point in order to make short and compact optical amplifiers. This effect might be partially achieved by the introduction of P_2O_5 or Al_2O_3 into the network. On the other hand, glass-ceramic systems may be used to reduce the concentration quenching among active ions and increase desired energy transfer efficiencies through selecting special local environment for rare earth ions.

The advantages of the sol–gel method are relative simplicity, mild reaction temperatures, purity, homogeneity, low cost, and availability of various metal precursors. The drawbacks in inorganic systems obtained by this route are mainly low reproducibility, lack of uniformity, layers cracking, and limited thickness that not exceeds a few hundred nanometres of a single layer. Deposition of several layers is possible but increases fabrication time and a tendency to crack during drying and sintering.

Unique materials obtained by the sol–gel process are ORMOSILs (organically modified silicates) – silica-based inorganic–organic hybrid polymers which combine inorganic and organic units and show good optical and dielectric ($\varepsilon = 3.3$ at 10 kHz) properties [25]. The combination in a single material of organic and inorganic counterparts permits to prevail over the limitations associated with each individual phase. Pure inorganic polymers are not universally suited for optical and optoelectronic devices since they have low mechanical flexibility, high brittleness and need to be processed at high temperatures. Organic polymer-based optical materials, such as polymethylmethacrylate (PMMA) or polystyrene, can overcome some of the disadvantages of inorganic ones. They exhibit a low heat resistance (PMMA ~80 °C) and poor adhesion on some substrates but show water resistance and can be processed at fairly low temperatures [26].

Another interesting aspect is that the ORMOSILs adhere very well on most substrates such as (metallized) Si wafers, inorganic glasses, and polymers. Their thermal stability is high enough for some of the technological applications (the decomposition temperature is about 270 °C). Their electrical and optical properties can be tailored by mixing various precursors [26].

Typically, the chemical synthesis of ORMOSILs is a two-step process, in which initial inorganic polycondensation of the silane molecules via sol–gel reaction is followed by crosslinking of the organic side chains (either photochemically or thermally, or both), resulting in three-dimensional duroplastic materials. Starting points are organo-alkoxysilanes. Depending on the number of sol–gel reactive substituents, mono-, di-, tri-, and tetrafunctional precursors can be distinguished. The precursors are linked up during the polymerization reactions to establish the organic–inorganic hybrid network. The sol–gel process offers a great flexibility in polymer synthesis by variation of catalyst, temperature, and alkoxysilane scaffold.

Using organic–inorganic hybrids, films with thicknesses of tens or even hundreds of microns can be obtained [27]. An ORMOSIL network is more flexible and less susceptible to cracking. The derived films are smooth and have more dense structure. Moreover, these materials were adapted for UV curing, and nowadays they can be directly patterned like other organic photoresists. The film patterning is possible, for instance, by direct laser writing. The exposure to laser irradiation causes a densification of the sample due to local heating [28] or photopolymerization reactions [29,30]. Various irradiation techniques allow one to fabricate such structures as diffractive gratings or stripe and channel waveguides. The hybrid materials may have wide application in micro-optical elements such as lenses, phase holograms, and gratings [30–32]. The simplicity, short processing time, low cost, and possibility of controlling the device parameters are the advantages of this process.

8.3 Flame hydrolysis deposition

FHD, or flame aerosol synthesis, is one of the most practical and frequently adopted methods to produce high-quality silica particles, planar films, and also glasses for optical planar devices.

FHD consists of three substances: oxygen–hydrogen flame generation, which provides high temperature, a liquid precursor, which evaporates in the flame and in the gas phase undergoes chemical reactions (hydrolysis, oxidation, etc.), and particles, which are formed during reaction and are deposited on a planar surface (usually silicon substrate) to form a layer with low stress [33].

Precursors undergo homogeneous reaction in the flame to form pyrogenic particles, which next grow by vapour scavenging and/or aggregation as they are transported (by mechanisms of convection, Brownian diffusion, and thermophoresis) to the substrate (Figure 8.1). Particles agglomeration starts to occur at the end

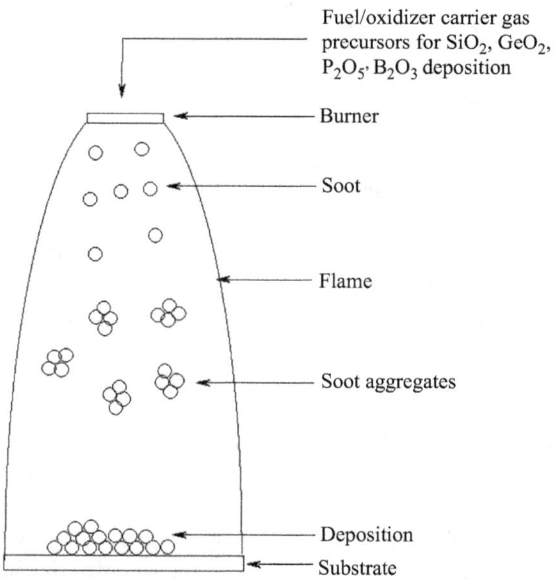

Figure 8.1 Scheme of flame hydrolysis deposition process

region of the flame and continues along the flame. The formed nanomaterial is highly dispersed and highly porous when collected on the substrates [34].

For silica layer deposition, inorganic or organometallic substances, such as silicon tetrachloride ($SiCl_4$) [35–37], silane (SiH_4) [38], tetraethoxysilane [34,39], octamethylcyclotetrasiloxane (D4) [40], or hexamethyldisiloxane (HMDSO) [41,42], are used as precursors. Silicon tetrachloride is corrosive and moisture sensitive and produces HCl gas as a by-product [39]. Therefore, it is going to be replaced by others, less hazardous and safe to handle compounds, such as siloxanes.

FHD allows the precise doping of GeO_2, P_2O_5, and B_2O_3 from their respective halide gases ($GeCl_4$, PCl_3, and BCl_3) or organometallics (D4, triethylphosphate, triethylborate, and tetraethoxygermane) within the formed SiO_2 films [36]. During the FHD process, each of the gaseous compounds is carried in a separate feedline so that different planar glass material systems are produced by either sequential deposition or co-deposition of the appropriate soots. The hydrolysis process of halide gases could be described by the following reactions [36]:

$$\begin{gathered}
SiCl_4(g) + 2H_2O(g) \rightarrow SiO_2(s) + 4HCl(g) \\
GeCl_4(g) + 2H_2O(g) \rightarrow GeO_2(s) + 4HCl(g) \\
2BCl_3(g) + 3H_2O(g) \rightarrow B_2O_3(s) + 6HCl(g) \\
2PCl_3(g) + 3H_2O(g) + O_2 \rightarrow P_2O_5(s) + 6HCl(g)
\end{gathered}$$

The reactions of organometallic precursors in the flame are as follows [33]:

$$\begin{gathered}
C_8H_{24}O_4Si_4 + 16O_2 \rightarrow 4SiO_2 + 8CO_2 + 12H_2O \\
C_8H_{20}O_4Ge_4 + 12O_2 \rightarrow GeO_2 + 8CO_2 + 10H_2O \\
2(OC_2H_5)_3PO + (37/2)O_2 \rightarrow P_2O_5 + 12CO_2 + 15H_2O \\
2B(OC_2H_5)_3 + 18O_2 \rightarrow B_2O_3 + 12CO_2 + 15H_2O
\end{gathered}$$

There are other diverse mechanisms of reaction which take place in the flame formed with various fuels [34,43,44].

Deposition of sequential different layers of SiO_2, or SiO_2 with dopants, leads to multiple layers with different refractive indices, which finally allows one to obtain buried optical waveguide structures [33]. Using masks during deposition allows one to prepare films with well-defined patterns [45]. Planar optical devices on silicon substrates require at least three layers with different compositions and thicknesses. Those are underclad, core, and overclad layers (Figure 8.2). The external, first (underclad) and third (overclad) layers have effective indices close to one another while the middle (second layer – core) has an effective index higher than that of adjoining layers [33]. The increase in core index is achieved by adding, for example, germanium, phosphorus, or boron precursors to the core layer. It is claimed that germanium oxide acts as the main agent for index enhancement [33].

The underclad layer is the bottom layer, and it must be the stiffest one, with the composition close to pure silica. Some amounts of phosphorus and boron are added to this layer to reduce glass roughness, which could cause germanium migration at the underclad–core interface. The overclad layer consists of glasses with high

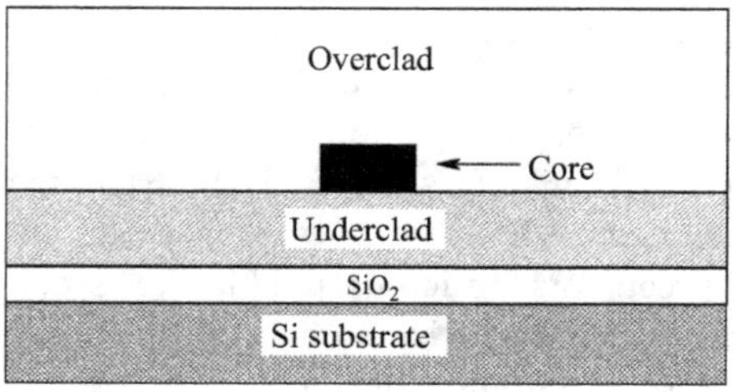

Figure 8.2 Scheme of multilayer structure for optical devices

levels of dopants (e.g. phosphorus or boron). The ratio of the dopants should be such that the overclad layer index is close to that of the underclad layer. Each layer of the planar device is deposited one over the other using the FHD process and consolidated before the next layer is deposited.

The deposited films are porous. In order to fabricate waveguides with low propagation losses, the porous SiO_2 thick films should be consolidated into a dense silica glass at high temperature (above 1,300 °C). Some authors have suggested that a small amount of PCl_3 and BCl_3 added to the precursor gas allows one to lower the melting point of the deposited films [14]. Sintering temperatures must be properly chosen such that the layer closer to the silicon substrate has higher viscosity for a given temperature, to avoid mixing of the layers [33]. Unfortunately, sintering temperatures are high enough to cause the dopants to volatilize at the surfaces of the deposit and create gradients in dopant concentration and inhomogeneity in composition in the densified glass layer, which further result in inferior optical properties. To minimize or even avoid this problem, layers could be deposited pre-sintered during laydown [33]. It is reported that the density of the film increases with increase in temperature because a denser film is formed during sintering [39]. Sintered materials have a higher value of refractive index. It was observed that the refractive index of the deposited film varies with the corresponding increase in precursor (TEOS) flow rate up to some values and then decreases. It is explained by the fact that, when the flow rate of TEOS increases, the flame temperature decreases and also the sintering temperature decreases, which leads to the less-dense films. With the increase in precursor flow rate, the film thickness increases. It is obvious because the increase in flow rate of precursor causes a higher concentration of reactants, so that the growth of the deposited film is faster.

The FHD process allows one to deposit layers with different compositions and thicknesses. Underclad and overclad layers of about 20 μm are usually prepared, whereas the core thickness varies from 5 to 7 μm [33]. But it is possible to obtain thicker films (about 10–100 μm), depending on the flow rate of $SiCl_4$ and O_2 and the deposition time [14].

The most considerable problem of gas phase methods is poor controllability in the growth of nanoparticles, especially when they are generated at the high concentrations which are needed for practical applications. Also, it is not easy in the FHD process to control the size and morphology of the SiO_2 droplets (SiO_2

particles-aggregates) formed in the flame, and thus also the quality of the final SiO_2 glass [46]. Most of flame-made materials consist of spherical primary particles with different degrees of agglomeration because, after formation in the gas phase, they are attached to one another with a bond strength that varies from weak van der Waals forces to significant necking. The agglomeration state is determined by the concentration of primary particles and their residence time (in other words, maximum temperature and cooling rate), which have influence on the competition between collision and coalescence processes [47]. The molecules collide to form particles which grow by a collision-coalescence mechanism. Collisions take place as a result of the particle Brownian motion, while coalescence is driven by the surface free energy excess of attached particles. Aggregates are formed when the collision of particles is faster than their coalescence [48]. In many cases, smaller and unagglomerated particles are desired to synthesize high-quality nanostructured materials. Understanding of interaction between flame and particle growth is important for the controlled formation of nanoparticles in flames.

Silica glass with high purity and high refractive index homogeneity is affected by size, size distribution, morphology, extent of aggregation, crystalline phase, and chemical composition of the silica droplets and the residue of radicals (e.g. OH groups) formed in the flame [44]. The formation and growth of silica droplets are connected with the temperature and species present in the flame. The quality of the flame-made silica droplets and the efficiency of the transition of these particles from the gas phase and melting onto the surface of the support are affected by their combustion process, expressed by the stability of the flame, particles concentration (controlled by burner configuration), precursor composition, oxidant/fuel flow rates, OH radical concentration, and temperature gradient near the surface of the support [45,47]. Especially critical are the uniformities of SiO_2 deposition and OH concentration on deposition surface. The homogeneity of SiO_2 glass is provided by maintaining a stable flow of air around the glass. Flow oscillation could result in changes in SiO_2 deposition and also OH concentration. It was observed that OH concentration decreases dramatically in the particle formation zone, which can be attributed to both the generation of hydrochloric acid (during hydrolysis processes) and the equilibrium shifts caused by the consumption of oxygen and water molecules during silica formation [49]:

$$SiCl_4 + O_2 \rightarrow SiO_2 + 2Cl_2$$
$$SiCl_4 + 2H_2O \rightarrow SiO_2 + 4HCl$$

The presence of OH groups is undesirable, even at low concentration, because absorption bands in the NIR would lead to strong optical losses, which are especially critical for optical fibres.

Flame synthesis of materials does not require an additional source of energy for precursor conversion such as plasmas, lasers, or electrically heated walls [45]. Gas phase methods generally produce purer nanoparticles than liquid-based processes since even the purest water contains traces of minerals. In the flame processes, energy is generated chemically in situ, driving reactions for particle formation. Then, the released energy is removed rapidly by radiation and convection, which is essential for the synthesis of nanostructured particles [45]. As a source of flame in the FHD, mixtures of oxygen/methane [33], H_2/O_2 [37,44,49,50], and H_2/air [51] are used. In all kinds of oxy-fuel flames, fractions of molecular O_2, H_2 and the reaction intermediates H, O, and OH exist at temperatures above 2,500K. Below this temperature, only OHs exist in considerable concentrations. H_2/O_2 flames burn at higher temperatures and have much higher concentrations of radicals and atoms compared to H_2/air flames of identical stoichiometry and meet requirements needed for homogeneous silica formation [52]. It was claimed that, as a carrier gas, the mixture of H_2 and O_2 gives the best results in the FHD synthesis of silica glass with high purity and refractive index [44]. The usage of only hydrogen caused a low deposition rate and incomplete reaction with $SiCl_4$. Complete reaction of this silica precursor and a high deposition rate were observed by replacing hydrogen by oxygen. In this case, $SiCl_4$ oxidation reaction occurs to form silica. In the presence of water vapour, the $SiCl_4$ hydrolysis reaction also takes place. Formation of SiO_2 is determined by the competition between $SiCl_4$ oxidation and hydrolysis reaction, if both oxygen and water vapour are present [43]. Hydrocarbon oxy-fuel flames tend to strong sooting and contamination of the silica [52]. It must be added that addition of a precursor (e.g. $SiCl_4$) to flames can significantly affect flame characteristics, including temperature and species concentrations [49,50].

An important factor affecting flame synthesis is flame temperature. The flame temperature has an influence on its structure, chemical reactions, and the process of particles growth, their transport and deposition. Formation and growth of particles in a typical flame occur in a high-temperature region, usually over 1,000K [49]. Generally, faster cooling rates and high temperatures reduce primary particle size and extent of aggregation, so, in these conditions, formation of less aggregated particles is privileged. It was observed that a high deposition temperature (2,283 °C) reduced the viscosity of SiO_2 aggregates, and thus, smaller particles (30 nm) were formed [46]. An extremely smooth structure of SiO_2 glass was formed at temperatures over 2,000K but below 3,000K (boiling temperature of amorphous SiO_2). Usually, the size of the primary particles in the combustion processes ranges from a few to several hundred nanometres in diameter, depending on material and process conditions.

The quality of final deposition product is also sensitive to the distance between the burner and the substrate. The final droplets which reach the substrate surface could consist of precursor droplets, precursors, or product vapours or product particles, all of them with different states of agglomeration [45]. The kind of formed droplets is influenced by the flame temperature. Droplets of unconverted precursor deposited on the substrate form a dense film. Product particles lead to porous films, which could collapse into a more compact structure during sintering process. The deposition step could be controlled by placing the substrate at an appropriate distance or by changing the gases flow [45].

The properties of materials are strongly connected with their crystal structure. Various polymorph forms of the same material can exhibit different physical properties. During synthesis, temperatures used in combustion processes determine the particle phase composition and have influence on the final structure. But until now controlling the crystal structure of flame-made materials is not well understood because there are no techniques to monitor the evolution of phase composition. There are no general rules for many materials to predict their crystal structures after flame process [45]. Sometimes, metastable phases and polymorphs that are not possible to obtain by conventional wet-phase and calcination techniques are formed in these combustion processes.

To understand the FHD process and the interacting effects of flame structure and silica formation on the morphology of the resulting deposit, a number of models and tools have been developed. The Euler–Lagrange multi-phase model has been applied to describe the formation and transport of silica droplets in an oxy-hydrogen diffusion flame with $SiCl_4$ [44]. In this model, the interphase exchanges of mass, momentum, and energy between the gas phase molecules and the liquid phase silica droplets during their formation and transportation were analysed. The commercial computational fluid dynamics (CFD)-code FLUENT was useful to analyse numerically the hydrogen combustion and SiO_2 particle formation in the premixed flame reactor [51]. Coherent anti-Stokes Raman spectroscopy (CARS) and planar laser-induced fluorescence (PLIF) were used to measure both the distributions of the flame temperature in silica-generating H_2/O_2 coflow diffusion flames and the concentration of OH radicals, respectively [49]. It was observed that addition of $SiCl_4$, even in a small amount, has an influence on thermal and chemical characteristics of H_2/O_2 diffusion flames, mainly because of the oxidation and hydrolysis reaction of this compound. As indicated by the measurements of CARS from nitrogen, the temperature in the non-reacting zone decreases, whereas the flame temperature in the zone where SiO_2 particles are formed increases. Both reactions, oxidation and hydrolysis of $SiCl_4$ forming SiO_2, have influence on OH radicals concentration (as shown by PLIF of OH radicals), which decreases dramatically in the particle formation zones due to the consumption of O_2 and H_2O and generation of HCl [49].

Generally, the FHD process is characterized by products (glass films) which could be deposited with high rates and form homogeneous films with low stresses and without cracks, a result difficult to achieve by classic wet coating techniques. The products formed from gaseous phase do not need drying processes for deposited films, so the layers are crack-free. The other advantage is the low optical loss of the fabricated structures.

8.4 Chemical vapour deposition

CVD is a method of thin-film formation on a substrate material by a chemical reaction of vapour phase precursors. Chemical reactions of precursors occur both in the gas phase and on the substrate. Reactions could be initiated by various ways (since there are different kinds of CVD methods):

- heat (thermal CVD)

- radiation, such as UV (photo-assisted CVD)
- low pressure (LPCVD)
- plasma (PECVD)

In a typical CVD process, the substrates placed in the reaction chamber are exposed to one or more volatile precursors which react on the substrate surface to form a thin film, till a desired thickness of the film is obtained (Figure 8.3). Chemical reactions start from islands forming, and, as the process continues, the islands grow and merge to create the desired film. By-products generated in these processes are removed by the gas flow through the reaction chamber.

This process is often used to produce thin films of polysilicone, silicon dioxide, silicon nitride, silicon carbide, silicon oxynitride, and some metals (W, Mo, Ti, and Ni) [53]. Among them, Si_3N_4, SiO_2, Si, silicon oxynitride (SiO_xN_y), and ZnO may be used as optical waveguides and have significant applications in IO.

Silane is a common compound for CVD of silicon-based thin films [6,54,55]. Reaction of its pyrolysis is as follows:

$$SiH_4(g) \leftrightarrow Si(s) + 2H_2(g)$$

and it is commonly used in the LPCVD systems (when a vacuum is used) operating at around 600 °C and 1–10 Torr to deposit polycrystalline Si [56]. Various materials, as liquid or gaseous precursors, that undergo decomposition, reduction, deprotonation, and other reactions can be used as well. For example, chlorosilane, which undergoes the following reaction,

$$SiCl_4(g) + 2H_2(g) \leftrightarrow Si(s) + 4HCl(g),$$

or dichlorosilane ($SiCl_2H_2$) [57], and organosilicon precursors, such as TEOS, hexamethyldisiloxane [58,59], and hexamethyldisilazane ($((CH_3)_3Si)_2NH$) [60], have been investigated. All these compounds are available in the liquid phase, so they can be easily vaporized at ambient temperatures and provide chemisorption in a controlled way. Unfortunately, organosilicon compounds cause significant impurities in formed films by carbon and hydrogen, and thus, the utility of these CVD precursors is limited [61].

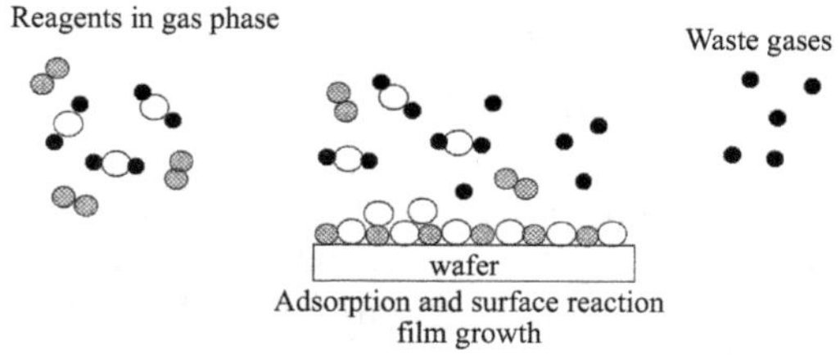

Figure 8.3 Schematic diagram of the chemical vapour deposition process

Various gases (diborane, phosphine, or arsine) can be added to reaction chamber for silicon doping. On the other hand, the CVD is also important in dielectric films deposition through different reactions:

$$SiH_4(g) + O_2(g) \rightarrow SiO_2(s) + 2H_2(g)$$
$$Si(C_2H_5O)_4(g) \rightarrow SiO_2(s) + 4C_2H_4(g) + 2H_2O(g)$$
$$3SiH_4(g) + 4NH_3(g) \rightarrow Si_3N_4(s) + 12H_2(g)$$
$$SiH_2Cl_2(g) + NH_3(g) + N_2O(g) \rightarrow SiON(s) + \text{by-products}$$

Oxygen and hydrogen are often used to oxidize or reduce precursors. Ammonia (NH_3) and nitrous oxide (N_2O) are the most widely used gases for the silicon oxynitride and silicon nitride [55,62] nano-metric regime thin-film deposition [6]. There are some examples of precursors which contain both silicon and nitrogen, for example tris(diethylamino)chlorosilane ($(Et_2N)_3SiCl$) [7], tris(dimethylamino) silane ($((CH_3)_2N)_3SiH$) [62], and silicon dimethylamido complexes. In the structure of these precursors, there are present direct single Si–N bonds and are absent Si–C bonds which minimize incorporation of carbon and hydrogen atoms into Si_3N_4 structure.

Using the CVD, silica glass can be made by hydroxylation. In this process, vapours of silicon tetrachloride ($SiCl_4$) react with steam (H_2O), causing silica deposition on cooler substrates [63]. When other vapours are added, such as phosphorus oxychloride ($POCl_3$) and germanium tetrachloride ($GeCl_4$), mixed oxide structures are fabricated with desired refractive index, according to the modified composition. Products are subsequently sintered to obtain a dense glass.

A special variant of this technique, known as epitaxy, represents a process in which a layer is created on top of another layer and inherits its crystal structure. The representative example of this process is the deposition of silicon (or gallium arsenide) on a wafer to produce an epitaxial growth of the crystal. Such films can be relatively thick (0.1 mm) and are commonly used for producing SOI substrates that lower the power requirements and speed the switching capabilities of complementary metal-oxide semiconductors (CMOSs) [64].

Among materials obtained by the CVD, ZnO is being paid special attention because it is considered a substitute for GaN, material used in short wavelength optoelectronic devices. ZnO, with its gap at 3.4 eV, is transparent in the near ultraviolet and may also have optoelectronic applications in this spectral region [65]. Two common zinc precursors, diethylzinc and dimethylzinc, are favourable for ZnO synthesis by CVD combined with metal-organic precursors. Both precursors are oxidized with O_2 or H_2O. Structure and optical properties of the ZnO films depend on the temperature and oxygen pressure. It was observed that ZnO films deposited at high O_2 content had better optical properties (showed intense UV emissions) than the same material prepared at O_2 deficient [66]. In this case, the material showed defect-related emission, along with UV luminescence. Sometimes, because of the aggressive gas-phase pre-reaction with oxygen and white powder deposition, alcohols (tertiary butanol) as oxidizing agents are used instead of O_2 [65].

ZnO shows n-type conduction, which can be enhanced by various (B, Al, Ga, In, N, or F donor) dopants. Some of them could by introduced by different CVD methods. Solid-source CVD was used to prepare N-doped p-type ZnO films [67]. In this method, solid zinc acetate dihydrate and solid ammonia acetate were used as the Zn and nitrogen sources, respectively. Gallium-doped ZnO layers prepared in a low-pressure system within a single CVD cycle can also be delivered [68].

In the plasma-enhanced CVD (PECVD), plasma is formed in a reaction chamber that transforms the gaseous precursors into more reactive species which react with faster rates and at lower temperatures. Commercially available systems allow one to process PECVD in a temperature range from room temperature to 350 °C, while in the CVD, much higher temperatures (600 °C–900 °C) are needed to develop thin films. A large variety of substrates, from wafers with a diameter up to 200 mm to parts loaded on carriers, can be processed, but this method tends to sacrifice uniformity of deposition, although it is relatively fast [64].

Structural and optical properties, such as uniformity of the refractive index and the optical loss, are strongly dependent on the layer composition which could be controlled by the parameters of the CVD process. These important parameters are flow rate and ratio of the process gases, the chamber pressure, and the substrate temperature. To improve layer parameters, a large variety of chamber designs have been used to minimize turbulence and to facilitate the removal of reaction products [56].

The CVD has two main disadvantages. The first is the difficulty of obtaining uniform films deposited on the substrate. It is caused by the conditions of this process (temperature, concentration, chemical composition, and velocity of the gas mixture) which are difficult to control to be the same along the length and across the reaction chamber. Moreover, the growth on horizontal surfaces might be different than on vertical surfaces; thus, the deposition does not conform. A high conformity can be achieved by high process temperatures, and the best results are obtained using low pressure systems. The second problem is related to the nature of used gases, which generally are dangerous because of their toxicity, tendency to explosion, or corrosive properties. There is also a probability that some undesirable N–H, Si–O, and Si–H groups will be incorporated in the layers and will increase the optical losses.

8.5 Atomic layer deposition

ALD belongs to the general class of the CVD techniques and is a technique which allows for a surface-controlled layer-by-layer deposition of a variety of thin-film materials from the vapour phase [69]. In this technique, two or more gaseous precursors are used to react with the substrate sequentially, one at a time. The deposition process is driven by the reaction between substrates functional groups and molecules of the incoming precursor (chemisorption). ALD process could be split up into two half reactions. The first is the reaction of the gaseous precursor containing the metal or nonmetal source atom, which then reacts with the substrate through a self-limiting process. In this way, only one monolayer is formed on the substrate surface. It is a 'half reaction' because it makes up only part of the materials synthesis. After this step, the excess of precursor is purged away. The second reaction could be reduction, oxidation, or nitridation. In this step, a reactant (e.g. oxygen for metal oxides or nitrogen for metal nitrides, water molecules) reacts with the surface groups of the first layer, and a film is deposited. Formed by-products and excess of second precursors are removed with an inert gas (N_2, Ar) introduced into the chamber. This is one cycle. All these processes could be repeated for each layer, until the desired composition and appropriate film thickness will be achieved [69,70]. The ALD is a modified CVD growth method in which a synthesis reaction is split into two surface reactions which are self-limited by separate sequential exposures of the substrate to the chemical precursors. A general ALD process is illustrated in Figure 8.4.

The cyclic nature of the ALD process causes stepwise growth of the desired material layer by layer, so the growth is rather per cycle than by growth rate, which allows for a thickness control of the formed layers [71]. The influence on the thickness of the film prepared in one cycle has also the size of the precursor molecules, especially ligands. Ligands form steric hindrance and limit the number of potential molecules which could adsorb on the surface. Great molecules of precursor block the active sites of the substrate and prevent them from reaction with precursors, so that the formed layer could not be homogeneous. Full monolayer growth per one cycle is possible, but only for small molecules.

General requirements for reactants used in the ALD are that they must quickly react with reactive sites of substrate surface (such as hydroxyl groups on oxide surfaces) or with a second precursor (H_2, O_2, H_2O, etc.) [70]. The ALD precursors may be highly reactive gases, volatile liquids, or solids and cannot decompose thermally at the ALD processing temperatures [72]. The vapour pressure of the precursors must be high enough for effective mass transportation. What is more, the gas–solid reactions of the reactants have to fulfil the criterion of self-termination [73,74].

The reactants used in the ALD can be divided into two main groups: inorganic and metalorganic. Metalorganic reactants can further be classified in those containing a direct metal–carbon bond (organometallic reactants), and those containing no direct metal–carbon bond. As an inorganic reactant, elements (e.g. transition metals, noble metals, and lanthanides) and halides (especially chlorides) have been used. The spectra of metalorganic reactants are broader and include compounds such as alkyls, cyclopentadienyls, alkoxides, β-diketonates, amides, and amidinates [73]. The group of nonmetal precursors in the ALD contains water, oxygen, ozone for oxygen production, ammonia, hydrazine, and amines as sources of nitrogen, hydrides for group V elements, and chalcogens [72]. Precursors should 'simply' have ligands, which will not create great steric hindrance and will not block the active sites. It is also desired that ligands will not pollute the product and will not form a high number of by-products.

The ALD technique developed in 1974 [75] has become of worldwide importance with the availability of commercial equipment for thin-film deposition. The group of ALD-grown materials includes many types of solid inorganic

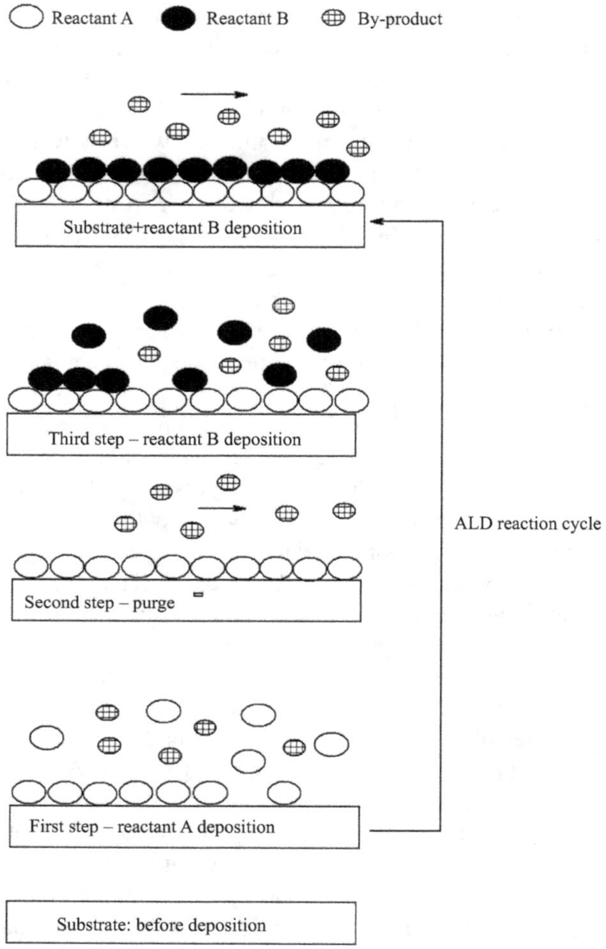

Figure 8.4 Scheme of atomic layer deposition process

materials, such as oxides, nitrides, sulfides, selenides, tellurides, arsenides, and pure elements [73], including metals (Al_2O_3, TiO_2, SiO_2, HfO_2, ZnO, ZrO, SiN_x, AlN_x, TaN_x, GaN, Pt, Ag, and Cu) [70]. For IO, significant are oxides (Al_2O_3, ZnO, others) or nitrides. Besides flat surfaces with dimensions up to around 500 mm, this technique allows one to cover 3D objects or powders.

Amorphous Al_2O_3 is studied as a waveguide platform for broad-spectrum-integrated photonics with transmission from the infrared to the ultraviolet [76]. Al_2O_3 thin films are often prepared using trimethylaluminium [$Al(CH_3)_3$] (TMA) [15,74]. Deposition of Al_2O_3 involves adsorption of TMA onto a hydroxylated surface, followed by the reaction of water with the resulting methylated surface [77]. It was observed that Al_2O_3 growth is linear with the number of cycles; typical measured growth rates are 1.1–1.2 Å per one full cycle. It is an efficient and self-limiting surface reaction in which strong Al–O bonds are formed. The overall reaction for Al_2O_3 formation in the ALD is [78]

$$3Al(CH_3)_3 + 3H_2O \rightarrow Al_2O_3 + 3CH_4$$

This reaction has one of the highest reaction enthalpies ($\Delta H = -376$ kcal) gained for any ALD reaction. The high exothermicity allows one to carry the ALD process at low temperatures, even at room temperature. It was noticed that Al_2O_3 films show a small decrease in density from 3.0 g/cm^3 at 177 °C to 2.5 g/cm^3 at 33 °C [79]. The refractive index also decreases slightly. Generally, Al_2O_3 thin films, even obtained at 33 °C, are smooth and show great conformality to the initial substrate. The ALD seems to be a good technique also for Al_2O_3 waveguides doped with erbium ions. Er^{3+} shows good solubility in Al_2O_3 due to its similarity in valency and lattice constant with Er_2O_3 [79]. In comparison to SiO_2, Al_2O_3 possesses relatively high refractive index ($n \sim 1.65$), which allows tighter light confinement in Er:Al_2O_3 waveguide structures. For these reasons, Al_2O_3 is claimed as one of the most common and reliable host materials for Er-based integrated waveguide devices useful for optical amplification or compensation for the losses present in passive integrated optical circuits [80].

ZnO is known as a semiconductor and a transparent conductive oxide with excellent optoelectronic properties. ZnO is usually prepared from diethyl zinc ($Zn(C_2H_5)_2$) precursor and water as an oxidant [81,82]. The thickness of the film formed by this method depends on the substrate temperature. The layer thickness varied between 14.1 and 33.4 nm when the substrate temperature increased from 100 °C to 300 °C. The maximum growth (33.4 nm) of ZnO thin films was observed at 200 °C, and for this temperature, also a remarkably high growth rate of 0.56 nm/cycle was noticed. The measured refractive index for film obtained in these conditions was 1.56 [81]. ZnO films could be also doped with other elements. Al-doped ZnO films of about 100 nm thickness with various Al doping treatments were prepared at 150 °C by ALD on quartz substrates. Al_2O_3 cycle was inserted after every ZnO cycle [83]. ZnO thin films prepared by ALD are characterized by high purity, ultra-thickness, and crystallinity, which influence the optical and electrical properties of this film. The ALD process was also used to prepare ZnO photonic crystals from diethyl zinc [84].

Silicon nitride (SiN_x) is one of the most significant materials for applications in integrated circuits and optical devices. Silicon nitride is utilized in several ways in manufacture of semiconductors and as a final passivation and mechanical protective layer in integrated circuits. Growth of SiN_x thin films via thermal ALD is often achieved using chlorosilanes ($SiCl_4$, SiH_2Cl_2, Si_2Cl_6, and Si_3Cl_8) as the silicon source; as to the nitrogen source, the most widely used reactant is ammonia [85]. Deposition of nitride thin films requires both a nitrogen source and reducing agent. Often NH_3 fulfils both roles. Sometimes, the more reactive hydrazine (N_2H_4) has been tested as a nitrogen source, but it is a very dangerous chemical to handle [86,87]. SiN_x films prepared from chlorosilanes are typically deposited at temperatures in the range 300 °C–650 °C. Chlorosilane-based precursors are necessary for thermal ALD SiN_x film deposition; other precursors have not given good products. Great variety of silicon precursors (chlorosilanes (SiH_3Cl, SiH_2Cl_2), tris(dimethylamino)silane ($SiH(N(CH_3)_2)_3$), trisilylamine ($N(SiH_3)_3$), bis(tertiary-butyl-amino)silane ($SiH_2(NH^tBu)_2$), tris(isopropylamino)silane ($C_9H_{25}N_3Si$), etc.) and plasma gases (N_2, N_2/H_2) are offered by plasma-enhanced ALD (PEALD) [85]. In this method, SiN_x films are deposited below 500 °C. N-rich silicon nitride films, with a wide range of lower values of refractive index (1.86–2.00) than the stoichiometric silicon nitride (2.01), were obtained by changing the deposition temperature [88]. It was reported that properties of the N-rich silicon nitride film are dominated by the deposition temperature, which has an influence on physical properties like N/Si ratio, etching rate, refractive index, residual stress, and the H concentration. N-rich silicon nitride films were prepared from dichlorosilane and ammonia in a PEALD process.

The ALD, unlike alternative deposition methods such as CVD and PVD, allows one to control the thickness of the deposited films at the angstrom or monolayer level, and their composition is accurately controlled as well [78]. By utilizing layer-by-layer deposition, the thickness of a formed film can be tailored by the number of cycles carried out. This is a great advantage of this method. The ALD process could be used to coat large substrates, even with extremely complex shapes. In this case, the process is limited only by the size of reaction chamber [72]. Gaseous precursors do not contact each other, and reactions are performed sequentially and are limited only to the surface. All this leads to producing a conformal material layer of high quality on the substrate.

The major limitation of the ALD is its slowness. It is a stepwise process and only one monolayer is deposited in one cycle. Long cycle times are involved in pulsing and purging precursors and by the layer-by-layer nature of the deposition. Typically, deposition rates are 100–300 nm/h [72]; therefore, this is not practical method for micrometre-thick films formation. But this process can run at temperatures below 350 °C [69]. In plasma-enhanced operation (PEALD), the plasma source enables homogenous and conformal coating of sensitive substrates (e.g. organic polymer foils), where layers at even lower temperatures (<100 °C) are obtained.

Another aspect connected with the ALD is chemistry. This is a chemical technique, and there is always a risk that the product (thin film) might be polluted with some residues of precursors or impurities present in carrier gases.

8.6 Other techniques

Electrodeposition is a method in which electric current is passed through a solution composed of a salt of the metal to be deposited. Here, metal ions adhere to the substrate-electrode and then, through oxidation or reduction processes, they are deposited on one of the electrodes immersed in the salt solution, in the form of metal or metal oxide. The thickness of the formed layer and its formation time are determined by the electrolysis laws, and the general rule for electrodeposition is that the higher the charge and the longer the current is connected, the thicker the deposition coating is obtained [2]. Anodization is used for the deposition of oxides (or less commonly for other compounds such as nitrides). In this technique, films are deposited on the anode of the parent metal, following a chemical reaction [2].

Ion exchange is not strictly a deposition technique, but it is probably one of the oldest and the most popular techniques for glass waveguide fabrication for integrated optical components. In an ion-exchange process, an ion A^+ in the glass, which represents an alkali oxide (usually Na^+ but also K^+, Li^+), is replaced by a B^+ ion with the same valence but of a larger size or higher polarizability, such as silver (Ag^+), potassium (K^+), caesium (Cs^+), thallium (Tl^+), and rubidium (Rb^+) [89]. The technique, therefore, is based on the following reaction [90]:

$$X\overline{A^+} + B^+ \leftrightarrow X\overline{B^+} + A^+$$

where X^- indicates the glass matrix. This exchange modifies locally the chemical composition of the glass surface.

In ion exchange, dopant atoms may be introduced into the substrate material through the usage of electric fields. The substrate, which is the anode, is submerged in a molten material and is placed inside an electric field generator. During the heating, the bonds in the substrate materials are loosed, and, due to the electric field, cations from substrate migrate towards the cathode. At the same time, dopant atoms of the molten material move into the substrate to fill the empty places open by the migrated atoms.

Combinations of different ion-exchange processes to fabricate waveguide structures on the same substrate have also been studied [89]. The dopant ions introduced in the glass take over the places in the glass structure left by the modifier ions, and because of this exchange of ions, the structure of glass is left unchanged.

The replacing process changes the optical properties of glass in its surface area. The changes involve optical properties of glass due to different electrical polarizabilities of the exchanged ions, difference in their size (ionic radii), and created mechanical stresses. The ion exchange also causes an increase of the refractive index in the diffused layer. The ion exchange process can be implemented after the creation of the glassy structure, e.g. after melting it and forming (as plates, bars, profiles, etc.) by an appropriate mechanical working [90]. Derived glasses should be characterized by low intrinsic losses and chemical durability, defined as the chemical resistance to the acids and other degradation agents [91]. But it is well known that glass has low chemical resistance to HF acid, which damages the Si–O bond. The chemical durability of a silica glass could be improved by adding CaO, MgO, ZnO, PbO, B_2O_3, or Al_2O_3 [91]. The glass must show resistance to thermal shocks. The temperature of vitreous transition (T_g) must be sufficiently high to envisage the fabrication of optical devices at temperatures of 300 °C–500 °C. The oxide glasses show high transparency in the visible and NIR bands, which could be disrupted by impurities such as transition elements: Cr, Mn, Fe, Co, Ni, or Cu.

Some specific glasses are developed for each type of exchanged ions. The two most widely used ions in ion-exchange glass waveguide research activities are Ag^+ and K^+. The radius of silver ion is close to that of sodium, due to which the mobility and the diffusion during the exchange is easy. The T_g of this kind of glass can be higher than 500 °C, whereas the ion-exchange process takes place at moderate temperatures (200 °C–350 °C) [91]. It is claimed that Ag^+ ions are also responsible for largest changes of refractive index [89]. The main disadvantage of silver-exchanged waveguides is the tendency of silver ions to reduce into neutral atoms which then aggregate into metallic clusters and form in the glass an absorbing colloid. All this causes high waveguide losses [92]. The key role in the reduction of silver to metallic form is played by the glass composition, especially by impurities such as elements Fe, As, and Sb which are present in the glass structure as polyvalent ions. The Ag^+ ions capture electrons from them and reduce. Another aspect connected with Ag^+ ions reduction is caused by composition of glass. In conventional glasses containing sodium, Na^+ ions are associated with nonbridging oxygen atoms in the glass network, and silver ions exchanged into glass extract electrons from these nonbridging oxygen atoms and reduce into silver atoms. The number of nonbridging oxygen sites is decreased when Al_2O_3 is added into glass, so that Ag^+ ions exchanged into such aluminosilicate glasses are quite stable [92].

8.7 Summary

The controlled synthesis of materials in the form of thin films is a fundamental step in many applications, including optical integrated circuits. Each deposition technique has particular advantages and disadvantages and has its own influence on film structure and properties. The decision which technique should be used depends on the desired application. The optoelectronic devices require coatings or waveguiding films of high quality, where good surface uniformity, high packing density, good adhesion, or low defect density must be ensured. Moreover, precise thickness control is demanded. Selected chemical techniques fulfil the requirements to deposit high-quality layers and allow one to fabricate components such as narrow bandpass filters, polarizing coatings, and integrated optical circuits.

Among the disadvantages in chemical methods of synthesis and deposition, one should mention a risk of product contamination with residues being left from the used precursors or carrier gases. The impurity content of final thin films depends on precursor's stability and on the completeness of the reaction. The carrier gas, too, could be a source of contamination. In comparison, physical methods assure much cleaner environment of processes, due to the high vacuum used and the lack of precursor compounds with unwanted ligands or elements. Besides, the evaporation distribution may not be geometrically uniform, and the properties of the coating can vary over the sample.

References

[1] Tien P.K. 'Integrated optics and new wave phenomena in optical waveguides'. *Rev. Mod. Phys.* 1977; **49**(2): 361–420.

[2] Smeeton T. and Humphreys C. 'Perspectives on electronic and photonic materials' in: Kasap S. and Capper P. (eds.) *Springer Handbook of Electronic and Photonic Materials*. (Cham, Springer Handbooks, 2017), pp. 1–15.

[3] DeCorby R. and Irannejad M. 'Glasses for photonic integration' in: Kasap S. and Capper P. (eds.) *Springer Handbook of Electronic and Photonic Materials*. (Cham, Springer Handbooks, 2017), pp. 1059–1080.

[4] Muñoz P., Micó G., Bru L.A., *et al.* 'Silicon nitride photonic integration platforms for visible, near-infrared and mid-infrared applications'. *Sensors* 2017; **17**: 2088.

[5] Marris-Morini D., Vakarin V., Ramirez J.M., *et al.* 'Germanium-based integrated photonics from near- to mid-infrared applications'. *Nanophotonic* 2018; **7**: 1781–1793.

[6] Sharma S.K., Barthwal S., Singh V., *et al.* 'PECVD based silicon oxynitride thin films for nano photonic on chip interconnects applications'. *Micron* 2013; **44**: 339–346.

[7] Liu B.T., Tang S.J., Yu Y.Y., and Lin S.H. 'High-refractive-index polymer/inorganic hybrid films containing high TiO_2 contents'. *Coll. Surf. A: Physicochem. Eng. Aspects* 2011; **377**: 138–143.

[8] Su H.W. and Chen W.C. 'High refractive index polyimide–nanocrystalline-titania hybrid optical materials'. *J. Mater. Chem.* 2008; **18**: 1139–1145.

[9] Liu B.T., Li P.S., Chen W.C., and Yu Y.Y. 'Ex situ synthesis of high-refractive-index polyimide hybrid films containing TiO_2 chelated by 4-aminobenzoic acid'. *Macromol. Nanotechnol.* 2014; **50**: 54–60.

[10] Kik P.G. and Polman A. 'Erbium doped optical waveguide amplifiers on silicon'. *MRS Bull.* 1998; **23**(4): 48–54.

[11] Sohler W. *Status and Prospects of Lithium Niobate Integrated Optics* [online]; 2012. Available from https://physik.unipaderborn.de/fileadmin/physik/Alumni/Sohler/2012/ECIO12_Sohler_Status_LN..._Plenary-2012.pdf [Accessed 26 Mar 2019].

[12] Bazzan M. and Sada C. 'Optical waveguides in lithium niobate: recent developments and applications'. *Appl. Phys. Rev.* 2015; **2**: 040603.

[13] Koch T.L., Leonberger F.J., and Suchoski P.G. 'Integrated optics' in: Bass M. (ed. in chief), *Handbooks of Optics Vol. II. Devices, Measurements, and Properties*. Optical Society of America, McGraw-Hill, Inc. 1995. pp. 6.1–6.42.

[14] Wu Y., Xing H., Zhang L., *et al.* 'Fabrication and properties of vitreous silica films prepared by flame hydrolysis deposition'. *Mat. Chem. Phys.* 2004; **84**: 234–237.

[15] Kumar P., Wiedmann M.K., Winter C.H., and Avrutsky I. 'Optical properties of Al_2O_3 thin films grown by atomic layer deposition'. *Appl. Opt.* 2009; **48**(28): 5407–5412.

[16] Smit M.K., Acket G.A., and van der Laan C.J. 'Al_2O_3 films for integrated optics'. *Thin Solid Films* 1986; **138**: 171–181.

[17] Orignac X., Barbier D., Du X.M., and Almeida R.M. 'Fabrication and characterization of sol-gel planar waveguides doped with rare-earth ions'. *Appl. Phys. Lett.* 1996; **69**: 895–897.

[18] Innocenzi P., Martucci A., Guglielmi M., *et al.* 'Optical and surface properties of inorganic and hybrid organic-inorganic silica-titania sol-gel planar waveguides'. *J. Non-Cryst. Solids* 1999; **259**: 182–190.

[19] Zampedri L., Righini G.C., Portales H., *et al.* 'Sol–gel-derived Er-activated SiO_2–HfO_2 planar waveguides for 1.5 μm application'. *J. Non-Cryst. Solids* 2004; **345–346**: 580–584.

[20] Lukowiak A., Dylewicz R., Patela S., Strek W., and Maruszewski K. 'Optical properties of SiO_2–TiO_2 thin film waveguides obtained by the sol–gel method and their applications for sensing purposes'. *Opt. Mater.* 2005; **27**: 1501–1505.

[21] Armelao L., Gross S., Obetti G., and Tondello E. 'Er^{3+}-doped SiO_2–Al_2O_3 thin films prepared by the sol–gel route'. *Surf. Coat. Technol.* 2005; **190**: 218–222.

[22] Aquino F.T., Ferrari J.L., Ribeiro S.J.L., Ferrier A., Goldner P., and Gonçalves R.R. 'Broadband NIR emission in novel sol–gel Er^{3+}-doped SiO_2–Nb_2O_5 glass ceramic planar waveguides for photonic applications'. *Opt. Mater.* 2013; **35**: 387–396.

[23] Lukowiak A., Zur L., Tran T.N.L., *et al.* 'Sol–gel-derived glass-ceramic photorefractive films for photonic structures'. *Crystals* 2017; **7**: 61.

[24] Berneschi S., Alombert-Goget G., Armellini C., *et al.* 'Rare-earth-doped silicate glass-ceramic thin films for integrated optical devices'. *Adv. Sci. Technol.* 2010; **71**: 6–15.

[25] Buestrich R., Kahlenberg F., Popall M., Dannberg P., Müller-Fiedler R., and Rösch O. 'ORMOCER®s for optical interconnection technology'. *J. Sol-Gel Sci. Technol.* 2001; **20**: 181–186.

[26] Houbertz R., Domann G., Cronauer C., *et al.* 'Inorganic–organic hybrid materials for application in optical devices'. *Thin Solid Films* 2003; **442**: 194–200.

[27] Yu W. and Yuan X.C. 'UV induced controllable volume growth in hybrid sol-gel glass for fabrication of a refractive microlens by use of a grayscale mask'. *Opt. Express* 2003; **11**: 2253–2258.

[28] Yang L., Saavedra S.S., Armstrong N.R., and Hayes J. 'Fabrication and characterization of low-loss, sol-gel planar waveguides'. *Anal. Chem.* 2015; **66**(8): 1254–1263.

[29] Coudray P., Etienne P., Moreau Y., Porque J., and Najafi S.I. 'Sol-gel channel waveguide on silicon: fast direct imprinting'. *Opt. Commun.* 1997; **143**: 199–202.

[30] Yu W.X., Yuan X.C., Ngo N.Q., Que W.X., Cheong W.C., and Koudriachov V. 'Single-step fabrication of continuous surface relief micro-optical elements in hybrid sol-gel glass by laser direct writing'. *Opt. Express* 2002; **10**: 443–448.

[31] Kärkkäinen A.H.O., Tamkin J.M., Rogers J.D., *et al.* 'Direct photolithographic deforming of organomodified siloxane films for micro-optics fabrication'. *Appl. Opt.* 2002; **41**: 3988–3998.
[32] Neiss E., Flury M., Mager L., *et al.* 'Multi-level diffractive optical elements produced by excimer laser ablation of sol–gel'. *Opt. Express* 2008; **16**: 14044–14056.
[33] Tandon P. and Boek H. 'Experimental and theoretical studies of flame hydrolysis deposition process for making glasses for optical planar devices'. *J. Non-Cryst. Solids* 2003; **317**: 275–289.
[34] Pitkänen A., Mäkelä J.M., Nurminen M., *et al.* 'Numerical study of silica particle formation in turbulent H_2/O_2 flame'. *IFRF Combust. J.* 2005; No. 200509.
[35] Huang Y., Zheng L., Zhang H., Zhang G., and Jiang L. 'Design and optimization of substrate placement for large-sized and high-quality fused silica glass by $SiCl_4$ flame hydrolysis deposition'. *Int. J. Heat Mass Transfer* 2017; **111**: 917–932.
[36] Ruano J.M., Benoit V., Aitchison J.S., and Cooper J.M. 'Flame hydrolysis deposition of glass on silicon for the integration of optical and microfluidic devices'. *Anal. Chem.* 2000; **72**: 1093–1097.
[37] Lee D. and Choi M. 'Coalescence enhanced synthesis of nanoparticles to control size, morphology and crystalline phase at high concentrations'. *Aerosol Sci.* 2002; **33**: 1–16.
[38] Kim Y.T., Cho S.M., Seo Y.G., Yoon H.D., Im Y.M., and Yoon D.H. 'Influence of hydrogen on SiO_2 thick film deposited by PECVD and FHD for silica optical waveguide'. *Cryst. Res. Technol.* 2002; **37**(12): 1257–1263.
[39] Bange J.P., Patil L.S., and Gautam D.K. 'Growth and characterization of SiO_2 films deposited by flame hydrolysis deposition system for photonic device application'. *Prog. Electromag. Res.* 2008; **3**: 165–175.
[40] Bange J.P., Patil L.S., and Gautam D.K. 'Effect of OMCTS flow rate on SiO_2 films grown by flame hydrolysis deposition'. *Optoelectron. Adv. Mater. Rapid Commun.* 2010; **4**(4): 584–587.
[41] Chagger H.K., Hainsworth D., Patterson P.M., Pourkashanian M., and Williams A. 'The formation of SiO_2 from hexamethyldisiloxane combustion in counterflow methane-air flames'. *Symp. (Int.) Combust.* 1996; **26**(2): 1859–1865.
[42] Wegner K. and Pratsinis S.E. 'Scale-up of nanoparticle synthesis in diffusion flame reactors'. *Chem. Eng. Sci.* 2003; **58**: 4581–4589.
[43] Huang Y., Lu Z., and Zheng L. 'Study of $SiCl_4/H_2/O_2$ chemical kinetics and its application to fused silica glass synthesis'. *Comb. Sci. Technol.* 2018; **190**(10): 1861–1885.
[44] Yao W., Zheng L., and Zhang H. 'Modeling analysis on the silica glass synthesis in a hydrogen diffusion flame'. *Int. J. Heat Mass Transfer* 2015; **81**: 797–803.
[45] Strobel R. and Pratsinis S.E. 'Flame aerosol synthesis of smart nanostructured materials'. *J. Mater. Chem.* 2007; **17**: 4743–4756.

[46] Lei W., Chuandong R., and Zhen'an G. 'Effects of flame temperature and $SiCl_4$ concentration on particle characteristics for synthetic silica glass'. *J. Wuhan Univ. Technol.-Mater. Sci. Ed.* 2010; **25**(13): 419–422.

[47] Lee J., Olfert J., Altman I.S., and Choi M. 'Determination of particle temperatures in a silica-generating counterflow flame via flame emission measurements'. *Int. J. Heat Mass Transfer* 2010; **53**: 564–567.

[48] Windeler R.S., Friedlander S.K., and Lehtinen K.E.J. 'Production of nanometer-sized metal oxide particles by gas phase reaction in a free jet. I: Experimental system and results'. *Aerosol Sci. Technol.* 1997; **27**(2): 174–190.

[49] Hwang J.Y., Gil Y.S., Kim J.I., Choi M., and Chung S.H. 'Measurements of temperature and OH radical distributions in a silica generating flame using CARS and PLIF'. *J. Aerosol Sci.* 2001; **32**: 601–613.

[50] Moore T., Brady B., and Martin L.R. 'Measurements and modeling of $SiCl_4$ combustion in a low-pressure H_2/O_2 flame'. *Combust. Flame* 2006; **146**: 407–418.

[51] Nasonova A., Park D.W., Charinpanitkul T., and Kim K.S. 'Numerical analysis on premixed combustion of H_2–$SiCl_4$–air system to prepare SiO_2 particles'. *J. Ind. Eng. Chem.* 2012; **18**: 509–512.

[52] Linow S., Schneider C., Geiss S., Janicka J., Hassel E.P., and Rüdiger F. 'Experimental study of the synthesis of fused silica by direct combustion synthesis'. *Exp. Fluids* 2002; **32**: 66–75.

[53] Tong X.C. *Advanced Materials for Integrated Optical Waveguides*. (Cham, Springer, 2014).

[54] Hussein M.G., Wörhoff K., Sengo G., and Driessen A. 'Reduction of hydrogen-induced optical losses of plasma-enhanced chemical vapor deposition silicon oxynitride by phosphorus doping and heat treatment'. *J. Appl. Phys.* 2007; **101**: 023517.

[55] Picciotto A., Bagolini A., Bellutti P., and Boscardin M. 'Influence of interfaces density and thermal processes on mechanical stress of PECVD silicon nitride'. *Appl. Surf. Sci.* 2009; **256**: 251–255.

[56] Arthur J.R. 'Physical and chemical methods for thin-film deposition and epitaxial growth' in: Czanderna A.W., Powell C.J., Madey T.E., Hercules D. M., and Yates J.T. (eds.). *Specimen Handling, Preparation, and Treatments in Surface Characterization. Methods of Surface Characterization*. (Boston, Springer, 2002), pp. 239–293.

[57] Korkin A.A., Cole J.V., Sengupta D., and Adams J.B. 'On the mechanism of silicon nitride chemical vapor deposition from dichlorosilane and ammonia'. *J. Electrochem. Soc.* 1999; **146**(11): 4203–4212.

[58] Xiao Z.G. and Mantei T.D. 'Plasma-enhanced deposition of hard silicon nitride-like coatings from hexamethyldisiloxane and ammonia'. *Surf. Coat. Technol.* 2003; **172**: 184–188.

[59] Jhansirani K., Dubey R.S., More M.A., and Singh S. 'Deposition of silicon nitride films using chemical vapor deposition for photovoltaic applications'. *Results Phys.* 2016; **6**: 1059–1063.

[60] Ohdaira K. and Matsumura H. 'Deposition of moisture barrier films by catalytic CVD using hexamethyldisilazane'. *Jpn. J. Appl. Phys.* 2014; **53**: 05FM03.

[61] Liu X., Pu X., Li H., Qiu F., and Huang L. 'Chemical vapor deposition of silicon nitride thin films from tris(diethylamino)chlorosilane'. *Mater. Lett.* 2005; **59**: 11–14.

[62] Levy R.A., Lin X., Grow J.M., Boeglin H.J., and Shalvoy R. 'Low pressure chemical vapor deposition of silicon nitride using the environmentally friendly tris(dimethylamino)silane precursor'. *J. Mater. Res.* 1996; **11**(6): 1483–1488.

[63] Varshneya A.K. *Industrial glass* [online]; 1999. Available from www.britannica.com/topic/glass-properties-composition-and-industrial-production-234890/Glassmaking-in-the-laboratory#ref76359 [Accessed 20 Feb 2019].

[64] Saint C. and Saint J.L. *Integrated circuit* [online]; 1999. Available from www.britannica.com/technology/integrated-circuit/Fabricating-ICs#ref837159 [Accessed 20 Feb 2019].

[65] Hahn B., Heindel G., Pschorr-Schoberer E., and Gebhardt W. 'MOCVD layer growth of ZnO using DMZn and tertiary butanol'. *Semicond. Sci. Technol.* 1998; **13**: 788–791.

[66] Hussain S., Khan Y., Khranovskyy V., Muhammad R., and Yakimova R. 'Effect of oxygen content on the structural and optical properties of ZnO films grown by atmospheric pressure MOCVD'. *Progr Nat. Sci.: Mat. Intl.* 2013; **23**(1): 44–50.

[67] Lu J., Ye Z., Wang L., Huang J., and Zhao B. 'Structural, electrical and optical properties of N-doped ZnO films synthesized by SS-CVD'. *Mater. Sci. Semicond. Proc.* 2003; **5**: 491–496.

[68] Ataev R.M., Bagamadova A.M., Djabrailov A.M., Mamedov V.V., and Rabadanov R.A. 'Highly conductive and transparent Ga-doped epitaxial ZnO films on sapphire by CVD'. *Thin Solid Films* 1995; **260**: 19–20.

[69] Johnson R.W., Hultqvist A., and Bent S.F. 'A brief review of atomic layer deposition: from fundamentals to applications'. *Mater. Today* 2014; **17**(5): 236–246.

[70] Knisley T.J., Kalutarage L.C., and Winter C.H. 'Precursors and chemistry for the atomic layer deposition of metallic first row transition metal films'. *Coord. Chem. Rev.* 2013; **257**: 3222–3231.

[71] Potts S.E. and Kessels W.M.M. 'Energy-enhanced atomic layer deposition for more process and precursor versatility'. *Coord. Chem. Rev.* 2013; **257**: 3254–3270.

[72] Leskelä M. and Ritala M. 'Atomic layer deposition chemistry: recent developments and future challenges'. *Angew. Chem. Int. Ed.* 2003; **42**: 5548–5554.

[73] Puurunen R.L. 'Surface chemistry of atomic layer deposition: a case study for the trimethylaluminum/water process'. *J. Appl. Phys.* 2005; **97**: 121301.

[74] Crowell J.E. 'Chemical methods of thin film deposition: chemical vapor deposition, atomic layer deposition, and related technologies'. *J. Vacuum Sci. Technol. A* 2003; **2**: S88–S95.
[75] Suntola T. and Antson J. 'Method for producing compound thin films'. US Patent 4,058,430. 1977.
[76] West G.N., Loh W., Kharas D., *et al.* 'Low-loss integrated photonics for the blue and ultraviolet regime'. *APL Photonics* 2019; **4**: 026101.
[77] Monsma D. and Becker J. 'The Savannah ALD System – an excellent tool for atomic layer deposition'. *Mater. Matters* 2006; **1.3**: 5.
[78] George S.M. 'Atomic layer deposition: an overview'. *Chem. Rev.* 2010; **110**: 111–131.
[79] Solehmainen K., Kapulainen M., Heimala P., and Polamo K. 'Erbium-doped waveguides fabricated with atomic layer deposition method'. *IEEE Photon. Technol. Lett.* 2004; **16**(1): 194–196.
[80] Rönn J., Karvonen L., Kauppinen C., *et al.* 'Atomic layer engineering of Er-ion distribution in highly doped Er:Al_2O_3 for photoluminescence enhancement'. *ACS Photonics* 2016; **3**: 2040–2048.
[81] Iqbal J., Jilani A., Hassan P.M.Z., Rafique S., Jafer R., and Alghamdi A. A. 'ALD grown nanostructured ZnO thin films: Effect of substrate temperature on thickness and energy band gap'. *J. King Saud Univ. – Sci.* 2016; **28**: 347–354.
[82] Guziewicz E., Kowalik I.A., Godlewski M., *et al.* 'Extremely low temperature growth of ZnO by atomic layer deposition'. *J. Appl. Phys.* 2008; **103**: 033515.
[83] Banerjee P., Lee W.J., Bae K.R., Lee S.B., and Rubloff G.W. 'Structural, electrical, and optical properties of atomic layer deposition Al-doped ZnO films'. *J. Appl. Phys.* 2010; **108**: 043504.
[84] Scharrer M., Wu X., Yamilov A., Cao H., and Chang R.P.H. 'Fabrication of inverted opal ZnO photonic crystals by atomic layer deposition'. *Appl. Phys. Lett.* 2005; **86**: 151113.
[85] Meng X., Byun Y.C., Kim H.S., *et al.* 'Atomic layer deposition of silicon nitride thin films: a review of recent progress, challenges, and outlooks'. *Materials* 2016; **9**: 1007.
[86] Morishita S., Sugahara S., and Matsumura M. 'Atomic-layer chemical-vapor-deposition of silicon-nitride'. *Appl. Surf. Sci.* 1997; **112**: 198–204.
[87] Alvarez D. Jr., Spiegelman J., Heinlein E., *et al.* 'Novel oxidants and sources of nitrogen for atomic layer deposition'. *ECS Trans.* 2016; **72**(4): 243–248.
[88] Jhang P.C., Lu C.P., Shieh J.Y., *et al.* 'Properties of N-rich silicon nitride film deposited by plasma-enhanced atomic layer deposition'. *Solid-State Electron.* 2017; **133**: 10–16.
[89] Yulianto M., Marzuki A., and Suryanti V, 'Fabrication and characterization planar waveguides of Na^+–Ag^+/K^+ by ion exchange and prism coupler'. *AIP Conf. Proc.* 2017; **1868**(1): 060006.
[90] Opilski A., Rogoziński R., Gut K., and Blahut M. 'Some aspects of application of ion exchange in glass'. *Proc. SPIE* 2000; **4239**.

[91] Rehouma F. and Aiadi K.E. 'Glasses for ion-exchange technology'. *Int. J. Commun.* 2008; **1**(4): 148–155.

[92] Tervonen A., Honkanen S., and West B.R. 'Exchanged glass waveguide technology: a review'. *Opt. Eng.* 2011; **50**(7): 071107.

Chapter 9

Photorefractive waveguides

Thi Ngoc Lam Tran[1,2,3], *Simone Berneschi*[4], *Gualtiero Nunzi Conti*[4] *and Maurizio Ferrari*[1]

In integrated optics, optical waveguides are fundamental elements for the development of integrated devices, including light sources, modulators, detectors, optical pathways, receivers and switches. For the fabrication of optical integrated waveguide components, grating couplers or other complex integrated structures realized on planar optical waveguides, materials with high photorefractivity are surely of great interest since they allow applying photorefractive direct-laser-writing techniques which are of lower cost and less time-consuming than conventional photolithography and etching processes. This chapter provides an overview of classes of photorefractive materials such as photorefractive glasses, glass-ceramics, crystals and polymers, which are promising candidates for integrated optics. Different aspects, including material preparation, photorefractivity investigation and the fabrication of photorefractive direct-laser-written micro/nano-optical structures such as channel waveguides and gratings in each type of photorefractive materials, are also discussed.

9.1 Fundamentals of photorefractive waveguides

This section provides a brief overview of photorefractive materials and the description of their photorefractive effects under certain irradiation conditions. The photorefractivity in different classes of materials, e.g. glasses, glass-ceramics, crystals and polymers is discussed. Some fundamentals of integrated optical waveguides are recalled, and the current fabrication techniques of optical integrated waveguides are summarized, emphasizing the advantages of photorefractive direct laser writing over other techniques.

[1]Institute of Photonics and Nanotechnologies (IFN) CSMFO Lab., Department of Physics Sciences and Matter Technologies (DSFTM), National Research Council (CNR), and FBK Photonics Unit, Trento, Italy

[2]Institute of Photonics and Nanotechnologies (IFN), Department of Physics Sciences and Matter Technologies (DSFTM), National Research Council (CNR), Milano, Italy

[3]Department of Materials Technology, Faculty of Applied Sciences, Ho Chi Minh City University of Technology and Education, Ho Chi Minh City, Vietnam

[4]"Nello Carrara" Institute of Applied Physics (IFAC) MiP Lab, National Research Council (CNR), Italy

9.1.1 Photorefractivity

In general, photorefractive materials are the special materials that exhibit a refractive index change when they are exposed to a light of suitable wavelength. The photorefractive effects are generated by different mechanisms in different materials, depending on their nature.

In centrosymmetric glasses, such as germanosilicate glasses, tin-dioxide-based glasses and glass-ceramics or other types of centrosymmetric materials, the photorefractivity, also specified as photosensitivity [1], is attributed to a change either in the polarizability or the density, leading to a refractive index change of the material. The polarizability change can be altered thanks to the photoinduced structural modification of the constituent molecules via photoconversion of optically active defects or via breakage of bonds [2]. The photosensitivity can also involve the creation of new phases [3,4]. For instance, in the works [5,6] on Nd^{3+}-doped and Ho^{3+}-doped strontium barium niobate [SBN: Nd_2O_3–SrO–BaO–Nb_2O_5–B_2O_3], respectively, the continuous wave (CW) argon laser irradiation was observed to induce local SBN nanocrystalline formation. Under the ultraviolet (UV) light irradiation, some materials were observed to change their density thanks to an undergone densification [7,8] or expansion [9,10], which in fact relate to the photoinduced rearrangement of the microscopic-scale structure of the materials. However, varied glass composition, fabrication history and irradiation condition, e.g. irradiation source, power and frequency, can induce varied photosensitive responses. The complexity and diversity of the photosensitive properties of glasses has led to challenges for defining the common mechanisms and microscopic origins that drive the photosensitivity of such materials.

In non-centrosymmetric photorefractive materials such as $LiNbO_3$, $LiTaO_3$, $Sn_2P_2S_6$ crystals [11] and polymers [12,13], under light irradiation, the refractive index change is induced by electric fields oscillating at optical frequencies, which can be considered as a manifestation of the classic electro-optic effect on a microscopic scale thanks to charge transportation [14,15]. If the optical frequency of the light is higher than the bandgap of the materials as in the works on $LiNbO_3$ [11,16] and $LiTaO_3$ [17,18], the refractive index change is originated by interband phototransition of intrinsic carriers in which the electron is promoted to the upper mobile state, leaving a hole in the lower state. A photorefractive space–charge field is thus generated, owing to the difference between the motions of electrons and holes, which leads to a refractive index change [14]. In other words, this effect involves the band-to-band electron excitation, and it is identified as interband photorefractivity [14]. In the doped crystals, the refractive index change can be induced by spatially inhomogeneous intensity illumination of a laser beam, which generates charge transport at the defect levels within the bandgap due to the presence of dopants; this photorefractive effect is named conventional photorefractivity. The magnitude of refractive index change in the case of interband photorefractivity is typically higher than in the other cases.

9.1.2 Integrated optical waveguides

Generally, optical waveguides are structures consisting of a dielectric medium with a higher refractive index than its surrounding dielectric media. Such structures allow confining electromagnetic waves in a specific volume and guiding them in a specific direction for signal transmission.

The optical waveguides are commonly classified, based on their geometries, as planar waveguides (e.g. strip or channel waveguides) and optical fibres. Together with the development of optical interconnects, optical fibres have long been the technical choice for long-distance data transfer [19]. For short-distance data processing, although fibre-based optical links have been developed and applied in printed circuit boards [20] and in-fibre integrated optical devices have been proposed [21], they are not sufficient to achieve high integration density, complex circuitry and scaling-down sizes at once for integrated optical devices. On the contrary, optical circuits based on two-dimensional planar waveguides show higher integration applicability for developing miniaturized and functionalized devices [22]. Therefore, they undoubtedly have more potential for miniaturized optical circuits than the fibre-based ones.

The simplest planar optical integrated circuit (OIC) consists of three basic components: light source, optical waveguide and photodetector [19]. However, for more functional OICs, other optical components such as switches and modulators can be integrated as well. If one only considers the basic three-component OICs, in terms of constituting materials, they can be classified into two types: either monolithic or hybrid. The features, pros and cons and several material options for each type of OIC are summarized in Table 9.1.

In comparison with the monolithic OICs, the hybrid OICs have been employed and developed dominantly since they have the advantages of both a larger material

Table 9.1 Two types of optical integrated circuits (OICs)

Type	Feature	Material options	Pros and cons
Monolithic OIC	A single substrate material is used for all the three basic components (light source, optical waveguide and photodetector)	Optically active materials such as compound semiconductors (GaAs, InP), active glasses and polymers [23]	P: Low cost and automated batch processing potential C: Limited available proper materials and unmatured technology at present [23]
Hybrid OIC	The basic components are made of two or more substrate materials bond together, each of them performing a specific optical function: generation, guiding and detection of light signal	Dielectric materials for waveguides and detector such as quartz, $LiNbO_3$, $LiTaO_3$, Si, Si_3N_4 and polymers; III–V semiconductors and doped active materials for light source	P: Possibility of exploiting the existing technologies for each functional material C: Assembling all the components is challenging [19]

choice and the availability of advanced manufacturing and processing technologies for each type of materials. However, the challenges arising from integration of all the component materials lead to high cost. In addition, among the possible drawbacks of the hybrid solution, it should also be mentioned that this approach generally offers greater coupling losses due to the index mismatch among the different materials constituting the integrated optical circuit. In other words, the problem of reducing coupling losses is more critical in the hybrid approach than that in the monolithic case, where the material on which the various devices – constituting the final integrated optical circuit – are fabricated is practically the same or almost the same [low index mismatch]. Even if the monolithic OICs at present are limited due to a lack of suitable optically active materials and technologies for constructing the three components at once from a single material substrate, they still show the potential of simultaneous bandwidth scaling, low cost and automated batching and processing [24]. Hence, there have been tremendous researches and efforts on making monolithic OICs commercially available by exploiting semiconductor materials [24,25].

Considering the optical waveguiding components for OIC applications, in order to obtain a strong light confinement and contemporarily optimize the device performance [e.g. in terms of losses], it is mandatory to guarantee a high refractive index contrast between the waveguide and the cladding, together with a single-mode propagation regime at the working wavelength. In other words, the size of optical waveguides is of the order of the wavelength of the propagating light, i.e. in micron-size. In addition, the design and implementation of proper optical coupling structures on the optical waveguides is highly demanded for interconnecting the circuit components. Micro/nano optical coupling elements such as micro-lenses [26], nanotapers [27] and gratings [28] are potential candidates for in-coupling and out-coupling light within integrated optical waveguides. However, achieving efficient optical couplers for the integrated optical circuits is still an ongoing challenge. Furthermore, to fabricate micro-optical components as optical waveguides and couplers for a board-level optical interconnect, a huge effort to develop advanced materials and manufacturing, micro-patterning and implementation technologies is still required.

Several technologies for such optical waveguide fabrication are currently available, such as film deposition, photolithography, moulding, etching, ion exchange, ion diffusion and direct laser writing. Among these technologies, a strategic place can be assigned to the direct band edge writing that exploits the photorefractive properties of suitable materials. This technique allows the fabrication of waveguides and components, such as directional couplers or Mach–Zehnder interferometers and grating couplers, for both hybrid and monolithic OICs, with less-complicated processing steps and consequently lower cost and less time, especially in comparison with the standard photolithography and chemical or physical etching patterning processes [8,11,29]. In addition, active photorefractive materials are novel candidates for the construction of monolithic OICs since they may offer solutions not only to the issue of the light source component but also to the fabrication of other micro/nano-optical components by using the direct-laser-writing techniques [23]. Therefore, this approach can combine the advantages of both the monolithic OICs themselves, as mentioned in Table 9.1, and the direct-laser-writing technique.

9.2 Photosensitive glasses and glass-ceramics

In integrated optics, besides several advanced waveguide platforms such as silicon [30], silicon-on-insulator (SOI) [31], silicon-on-sapphire [32], AlN on sapphire [33,34], SiN [35] and indium phosphide [29], silica-based materials are other potential candidates for constructing OICs [36–38]. Silica-based planar circuits mostly rely on the mature technology of guided wave optics and silica fibres. Their advantages for the development of optical interconnects include reliability, low losses, integration capability and high efficiency in coupling to single-mode optical fibres [39]. In addition, silica-glass waveguides can be fabricated on large-size silicon wafers with low cost and large scale [36]. Moreover, the high photosensitivity of some silica-based materials makes them even more convenient for integrated optical circuits since they enable the robust direct UV-writing technique for fabricating optical components. Germanosilicate glasses have been by far the most commonly used photosensitive silica-based materials for constructing planar lightwave circuit (PLC) devices [36]. Ion-exchanged phosphate glasses were also reported to have high photosensitivity [3,35]. Recently, studies on tin-dioxide-based glasses [40] and glass-ceramics [10,41] have shown the potential of these binary systems for integrated optical circuits thanks to their high photosensitivity.

9.2.1 Germanosilicate glasses

9.2.1.1 Photosensitivity investigation

Starting from the first demonstration of UV photoinduced Bragg gratings in Ge-doped fibre cores [42], a great attention has been paid to germanosilicate glasses for integrated optics and photonics. As summarized in [38], the photosensitivity of germanosilicate glasses can be explained by the photoinduced electronic transitions of localized defects or dopants, such as germanium–oxygen-deficient centres, which lead to a strong UV absorption [43]. However, another contribution to photosensitivity of germanosilicate glasses is reported to come from photoinduced volume densification [8,44].

To enhance the photosensitivity, a huge effort has been made to increase or create defect centres in these glasses through different methods, including increasing germanium concentration in the glasses [38,45], subsequent sensitization processes such as low-temperature D_2 [38] or H_2 loading [46] and codoping with elements such as B [47,48] and Sn [49]. Concerning the fabrication of photosensitive germanosilicate glasses and films, several techniques have been employed, such as radio-frequency-magnetron-sputtering deposition (RF-sputtering) [8,50,51], plasma-enhanced chemical vapour deposition (PECVD) [52,53], sol–gel [45] and flame hydrolysis deposition (FHD) [54,55].

For the photosensitivity investigation and direct laser writing of micro-optical waveguides and gratings, the popularly used UV lasers are the pulsed 193 nm ArF and 248 nm KrF excimer lasers, the 244 nm CW frequency-doubled argon-ion (FDA) laser and the 266 nm CW diode-pumped all-solid-state laser [56]. Table 9.2 summarizes the properties of a few reported germanosilicate glasses prepared by different methods. One can see that the photosensitive responses scatter, since they depend not only on composition and sample preparation but also on laser irradiation conditions.

9.2.1.2 Direct UV-laser-written optical structures

Due to the high photosensitivity, germanosilicate glasses have been intensively employed for developing silica-based guided wave circuits via the robust and effective direct UV processing, which can include trimming/tuning of the device characteristics, thus providing new functionalities. One can find the potentials and current status of both the photosensitive UV laser processing and the development of silica-based PLCs for integrated optics in some review papers [29,36,38,57]. Some examples of photosensitivity investigation and direct writing of micro-optical structures (i.e. waveguides or gratings) on germanosilicate glasses are summarized in Table 9.3.

9.2.2 Tin dioxide–silica glasses and glass-ceramics

The photosensitivity potential of tin-dioxide-based transparent materials was first realized in tin-codoped germanosilicate [49] and phosphosilicate [60,61] optical fibres. By introducing tin in such silicate glasses, strong photosensitive gratings with enhanced refractive index change up to ~1.2–1.4 $\times$ 10^{-3} were obtained. Thanks to Sn presence, the refractive index change was even three times larger than in the case of pure germanosilicate [49]. All these aspects have led to numerous researches on the development of binary tin–silicon dioxides and on the exploitation of their photorefractivity.

According to [62], two different tin-silica systems could be obtained, based on SnO_2 content: glasses (with tin content up to 0.5 mol%) and glass-ceramics (with tin content higher than 0.5 mol%). The two materials have different photosensitive responses after the UV irradiation: the tin–silica glasses facilitate positive refractive change while the glass-ceramics have negative index change with higher magnitude. Increasing SnO_2 content leads to higher refractive index change [63] in the tin dioxide–silica binary systems [41,63]. Several methods have been applied for the preparation of SnO_2:SiO_2 materials, such as modified chemical vapour deposition (MCVD) [64], melt-quenching method [65], ion-implantation method [66], helicon-activated reaction evaporation (HARE) process [63] and sol–gel [41,67]. Table 9.4 reports the achievable SnO_2 content in SiO_2 materials prepared by different techniques. Among these reported methods, sol–gel [41,67] and HARE [63] are the ones that provide higher SnO_2 contents in the binary system material. Concerning the sol–gel-derived SiO_2–SnO_2 binary systems, proper thermal treatment processes are mandatory to obtain densified glass-ceramics: 900°C for 100 h for the SiO_2–SnO_2 glass-ceramic monoliths and 1,000°C for 1 h for the SiO_2–SnO_2 glass-ceramic planar waveguides [41,68].

Table 9.2 Germanosilicate glasses (SiO_2–GeO_2) prepared by different methods

Method	Form	GeO_2 content (x) (mol%)	Reported photosensitive response	Ref.
RF-sputtering	Thin film	5–60	After 193 nm ArF laser exposure: $x > 25$ mol%: negative refractive index change $x < 25$ mol%: positive refractive index change	[51]
Sol–gel	Thin film	10–45	After 248 nm KrF excimer laser exposure: higher GeO_2 content, higher positive refractive index change	[3,45]
MCVD	Fibre	Up to 75	After 244 nm FDA laser exposure: irradiation fluence <10 kJ/cm^2: positive refractive index change irradiation fluence >10 kJ/cm^2: negative refractive index change	[45]

Table 9.3 Several examples of photosensitivity investigation and direct writing of micro-optical structures on germanosilicate glasses (SiO_2–GeO_2)

Materials	Film deposition technique	Irradiation condition	Refractive index change	Laser-written structures	Refs.
SiO_2–GeO_2 buried layer	PECVD	244 nm CW FDA laser Fluence: 10 kJ/cm^2	$\Delta n = +2.5 \times 10^{-3}$	Channel waveguides	[53]
H_2-loaded SiO_2–GeO_2 buried layer	FHD	244 nm CW FDA laser Fluence: 12 kJ/cm^2	$\Delta n = +0.8 \times 10^{-3}$	Channel waveguides with integral Bragg gratings	[58,59]
SiO_2–GeO_2 planar wave-guides	RF-sputtering	248 nm KrF excimer laser Cumulative fluence: 20 kJ/cm^2	$\Delta n = +3.0 \times 10^{-3}$	Bragg gratings	[8]
SiO_2–GeO_2 thin films	Sol–gel	248 nm KrF excimer laser Cumulative fluence: N/A	$\Delta n = +6.0 \times 10^{-5}$	N/A	[45]

Table 9.4 Tin-dioxide-based glasses and glass-ceramics (SiO_2–SnO_2) prepared by different methods

Method	Form	Maximum SnO_2 content (mol%)	Feature comments	Ref.
MCVD	Fibre	0.15	Volatility of SnO_2 due to high fabrication temperature is limiting SnO_2 content	[51]
Melt quenching	Bulk	3		[65]
Ion implantation	Bulk and film	N/A	α-Sn nanoclusters and non-stoichiometric SnO_x nanoparticles	[66]
Sol–gel	Bulk	10	Low-temperature and melt-free synthesis, multicomponent materials	[67]
	Film	30		[68]
HARE	Film	25	N/A	[63]

Table 9.5 Photosensitivity investigation of tin-doped silica glasses

Material	Fabrication method	Irradiation condition	Refractive index change	Ref.
SiO_2:SnO_2 optical fibres	MCVD	248 nm KrF excimer laser Cumulative fluence: 20 kJ/cm^2	$\Delta n = +3.0 \times 10^{-4}$ for 0.15 mol% SnO_2 sample	[64]
SiO_2:SnO_2 bulks	Sol–gel	266 nm pulsed fourth harmonic Nd–YAG laser Cumulative fluence: 0.17 kJ/cm^2	$\Delta n = +4.0 \times 10^{-4}$ for 0.4 mol% SnO_2 sample	[70]

9.2.2.1 Photosensitivity of tin-doped silica glasses

Concerning the photosensitivity of tin dioxide–silica glasses in which a very small content of tin (between 0.4 and 0.5 mol%) was introduced into silica, the refractive index change was demonstrated to relate to laser-induced bleaching of the Sn–oxygen-deficient centres absorption [64], photoinduced material expansion [9,69] and Sn–$(SiO_4)_n$ rings structural units of which the dimensions were reduced during UV exposure [40,70–72]. In the paper by Chiodini *et al.* [40], the power dependence of the refractive index change demonstrated the contribution of two photoinduced processes: one- and two-photon absorption. During the first few tens of pulses, the refractive index change (Δn) was proportional to the square of the laser power density per pulse P: $\Delta n \sim P^2$, which refers to two-photon absorption. On the contrary, the one-photon absorption was observed after a certain number of pulses when the value of Δn increased linearly with respect to the power density P: $\Delta n \sim P$. The corresponding results are listed in Table 9.5. The reported refractive index change of the substitutional tin-doped silica glasses is typically in the order of 10^{-4}.

9.2.2.2 Photosensitivity of tin-dioxide-based glass-ceramics

Contrary to the tin-doped silica glasses, the photosensitivity of tin-dioxide-based glass-ceramics shows a negative refractive index change. The mechanism was suggested to involve two factors: (i) surface modification of SnO_2 nanoclusters and consequent decrease of their crystalline size [62] and (ii) the breaking Sn-related network bonds [41]. The reported refractive index changes of such tin-dioxide-based glass-ceramics are tabulated in Table 9.6.

The values of refractive index change of the sol–gel-derived tin-dioxide-based glass-ceramics are one order of magnitude higher than the ones of tin-doped silica glasses, and they increase with the increasing SnO_2 content in SiO_2 matrix as one can see from Figure 9.1.

9.2.2.3 Direct UV-laser-written micro-optical structures in SiO_2–SnO_2

Exploiting the refractive index change of tin-doped silica glasses in the order of 10^{-4}, UV-written Bragg gratings were first fabricated in SiO_2–SnO_2 optical glass fibres [65,73]. In the paper by Brambilla *et al.* [64], a detailed comparison between SiO_2–GeO_2 and SiO_2:SnO_2 fibres suggested that, under similar UV intensity, the same photosensitivity effects could be obtained with the fibres containing very low SnO_2 contents, almost two orders of magnitude less than GeO_2. Moreover, tin doping helped avoiding hydrogen loading, which is time-consuming and induces high losses in the 1.5 μm region. In addition, SiO_2:SnO_2 fibres potentially provide a lower numerical aperture, and the fabricated fibre Bragg gratings on such glasses have higher thermal stability in comparison with germanosilicate, boro-germanosilicate, H_2-loaded telecom fibres [73].

Table 9.6 Photosensitivity investigation of tin-dioxide-based glass-ceramics

Materials	Fabrication method	Irradiation condition	Refractive index change	Ref.
SiO_2–SnO_2 films	HARE	248 nm KrF excimer laser Cumulative fluence: 2.0 kJ/cm^2	$\Delta n = -2.7 \times 10^{-3}$ for 25 mol% SnO_2 sample	[63]
SiO_2–SnO_2 bulks	Sol–gel	266 nm pulsed fourth harmonic Nd–YAG laser Cumulative fluence: 0.15 kJ/cm^2	$\Delta n = -6.0 \times 10^{-4}$ for 5 mol% SnO_2 sample	[70]
SiO_2–SnO_2 films doped with Eu^{3+}	Sol–gel	248 nm KrF excimer laser Cumulative fluence: 0.3 kJ/cm^2	$\Delta n = -1.6 \times 10^{-3}$ for 25 mol% SnO_2 sample	[9,69]
SiO_2–SnO_2 guiding films doped with Er^{3+}	Sol–gel	248 nm KrF excimer laser Cumulative fluence: 0.3 kJ/cm^2	$\Delta n = -2.8 \times 10^{-3}$ for 30 mol% SnO_2 sample $\Delta n = -2.0 \times 10^{-3}$for 20 mol% SnO_2 sample	[41]

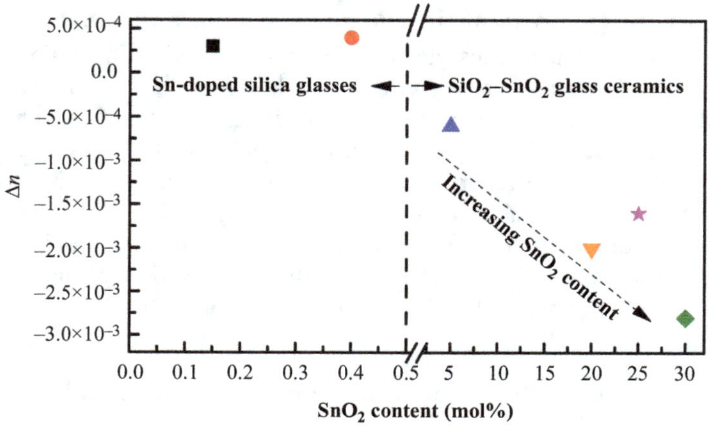

Figure 9.1 The refractive index change under UV irradiation of the sol–gel-derived tin-dioxide-based glass-ceramics and the tin-doped silica glasses as a function of SnO_2 content: 0.15 [64], 0.4 [70], 5 [70], 20 [41], 25 [9,69] and 30 mol% [41]

Referring to the SiO_2–SnO_2 glass-ceramics, in a recent work [41], the optical gratings were fabricated for the first time in the highly photosensitive SiO_2–SnO_2 glass-ceramic planar waveguides (with 30 mol% SnO_2) activated by Er^{3+}, by employing a CW frequency-doubled argon laser at $\lambda = 244$ nm with a writing fluence of 1 kJ/cm^2. Using a ~5 μm MFD (mode-field diameter) spot, the gratings were formed on a small piece of the sample, with 5 μm pitch raster pattern over 4×4 mm^2 area, as revealed by the image in Figure 9.2. A high average refractive index change of 4.3×10^{-3} was measured via a prism coupling technique at both 633 and 1,550 nm. In fact, this high refractive index change is sufficient for the UV direct writing of channel waveguides.

It is noteworthy to emphasize the energy-efficient aspect of the grating's fabrication on the SiO_2–SnO_2:Er^{3+} glass-ceramic planar waveguides in comparison with hydrogen-loaded germano borosilicate glasses. To obtain the same value of refractive index change, the irradiation fluence needed for the germanosilicate glasses is one order of magnitude higher than an SiO_2–SnO_2:Er^{3+} glass-ceramic film [41]. The high photosensitivity and the mentioned advantages of UV-direct-written gratings in SiO_2–SnO_2 glass-ceramics over germanosilicate glasses make them a promising alternative for constructing integrated optic circuits. Moreover, SiO_2–SnO_2 glass-ceramics activated by rare-earth (RE) ions possess another outstanding property, namely the role of SnO_2 as efficient RE luminescence sensitizer [41,67,69,74]. This can be ideal for the development of active optical integrated components, e.g. light sources and monolithic OICs based on RE-doped SiO_2–SnO_2 glass-ceramics.

Figure 9.2 *Image of the 4 × 4 mm^2 area UV-direct-written gratings [5 μm pitch raster pattern] on SiO_2–SnO_2 glass-ceramic planar waveguides fabricated by CW frequency-doubled argon laser at $\lambda = 244$ nm with a writing fluence of 1 kJ/cm^2. One can see the colours due to the interference patterns from the gratings*

9.2.3 Chalcogenide glasses

Chalcogenide glasses (ChGs) are a class of amorphous semiconductors majorly constituted by the VI A-group (now group 16) chalcogen elements such as S, Se and Te, which are covalently bonded to network formers made of As, Ge, Sb, Ga, Si or P [75]. Thanks to the special chemistry and glass formation from covalently bonded heavy elements, ChGs possess unique properties for applications in solar cells, sensors, photonics and integrated optics, such as infrared transparency, bandgap in visible or NIR, low optical attenuation, large refractive index and high optical non-linearity [76]. Moreover, ChGs also exhibit photosensitivity when they are exposed to a light of a wavelength near the band edge, i.e. in visible or NIR range depending on their composition [75], which makes them viable for the photorefractive direct-laser writing of optical components for fibre- or film-based integrated photonic devices [77]. Many of the photosensitivity properties of ChGs have been observed to be more pronounced in thin films than in bulks, thanks to their large number of defective bonds and topologies. This makes the planar format of ChGs ideal for direct laser patterning and constructing OICs. Concerning the preparation of infrared optical ChG fibres, their preforms have been made by several methods such as core drilling [78,79], rotational casting [80,81], extrusion [82] and double-crucible technique [83]. Several techniques have been employed to prepare the ChG thin films, including sputtering [84], pulsed laser depositions [85], chemical vapour deposition [86] and thermal vapouring [87].

Table 9.7 Effect of composition on the photosensitivity of As_xS_y-based ChG films [78,90] prepared by thermal evaporation after exposure to 10^7 femtosecond laser pulses

Materials	Composition	Refractive index change	Photosensitivity
As_xS_y	$As_{42}S_{58}$	$\Delta n = +0.060$	–
Ge substitution for As	$As_{36}Ge_6S_{58}$	$\Delta n = +0.090$	Increase
Sb substitution for As	$As_{36}Sb_6S_{58}$	$\Delta n = +0.005$	Decrease

9.2.3.1 Photosensitivity investigation

The photosensitivity, as well as the photoinduced refractive index change, in ChGs is basically attributed to the photogenerated electron–hole pairs which change the valence and chemical bonding of their nearby atoms and create defects [88], and this phenomenon is known as photo-darkening [89]. A broad range of glass-forming systems and the large composition choices of ChGs lead to a huge diversity and complexity of their photosensitive responses in comparison with silica-based glasses discussed ahead [38]. Different ChG families exhibit different photosensitive responses, which represent a further advantage of ChGs since their photosensitivity can be tailored and engineered by changing their chemical composition and glass formation [77]. For instance, in the works [78,90], when the authors modified the composition of the binary $As_{42}S_{58}$ ChG films by substituting a portion of As atoms by Ge ($As_{36}Ge_6S_{58}$) or Sb ($As_{36}Sb_6S_{58}$), they found two different trends of the refractive index change of the ChGs under 795 nm Ti:Sapphire laser irradiation. The change of refractive index with respect to the original films increases in the case of Ge substitution and decreases in the case of Sb substitution, as summarized in Table 9.7.

The effect of composition on the photosensitivity of the ternary ChGs family $Ge_xAs_ySe_{1-x-y}$ was comprehensively demonstrated in the paper by Su *et al.* [91], looking at their 12 glass compositions. By changing the mean coordination number (MCN), i.e. the mean number of bonded neighbours per atom, the photosensitivity of such glasses changes: the films with an MCN in the range of 2.45–2.50 are photostable, while films beyond this MCN range are highly photosensitive. Optimization of chalcogenide compositions and fabrication methodologies tailoring certain desired properties, including the photosensitivity of ChGs, is one of the main research areas of chalcogenide materials and applications at present [92]. In a recent study on the amorphous Ge–Sb–Se thin films [93] fabricated by an RF-magnetron co-sputtering technique, the authors demonstrated that the optical bandgap in a range of 1.35–2.08 eV and the corresponding refractive index ranging from 3.33 to 2.36 of the selenide-sputtered films could be modified by a near-bandgap irradiation under Ar atmosphere. The photobleaching effect in the Ge-rich films was observed, which tends to decrease with increasing Sb content. The glass composition-dependent behaviour of the photosensitivity at 1.55 μm in a femtosecond regime of such ternary Ge–Sb–Se ChGs was also identified in the slab waveguides synthesized by the conventional melting and quenching technique [94].

9.2.3.2 Direct-laser-written micro-optical structures in ChGs

The high photosensitivity of ChGs is important for applications in infrared integrated optics since it enables the direct laser writing of not only effective Bragg gratings but also of channel waveguides, as demonstrated in several works [88,93–96]. More effort has been made to investigate and exploit ChGs for infrared photonic applications, including integrated optics [92]. For instance, chalcogenide-assisted SOI circuits were realized on As_2S_3 ChGs [97], showing a potential solution for trimmable photonic integrated circuits. For the development of active infrared optical components, e.g. infrared light sources, the RE-doped mid-infrared fibre lasers [98] and microsphere lasers based on ChGs are viable for both hybrid and monolithic OICs working in an infrared region. Exploiting the photosensitivity of As_2S_3 ChG glasses, post-fabrication trimming techniques were developed to compensate for fabrication imperfections of integrated chalcogenide waveguide resonators [99]. UV-direct-written channel waveguides with a refractive index change of 10^{-3} were achieved on gallium lanthanum sulphide (NdQC–Ga:La:S) ChG using a 244 nm CW FDA laser with writing fluence of 10.8 J/cm^2 [100]. In their work, Viens *et al.* [96] reported the fabrication of channel waveguides on the photosensitive As–S–(Se)-based ChG using a direct laser writing at 514 nm, which closely matches the bandgap of the material. A review paper on the progress of ChGs optical waveguide fabrication [89] indicated that the band edge irradiation writing is useful for fabricating Bragg gratings and low loss channel waveguides on ChGs.

9.2.4 Other photosensitive glasses

9.2.4.1 Pyrex borosilicate glasses

Pyrex borosilicates are commercially available, inexpensive and with customizable composition glasses. Milton *et al.* [101] investigated the photosensitivity of Pyrex borosilicate glasses ($82SiO_2$, $3.5Na_2O$, $2.5Al_2O$, $0.7K_2O$) and obtained a maximum refractive index change of 10^{-3} by using a 244 nm CW FDA laser. To obtain such refractive index change for fabricating the channel waveguides, they employed a fluence of 206 kJ/cm^2, which is much higher than in other glasses mentioned earlier.

9.2.4.2 Nd-doped fluoroaluminate glasses

Channel waveguides and a waveguide laser were produced in an Nd-doped fluoroaluminate glass fabricated by dip casting by using a 244 nm CW FDA laser with a laser power of 200 mW, a scan speed of 10 mm/min and a spot diameter of 10 μm [102]. The reported negative refractive index change was 10^{-2}. The lasing action at 1,317 nm was achieved in the channel waveguide; the lateral confinement was provided by the negative index change induced by the photothermal expansion of the glass.

9.2.4.3 Bismuth-based silicate glasses

Using a pulsed 248 nm KrF excimer laser, the photosensitivity of bismuth-based silicate glasses prepared by melt-quenching method was investigated in the work by Yang *et al.* [103]. After the UV irradiation with a pulse intensity of 30 mJ/cm^2

for 10 min, the maximum UV-induced refractive index change of such glasses varied depending on the material preparation condition: $\Delta n = 2.1 \times 10^{-4}$ for the glass prepared under oxidizing melting conditions and $\Delta n = 6 \times 10^{-4}$ for the glass prepared in a nitrogen atmosphere.

9.2.4.4 Lead silicate glasses

The photosensitivity of lead silicate glasses (SiO_2–PbO–Na_2O–K_2O–As_2O_3) was studied by Contardi *et al.* [104] under two irradiation conditions, i.e. using: (i) a pulsed 248-nm KrF excimer laser and (ii) a CW FDA laser at 244 nm. A large refractive index change, up to 2.9×10^{-2}, was achieved by direct 244 nm UV writing: channel waveguides were produced without the use of a phase mask. The thermal stability of the channel waveguides was reported to be up to 300°C.

9.3 Photorefractive crystals

9.3.1 Material preparation

Photorefractive crystals such $LiNbO_3$, $LiTaO_3$, $BaTiO_3$ and $KNbO_3$ are an important class of materials for applications in integrated optics [15]. The amount of their refractive index change allows the direct writing of optical integrated components such as permanent holographic gratings [105]. Although channel waveguides can be directly laser-written in such photorefractive crystals [106], most researches have focused on applying the direct laser writing for the fabrication of gratings in the photorefractive waveguides already fabricated by other methods [15]. To fabricate planar waveguides or channel waveguides in crystals such as $LiNbO_3$ and $LiTaO_3$, different methods have been used:

1. diffusion of metals such as Ti [107,108] and Zn [109,110];
2. ion exchange, such as proton exchange [111,112], combined proton and copper exchange [113,114] and metal ion exchange such as Cu–Ti exchange [115] and
3. ion implantation of elements such as H^+ and He^+ [116].

For the preparation of a crystalline thin film, different techniques can be employed, including pulsed laser deposition [117], sputtering [118,119], molecular-beam epitaxy [118], liquid-phase epitaxy [120], metal organic chemical vapour deposition [121] and sol–gel [122].

9.3.2 Photorefractive effects

In crystals such as $LiNbO_3$, $LiTaO_3$ and $Sn_2P_2S_6$ [11], the photoinduced refractive index change can be altered just by inhomogeneous intensity illumination of a coherent laser beam, due to the classic electro-optic effect on a microscopic scale thanks to charge transportation [14,15]. Therefore, the photorefractive effects in crystals have two aspects: (i) it is of great interest for the development of direct-laser-written optical components for integrated optics and (ii) it is feared as optical damage which can degrade the performance of integrated optical devices [11]. Therefore, optical damage resistance is one of the considered factors for applying such photorefractive crystals to integrated optics [15]. The determination of the photorefractive properties, such as saturated refractive index changes, photoconductivity, holographic sensitivity and temporal stability of refractive index profiles of the crystals at the working wavelength, is also important to define the performance of the photorefractive crystal waveguides. Such photorefractive effect investigations can be carried out by two principal holographic methods: by a two-beam interferometry set-up, or by monitoring the output intensity and the beam shape of a single guided beam [15].

9.3.3 Some examples of photorefractive crystal waveguides

9.3.3.1 $LiNbO_3$ waveguides

Among different photorefractive oxide crystals, $LiNbO_3$ is one of the most important technical and industrial photonic materials thanks to its combination of functional properties and commercial availability; in fact, most of the integrated optical modulators or non-linear devices available today are based on titanium-diffused $LiNbO_3$ waveguides [15]. The photorefractivity of $LiNbO_3$ is dependent on the transition metal dopant which drives the charge transport under specific illumination condition of CW laser and pulsed laser [123]. Different $LiNbO_3$-based optical waveguides exhibit different photorefractive properties, including saturated refractive index changes and photorefractive sensitivity depending on the fabrication methods and dopants [124–126]. In general, the photorefractive sensitivity and refractive index changes of doped crystals are high, allowing the laser-induced fabrication of channel waveguides, as demonstrated in the work by Matoba *et al.* [127] with the achieved refractive index change of 1.24×10^{-3}.

In [128], photo-written waveguides in Fe-doped lithium niobate crystal were also demonstrated by employing binary optical masks. Micro-Bragg reflectors have also been realized in Cu:H:$LiNbO_3$:Mg waveguides fabricated by the combined proton and copper exchange [129]. It was shown that the holographic recording of the Bragg gratings with guided beams can be more effective than recording with external beams.

9.3.3.2 $LiTaO_3$ waveguides

Besides $LiNbO_3$, $LiTaO_3$ is another potential photorefractive oxide crystal for integrated optics. $LiTaO_3$ has some advantageous properties over $LiNbO_3$, such as transparency and less susceptibility to optical damage in blue spectral region, and better mechanical properties for applications in integrated optics [15,18]. Exploiting the interband photorefractivity of Mg-doped $LiTaO_3$, dynamic waveguides were written beneath the surface of the crystals by using the 257 nm laser beam from an FDA laser [130]. Holographic gratings in pure and Mg-doped near-stoichiometric $LiTaO_3$ were also fabricated by a deep-UV 257-nm laser beam thanks to interband photorefractivity [131].

9.3.3.3 Other photorefractive crystal waveguides

For applications in integrated optics, there are also other viable crystals which exhibit photorefractivity properties, including the high interband photorefractive effects, such as $KNbO_3$ [132] and $Sn_2P_2S_6$ [133]. Interband photorefractive laser-written gratings on crystals waveguides were also demonstrated in the case of $Sn_2P_2S_6$ [134] and of $KNbO_3$ [132].

9.4 Photorefractive polymers

Polymer optical waveguide devices have recently emerged and also played an important role in the development of certain optical components for hybrid OICs [135]. In comparison with other discussed inorganic material classes, i.e. glasses, glass-ceramics and crystals, polymeric materials offer some attractive aspects in terms of rapid and easy processing, cost-effectiveness, high yields, high performance and low optical loss for the fabrication of simple, precise, low-cost and flexible PLC devices [135–137]. Polymer optical waveguides devices find several photonic and optic applications, including low-cost all-polymer integrated circuits [138], hybrid photonic and optical integrated devices such as tuneable filters [139], switches [137] and optical amplification [137]. Furthermore, the photorefractive polymers have been considered as a technological potential since they offer high efficiency and capability of manufacturing large-area films for integrated optical boards at low cost [140].

9.4.1 Photorefractive effects

Similar to the case of photorefractive crystals, the photorefractivity of polymers involves photoinduced charge generation, transport and trapping to new sites [12,13]. The photorefractivity of polymers can be easily tailored by changing their composition, i.e. their chemical and structural nature [12]. The relation between the molecular structures of the polymers and their photorefractivity was discussed in a review paper about several polymeric molecular structures largely studied over past two decades [141]. Concerning the photorefractive response of such polymers, PTAA (poly(bis(4-phenyl) (2,4,6-trimethylphenyl)-amine)) composites, for example, exhibit a photorefractive response at least one order of magnitude faster than other conventional polymers for the asymmetric energy transfer and optical diffraction [141–143]. For tailoring the photorefractivity of the polymers, the introduction of organic sensitizers such as graphene [144], PBI (perylene bisimide) [145], DiPBI (di(perylene bisimide)) [146–147] and fullerene derivative of PCBM (phenyl-C61-butyric acid methyl ester) [148] into such polymers is also crucial [149]. These examples of relation between the molecular structure and the photorefractive response show the versatility and potential of polymers in term of easily tuning their properties, including photorefractivity, for achieving the desired functions of the devices. The polymers typically exhibit a refractive index change in the order of 10^{-2} [150–151].

9.4.2 Photorefractive polymer waveguides

The direct-written micro-optical structures can be fabricated in photorefractive polymers by simple techniques employing only a UV lamp, which is of great interest for low cost and easy fabrication of OIC components [151]. As an example, a simple UV lamp writing technique was proposed to fabricate buried waveguides and devices in a commercial benzocyclobutene thin films with a refractive index increase of 1.2×10^{-2} [150]. Moreover, polymers are potential candidates for long-period gratings which can be used as band-rejection filters [152] and gain flatteners [153] for integrated lasers. Long-period polymer waveguide gratings have also been demonstrated to be widely tuneable and polarization-insensitive resonance wavelength, which is undoubtedly of great interest for integrated optic applications [154]. Some photorefractive polymer waveguides can also be utilized in the infrared region for optical interconnect applications [140].

9.5 Summary

Looking at the state of the art of optical interconnects and communications, the development of OICs is crucial for the broadband data transformation and miniaturization, thanks to their advantages over electrical integrated circuits, such as large bandwidth, expanded frequency-division multiplex, low-loss, small size, lightweight and lower power consumption. On the other hand, the basic element of any optical circuit is the optical waveguide, which governs the optical connections between optical components and devices. To have a high propagating light confinement, the integrated optical waveguides should have high refractive index contrast with their claddings, and sizes in the micron-scale. The development and construction of micro-optical components, integrated circuits and devices urge the intensive investigation of advanced materials, fabrication and integration technologies.

Materials with high photorefractivity are potential candidates for OICs since they not only provide the solution for fabricating micro-optical structures, such as channel waveguides and grating couplers, but also reduce time-consuming and costly processes by exploiting the robust photorefractive direct-laser writing. To provide a brief overview of photorefractive waveguides as materials of choice for integrated optics, this chapter summarized some photorefractive properties of different classes of optical materials, including glasses, glass-ceramics, crystals and polymers.

The photorefractive silica-based glasses and glass-ceramics have the advantage of exploiting and matching the already mature guided wave optics and technology of optical communications silica fibres. ChGs are currently promising for integrated optical circuits operating in the mid-infrared region. With a large diversity of composition choices, leading to different numbers of defective bonds and topologies, the photosensitivity of ChGs can be tailored and engineered depending on the desired functions to be performed by the integrated optical devices. The photorefractive crystals such as $LiNbO_3$, $LiTaO_3$, $BaTiO_3$ and $KNbO_3$ are an important class of materials, too. In particular, titanium-diffused $LiNbO_3$ waveguides have been already applied in the commercial integrated optical devices.

Photorefractive polymer optical waveguides have recently emerged; they also played an important role in the development of certain optical components for hybrid OICs and especially for simple, precise, low-cost and flexible PLC devices.

Finally, the current status of development of optical guided wave components fabricated by photorefractive direct-laser-writing techniques, mostly channel waveguides and gratings, in the above-listed materials has been shortly reviewed.

References

[1] Borrelli NF. *Photosensitive Glass and Glass-Ceramics*. 1st ed. (Boca Raton, FL, CRC Press, 2016).

[2] Chen GH, Li YG, Liu LY, He YJ, Xu L, and Wang WC. 'The photosensitivity and ultraviolet absorption change of Sn-doped silica film fabricated by modified chemical vapor deposition'. *Journal of Applied Physics*. 2004;**96**(11):6153–6158.

[3] Yliniemi S, Albert J, Wang Q, and Honkanen S. 'UV-exposed Bragg gratings for laser applications in silver-sodium ion-exchanged phosphate glass waveguides'. *Optics Express*. 2006;**14**(7):2898–2903.

[4] Montero C, Gomez-Reino C, and Brebner JL. 'Planar Bragg gratings made by excimer-laser modification of ion-exchanged waveguides'. *Optics Letters*. 1999;**24**(21):1487–1489.

[5] González-Pérez S, Arbelo-Jorge E, Cáceres JM, Núñez P, Martín IR, and Haro-González P. 'Laser irradiation in Nd^{3+} doped strontium barium niobate glass'. *Journal of Applied Physics*. 2008;**104**(1):013112–013115.

[6] Haro-González P, Martín IR, and Creus AH. 'Nanocrystals distribution inside the writing lines in a glass matrix using Argon laser irradiation'. *Optics Express*. 2010;**18**(2):582–590.

[7] Conti GN, Berneschi S, Brenci M, *et al.* 'UV photoimprinting of channel waveguides on active SiO_2-GeO_2 sputtered thin films'. *Applied Physics Letters*. 2006;**89**(12):121102.

[8] Sebastiani S, Conti GN, Pelli S, *et al.* 'Characterization of a highly photorefractive RF-sputtered SiO_2-GeO_2 waveguide'. *Optics Express*. 2005;**13**(5): 1696–1701.

[9] Berneschi S, Bhaktha BNS, Chiappini A, *et al.* 'Highly photorefractive Eu^{3+} activated sol-gel SiO_2-SnO_2 thin film waveguides'. *Proceedings of SPIE*. 2010;**7604**:76040Z.

[10] Lukowiak A, Zur L, Tran TNL, *et al.* 'Sol–gel-derived glass-ceramic photorefractive films for photonic structures'. *Crystals*. 2017;**7**(2):61.

[11] Kip D and Wesner M. 'Photorefractive waveguides' in Gunter P, Huignard JP (eds.). *Photorefractive Materials and Their Applications – Volume 1: Basic Effects*. (New York, NY, Springer-Verlag, 2006), pp. 289–310.

[12] Moerner WE and Silence SM. 'Polymeric photorefractive materials'. *Chemical Reviews*. 1994;**94**(1):127–155.

[13] Ducharme S, Scott JC, Twieg RJ, and Moerner WE. 'Observation of the photorefractive effect in a polymer'. *Physical Review Letters*. 1991;**66**(14): 1846.

[14] Montemezzani G, Dittrich P, and Gunter P. 'Band-to-band photorefraction' in Gunter P, Huignard JP (eds.). *Photorefractive Materials and Their Applications – Volume 1: Basic Effects*. (New York, NY, Springer-Verlag, 2006), pp. 203–230.

[15] Kip D. 'Photorefractive waveguides in oxide crystals: fabrication, properties, and applications'. *Applied Physics B: Lasers and Optics*. 1998;**67**(2): 131–150.

[16] Volk T and Wöhlecke M. *Lithium Niobate: Defects, Photorefraction and Ferroelectric Switching*. (Bach, Springer, 2008).

[17] Mosimann R, Haertle D, Jazbinsek M, Montemezzani G, and Günter P. 'Determination of the absorption constant in the interband region by photocurrent measurements'. *Applied Physics B: Lasers and Optics*. 2006;**83**(1): 115–119.

[18] Bernasconi P, Montemezzani G, Günter P, Furukawa Y, and Kitamura K. 'Stoichiometric $LiTaO_3$ for ultraviolet photorefraction'. *Ferroelectrics*. 1999;**223**(1–4):373–379.

[19] Tong XC. *Advanced Materials for Integrated Optical Waveguides*. (New York, NY, Springer International Publishing, 2014).

[20] Schneider M, Kuhner T, Mohr J, and Maas D. 'Fibers in printed circuit boards with passively aligned coupling'. *Journal of Lightwave Technology*. 2010;**28**(15):2121–2128.

[21] Yuan L. 'In-fiber integrated optic devices for sensing applications'. *Proceedings of SPIE*. 2012;**8421**:84211D.

[22] Miller DAB. 'Rationale and challenges for optical interconnects to electronic chips'. *Proceedings of the IEEE*. 2000;**88**(6):728–749.

[23] Hunsperger RG. *Integrated Optics: Theory and Technology*. (New York, NY, Springer-Verlag, 2009).

[24] Nagarajan R, Doerr CR, Kish FA, Kaminow I, and Li T. 'Semiconductor photonic integrated circuit transmitters and receivers'. *Optical Fiber Telecommunications* 2013;**6**(A):25–98.

[25] Doerr CR. 'Highly integrated monolithic photonic integrated circuits'. *Proceedings of 39th European Conference and Exhibition on Optical Communication (ECOC 2013)*. IET; 2013.

[26] Thual M, Rochard P, Chanclou P, and Quetel L. 'Contribution to research on micro-lensed fibers for modes coupling'. *Fiber and Integrated Optics*. 2008;**27**(6):532–541.

[27] Liu J, Raja AS, Pfeiffer MHP, *et al*. 'Double inverse nanotapers for efficient light coupling to integrated photonic devices'. *Optics Letters*. 2018;**43**(14): 3200–3203.

[28] Pernice WHP, Li M, Hochberg M, Xiong C, Tang HX, and Baehr-Jones T. 'Harnessing optical forces in integrated photonic circuits'. *Nature*. 2008;**456** (7221):480–484.

[29] Righini GC and Chiappini A. 'Glass optical waveguides: a review of fabrication techniques'. *Optical Engineering*. 2014;**53**(7):071819.

[30] Soref R. 'The past, present, and future of silicon photonics'. *IEEE Journal of Selected Topics in Quantum Electronics*. 2006;**12**(6):1678–1687.

[31] Mäkinen J and Suni T. 'Thick-film SOI wafers: preparation and properties' in *Handbook of Silicon Based MEMS Materials and Technologies*. (Oxford, UK, William Andrew Publishing, 2015), pp. 206–237.

[32] Li F, Jackson S, Magi E, *et al*. 'Low propagation loss silicon-on-sapphire integrated waveguides for the mid-infrared'. *Optics Express*. 2013;**19**(16): 15212–15220.

[33] Troha T, Rigler M, Alden D, *et al*. 'UV second harmonic generation in AlN waveguides with modal phase matching'. *Optical Materials Express*. 2016; **6**(6):2014–2023.

[34] Li Y, Bhattacharyya A, Thomidis C, Moustakas TD, and Paiella R. 'Nonlinear optical waveguides based on near-infrared intersubband transitions in GaN/AlN quantum wells'. *Optics Express*. 2007;**15**(9):5860–5865.

[35] Carlson DR, Hickstein DD, Lind A, *et al*. 'Self-referenced frequency combs using high-efficiency silicon-nitride waveguides'. *Optics Letters*. 2017; **42**(12):2314–2317.

[36] Takahashi H. 'Planar lightwave circuit devices for optical communication: present and future'. *Proceedings of SPIE*. 2003;**5246**:520–531.

[37] Politi A, Cryan MJ, Rarity JG, Yu S, and O'brien JL. 'Silica-on-silicon waveguide quantum circuits'. *Science*. 2008;**320**(5876):646–649.

[38] Ebendorff-Heidepriem H. 'Laser writing of waveguides in photosensitive glasses'. *Optical Materials*. 2004;**25**(2):109–115.

[39] Thompson MG, Politi A, Matthews JC, and O'Brien JL. 'Integrated waveguide circuits for optical quantum computing'. *IET Circuits, Devices & Systems*. 2011;**5**(2):94–102.

[40] Chiodini N, Ghidini S, and Paleari A. 'Mechanisms responsible for the ultraviolet photosensitivity of SnO_2-doped silica'. *Physical Review B*. 2001;**64**(7):073102.

[41] Tran LTN. *Tin dioxide-based photonic glass-ceramics*. PhD thesis, University of Trento (Trento, 2019). Available on-line: http://eprints-phd.biblio.unitn.it/3571/1/ _Tran_Thi_Ngoc_Lam.pdf (Accessed 20 Feb 2020).

[42] Hill KO, Fujii Y, Johnson DC, and Kawasaki BS. 'Photosensitivity in optical fibre waveguides: application to reflection filter fabrication'. *Applied Physics Letters*. 1978;**32**(10):647–649.

[43] Skuja L. 'Optically active oxygen-deficiency-related centers in amorphous silicon dioxide'. *Journal of Non-crystalline Solids*. 1998;**239**(1–3):16–48.

[44] Nishiyama H, Kintaka K, Nishii J, Sano T, Ohmura E, and Miyamoto I. 'Thermo-and photo-sensitive GeO_2–B_2O_3–SiO_2 thin glass films'. *Japanese Journal of Applied Physics*. 2003;**42**(2R):559.

[45] Simmons KD, Stegeman GI, Potter Jr BG, and Simmons JH. 'Photosensitivity in germano-silicate sol-gel thin films'. *Journal of Non-crystalline Solids*. 1994;**179**:254–259.

[46] Malo B, Albert J, Hill KO, Bilodeau F, and Johnson DC. 'Effective index drift from molecular hydrogen diffusion in hydrogen-loaded optical fibres and its effect on Bragg grating fabrication'. *Electronics Letters*. 1994;**30**(5): 442–444.

[47] Williams DL, Ainslie BJ, Armitage JR, Kashyap R, and Campbell R. 'Enhanced UV photosensitivity in boron codoped germanosilicate fibres'. *Electronics Letters*. 1993;**29**(1):45–47.

[48] Medvedkov OI, Vasiliev SA, Gnusin PI, and Dianov EM. 'Photosensitivity of optical fibres with extremely high germanium concentration'. *Optical Materials Express*. 2012;**2**(11):1478–1489.

[49] Dong L, Cruz JL, Reekie L, Xu MG, and Payne DN. 'Enhanced photosensitivity in tin-codoped germanosilicate optical fibres'. *IEEE Photonics Technology Letters*. 1995;**7**(9):1048–1050.

[50] Nishii J, Yamanaka H, Hosono H, and Kawazoe H. 'Preparation of Bragg gratings in sputter-deposited GeO_2–SiO_2 glasses by excimer-laser irradiation'. *Optics Letters*. 1996;**21**(17):1360–1362.

[51] Hosono H and Nishii J. 'High photosensitivity and nanometer-scale phase separation in GeO_2–SiO_2 glass thin films'. *Optics Letters*. 1999;**24**(19): 1352–1354.

[52] Takahashi M, Sakoh A, Ichii K, Tokuda Y, Yoko T, and Nishii J. 'Photosensitive GeO_2–SiO_2 films for ultraviolet laser writing of channel waveguides and Bragg gratings with Cr-loaded waveguide structure'. *Applied Optics*. 2003;**42**(22):4594–4598.

[53] Svalgaard M, Poulsen CV, Bjarklev A, and Poulsen O. 'Direct UV writing of buried singlemode channel waveguides in Ge-doped silica films'. *Electronics Letters*. 1994;**30**(17):1401–1403.

[54] Tandon P and Boek H. 'Experimental and theoretical studies of flame hydrolysis deposition process for making glasses for optical planar devices'. *Journal of Non-crystalline Solids*. 2003;**317**(3):275–289.

[55] Turvey MT, Mennea PL, Holmes C, Smith PGR, and Gates JC. 'Direct UV written waveguides in phosphogermanate planar glass layers fabricated via flame hydrolysis deposition'. *Proceedings of Advanced Photonics [IPR, NOMA, Sensors, Networks, SPPCom, SOF]*. Washington, D.C., USA, OSA Publishing; 2006:ITu3A.3.

[56] Jensen JB, Varming P, Liu B, and Gries W. 'Comparison of photosensitivity in germanium doped silica fibres using 244 nm and 266 nm continuous-wave lasers'. *Proceedings of Optical Fibre Communication Conference and Exhibit. Technical Digest Postconference Edition [IEEE Cat. 01CH37171]*. Washington, D.C., USA, OSA Publishing; 2001; 3:WDD90.

[57] Nasu Y, Abe M, and Kohtoku M. 'Developments in laser processing for silica-based planar lightwave circuits'. *Proceedings of SPIE*. 2006;**6107**: 61070B.

[58] Emmerson GD, Gawith CB, Williams RB, *et al.* 'Ultra-wide detuning through direct grating writing of planar Bragg structures'. *Proceedings of*

Bragg Gratings, Photosensitivity, and Poling in Glass Waveguides. Optical Society of America; 2003:TuA3.

[59] Sima C, Gates JC, Rogers HL, *et al.* 'Ultra-wide detuning planar Bragg grating fabrication technique based on direct UV grating writing with electro-optic phase modulation'. *Optics Express*. 2013;**21**(13):15747–15754.

[60] Dong L, Cruz JL, Tucknott JA, Reekie L, and Payne DN. 'Strong photosensitive gratings in tin-doped phosphosilicate optical fibres'. *Optics Letters*. 1995;**20**(19):1982–1984.

[61] Reekie L and Dong L. 'Material considerations for Bragg fibre gratings'. *Proceedings of SPIE*. 1997;**2998**:2–10.

[62] Chiodini N, Paleari A, Brambilla G, and Taylor ER. 'Erbium doped nanostructured tin–silicate glass–ceramic composites'. *Applied Physics Letters*. 2002;**80**(23):4449–4451.

[63] Gaff K, Durandet A, Weijers T, Love J, and Boswell R. 'Strong photosensitivity in tin-doped silica films'. *Electronics Letters*. 2000;**36**(9): 842–843.

[64] Brambilla G, Pruneri V, and Reekie L. 'Photorefractive index gratings in SnO_2: SiO_2 optical fibres'. *Applied Physics Letters*. 2000;**76**(7):807–809.

[65] Zatsepin AF, Buntov EA, Kortov VS *et al.* 'Low-temperature photoluminescence of ion-implanted SiO_2:Sn^+ films and glasses'. *Journal of Surface Investigation. X-ray, Synchrotron and Neutron Techniques*. 2012; **6**(4):668–672.

[66] Granger G, Restoin C, Roy P *et al.* 'Nanostructured optical fibres in the SiO_2/SnO_2 system by the sol-gel method'. *Materials Letters*. 2014;**120**:292–294.

[67] Tran L, Massella D, Zur L, *et al.* 'SiO_2-SnO_2: Er^{3+} glass-ceramic monoliths'. *Applied Sciences*. 2018;**8**(8):1335.

[68] Zur L, Tran LT, Massella D, *et al.* 'SiO_2-SnO_2 transparent glass-ceramics activated by rare earth ions'. *Proceedings of SPIE*. 2019;**10914**:1091411.

[69] Bhaktha BS, Berneschi S, Conti GN, *et al.* 'Spatially localized UV-induced crystallization of SnO_2 in photorefractive SiO_2-SnO_2 thin film'. *Proceedings of SPIE*. 2010;**7719**:77191B.

[70] Chiodini N, Paleari A, Spinolo G, *et al.* 'Photosensitive erbium doped tin-silicate glass'. *Journal of Non-Crystalline Solids*. 2002;**311**(3):217–222.

[71] Chiodini N, Paleari A, and Spinolo G. 'Photorefractivity in nanostructured tin-silicate glass ceramics: a radiation-induced nanocluster size effect'. *Physical Review Letters*. 2003;**90**(5):055507.

[72] Chiodini N, Paleari A, Spinolo G, and Crespi P. 'Photorefractivity in SiO_2: SnO_2 glass-ceramics by visible light'. *Journal of Non-crystalline Solids*. 2003;**322**(1–3):266–271.

[73] Brambilla G and Rutt H. 'Fibre Bragg gratings with enhanced thermal stability'. *Applied Physics Letters*. 2002;**80**(18):3259–3261.

[74] Fernández J, García-Revilla S, Balda R, and Cascales C. 'Rare-earth-doped wide-bandgap tin-oxide nanocrystals: pumping mechanisms and spectroscopy'. *Proceedings of SPIE*. 2018;**10528**:1052805.

[75] Eggleton BJ, Luther-Davies B, and Richardson K. 'Chalcogenide photonics'. *Nature Photonics*. 2011;**5**(3):141.

[76] Sanghera J and Gibson D. 'Optical properties of chalcogenide glasses and fibres'. *Proceedings of Chalcogenide Glasses*. Sawston, Cambridge, UK, Woodhead Publishing; 2014:113–138.

[77] Musgraves JD, Carlie N, Guery G, *et al.* 'Chalcogenide glasses and their photosensitivity: engineered materials for device applications'. *Proceedings of Bragg Gratings, Photosensitivity, and Poling in Glass Waveguides*. Optical Society of America; 2010:BWD1.

[78] Schimmel RC, Faber AJ, De Waardt H, Beerkens RG, and Khoe GD. 'Development of germanium gallium sulphide glass fibres for the 1.31 μm praseodymium-doped fibre amplifier'. *Journal of Non-Crystalline Solids*. 2001;**284**:188–192.

[79] McNamara P, Lancaster DG, Bailey R, Hemming A, Henry P, and Mair RH. 'A large core microstructured fluoride glass optical fibre for mid-infrared single-mode transmission'. *Journal of Non-Crystalline Solids*. 2009;**355**(28–30):1461–1467.

[80] Nishii J, Morimoto S, Inagawa I, Iizuka R, Yamashita T, and Yamagishi T. 'Recent advances and trends in chalcogenide glass fibre technology: a review'. *Journal of Non-Crystalline Solids*. 1992;**140**:199–208.

[81] Sanghera JS, Shaw LB, Busse LE, *et al.* 'Development and infrared applications of chalcogenide glass optical fibres'. *Fibre & Integrated Optics*. 2000;**19**(3):251–274.

[82] Savage SD, Miller CA, Furniss D, and Seddon AB. 'Extrusion of chalcogenide glass preforms and drawing to multimode optical fibres'. *Journal of Non-Crystalline Solids*. 2008;**354**(29):3418–3427.

[83] Sanghera JS, Florea CM, Shaw LB, *et al.* 'Non-linear properties of chalcogenide glasses and fibres'. *Journal of Non-Crystalline Solids*. 2008;**354**(2–9):462–467.

[84] Tan WC, Solmaz ME, Gardner J, Atkins R, and Madsen C. 'Optical characterization of a-As_2S_3 thin films prepared by magnetron sputtering'. *Journal of Applied Physics*. 2010;**107**(3):033524.

[85] Youden KE, Grevatt T, Eason RW, Rutt HN, Deol RS, and Wylangowski G. 'Pulsed laser deposition of Ga-La-S chalcogenide glass thin film optical waveguides'. *Applied Physics Letters*. 1993;**63**(12):1601–1603.

[86] Huang CC, Hewak DW, and Badding JV. 'Deposition and characterization of germanium sulphide glass planar waveguides'. *Optics Express*. 2004;**12**(11):2501–2506.

[87] Hu J, Tarasov V, Agarwal A, *et al.* 'Fabrication and testing of planar chalcogenide waveguide integrated microfluidic sensor'. *Optics Express*. 2007;**15**(5):2307–2314.

[88] Shimakawa K, Kolobov A, and Elliott SR. 'Photoinduced effects and metastability in amorphous semiconductors and insulators'. *Advances in Physics*. 1995;**44**(6):475–588.

[89] Gai X, Han T, Prasad A *et al.* 'Progress in optical waveguides fabricated from chalcogenide glasses'. *Optics Express*. 2010;**18**(25):26635–26646.
[90] Anderson T, Hu J, Ramme M, *et al.* 'Integrating optics and micro-fluidic channels using femtosecond laser irradiation'. *Proceedings of Commercial and Biomedical Applications of Ultrafast Lasers IX*. International Society for Optics and Photonics; 2009;7203:72030I.
[91] Su X, Wang R, Luther-Davies B, and Wang L. 'The dependence of photo-sensitivity on composition for thin films of $Ge_xAs_ySe_{1-x-y}$ chalcogenide glasses'. *Applied Physics A*. 2013;**113**(3):575–581.
[92] Eggleton BJ. 'Chalcogenide photonics: fabrication, devices and applications'. *Optics Express*. 2010;**18**(2):26632–26634.
[93] Halenkovič T, Gutwirth J, Němec P, *et al.* 'Amorphous Ge-Sb-Se thin films fabricated by co-sputtering: properties and photosensitivity'. *Journal of the American Ceramic Society*. 2018;**101**(7):2877–2887.
[94] Olivier M, Tchahame JC, Němec P, *et al.* 'Structure, nonlinear properties, and photosensitivity of $(GeSe_2)_{100-x}(Sb_2\ Se_3)_x$ glasses'. *Optical Materials Express*. 2014;**4**(3):525–540.
[95] Meneghini C and Villeneuve A. 'As_2S_3 photosensitivity by two-photon absorption: holographic gratings and self-written channel waveguides'. *Journal of the Optical Society of America B*. 1998;**15**(12):2946–2950.
[96] Viens JF, Meneghini C, Villeneuve A, *et al.* 'Fabrication and characterization of integrated optical waveguides in sulfide chalcogenide glasses'. *Journal of Lightwave Technology*. 1999;**17**(7):1184.
[97] Canciamilla A, Morichetti F, Grillanda S, *et al.* 'Photo-induced trimming of chalcogenide-assisted silicon waveguides'. *Optics Express*. 2012;**20**(14): 15807–15817.
[98] Seddon AB, Tang Z, Furniss D, Sujecki S, and Benson TM. 'Progress in rare-earth-doped mid-infrared fibre lasers'. *Optics Express*. 2010;**18**(25): 26704–26719.
[99] Carlie N, Musgraves JD, Zdyrko B, *et al.* 'Integrated chalcogenide waveguide resonators for mid-IR sensing: leveraging material properties to meet fabrication challenges'. *Optics Express*. 2010;**18**(25):26728–26743.
[100] Mairaj AK, Chardon AM, Shepherd DP, and Hewak DW. 'Laser performance and spectroscopic analysis of optically written channel waveguides in neodymium-doped gallium lanthanum sulphide glass'. *IEEE Journal of Selected Topics in Quantum Electronics*. 2002;**8**(6):1381–1388.
[101] Sheridan AK, Gawith CB, Emmerson GD, Milton JA, Smith PG, and Wilkinson JS. 'Channel waveguides in ion-exchanged pyrex by direct UV writing'. *Optics Communications*. 2004;**242**(1–3):109–114.
[102] Harwood DWJ, Fu A, Taylor ER, Moore RC, West YD and Payne DN. 'A 1317nm neodymium doped fluoride glass waveguide laser'. *Procceedings 26th European Conference on Optical Communication 2020*. VDE-VERLAG Berline; **2**:191–192.

[103] Yang Z, Zhang Q, and Jiang Z. 'Photo-induced refractive index change of bismuth-based silicate glass'. *Journal of Physics D: Applied Physics.* 2005;**38**(2):1461–1463.

[104] Contardi C, Taylor ER, and Fu A. 'Study of UV-written channels in lead silicate glasses'. *Journal of Non Crystalline Solids.* 2001;**291**(1–2):113–120.

[105] Bazzan M and Sada C. 'Optical waveguides in lithium niobate: recent developments and applications'. *Applied Physics Reviews.* 2015;**2**(4): 040603.

[106] Zozulya AA and Anderson DZ. 'Propagation of an optical beam in a photorefractive medium in the presence of a photogalvanic nonlinearity or an externally applied electric field'. *Physical Review A.* 1995;**51**(2):1520.

[107] Schmidt RV and Kaminow IP. 'Metal-diffused optical waveguides in $LiNbO_3$'. *Applied Physics Letters.* 1974;**25**(8):458–460.

[108] Atuchin VV, Ziling KK, and Shipilova DP. 'Investigation of optical waveguides fabricated by titanium diffusion in $LiTaO_3$'. *Soviet Journal of Quantum Electronics.* 1984;**14**(5):671.

[109] Florencio R, Boix RR, Carrasco E, Encinar JA, Barba M, and Pérez-Palomino G. 'Broadband reflectarrays made of cells with three coplanar parallel dipoles'. *Microwave and Optical Technology Letters.* 2014;**56**(3): 748–753.

[110] Espeso-Gil O, Garcıa G, Agulló-López F, *et al.* 'Characterization of surface layers in Zn-diffused $LiNbO_3$ waveguides by heavy ion elastic recoil detection'. *Applied Physics Letters.* 2002;**81**(11):1981–1983.

[111] Rice CE and Veselka JJ. 'Proton exchange in $LiNbO_3$'. *Ferroelectrics.* 1983;**50**(1):165–170.

[112] Ganshin VA, Korkishko YN, Morozova TV, and Saraikin VV. 'The study of proton exchange in lithium tantalate crystals'. *Physica Status Solidi A.* 1989;**114**(2):457–465.

[113] Kostritskii SM, Kip D, and Krätzig E. 'Improvement of photorefractive properties of proton-exchanged $LiTaO_3$ waveguides'. *Applied Physics B.* 1997;**65**(4–5):517–522.

[114] Rickermann F, Kip D, Gather B, and Krätzig E. 'Characterization of photorefractive $LiNbO_3$ waveguides fabricated by combined proton and copper exchange'. *Physica Status Solidi (A).* 1995;**150**(2):763–772.

[115] Caccavale F, Sada C, Segato F, *et al.* 'Copper–lithium ion exchange in $LiNbO_3$'. *Journal of Materials Research.* 2000;**15**(5):1120–1124.

[116] Destefanis GL, Gailliard JP, Ligeon EL, *et al.* 'The formation of waveguides and modulators in $LiNbO_3$ by ion implantation'. *Journal of Applied Physics.* 1979;**50**(12):7898–7905.

[117] Hu WS, Liu ZG, Lu YQ, Zhu SN, and Feng D. 'Pulsed-laser deposition and optical properties of completely [001] textured optical waveguiding $LiNbO_3$ films upon SiO_2/Si substrates'. *Optics Letters.* 1996;**21**(13): 946–948.

[118] Akazawa H and Shimada M. 'Electron cyclotron resonance plasma sputtering growth of textured films of *c*-axis-oriented $LiNbO_3$ on Si [100] and Si [111] surfaces'. *Journal of Vacuum Science & Technology A: Vacuum, Surfaces, and Films*. 2004;**22**(4):1793–1798.

[119] Hewig GM, Jain K, Sequeda FO, Tom R, and Wang PW. 'RF sputtering of $LiNbO_3$ thin films'. *Thin Solid Films*. 1982;**88**:67–74.

[120] Kawaguchi T, Yoon DH, Minakata M, Okada Y, Imaeda M, and Fukuda T. 'Growth of high crystalline quality $LiNbO_3$ thin films by a new liquid phase epitaxial technique from a solid-liquid coexisting melt'. *Journal of Crystal Growth*. 1995;**152**(1–2):87–93.

[121] Bihari B, Kumar J, Stauf GT, Van Buskirk PC, and Hwang CS. 'Investigation of barium titanate thin films on MgO substrates by second-harmonic generation'. *Journal of Applied Physics*. 1994;**76**(2):1169–1174.

[122] Ohya Y, Ito T, and Takahashi Y. 'Dielectric properties of multilayered ferroelectric thin films fabricated by sol-gel method'. *Japanese Journal of Applied Physics*. 1994;**33**(9S):5272.

[123] Buse K. 'Light-induced charge transport processes in photorefractive crystals II: Materials'. *Applied Physics B*. 1997;**64**(4):391–407.

[124] Yamada A, Tamada H, and Saitoh M. Photorefractive damage in $LiNbO_3$ thin-film optical waveguides grown by liquid phase epitaxy. *Journal of Applied Physics*. 1994;**76**(3):1776–1783.

[125] Gericke V, Hertel P, Krätzig E, Nisius JP, and Sommerfeldt R. 'Light-induced refractive index changes in $LiNbO_3$: Ti waveguides'. *Applied Physics B*. 1987;**44**(3):155–162.

[126] Kondo Y, Miyaguchi S, Onoe A, and Fujii Y. 'Quantitatively measured photorefractive sensitivity of proton-exchanged lithium niobate, proton-exchanged magnesium oxide-doped lithium niobate, and ion-exchanged potassium titanyl phosphate waveguides'. *Applied Optics*. 1994;**33**(16): 3348–3352.

[127] Matoba O, Itoh K, and Ichioka Y. 'Analysis of photo-induced waveguide in lithium niobate crystal'. *Optical Review*. 1994;**1**(1):73–75.

[128] Zhang P, Yang DX, Zhao J, and Wang M. 'Photo-written waveguides in iron-doped lithium niobate crystal employing binary optical masks'. *Optical Engineering*. 2006;**45**(7):074603.

[129] Kostritskii SM, Korkishko YN, Fedorov VA, and Sevostyanov OG. 'Fabrication of micro-Bragg reflectors by guided beams in photorefractive Cu: H: $LiNbO_3$: Mg waveguides'. *Proceedings of 2011 International Workshop on Nonlinear Photonics*. IEEE; 2011:1–2.

[130] Juvalta F, Koziarska-Glinka B, Jazbinšek M, Montemezzani G, Kitamura K, and Günter P. 'Deep UV light induced, fast reconfigurable and fixed waveguides in Mg doped $LiTaO_3$'. *Optics Express*. 2006;**14**(18): 8278–8289.

[131] Juvalta F, Dittrich P, Montemezzani G, *et al*. 'Holographic gratings in pure and Mg-doped near-stoichiometric $LiTaO_3$ induced by deep-ultraviolet light'. *Proceedings of SPIE*. 2006;**6252**:62520A.

[132] Medrano C, Zgonik M, Liakatas I, and Günter P. 'Infrared photorefractive effect in doped $KNbO_3$ crystals'. *Journal of the Optical Society of America B*. 1996;**13**(11):2657–2661.

[133] Jazbinsek M, Haertle D, Bach T, *et al.* '$Sn_2P_2S_6$ crystals for fast near-infrared photorefraction'. *Ferroelectrics*. 2005;**318**(1):89–94.

[134] Juvalta F, Mosimann R, Jazbinšek M., and Günter P. 'Fast dynamic waveguides and waveguide arrays in photorefractive Sn2P2S6 induced by visible light'. *Optics express*. 2009;**17**(2):379–384.

[135] Ma H, Jen AY, and Dalton LR. 'Polymer-based optical waveguides: materials, processing, and devices'. *Advanced Materials*. 2002;**14**(19): 1339–1365.

[136] Jeong MY. 'Replication of precise optical waveguide for an optical device'. *Proceedings of SPIE*. 2006;**6124**:61241H.

[137] Oh MC, Lee HJ, Lee MH, Ahn JH, Han SG, and Kim HG. 'Tunable wavelength filters with Bragg gratings in polymer waveguides'. *Applied Physics Letters*. 1998;**73**(18):2543–2545.

[138] Drury CJ, Mutsaers CM, Hart CM, Matters M, and De Leeuw DM. 'Low-cost all-polymer integrated circuits'. *Applied Physics Letters*. 1998;**73**(1): 108–110.

[139] Zhang X, Hosseini A, Lin X, Subbaraman H, and Chen RT. 'Polymer-based hybrid-integrated photonic devices for silicon on-chip modulation and board-level optical interconnects'. *IEEE Journal of Selected Topics in Quantum Electronics*. 2013;**19**(6):196–210.

[140] Kippelen B, Marder SR, Hendrickx E, *et al.* 'Infrared photorefractive polymers and their applications for imaging'. *Science*. 1998;**279**(5347): 54–57.

[141] Tsutsumi N. 'Molecular design of photorefractive polymers'. *Polymer Journal*. 2016;**48**(5):571.

[142] Kinashi K, Shinkai H, Sakai W, and Tsutsumi N. 'Photorefractive device using self-assembled monolayer coated indium-tin-oxide electrodes'. *Organic Electronics*. 2013;**14**(11):2987–2993.

[143] Tsutsumi N, Kinashi K, Masumura K, and Kono K. 'Photorefractive dynamics in poly(triarylamine)-based polymer composites'. *Optics Express*. 2015;**23**(19):25158–25170.

[144] Zamiri G and Bagheri S. 'Fabrication of green dye-sensitized solar cell based on ZnO nanoparticles as a photoanode and graphene quantum dots as a photo-sensitizer'. *Journal of Colloid and Interface Science*. 2018;**511**:318–324.

[145] Schemme T, Ditte K, Travkin E, Jiang W, Wang Z, and Denz C. 'Perylene bisimide derivatives as innovative sensitizers for photorefractive composites'. *Proceedings of SPIE*. 2012;**8435**:843526.

[146] Zhen Y, Qian H, Xiang J, Qu J, and Wang Z. 'Highly regiospecific synthetic approach to monobay-functionalized perylene bisimide and di(perylene bisimide)'. *Organic Letters*. 2009;**11**(14):3084–3087.

[147] Ditte K, Jiang W, Schemme T, Denz C, and Wang Z. 'Innovative sensitizer DiPBI outperforms PCBM'. *Advanced Materials*. 2012;**24**(16):2104–2108.

[148] Köber S, Gallego-Gomez F, Salvador M, *et al.* 'Influence of the sensitizer reduction potential on the sensitivity of photorefractive polymer composites'. *Journal of Materials Chemistry*. 2010;**20**(29):6170–6175.

[149] Christenson CW, Thomas J, Blanche PA, *et al.* 'Grating dynamics in a photorefractive polymer with Alq_3 electron traps'. *Optics Express*. 2010; **18**(9):9358–9365.

[150] Lor KP, Chiang KS, Liu Q, and Chan HP. 'Ultraviolet writing of buried waveguide devices in epoxy-coated benzocyclobutene'. *Optical Engineering*. 2009;**48**(4):044601.

[151] Chiang KS, Lor KP, Liu Q, and Chan HP. 'UV-written buried waveguides in benzocyclobutene'. *Proceedings of SPIE*. 2006;**6351**:63511J.

[152] Vengsarkar AM, Lemaire PJ, Judkins JB, Bhatia V, Erdogan T, and Sipe JE. 'Long-period fibre gratings as band-rejection filters'. *Journal of Lightwave Technology*. 1996;**14**(1):58–65.

[153] Vengsarkar AM, Pedrazzani JR, Judkins JB, Lemaire PJ, Bergano NS, and Davidson CR. 'Long-period fibre-grating-based gain equalizers'. *Optics Letters*. 1996;**21**(5):336–338.

[154] Chiang KS, Chow CK, Chan HP, Liu Q, and Lor KP. 'Widely tunable polymer long-period waveguide grating with polarisation-insensitive resonance wavelength'. *Electronics Letters*. 2004;**40**(7):422–424.

Chapter 10

Integrated optics using liquid crystals

Rita Asquini[1] *and Antonio d'Alessandro*[1]

Liquid crystals (LCs) are well-known for the worldwide commercial success of flat-panel displays in high and ultra-high-resolution TV large area screens, computer monitors, cellular phones and many other daily used consumer electronics products. LCs were discovered in 1888 by the Austrian botanic Friedrich Reinitzer observing two melting points and the birefringence properties of cholesteryl benzoate and then named 'flussigkristalle' by the German physicist Otto Lehmann [1]. In fact, LCs are materials characterized by an intermediate phase, called 'mesophase', between solid and liquid, since their molecular order is higher than in a liquid and at the same time are fluid, requiring a confinement geometry.

It took about 80 years to demonstrate the first application of these delicate and fascinating materials, when George Heilmeier fabricated the first LC display at RCA Labs in 1964 [2–4]. Since then, LCs have been studied and synthesized to develop flat screens based on their ability to change light polarization. Moreover, the linear and non-linear optical properties of LCs were deeply studied, opening a new scenario of optoelectronic and photonic device applications beyond displays. They encompass spatial light modulators for image processing, adaptive lenses, light valves, smart windows, tuneable gratings, tuneable lasers and so on. They are used to make metamaterials, photonic bandgap structures and structures to obtain optical vortexes, photon-induced phase transformations. This chapter shows how LC and their optical properties can be used to make reconfigurable optical waveguides, electro-optical and all-optical devices integrated on various substrate materials such as silicon, silicon nitride, polymers and glass.

10.1 Optical properties of liquid crystals

LCs are materials able to show several mesophases between solid and liquid phase according to either a solution concentration in a solvent, in this case, they are referred to as lyotropic LCs or according to temperature, in this case, LCs are referred to as thermotropic LCs [5]. The formers are mostly used in cosmetics, an

[1]Department of Information Engineering, Electronics and Telecommunications, Sapienza University of Rome, Rome, Italy

example is represented by a mixture of soap and water, or they appear in some biological structures, whereas the latters, made of rod-like molecules, are typically used in most optoelectronic applications.

LC molecules are chemical compounds characterized by ring systems, with linking groups, substituents and terminal groups, as shown in Figure 10.1, to optimize physical and optical properties. They are often the results of mixtures of different compounds to optimize their electro-optical performance. LCs are transparent materials from ultraviolet (UV) to infrared (IR), with scattering losses decreasing as $\lambda^{-2.34}$. For this reason, they are attractive to make photonic devices [6].

In the nematic phase, the rod-like molecules are oriented along an average direction indicated by a unity vector referred to as director **n**. At lower temperatures, smectic mesophases can be formed, in which a higher molecular order is characterized by smectic planes. In smectic A phase, molecules are perpendicular to the smectic planes while tilted by a small angle in the smectic C mesophase. In some LC mixtures with chiral compounds, molecules are organized to form a helix forming the so-called cholesteric mesophase.

10.1.1 Optical anisotropy and refractive index

LCs are characterized by optical anisotropy both at molecular and at macroscopic levels, when molecules are oriented like in the nematic phase. In this case, nematic LCs (NLCs) behave as uniaxial materials with the optical axis parallel to the director, as sketched in Figure 10.2. At molecular level, a refractive index $n_{\|}$ along the molecular axis and a refractive index $n_{\perp}$ perpendicular to the molecular axis can be defined. At macroscopic level, an extraordinary refractive index n_e and an ordinary refractive index n_o can be defined along and perpendicularly to the director **n**, respectively. LCs can have a high birefringence: more than 0.5 at visible and more than 0.15 at near-IR wavelengths.

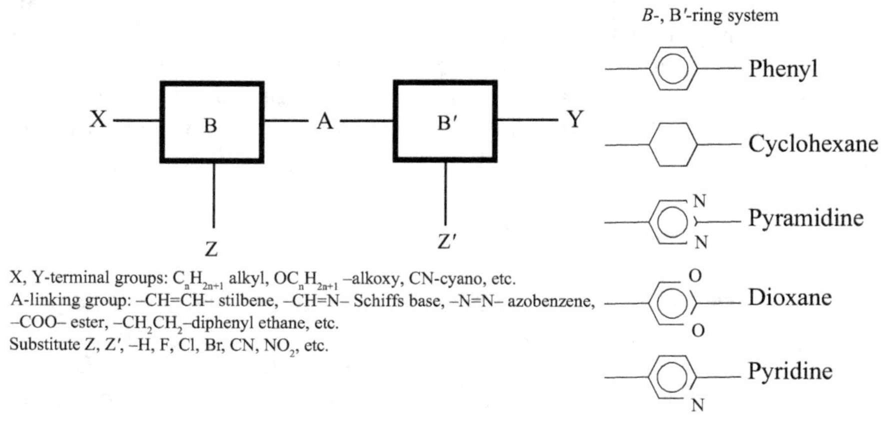

Figure 10.1 Chemical structure of LC molecules

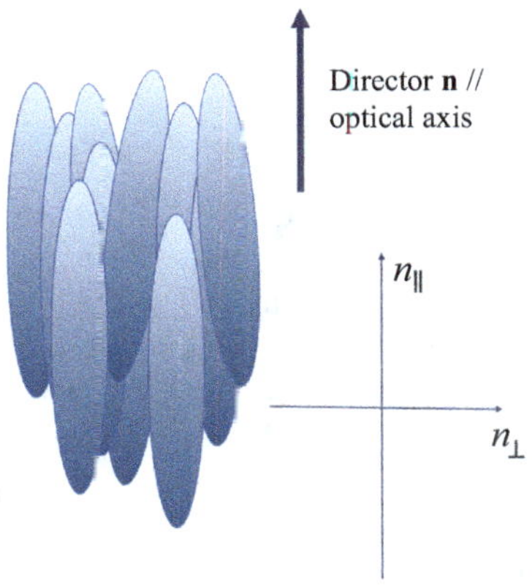

Figure 10.2 Optical uniaxiality of a nematic LC with the optical axis parallel to the molecular director

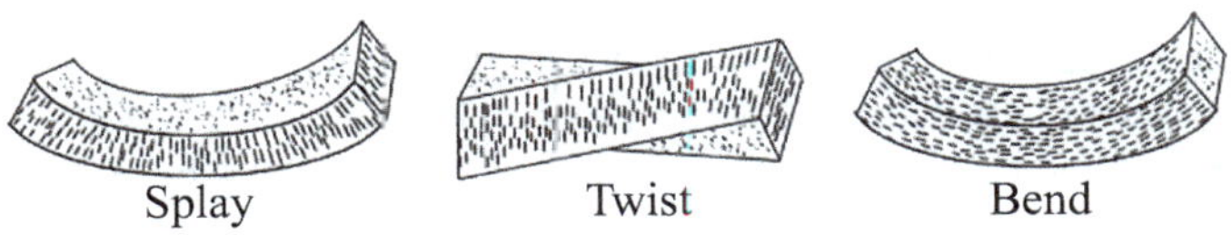

Figure 10.3 Schematic of basic elastic deformations of LC molecules

Birefringence decreases as temperature increases, vanishing when the liquid isotropic phase is reached. LCs need to be confined and their interaction with boundary surfaces is crucial for their reorientation under an applied electric field.

10.1.2 Electro-optic effect in LC and molecular reorientation

An external electric field induces a molecular reorientation of LC molecules, which assume a new configuration depending on the elastic and dielectric properties. The LC molecular configuration under the action of an external electric field is the result of the balance between the electric torque and the elastic restoring. The deformation pattern determined by the elastic torque is the result of a combination of three basic deformations: splay, twist and bend, as sketched in Figure 10.3.

The orientation of the LC corresponds to the minimum of the free energy $F = F_{el} - F_{diel}$, which includes the elastic term F_{el} and the dielectric term F_{diel}. The elastic energy according to the Oseen–Frank theory is expressed by

$$F_{el} = \frac{1}{2}\iiint \left[k_{11}(\nabla \cdot \mathbf{n})^2 + k_{22}(\mathbf{n} \cdot \nabla\times\mathbf{n})^2 + k_{33}(\mathbf{n}\times\nabla\times\mathbf{n})^2\right] dv \tag{10.1}$$

where k_{11}, k_{22} and k_{33} are the elastic constants corresponding to the splay, twist and bend, respectively, and dv is the elementary integration volume. The dielectric energy is given by

$$F_{\text{diel}} = \frac{1}{2} \iiint \left[\varepsilon_0 \varepsilon_\perp |\mathbf{E}|^2 + \varepsilon_0 \Delta\varepsilon (\mathbf{E} \cdot \mathbf{n})^2 \right] dv \tag{10.2}$$

where ε_0 is dielectric permittivity in vacuum, $\varepsilon_\perp$ is the dielectric permittivity along the optical axis, $\Delta\varepsilon$ is the dielectric anisotropy and $\mathbf{E}$ is the applied electric field. The minimization of F is achieved by solving the Euler–Lagrange equation. This approach is simpler than global or relaxation approaches, and, therefore, more adapted for large three-dimensional (3D) problems of typical integrated optical structures. To solve the partial derivative equation, a finite element method is well suited since it allows the implementation of the weak form of the Euler–Lagrange equation.

The minimization of F is coupled with the solution of the Poisson equation for the distribution of the electric potential V in the LC structure:

$$\nabla \cdot [\varepsilon_\perp \nabla V + \Delta\varepsilon (\nabla V \cdot \mathbf{n})\mathbf{n}] = 0 \tag{10.3}$$

The resulting output data are the spatial distribution of the LC director orientation, from which the refractive index profile can be derived. The possibility to have a refractive index distribution allows to design every optical device, and this is more crucial in LC waveguide (LCW)-based integrated optic devices. The advantage of this model is that we do not make any hypothesis or simplification of the director and electric potential distributions. Therefore, this fully consistent model may be used for every type of geometry, either two-dimensional (2D) or 3D, provided the boundary conditions are correctly stated. Alternatively, a study of the molecular director distribution can be obtained by means of Monte Carlo simulation techniques [7,8].

10.2 Switchable optical waveguides with liquid crystal core in silicon

The basic element of any integrated optics device is the optical waveguide. LC offers the possibility to design and fabricate electrically controllable optical waveguides by confining LC in a capillary or in a channel. A technology solution is to embed an NLC in a silicon V-groove [9], in which electrodes can be integrated to control propagation in the waveguide LC core. The main advantage of using Si is the potential integration of both optical and electronic circuits and devices in the same substrate.

10.2.1 Fabrication technology

A silicon V-groove can be obtained by exploiting the preferential etching of silicon. Since LC has refractive index in the range of 1.45–1.6, a low index cladding is necessary. Hence, a native thermal oxide SiO_2 is obtained in an oven at about

1,100°C. The V-groove is obtained by using the typical micromachining process in which an initial sacrificial thermal oxide layer of about 300 nm is grown to be used as a mask. A strip of oxide, with a groove width typically between 5 and 15 μm, is etched in buffered HF, and the next anisotropic etching of Si by KOH at 80°C allows to obtain the desired V-groove in the Si wafer. The sacrificial SiO_2 is removed and a uniform SiO_2 layer is thermally grown up to a 1.5-μm thickness to act as substrate for the final waveguide. The anisotropic etching allows to obtain a V-groove with an angle $\alpha = 54.7°$ using (1 0 0) Si wafer. An interesting characteristic of SiO_2/Si is the smoothness of the surfaces, which prevents defects in the LC core, thus reducing scattering of light in the waveguide. A glass cover of 0.5 mm is then placed on top of the V-groove [10]. The glass type is borosilicate with a refractive index $n_{D263} = 1.516$ (at the wavelength $\lambda = 1.55$ μm), intermediate between the values of the extraordinary and the ordinary refractive index of the LC used, like E7 ($n_e = 1.69$, $n_c = 1.5$ at wavelength $\lambda = 1.55$ μm) to obtain an optical switch [11]. The inner face of the glass is covered by a 100-nm thick layer of ITO (indium tin oxide), a transparent conductive oxide used as electrode, and on top of it, an alignment layer is deposited, made of spun and rubbed Nylon 6 to orient LC molecules along the groove direction. The ITO sheet resistance is about 80 Ω/square. NLC E7 is infiltrated in the covered V-groove by capillarity in vacuum at 80°C, when the NLC is in the isotropic phase, and then cooled down at room temperature. Figure 10.4 shows the final waveguide structure in which a voltage can be applied between the ITO and the n-Si substrate.

LC waveguides in SiO_2/Si grooves, referred to as LCW, can be butt-coupled to single-mode optical fibres with a coupling loss of 4.5 dB or better by using LC-NOA61 interface. Such interface, obtained by infiltrating NOA61 UV curable polymer at both edges of the channel, reduces LC molecular random orientation responsible for high scattering at the input and output faces of the LCW. Propagation losses are 6 dB/cm, which can be reduced by using photoalignment techniques of the LC molecules [12].

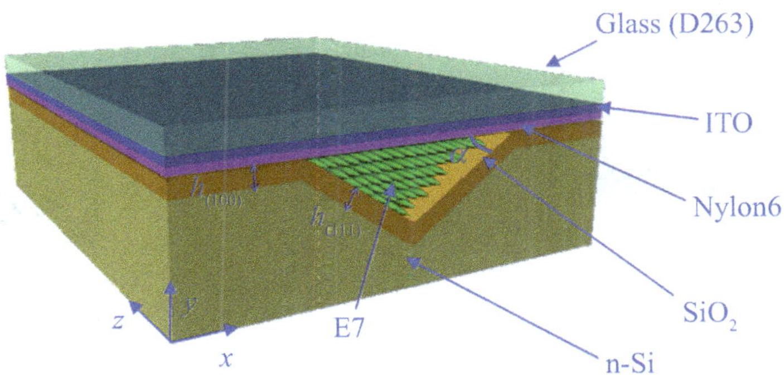

Figure 10.4 Schematic of an optical waveguide made of LC infiltrated in an SiO_2/Si V-groove. © 2010 IEEE. Reprinted, with permission, from Reference [11]

10.2.2 Electro-optical and all-optical switching

Light propagation can be controlled in the SiO_2/Si LCW by either electro-optic or non-linear optical effect, or both. Since LC molecules can be reoriented along an applied electric field, a square wave voltage is applied between the ITO electrode and the n-type Si substrate acting as a second electrode. An AC field is required to avoid electrolysis of the LC. Figure 10.5 shows the working principle of an LCW acting as an on–off switch. When a voltage is applied, the LC molecules are mostly reoriented perpendicularly to the propagation direction and the glass plates. A vertically polarized light signal sees an increasing refractive index of the LC, approaching the value of n_e of the LC along the molecular director, as the voltage increases.

When the refractive index of the LC seen by the light beam becomes higher than the refractive index both of SiO_2 and of the borosilicate glass of the cover, light can be confined and propagates in the LC core. If no voltage is applied, the polarized light beam sees the ordinary refractive index n_o of the LC which is lower than the glass refractive index. The device can behave not only as a switch but also as a variable attenuator. In fact, as shown in Figure 10.6, when a voltage is higher than just 2 V, light starts to propagate, and transmission increases with an on–off extinction ratio (ER) above 40 dB at a voltage of about 8 V. The switching time is related to reorientation dynamics of the LC in the range of a few ms, depending on the viscosity and elastic properties of the LC mixture. As the voltage amplitude increases, the number of supported modes also increases.

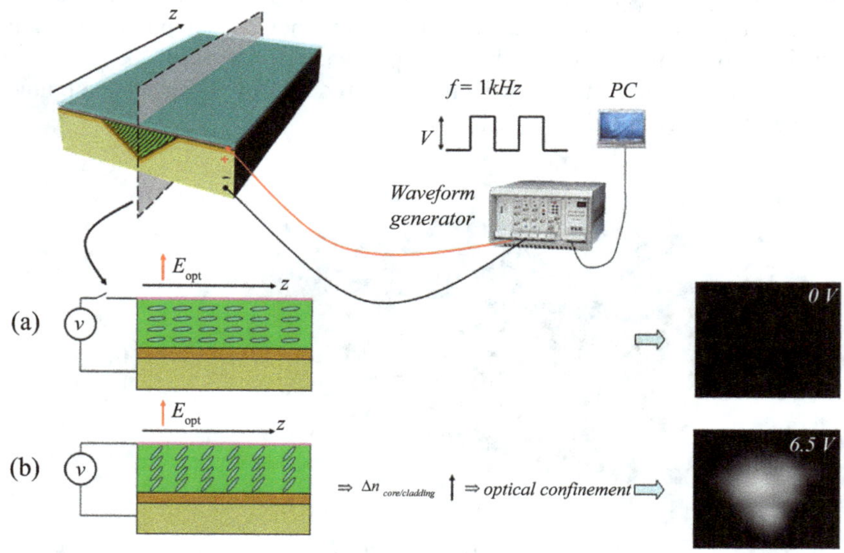

Figure 10.5 Working principle of an electrically controlled LCW without (a) and with (b) applied electric field. © [2010] IEEE. Reprinted with permission from Reference [11]

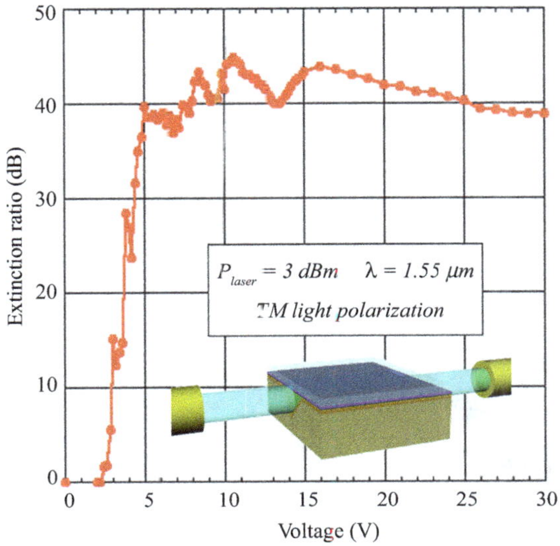

Figure 10.6 Extinction ratio versus applied voltage showing the switching characteristic of SiO_2/Si LCW. © 2010 IEEE. Reprinted, with permission, from Reference [11]

The LCW propagation characteristics can be theoretically modelled by minimizing the free energy F in the waveguide core and by solving the Poisson equation as described in Section 10.1. This approach allows to obtain the refractive index distribution of the LC inside the SiO_2/Si which can be implemented in a beam propagator able to simulate the guided light characteristics in the LCW.

LC molecules can also be reoriented by means of the intensity of the optical field confined in the LCW [13]. An optically controlled modulator/switch can be obtained by using the same structure. Even in this case, modelling of the all-optical device can be obtained by adding a third term F_{opt} to F_{el} and F_{diel} of (10.1) and (10.2), respectively, in the free energy expression related to the action of the optical electric field on the LC molecules:

$$F_{opt} = \frac{1}{2}\iiint \left[\varepsilon_0 \varepsilon_{\perp,opt} \left| \mathbf{E_{opt}} \right|^2 + \varepsilon_0 \Delta\varepsilon_{opt} \left(\mathbf{E_{opt}} \cdot \mathbf{n} \right)^2 \right] dv \tag{10.4}$$

which will contribute to control the refractive index distribution in the LCW channel.

Figure 10.7 shows a comparison of both the theoretical, shown in Figure 10.7(a), and experimental, shown in Figure 10.7(b), plots of P_{out} influenced by P_{in} intensity due non-linear optical reorientation. Simulations have been obtained by minimizing $F = F_{el} - F_{diel} - F_{opt}$. Agreement between simulation and experimental data can be observed, including a pretilt of LC molecules required by applying a small voltage between 6 and 8 V.

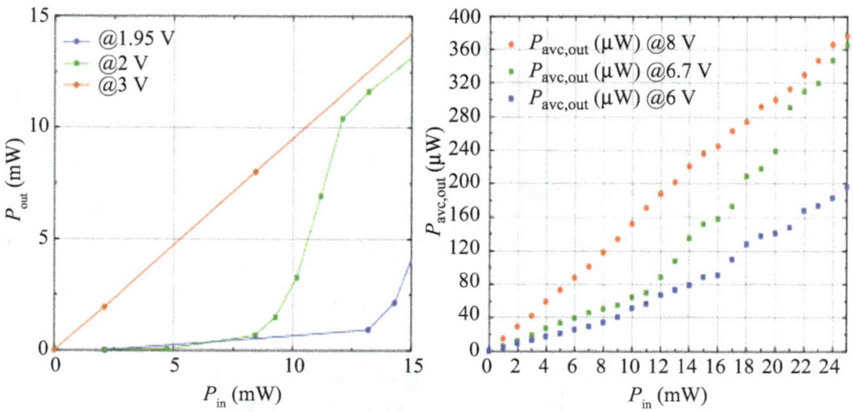

Figure 10.7 Non-linear optical characteristics of SiO_2/Si LCW: (a) theory and (b) experiment

Plots at maximum pretilt voltages of 1.95 V in Figure 10.7(a) and of 8 V in Figure 10.7(b) show a linear behaviour since, in this case, molecules are completely tilted by the applied electric field and the optical field intensity is not able to reorient them further.

This effect was used to demonstrate the concept of an all-optical switch. A probe beam transmission at the wavelength of 1,510 nm can be switched on and off in an LCW by controlling the power of a pump beam at the wavelength of 1,560 nm. An on–off ER of 10 dB was measured.

The difference of pretilt voltages between the corresponding plots of Figure 10.7 (a) (1.95, 2, 3 V) and 10.7(b) (6, 6.7, 8 V) is determined by two main reasons. The first one is related in part to the presence of unavoidable small defects in the groove walls perturbing the LC molecular orientation. The second reason is due to the voltage drop at the electrodes not considered in the model. Further reduction of driving power in all-optical devices can be obtained by using doped LC mixtures. Simulations with a dopant methyl red (MR) indicate that the waveguide can be switched on by an optical signal with a power of just 5 mW. Novel azo dye LC mixtures can be also used to obtain faster response below microsecond regime. A limitation of LCW in SiO_2/Si V-groove based on preferential anisotropic etching of Si is that bends cannot be obtained, and a different technological platform must be used to obtain more complex devices than just straight switchable channels.

10.3 Photonic devices with liquid crystal core in polydimethylsiloxane

Optical waveguides with an LC core and a polydimethylsiloxane (PDMS) cladding, referred to as LC:PDMS waveguides, have been demonstrated. They can be fabricated by filling PDMS channels with LC by capillarity. PDMS has been becoming an interesting material for microfluidic applications [14]. PDMS is also considered a very good material for photonic applications for its high optical quality and transparency. It is characterized by a low surface energy and low dielectric constant; it has a low Young's modulus and is biocompatible. Moreover, it is particularly suitable for fabrication processes, in fact it is an effective, reliable and cheap material for soft lithography to make both microfluidic and micro-optical devices on flexible substrates [15,16].

10.3.1 Fabrication of LC:PDMS waveguides

PDMS empty channels with different dimensions can be obtained by the well-known and reliable cast and moulding technique described in Figure 10.8.

A mould is obtained by photolithographically patterned SU-8 photoresist on an Si wafer forming a negative master of the final channels. PDMS casting is obtained by pouring liquid PDMS on the mould and cured at 80°C. A cooled PDMS layer, which includes the channels, is peeled away and deposited on another thin PDMS layer on top of a glass substrate used as holder. The empty channels are then filled with an NLC. Due to the low surface energy of PDMS, the LC molecules orient homeotropically, namely perpendicular to the inner hydrophobic surfaces of PDMS channels with a minimum contact surface, as schematically shown on the left-hand side of Figure 10.9. A typical chip, including some LC:PDMS waveguides, is reported on the right-end side of the same figure.

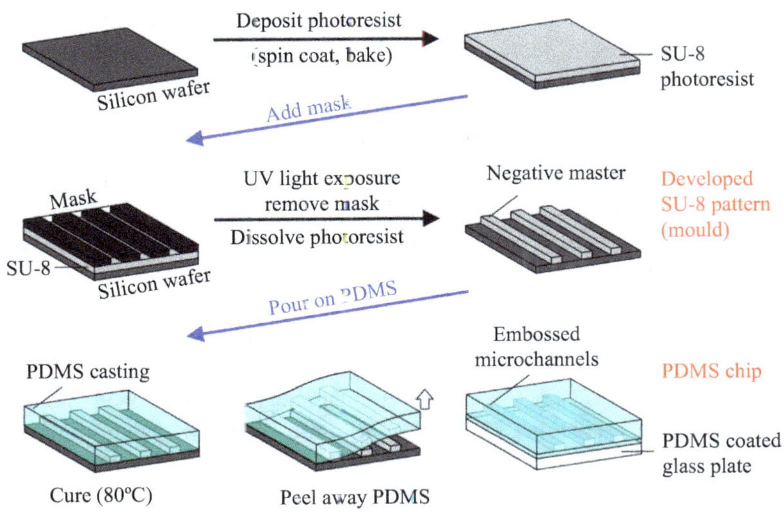

Figure 10.8 Fabrication steps of PDMS channels by using casting and moulding technique

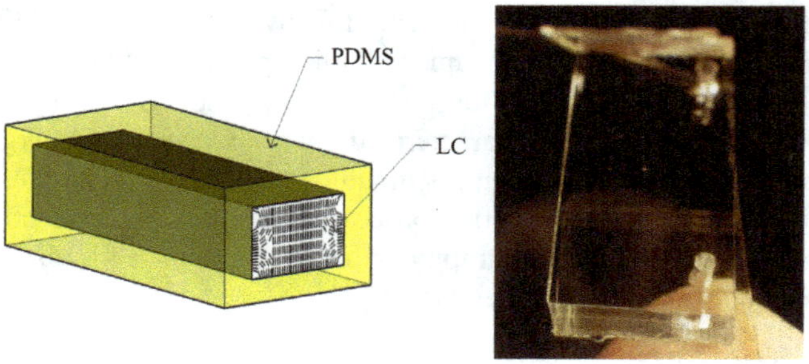

Figure 10.9 LC:PDMS waveguides. Oriented LC molecules inside the PDMS channel on the left. A PDMS chip with optical waveguides on the right

The homeotropic alignment of the LC inside the PDMS channel creates a distribution of the LC refractive index which allows polarization-independent propagation of light at both visible and near-IR wavelengths [17].

10.3.2 LC:PDMS-based photonic switches and demultiplexers

Several integrated optical devices can be made by using LC:PDMS waveguides. The design can be carried out by using the beam propagation technique in which the refractive index distribution can be implemented. As described in Section 10.1, the refractive index of LC seen by polarized light can be obtained from the LC director distribution as a result of the minimization of the free energy of the LC molecules.

An optical switch can be designed by using a directional coupler based on two LC:PDMS waveguides. Coupling condition depends on the LC refractive index controlled by in-plane voltage applied to coplanar electrodes. An ER of 20 dB, as the ratio of output optical power, can be obtained with a coupling length of 500 μm for a gap of 1 μm between two LC:PDMS waveguides using NLC E7. Since a 1-μm gap can be critical to fabricate, a more feasible zero-gap directional coupler switch can be designed, and it is sketched in Figure 10.10 [18].

In this case, waveguide coupling is determined by the interference between the two modes propagating in the central waveguide controlled by the voltage applied to the coplanar electrodes. Light can be switched to either of the two output waveguides with a contrast higher than 16 dB by applying only 1.62 V to route light to the right-hand output waveguide (P_{out} 2) and 1.76 V to route light to the left-hand output waveguide (P_{out}1). A similar coupler can also be designed to operate as demultiplexer for wavelengths 980 and 1,550 nm by separating them into the two different output ports with a contrast better than 10 dB. Coplanar electrodes of different metals can be deposited on PDMS by several techniques, including sputtering, e-beam evaporation or electroplating. Coplanar gold electrodes deposited by electroplating are able to switch LCs in a PDMS channel by applying only 2.85 V.

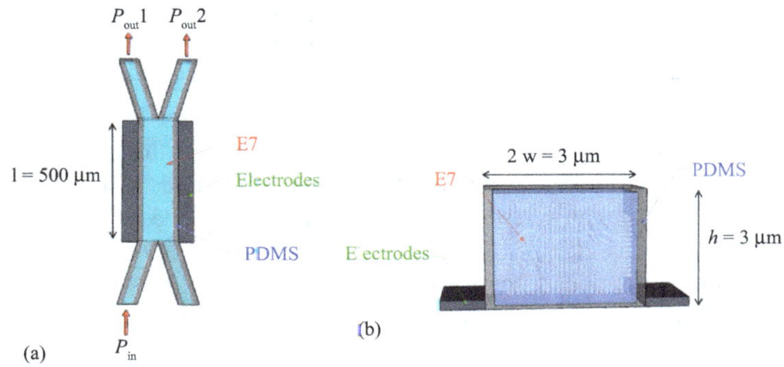

Figure 10.10 Optical switch based on a zero-gap directional coupler of LC: PDMS waveguides: (a) top view of the device and (b) front view of the bimodal central waveguide

10.4 Bragg reflectors based on liquid crystals

A Bragg grating consists of an optical structure with a periodic modulation of the refractive index able to reflect by diffraction a specific wavelength, which can be adjusted by either modifying the refractive index distribution or changing the grating period [19]. Bragg reflectors are widely used as tuneable optical filters, thanks to their narrow bandwidth and high wavelength selectivity.

A wide variety of materials and structures have been employed to make integrated optical Bragg gratings, including polymers, silicon-on-insulator (SOI), hollow capillaries, lithium niobate, metal–insulator–metal, silica and LCs [20–47]. As shown in the previous sections, LC and doped-LC novel materials are excellent options for the realization of low-power and low-cost optoelectronic devices, due to their large electro-optic response and non-linear optical properties [48–51]. Moreover, a low input power is able to change the refractive index by driving molecular reorientation for different applications, including photonic switching [50–57] and optofluidics [58–60].

10.4.1 Tuneable optical filters using composite gratings on glass

POLICRYPS, acronym for POlymer LIquid CRYstal Polymer Slices, are very well-known microstructures made of alternated microstripes of UV curable polymers and NLCs [25,26,33,61], which can produce efficient optical diffraction of light propagating in free space. Many applications were demonstrated in which POLICRYPS gratings can be used as electro-optic and all-optical free-space devices. An induced reorientation of the LC molecules can modulate indeed the periodic variation of the refractive index of the grating. Integrating POLICRYPS with channel waveguides makes possible to obtain tuneable waveguide filters, which can be driven either electrically or optically. In such a configuration, a

POLICRYPS grating is fabricated on top of a channel waveguide perpendicular to the orientation of polymer/LC slices. The POLICRYPS microstructure acts as a periodic overlayer, which induces a tuneable modulation for the effective refractive index of the waveguide. When the LC refractive index 'seen' by a TE-like polarized waveguide beam matches the refractive index of the polymer in the POLICRYPS, it basically hinders the action of the grating, and, in this case, a white light signal can propagate unperturbed through the waveguide. When a mismatch between the refractive index of the polymer and the one of the LCs is either electrically or optically induced by mean of the LC molecular reorientation, the POLICRYPS grating produces a Bragg reflector for a particular wavelength propagating in the optical waveguide.

The refractive index change implies a tuning of the Bragg back-reflected wavelength:

$$\lambda_B = \frac{2\Lambda n_{\text{eff}}}{m} \tag{10.5}$$

where n_{eff} is the effective refractive index of the waveguide affected by the grating index modulation, being Λ the pitch of the grating and m is the grating diffraction order.

Figure 10.11 shows, for comparison, an electrically driven (a) [33] and an optically driven (b) [41] integrated optical tuneable filter [62]. In both devices, a single-mode double ion-exchange K^+–Na^+/Ag^+–Na^+ waveguide in a BK7 substrate is used to confine light beams, the wavelengths of which can be selected by the Bragg diffraction due to the overlaying POLICRYPS grating. In both devices, the grating is embedded between the substrate, which includes the waveguide, and a cover made of the same glass BK7. Both gratings use Norland Optical Adhesive (NOA) 61 polymeric stripes alternated to the NLC E7 in the electro-optic version and to the MR-doped E7 (MR:E7) in the all-optical version of the grating.

In the examples presented in Figure 10.11, the grating pitch is 2.55 μm in the electro-optically controlled filter and 1.55 μm in the optically driven version. In

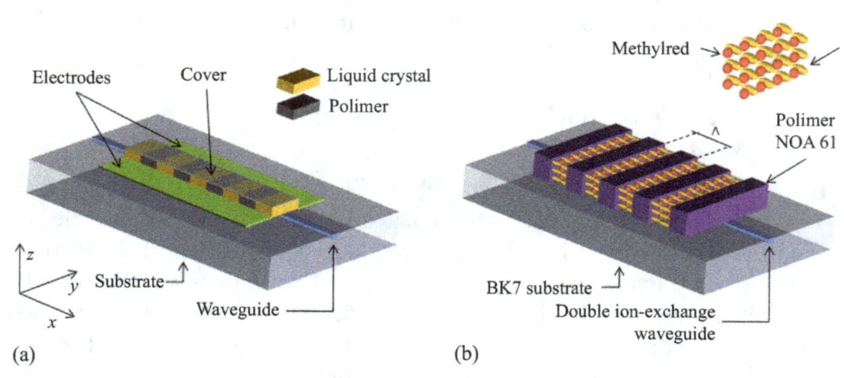

Figure 10.11 POLICRYPS-based electrically (a) and optically tuneable (b) integrated optical filters using double ion-exchange waveguides in BK7 glass

both cases, the grating thickness is 1.1 μm. The devices are designed to operate in the 1.55 μm region.

10.4.1.1 Electro-optic filter

In the electro-optic type filter, the tuning is provided by the molecular reorientation induced electro-optically by the electric field applied to the coplanar electrodes [33,62]. The transmitted and the corresponding reflected spectra are shown in Figure 10.12(a) and (b), respectively. The maximum attenuation of transmitted spectrum is about 20 dB at the wavelength of about 1,552 nm.

When a variable voltage between 0 and 40 V is applied to the coplanar electrodes of the electro-optical filter, a continuous tuning of about 4 nm is obtained, as shown in Figure 10.13, with overall sub-milliwatt driving power due to the negligible current absorption.

10.4.1.2 All-optical filter

In the all-optical filter, no electrodes are required since the MR:E7 can be reoriented by using green light (at the wavelength of 533 nm). The device behaves as a Bragg filter because a TE-like optical beam launched in the channel waveguide experiences the overlaying phase grating [41,62]. At thermal equilibrium, the MR is in its elongated trans form and the long axes of both the LC molecules and trans-MR are aligned perpendicularly to the polymer slices (i.e. parallel to z axis in Figure 10.11). The phase grating in this case is due to the mismatch between the trans MR:E7, the ordinary refractive index is about 1.5 at 1,550 nm (approximately equal to the E7 LC ordinary refractive $n_{\perp} = 1.5$ at 1,550 nm because of the low concentration of MR in E7 LC), and the NOA 61 whose refractive index is 1.5419 at 1,550 nm. The period of the grating is estimated to be $\Lambda = 1.50–1.55$ μm. Considering the third diffraction order ($m = 3$) in (10.5), the achieved Bragg wavelength λ_B falls in the telecom range of 1,520–1,570 nm.

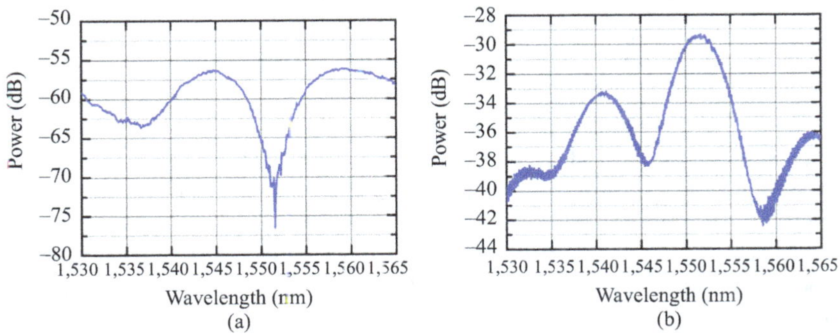

Figure 10.12 Transmitted (a) and reflected (b) spectra of the electrically tunable integrated optical filter. © 2008 OSA. Reprinted, with permission, from Reference [33]

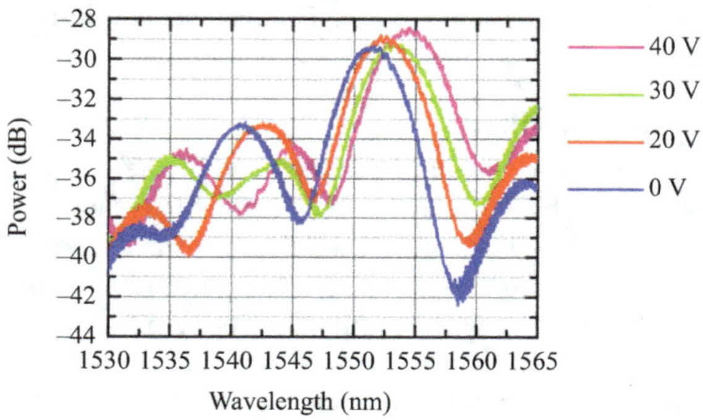

Figure 10.13 Overlapped transmitted spectra obtained by varying the applied voltage. © 2008 OSA. Reprinted, with permission, from Reference [33]

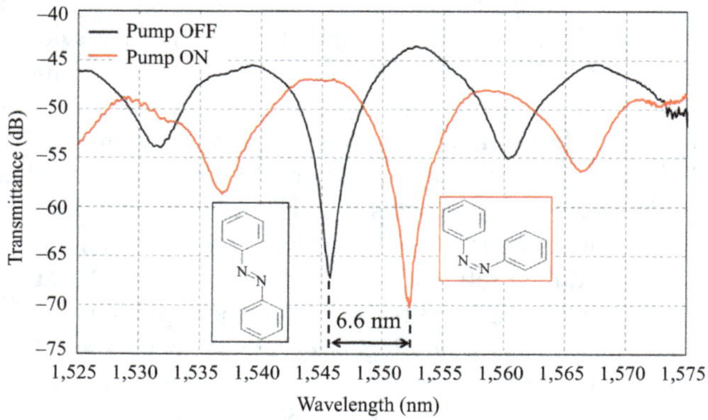

Figure 10.14 Optical transmission spectra of the NOA61/MR:E7 POLICRYPS-based optical filter when no driving signal is applied (solid line), and when a 45-mW laser beam at 532 nm impinges on the grating (dotted line). © 2011 OSA. Reprinted, with permission, from Reference [41]

When the grating is lighted up by a pump laser source at $\lambda = 532$ nm, the MR turns into the spherical *cis* form, changing the directional order of the MR: E7 mixture. Accordingly, there is a change in the refractive index mismatch of the grating structure, resulting in a variation of the back-reflected Bragg wavelength λ_B depending on a variation of n_{eff} in (10.1). The filter transmittance peak wavelength λ_B can be controlled by switching on and off the pump signal. Therefore, the device can be used as all-optical tuneable filter.

Figure 10.14 shows the transmission optical band of the all-optical filter. The spectrum is tuned by 6.6 nm when a 45-mW laser beam at 532 nm is applied. In this case, the controlling green light beam forces a reorientation of the NLC, homogeneous alignment of which, perpendicular to the polymer stripes, is broken by the MR molecular isomerization. The optically induced conformational molecular transition of MR molecules transforms the initial elongated *trans* molecular phase to isotropic *cis* form, leading to the shift of the optical filter response.

By exposing the grating to the pump signal, the shifted spectrum preserved its shape with a slightly deeper notch of about −22 dB. The bandwidth of the transmitted notch, at −3 dB with respect to the transmittance minimum, is 3.3 nm without an active pump signal, while the bandwidth is 2.7 nm when the pump illuminates the grating. Another interesting feature of this filter with respect to its electrically controlled counterpart described in the previous section is its polarization-independent operation since upon illumination the NLC is randomly orientated and light ‘senses’ an average refractive index of the doped E7 NLC.

The switching times for both filters are in the range of few tens of ms, as typically measured for E7 NLC standard cell with the same thickness. Temperature tuning can be also obtained in both filters. As an example, by changing the temperature of 4°C, a tuning range over 15 nm can be measured for the filter using the NOA61/MR:E7 grating, as expected for the thermo-optical properties of E7.

10.4.2 Guided-wave tuneable Bragg gratings

Guided-wave tuneable planar Bragg reflectors can be simply created with an LC layer sandwiched between two glass plates in which wavelength tuning is obtained by varying the LC refractive index through an external applied voltage [39,62]. In such devices, a periodic LC molecular reorientation is induced to obtain a Bragg grating directly inside the waveguide LC layer by means of a voltage applied to periodical patterns of electrodes. Estimation of the reorientation of the LC molecular director **n** can be obtained by coupling the solution of the Poisson equation, which describes the distribution of the electric field, with the minimization of the free energy F, which involves an elastic term and an electrostatic term calculated by means of the Oseen–Frank expression, as described in Section 10.1.2. Different electrode configurations and patterns lead to different performances.

10.4.2.1 Coplanar electrode configuration

An optical slab LCW layer can be controlled electrically by using several possible electrode patterns. Figure 10.15 shows a device configuration in which a periodic

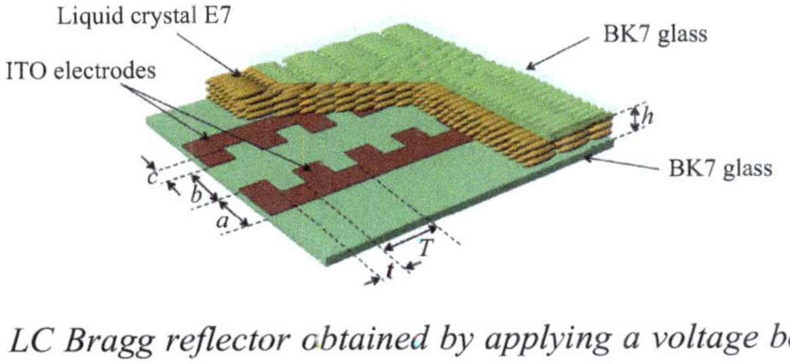

Figure 10.15 LC Bragg reflector obtained by applying a voltage between comb-like electrodes

LC molecular reorientation can induce a Bragg grating inside the waveguiding LC layer by applying a voltage to periodically patterned coplanar ITO electrodes [37]. In fact, the different distance b between the two comb-like patterned electrodes creates a periodic value of the in-plane electric field.

The device behaviour and simulation can be obtained by minimizing the Frank–Oseen free energy of the LC, including the optical anisotropy. Both dielectric and elastic energy contributions are included in the calculations. The LC orientation and the related refractive index spatial distributions can be obtained by solving the Poisson equation.

The design of the device is optimized, including the LC mixture E7 infiltrated between ITO patterned BK7 glass substrates, and refers to the following design values: $h = 1$ µm, $t = 250$ nm, $T = 500$ nm, $a = b = 500$ nm, $c = 250$ nm.

The refractive index modulation resulting from such a device is displayed in Figure 10.16(a). Results were evaluated 200 nm above the electrodes and in the symmetry axis between them ($y = 0$) and achieved for applied voltages between 2.8 V (bottom line) and 4.5 V (top line) in 0.1 V steps. The figure highlights the sinusoidal behaviour of the refractive index, with a period of 500 nm, the modulation amplitude depending on the applied voltage and, for each curve, the maximum value corresponding to the minimum electrode spacing. The resulting index contrast versus voltage between 0 and 13 V is reported in Figure 10.16(b). The index modulation is almost negligible up to 2.4 V and increases rapidly after this threshold, due to the NLC reorientation effect. At the voltage of 5 V, the NLC molecules are completely reoriented, reaching their maximum twist $\phi = 90°$, parallel to y, in the regions with the minimum electrode separation but partially reoriented where the distance between the electrodes is higher (distance $b + 2c$).

When the voltage increases over 5 V, molecular reorientation takes place only in the regions which are not completely reoriented yet, yielding a decrease in modulation. Above 12.4 V, almost all of the NLC molecules are reoriented parallel to y and the index contrast is zero.

Such a device configuration shows an impressive performance in terms of tuning range. In Figure 10.17, the back-reflected power (spectral reflectivity) is

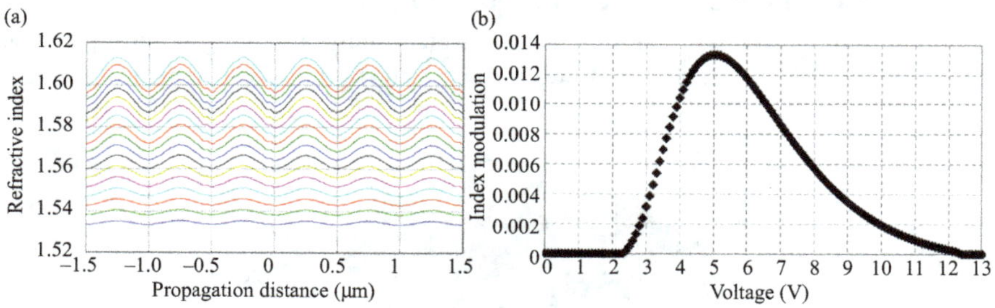

Figure 10.16 (a) Refracting index modulation for applied voltages between 2.8 V (bottom line) and 4.5 V (top line) in 0.1 V step. (b) Longitudinal modulation versus applied voltage. © 2010 OSA. Reprinted, with permission, from Reference [37]

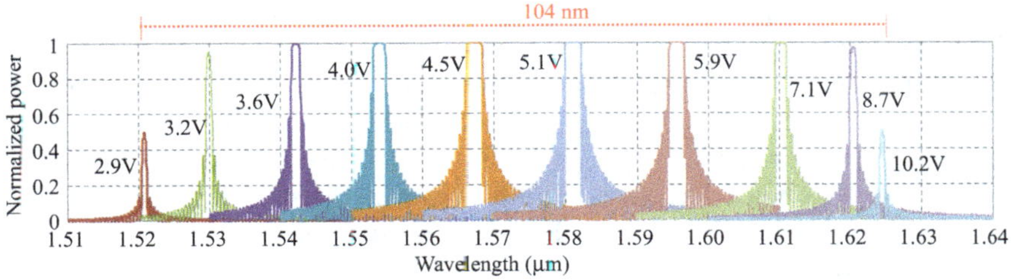

Figure 10.17 Spectral reflectivity for propagation over 1.5 mm (3,000 periods) and different voltages. © 2010 OSA. Reprinted, with permission, from Reference [37]

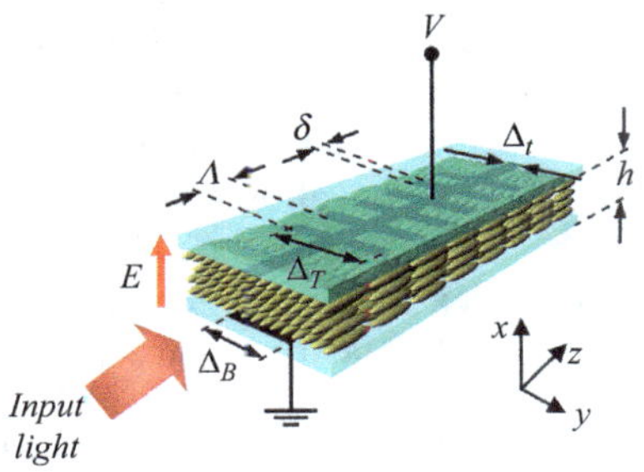

Figure 10.18 Three-dimensional view of the tuneable Bragg reflector with top–bottom electrode configuration. © 2009 OSA. Reprinted, with permission, from Reference [36]

reported as a function of the resonant wavelength, evaluated by using coupled-mode theory and considering the refractive index distribution for a device length of 1.5 mm, corresponding to 3,000 periods. A tuning range of 98 nm can be obtained with reflectivity above 90% and voltages between 3.2 and 9 V. An extended tuning range of about 104 nm in the near-IR regions between 1,521 nm and about 1,625 nm can be achieved by varying the applied voltage between 2.9 and 10.2 V, keeping the back reflection above 50%, with a good spectral selectivity.

10.4.2.2 Top–bottom electrode configuration

A second tuneable Bragg reflector affects the TM mode, instead of the TE mode as in the previous device. The device structure, shown in Figure 10.18, consists of a planar glass cell with an external voltage applied to the NLC layer by a set of top and bottom transparent electrodes [36]. The LC molecules inside the device are arranged with their director **n** parallel to z without an applied voltage.

The thickness of the NLC layer and its refractive index are assumed to be able to support the propagation of TM guided waves in the planar slab. In order to meet the highest coupling between the TM optical field and the induced Bragg grating for a given wavelength of operation and NLC, the electrode overall width, periodicity and duty cycle along z and layer thickness along x have to be selected. The application of a suitable voltage between the top electrode and the ground plane at the opposite boundary allows a 2D optical confinement. The top electrode is periodic along z, in order to define the one-dimensional photonic lattice entailing Bragg reflection via distributed feedback. An effective periodic modulation of the waveguide refractive index results in a Bragg resonance, despite the inherent non-locality in the electro-optic response.

Applying an appropriate voltage, it is possible to modulate the refractive index along the device. Figure 10.19(a) shows the refractive index modulation that can be experienced 100 nm below the top electrode by the TM00 mode, as a function of the propagation distance along z, for applied voltages ranging from 2 to 3 V in steps of 0.2 V. For the scope, an elementary electro-optic distributed feedback waveguide segment with only four periods of the top electrode was considered, and the device structure was chosen in order to have single-mode propagation in the aforesaid voltage range. The modulation effect can be appreciated as the voltage increases, doubling the index contrast from 0.007 to 0.015 when the voltage increases from 2.4 to 3.0 V.

The variation of the applied voltage allows also the tuning of filter. The filter reflectivity and the full width at half maximum (FWHM) of the reflection, for a given number of periods and index modulation depth, can be computed by using the coupled-mode theory. Figure 10.19(b) presents the normalized Bragg reflectivity versus wavelength for three different applied voltages. It is worthwhile to highlight that a 14 nm (from 1,536 to 1,550 nm) tuning of the voltage-controlled filter is achievable with applied voltages ranging from 2.5 to 3.0 V, being met in all cases a reflectivity always above 80% in single-mode operation. In particular, the filter reflectivity at 3 V has an FWHM of 0.38 nm with a device length of 5 mm achieving almost 100% reflection.

Such results are obtained considering NLC E7, with parameters $k = k_{22} = 7.3$ pN (assuming single constant approximation with pure twist deformation), $\varepsilon_{\perp} = 7$,

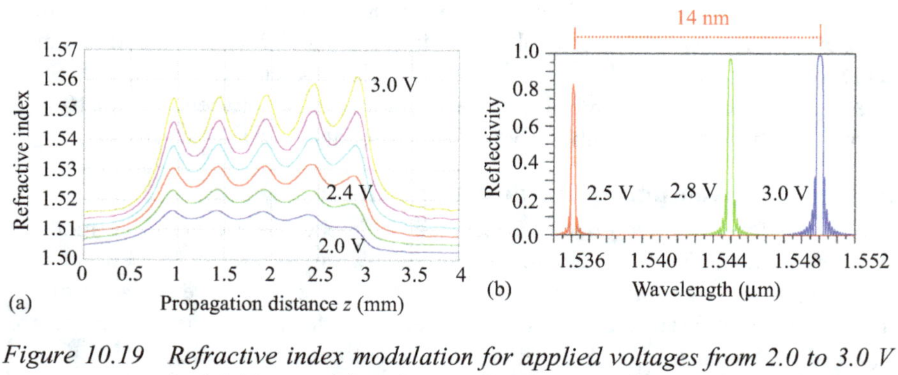

Figure 10.19 Refractive index modulation for applied voltages from 2.0 to 3.0 V in 0.2 V steps (a) and normalized grating filter response for 3 different values of applied voltage (b). © 2009 OSA. Reprinted, with permission, from Reference [36]

$\varepsilon_{\parallel} = 20$, $n_{\perp} = 1.50$, $n_{\parallel} = 1.689$, and a planar anchoring for the NLC with a rubbed Nylon 6 polymeric layer with a pretilt of 1° to eliminate the Fréedericksz threshold. As mentioned, NLC thickness $h = 1$ μm is assumed to obtain a single-mode propagation for wavelengths around 1,550 nm. The bottom straight strip electrode (along z in Figure 10.15) is $\Delta_B = 0.5$ μm wide (along y), and the top electrode includes a strip $\Delta_t = 0.1$ μm wide and a central periodic and symmetric comb structure, with a pitch $\Lambda = 0.5$ μm and duty cycle 3:20. Moreover, the top electrode had transverse segments $\Delta_T = 3$ μm wide along y and $\delta = 75$ nm along z.

10.5 Integrated optic devices based on a liquid crystal overlayer

In narrow and high-contrast waveguides, such as the SOI, a significant fraction of the light is carried in the cladding; it is thus possible to tune the waveguide's effective index through the cladding, by replacing it with an LC. The capacity to vary the refractive index of LC by means of a low power driving voltage is a feature widely exploited in the fabrication of several integrated optic devices.

Ring resonators are resonant optical structures, able to filter narrow wavelength bands from a broad spectrum. The high refractive index contrast available on SOI ring resonators enables low loss and high-Q filters, achieved with radii of a few nm. These structures can be used as notch filter for adding or dropping channels in dense photonic networks. The optical path length of microring resonators can be controlled by either adjusting the physical dimensions (in particular their circumference) or the refractive indices of the constituent materials of the resonator. Dynamic tuning can efficiently be achieved by using a layer of NLC as cladding on an SOI ring resonator and changing its refractive index as a consequence of the orientation of director of the NLC [63]. The resonances of a ring correspond to those frequencies which are integer multiple of 2π upon one round trip in the resonator:

$$\lambda_m = 2\pi R \cdot \frac{n_{\text{eff}}}{m} \tag{10.6}$$

where λ_m is the wavelength of the mth resonator mode, R is the resonator radius and n_{eff} is the effective refractive index of waveguide mode, being the latter dependent on the refractive index of the cladding. To characterize the tuning efficiency of the device, the tuning range of resonance wavelength can be defined as

$$\Delta\lambda_{\text{Res}} = \frac{\lambda_m \cdot \Delta n_{\text{eff}}}{n_{\text{eff}}} \tag{10.7}$$

which describes the dynamic tuning range, and Δn_{eff} is the corresponding maximum change value of the effective refractive index.

By electrically modulating the refractive index, Maune *et al.* have demonstrated a resonance shift of about 0.22 nm (27.5 GHz at 1.55 μm) by applying 20 V [64]. Moreover, a tuning of 31 nm has been demonstrated for ring resonators guiding the TM mode, almost covering the entire C-band of optical communications [65]. Recent optimizations of the resonator's geometry led to an ultra-wide tuning range of 56 nm with a driving voltage of only 5 V [66].

To obtain a phase shift, a promising research stream proposes a silicon-organic hybrid approach, combining waveguides with organic LC cladding: it supersedes certain limitations intrinsic to the traditional SOI approach. For example, adopting the so-called strip-loaded slot waveguide, a structure in which two parallel high index SOI rails are spaced by a narrow slot and are immersed in LC cladding, a large phase shift, up to 35π, can be achieved with short devices (1.7 mm) and little driving voltage (5 V) [63]. Using LC cladding helps also power dissipation, which is about six orders of magnitude smaller than that of state-of-art thermo-optic devices, allowing its integration in high-density photonic integrated circuits.

A broadband LC cladding switch integrated with polymer waveguides on PCB seems an interesting alternative to high-speed multi-gigabit electrical connections. In [67], the development of such a device is reported, which provides a switching contrast of 15 dB with a low on-state excess loss of approximately 0.5 dB. The switch is controlled directly by the PCB electrical layer.

10.6 Conclusions

A large variety of integrated optic devices can be made by using LC, either as core material to confine light beams or as overlayer to control the devices made on the substrate. In this chapter, examples of optical switches with high contrast over 40 dB and of Bragg gratings to obtain optical filters with a wide tuning range over 100 nm have been shown. Both electrical and optical control can be implemented on active devices, based on either the effective electro-optical or the non-linear optical effects.

One of the main attracting features of LC-based integrated optic devices is that a low power, either electrical or optical, is required to drive photonic devices independently of the employed material platforms and the corresponding technologies. It has also been shown that LC can be processed on several kinds of substrate materials, including silicon, glass, silicon nitride and polymers like PDMS. This opportunity allows to envisage the possibility to make low-cost hybrid integrated optic devices combined with electronic circuits fabricated on the same chip for effective optical interconnections and sensing applications.

References

[1] Kawamoto H. 'History of liquid-crystal displays'. *Proc. IEEE* 2002, vol. 90(4), pp. 460–500.

[2] Heilmeier G.H. and Zanoni L.A. 'Guest–host interactions in nematic liquid crystals. A new electro-optic effect'. *Appl. Phys. Lett.* 1968, vol. 13(3), pp. 91–2.

[3] Heilmeier G.H., Zanoni L.A., and Barton L.A. 'Dynamic scattering: a new electrooptic effect in certain classes of nematic liquid crystals'. *Proc. IEEE* 1968, vol. 56, pp. 1162–71.
[4] Heilmeier G.H. 'Liquid crystal displays: an experiment in interdisciplinary research that worked'. *IEEE Trans. Electron Devices* 1976, vol. 23(7), pp. 780–5.
[5] De Gennes P.G. and Prost J. *The physics of liquid crystals*. 2nd ed. Oxford, UK: University Press; 2003.
[6] Khoo I.C. *Liquid crystals*. 2nd ed. Hoboken: Wiley; 2007.
[7] Lebwohl P.A. and Lasher G. 'Nematic-liquid-crystal order—a Monte Carlo calculation'. *Phys. Rev. A* 1972, vol. 6, p. 426.
[8] Pasini P., Chiccoli C., and Zannoni C. *Advances in the computer simulations of liquid crystals*. Dordrecht: Kluwer; 2000.
[9] Bellini B., Larchanché J.-F., Vilcot J.-P., Decoster D., Beccherelli R., and d'Alessandro A. 'Photonic devices based on preferential etching'. *Appl. Opt.* 2005, vol. 44(33), pp. 7181–6.
[10] d'Alessandro A., Bellini B., Donisi D., Beccherelli R., and Asquini R. 'Nematic liquid crystal optical channel waveguides on silicon'. *IEEE J. Quant. Electron.* 2006, vol. 42(10), pp. 1084–90.
[11] Donisi D., Bellini B., Beccherelli R., *et al.* 'A switchable liquid-crystal optical channel waveguide on silicon'. *IEEE J Quantum Electron.* 2010, vol. 46(5), pp. 762–8.
[12] Chen C.-H., Lee C.-H., and Lin T.-H. 'Loss-reduced photonic liquid-crystal fiber by using photoalignment method'. *Appl. Opt.* 2010, vol. 49, pp. 4846–50.
[13] d'Alessandro A., Asquini R., Trotta M., Gilardi G., Beccherelli R., and Khoo I.C. 'All-optical intensity modulation of near infrared light in a liquid crystal channel waveguide'. *Appl. Phys. Lett.* 2010, vol. 97(9), p. 093302.
[14] Duffy D.C., McDonald J.C., Schueller O.J.A., and Whitesides G.M. 'Rapid prototyping of microfluidic systems in poly(dimethylsiloxane)'. *Anal. Chem.* 1998, vol. 70, pp. 4974–84.
[15] McDonald J.C., Duffy D.C., Anderson J.R., *et al.* 'Fabrication of microfluidic systems in poly(dimethylsiloxane)'. *Electrophoresis* 2000, vol. 21, pp. 27–40.
[16] Chang-Yen D.A., Eich R.K., and Gale B.K. 'A monolithic PDMS waveguide system fabricated using soft-lithography techniques'. *J. Lightwave Technol.* 2005, vol. 23(6), pp. 2088–93.
[17] d'Alessandro A., Martini L., Gilardi G., Beccherelli R., and Asquini R. 'Polarization independent nematic liquid crystal waveguides for optofluidic applications'. *IEEE Photon. Technol. Lett.* 2015, vol. 27(16), pp. 1709–12.
[18] Asquini R., Chiccoli C., Pasini P., Civita L., and d'Alessandro A. 'Low power photonic devices based on electrically controlled nematic liquid crystals embedded in poly(dimethylsiloxane)'. *Liq. Cryst.* 2018, vol. 45, pp. 2174–83.
[19] Yariv A. and Nakamura M. 'Periodic structures for integrated optics'. *IEEE J. Quantum Electron.* 1977, vol. 13(4), pp. 233–53.

[20] Sutherland R.L., Natarajan L.V., Tondiglia V.P., and Bunning T.J. 'Bragg gratings in an acrylate polymer consisting of periodic polymer-dispersed liquid crystal planes'. *Chem. Mater.* 1993, vol. 5, pp. 1533–8.

[21] Sutherland R.L., Tondiglia V.P., Natarajan L.V., Bunning T.J., and Adams W.W. 'Electrically switchable volume gratings in polymer-dispersed liquid crystals'. *Appl. Phys. Lett.* 1994, vol. 64, pp. 1074–6.

[22] Bowley C.C., Kossyrev P.A., Crawford G.P., and Faris S. 'Variable-wavelength switchable Bragg gratings formed in polymer-dispersed liquid crystals'. *Appl. Phys. Lett.* 2001, vol. 79(1), pp. 9–11.

[23] Sirleto L., Coppola G., Breglio G., Abbate G., Righini G.C., and Oton M.J. 'Electro-optical switch and continuously tunable filter based on a Bragg grating in a planar waveguide with a liquid crystal overlayer'. *Opt. Eng.* 2002, vol. 41, pp. 2890–8.

[24] Qi J., Xianyu H., Jianhua L., and Crawford G.P. 'Active U-turn electrooptic switch formed in patterned holographic polymer-dispersed liquid crystals'. *IEEE Photonics Technol. Lett.* 2003, vol. 15(5), pp. 685–7.

[25] Asquini R., d'Alessandro A., Gizzi C., *et al.* 'Optical characterization at wavelengths of 632.8 nm and 1549 nm of POLICRYPS switchable diffraction gratings'. *Mol. Cryst. Liq. Cryst.* 2003, vol. 398, pp. 223–33.

[26] d'Alessandro A., Asquini R., Gizzi C., *et al.* 'Electro-optic properties of switchable gratings made of polymer and nematic liquid crystal slides'. *Opt. Lett.* 2004, vol. 29, pp. 1405–7.

[27] Fujieda I., Mikami O., and Ozawa A. 'Active optical interconnect based on liquid-crystal grating'. *Appl. Opt.* 2005, vol. 42, pp. 1520–5.

[28] Bellini B., Donisi D., Asquini R., and d'Alessandro A. 'A tuneable waveguided optical filter made of polymer and liquid crystal slices operating in C-band: analysis of transmission and reflection properties'. *Mol. Cryst. Liq. Cryst.* 2005, vol. 429, pp. 265–76.

[29] Asquini R., D'Angelo J., and d'Alessandro A. 'A switchable optical add-drop multiplexer using ion-exchange waveguides and a POLICRYPS grating overlayer'. *Mol. Cryst. Liq. Cryst.* 2006, vol. 450, pp. 203–14.

[30] Adikan F.R.M., Gates J.C., Dyadyusha A., *et al.* 'Demonstration of 100 GHz electrically tunable liquid-crystal Bragg gratings for application in dynamic optical networks'. *Opt. Lett.* 2007, vol. 32, pp. 1542–4.

[31] Donisi D., d'Alessandro A., Asquini R., and Beccherelli R. 'Theoretical performance analysis of an integrated optic filter made of glass waveguides and POLICRYPS holographic gratings'. *Mol. Cryst. Liq. Cryst.* 2007, vol. 465, pp. 227–37.

[32] Sutherland R.L., Tondiglia V P., Natarajan L.V., *et al.* 'Liquid crystal Bragg gratings: dynamic optical elements for spatial light modulators'. *Proc. SPIE* 2007, vol. 6487, 64870V.

[33] d'Alessandro A., Donisi D., De Sio L., *et al.* 'Tunable integrated optical filter made of a glass ion-exchanged waveguide and an electro-optic composite holographic grating'. *Opt. Express* 2008, vol. 16(13), pp. 9254–60.

[34] Donisi D., d'Alessandro A., Asquini R., *et al.* 'Realization of an optical filter using POLICRYPS holographic gratings on glass waveguides'. *Mol. Cryst. Liq. Cryst.* 2008, vol. 486, pp. 31–7.

[35] Liu Y.J., Zheng Y.B., Shi J., Huang H., Walker T.R., and Huang T.J. 'Optically switchable gratings based on azo-dye-doped, polymer-dispersed liquid crystals'. *Opt. Lett.* 2009, vol. 34, pp. 2351–3.

[36] Donisi D., Asquini R., d'Alessandro A., and Assanto G. 'Distributed feedback grating in liquid crystal waveguide: a novel approach'. *Opt. Express* 2009, vol. 17, pp. 5251–6.

[37] Gilardi G., Asquini R., d'Alessandro A., and Assanto G. 'Widely tunable electro-optic distributed Bragg reflector in liquid crystal waveguide'. *Opt. Express* 2010, vol. 18(11), pp. 11524–9.

[38] Donisi D., Asquini R., d'Alessandro A., *et al.* 'Integration and characterization of LC/polymer gratings on glass and silicon platform'. *Mol. Cryst. Liq. Cryst.* 2010, vol. 516, pp. 152–8.

[39] Asquini R., Gilardi G., d'Alessandro A., and Assanto G. 'Integrated Bragg reflectors in low index media: enabling strategies for wavelength tunability in electro-optic liquid crystals'. *Opt. Eng.* 2011, vol. 50(7), pp. 071108(1)–(9).

[40] Gilardi G., Asquini R., d'Alessandro A., and Assanto G. 'An electro-optically tunable Bragg reflector based on liquid crystals'. *Mol. Cryst. Liq. Cryst.* 2011, vol. 549, pp. 62–8.

[41] Gilardi G., De Sio L., Beccherelli R., Asquini R., d'Alessandro A., and Umeton C. 'Observation of tuneable optical filtering in photosensitive composite structures containing liquid crystals'. *Opt. Lett.* 2011, vol. 36(24), pp. 4755–7.

[42] Gilardi G., Asquini R., d'Alessandro A., Beccherelli R., De Sio L., and Umeton C. 'All-optical and thermal tuning of a Bragg grating based on photosensitive composite structures containing liquid crystals'. *Mol. Cryst. Liq. Cryst.* 2012, vol. 558(1), pp. 64–71.

[43] Kato A., Nakatsuhara K., and Nakagami T. 'Wavelength tunable operation in Si waveguide grating that has a ferroelectric liquid crystal cladding'. *J. Lightwave Technol.* 2013, vol. 31, pp. 349–54.

[44] Lee Y.-H., Yin K., and Wu S.-T. 'Reflective polarization volume gratings for high efficiency waveguide-coupling augmented reality displays'. *Opt. Express* 2017, vol. 25, pp. 27008–14.

[45] Xiang X., Kim J., and Escuti M.J. 'Bragg polarization gratings for wide angular bandwidth and high efficiency at steep deflection angles'. *Sci. Rep.* 2018, vol. 8, p. 7202.

[46] Weng Y., Zhang Y., Cui J., *et al.* 'Liquid-crystal-based polarization volume grating applied for full-color waveguide displays'. *Opt. Lett.* 2018, vol. 43, pp. 5773–6.

[47] Lee Y.-H., He Z., and Wu S.-T. 'Optical properties of reflective liquid crystal polarization volume gratings'. *J. Opt. Soc. Am. B* 2019, vol. 36, pp. D9–12.

[48] Khoo I.C. and Wu S.T. *Optics and nonlinear optics of liquid crystals*. Singapore: World Scientific; 1993.
[49] Khoo I.C. 'Nonlinear optics of liquid crystalline materials'. *Phys. Rep.* 2009, vol. 471(5–6), pp. 221–67.
[50] Piccardi A., Trotta M., Kwasny M., *et al.* 'Trends and trade-offs in nematicon propagation'. *Appl. Phys. B: Lasers Opt.* 2011, vol. 104(4), pp. 805–11.
[51] Henninot J.F., Debailleul M., Asquini R., d'Alessandro A., and Warenghem M. 'Self-waveguiding in an isotropic channel induced in dye doped nematic liquid crystal and a bent self-waveguide'. *J. Opt. A: Pure Appl. Opt.* 2004, vol. 6(4), pp. 315–23.
[52] d'Alessandro A. and Asquini R. 'Liquid crystal devices for photonic switching applications: state of the art and future developments'. *Mol. Cryst. Liq. Cryst.* 2003, vol. 398, pp. 207–21.
[53] Beeckman J., Neyts K., and Vanbrabant P.J.M. 'Liquid-crystal photonic applications'. *Opt. Eng.* 2011, vol. 50(8), p. 081202.
[54] Zografopoulos D.C., Asquini R., Kriezis E.E., d'Alessandro A., and Beccherelli R. 'Guided-wave liquid-crystal photonics'. *Lab Chip* 2012, vol. 12(19), pp. 3598–610.
[55] Asquini R., Donisi D., Trotta M., *et al.* 'Realization of a liquid crystal electrically controlled optical waveguide on micromachined silicon'. *Mol. Cryst. Liq. Cryst.* 2009, vol. 500, pp. 23–30.
[56] Asquini R., Fratalocchi A., d'Alessandro A., and Assanto G. 'Electro-optic routing in a nematic liquid-crystal waveguide'. *Appl. Opt.* 2005, vol. 44(19), pp. 4136–43.
[57] Fratalocchi A., Asquini R., and Assanto G. 'Integrated electro-optic switch in liquid crystals'. *Opt. Express* 2005, vol. 13, pp. 32–7.
[58] De Sio L., Cuennet J.G., Vasdekis A.E., and Psaltis D. 'All-optical switching in an optofluidic polydimethylsiloxane: liquid crystal grating defined by cast-molding'. *Appl. Phys. Lett.* 2010, vol. 96, 131112.
[59] Cuennet J.G., Vasdekis A.E., De Sio L., and Psaltis D. 'Optofluidic modulator based on peristaltic nematogen microflows'. *Nat. Photonics* 2011, vol. 5, pp. 234–8.
[60] Liu Y., Cheng D., Lin I.H., Abbott N.L., and Jiang H. 'Microfluidic sensing devices employing in situ-formed liquid crystal thin film for detection of biochemical interactions'. *Lab Chip* 2012, vol. 12(19), pp. 3746–53.
[61] Caputo R., De Luca A., De Sio L., *et al.* 'POLCRYPS: a liquid crystal composed nano/microstructure with a wide range of optical and electro-optical applications'. *J. Opt. A: Pure Appl. Opt.* 2009, vol. 11, p. 024017.
[62] Asquini R. and d'Alessandro A. 'Tunable photonic devices based on liquid crystals and composites'. *Proc. SPIE* 2013, vol. 8828, 88280T.
[63] Pfeifle J., Alloatti L., Freude W., Leuthold J., and Koos C. 'Silicon-organic hybrid phase shifter based on a slot waveguide with a liquid-crystal cladding'. *Opt. Express* 2012, vol. 20, pp. 15359–76.

[64] Maune B., Lawson R., Gunn C., Scherer A., and Dalton L. 'Electrically tunable ring resonators incorporating nematic liquid crystals as cladding layers'. *App. Phys. Lett.* 2003, vol. 83, p. 23.
[65] Cort W.D., Beeckman J., Claes T., Neyts K., and Baets R. 'Wide tuning of silicon-on-insulator ring resonators with a liquid crystal cladding'. *Opt. Lett.* 2011, vol. 36, pp. 3876–8.
[66] Dai J., Zhang M., Zhou F., and Liu D. 'Highly efficient tunable optical filter based on liquid crystal micro-ring resonator with large free spectral range'. *Optoelectronics* 2016, vol. 9, pp. 112–20.
[67] Beals J., Hadeler O., Morris S.M., Wilkinson T.D., Penty R.V., and White I. H. 'Electro-optic integration of liquid crystal cladding switch with multi-mode passive polymer waveguides on PCB', in *Conference on Lasers and Electro-Optics/International Quantum Electronics Conference*, OSA Technical Digest (CD) (Optical Society of America, 2009), paper CFV6.

Chapter 11

Silicon nitride integrated optics*

Jonathan D.B. Bradley[1]*, Renjie Wang*[1]*, Henry C. Frankis*[1]*, Dawson B. Bonneville*[1]*, Khadijeh Miarabbas Kiani*[1] *and Hamidu M. Mbonde*[1]

Silicon nitride has emerged as a highly versatile integrated optics material. Its ultra-low losses and mature processing methods provide advantages for passive, active and non-linear optical devices and microsystems operating in the visible, near-infrared (NIR) and mid-infrared (MIR) wavelength ranges. This chapter reviews progress in silicon nitride waveguide technology, devices and photonic integrated circuits (PICs) and recent efforts to build novel materials and devices onto silicon nitride platforms.

11.1 Introduction

Integrated optics involves the development of optical materials, waveguides and devices on a small chip in order to build compact and mass-producible photonic microsystems, analogous to microelectronic circuits. Silicon nitride (SiN) has long been of interest as an integrated optics material because of its complementary metal–oxide–semiconductor (CMOS) compatibility, thus maturity through extensive parallel development as a dielectric layer for integrated silicon electronics, and excellent optical properties, including its wide transparency from the visible to MIR. This interest is evidenced by early pioneering developments on silicon nitride and silicon oxynitride waveguides by numerous commercial and academic research groups in the 1970s, 1980s and 1990s [1–7]. Over time, the distinct advantages of silicon nitride have emerged, which can be conveniently divided into its fabrication and optical properties. Its fabrication advantages include low cost, SiN thin-film deposition and waveguide etching that are carried out using mature, industry-standard wafer-scale methods allowing for mass-producible PICs, and the patterning methods employed (deep-ultraviolet stepper and electron-beam lithography) enable high-resolution, high-quality nanoscale features. Critically, SiN waveguide

*All authors contributed equally to this chapter.

[1]Department of Engineering Physics, McMaster University, Hamilton, Canada

fabrication steps are also fully compatible with state-of-the-art silicon photonics platforms, which have matured in parallel over the last two decades, providing access to all the active functionalities (high-speed modulation, switching and detection) available on those platforms. Silicon nitride's optical advantages include its ultra-low losses, low thermal sensitivity, high power handling, high non-linearity, low non-linear absorption enabling efficient Kerr non-linear devices, moderately high refractive index allowing for highly compact devices and wide transparency from visible to the MIR enabling access to traditional tele-communications applications and specialized applications such as biosensing. In the development of SiN technology, scientists and engineers have faced persistent challenges, including material considerations like hydrogen-related absorption and stress in thicker layers and a lack of active functionalities (switching, modulation, detection, amplification and light emission). In recent years, however, these challenges have gradually been addressed as new fabrication methods, and materials and devices have been introduced onto silicon nitride platforms. Over the last two decades, silicon nitride technology has moved steadily from a research activity to a robust, mature platform for commercial PICs and their many current and emerging applications in communications, sensing, ranging, metrology and information processing. As such, in recent years, several excellent surveys on silicon nitride photonics have emerged, including, from the perspective of SiN platforms, PICs and applications [8–10], non-linear optics [11,12] and silicon photonics [13–16]. Silicon nitride technology is well established, with many foundry services now available for researchers and industry to leverage for exploration of fundamental science or development of commercial devices, and its importance can only be expected to grow in the years to come. In this chapter, we review silicon nitride waveguide technology, devices, PICs and applications, with particular emphasis on recent efforts to build new materials, devices and functionalities onto this flexible integrated optics platform.

11.2 Silicon nitride waveguide technology

Silicon nitride technology is deceptively simple, but the material properties can vary widely, and different types and platforms optimized for varying passive, active, and non-linear applications have emerged. Low-loss wafer-scale SiN platforms have been developed with different properties in terms of thin-film material development, waveguide type, substrate type and photonic integration capabilities, complementing the capabilities of silicon-on-insulator (SOI) and III–V technology. SiN technology makes extensive applications possible with its high design flexibility and transparency over a broad spectral range, and various photonic device building blocks with small footprints have been realized. In this section, we briefly review silicon nitride thin films and waveguides and important considerations in their fabrication and design, as well as some of the platforms that have emerged, which researchers and industry can now access and make SiN technology highly accessible.

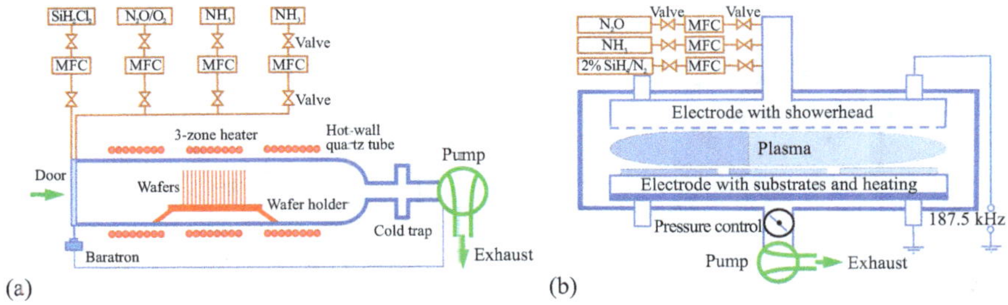

Figure 11.1 *(a) Schematic of an LPCVD set-up for silicon nitride and oxynitride growth. (b) Schematic diagram of a parallel plate PECVD reactor for silicon nitride and oxynitride growth. The diameter of the electrodes is 300 mm [27]. (a) © 1997 SPIE. Reproduced, with permission, from Reference [18]. (b) © 1999 IEEE. Reproduced, with permission, from Reference [19]*

11.2.1 Silicon nitride thin films

Dielectric SiN thin films have long received attention for PICs, because of their high transparency over a wide wavelength range (0.4–6.7 μm) [8], flexible composition and dimension control, relatively high refractive index and good interface properties [2,17]. Moreover, SiN technology is compatible with standard silicon integrated circuit processing or CMOS processing [13–16], and it enables cost-efficient packaging [8].

The achievement of high-quality and suitable SiN thin films on thermally oxidized Si or SOI wafers [7] lays a solid foundation for the development of SiN technology and its use in extensive photonic applications. While many deposition techniques have been explored, low-pressure chemical vapour deposition (LPCVD) and plasma-enhanced chemical vapour deposition (PECVD) are the two most popular methods used to deposit SiN films within state-of-the-art SiN platforms. Schematics of LPCVD [18] and PECVD set-ups [19] from literature for SiN and silicon oxynitride (SiON) film growth are presented in Figure 11.1(a) and (b), respectively.

Because of the precursors used in CVD processes, SiN-based films suffer from the incorporation of hydrogen content [20,21], including N–H and Si–H bonds, which have been investigated using Fourier-transform infrared spectroscopy (FTIR) studies [22]. The absorption due to the vibrational overtones of the hydrogen bonds, especially the N–H bonds, leads to optical loss in the telecommunications spectrum around 1,460–1,590 nm [18]. Figure 11.2(a) summarizes the loss of LPCVD and PECVD SiN thin films in the wavelength range from 1,500 to 1,600 nm. The loss measurements of the thin films were conducted using the prism coupling method. LPCVD SiN thin films have the advantages of low H content and low optical loss compared to PECVD films which are generally deposited at low temperature (200°C–500°C) but with higher H content [23,24]. As shown in

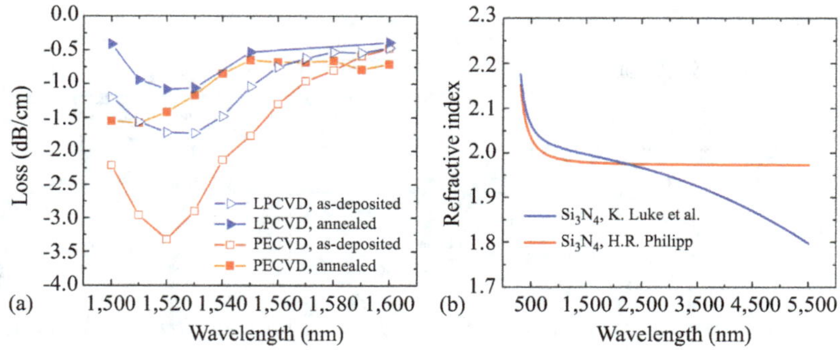

Figure 11.2 (a) Measured optical loss of LPCVD and PECVD silicon nitride thin films vs. wavelength. As-deposited silicon nitride thin films are annealed at high temperature (~1,050 °C) to reduce Si–H and N–H absorption. (b) Refractive index of silicon nitride extrapolated from the Sellmeier equations of Luke et al. [35] and Philipp [1]

Figure 11.2(a), the lower loss at the wavelength around 1,520 nm for both annealed LPCVD and PECVD films indicates that the annealing process at 1,050 °C can largely reduce the H content. It is worthwhile mentioning that the reduction in loss is more dramatic in the PECVD film compared to that in the LPCVD film, due to higher H content in the as-deposited PECVD film. However, the high deposition temperature of LPCVD SiN-based thin films (700 °C–800 °C) [2,25] limits their inclusion in active silicon photonics platforms via back-end-of-line (BEOL) or post-processing steps, which require temperatures <~400 °C. In addition, LPCVD SiN films also encounter the high tensile stress problem in thicker (>~400 nm) films [2,25]. Efforts have been made towards reducing the residual stress in thick SiN films, such as optimizing the precursor gas ratio [26]. A decrease of the H content to less than 1 at.% is desirable to achieve low-loss SiN films [27]. In addition to the efforts to further improving LPCVD films, techniques and methods are being developed to reduce the H content in the PECVD films considering its advantage of low deposition temperature. Besides the approach of post-PECVD heat treatment to reduce loss [28,29], the suppression of H incorporation can be achieved through PECVD chemistries optimization by implementing N_2 instead of NH_3 [20,30] or deuterated silane (SiD_4) instead of conventional silane (SiH_4) [31,32], or using electron cyclotron resonance PECVD for its low-pressure advantage [33]. Losses in the visible and in the NIR range below the silicon absorption edge (~1.1 µm) are significantly less dependent on H content and largely affected by scattering on small material and surface imperfections [27] and by the silicon content in non-stoichiometric SiN films [12].

Figure 11.2(b) shows the refractive index of stoichiometric Si_3N_4 extrapolated from the Sellmeier equation. By using Sellmeier coefficients derived from the ellipsometry measurements in the NIR range, the optical properties of Si_3N_4 have been successfully characterized by Philipp [1] and Bååk [34] in the visible and NIR range. Recently, Luke *et al.* [35] derived new Sellmeier coefficients based on ellipsometry measurements over the 193–1,690 nm and 1.7–33 μm wavelength ranges. As shown in Figure 11.2(b), the Sellmeier coefficients extrapolated from NIR and MIR data [35] are significantly different from the ones extrapolated from only NIR data [1] which exclude the influence of MIR absorption of Si_3N_4.

The flexible and widely accessible composition of SiN-based materials is one attractive advantage of SiN technology for its extensive integration applications. Besides stoichiometric Si_3N_4 thin films, SiON, silicon-rich SiN thin films and doped SiN have attracted intense interest and research efforts.

SiON thin films enable large flexibility in choosing the refractive index between SiO_2 (1.46)–Si_3N_4 (2.0)–amorphous silicon (theoretically up to 3.7) [12]. However, a suitable deposition technique should be chosen to obtain the desirable index of SiON thin films between low-index-contrast and high-index-contrast systems. Earlier studies showed that accurate control and good uniformity can be achieved by using PECVD for a refractive index <1.7 [7] or using LPCVD for a refractive index >1.7 [7,18]. However, more recently, it has been shown in many studies that inductively coupled plasma (ICP)-CVD and PECVD deposition can be used to obtain high-quality and high-refractive-index stoichiometric and silicon-rich SiN_x films [12]. The deposition rate of PECVD SiON, which depends on the flow ratio N_2O/SiH_4, is reported to be in the range of 20–50 nm/min [7]. In contrast to that, the deposition rate of LPCVD SiON, which varies with O_2 flow rate, is reported to be in the range of 2–12 nm/min [7]. SiON and SiN thin films can also be obtained by using reactive sputtering deposition in an oxygen and nitrogen ambient [36,37].

Bandgap-engineered silicon-rich SiN enables dispersion engineering of SiN technology for non-linear applications, such as optical pulse compression [38] and supercontinuum generation (SCG) [39]. In contrast to silicon, which exhibits strong two-photon absorption (TPA) in the telecom band around 1.5 μm [40], silicon-rich SiN exhibits negligible TPA [41] while having a high refractive index for highly compact devices [42,43] with tighter modal confinement compared to stoichiometric Si_3N_4 [12,44]. The silicon/nitrogen ratio can be tailored using ICP-CVD [42], PECVD [45] and LPCVD [46], and it is desirable to avoid the use of NH_3 to reduce the loss [20,47].

The introduction of dopant ions into SiN films makes stimulated emission possible in compact on-chip devices [48–50], such as high-*Q* whispering gallery microdiscs [50]. Si nanocrystals [51] can be used as efficient sensitizers for rare-earth ions in SiN films, and emission efficiency is found to be highly dependent on annealing temperature in SiN films with varying Si content [52]. The typical deposition method of doped SiN films is reactive co-sputtering deposition [50,51,53,54]. The lifetime of erbium (Er) in SiON is reported to be 3.8 ms and depends on the concentration of silicon dangling bonds [53]. The energy-transfer time from SiN to neodymium (Nd) ions is reported in the range of 1–10 ns and photoluminescence (PL) lifetime in the range of 40–70 μs is affected by the post-annealing temperature [49,50]. The decay times of terbium (Tb) ions in SiON are reported in the range of ~0.2–0.8 ms and are affected by nitrogen flow and post-annealing temperature during their deposition [54]. Although there has been progress in developing rare-earth-doped SiN materials, rare-earth solubilities and emission efficiencies remain low relative to other rare-earth-doped oxide-based materials. This has motivated the development of other materials (including rare-earth-doped Al_2O_3 and TeO_2, discussed in other sections in this chapter) into the silicon nitride platform for active devices such as amplifiers and lasers.

11.2.2 Silicon nitride optical waveguides

The most basic component of a PIC is the waveguide, the integrated optical equivalent of an optical fibre, used to confine and guide light across the circuit. Silicon nitride waveguides are typically fabricated by the deposition of silicon nitride films onto thermally oxidized silicon wafers with a 6–15 μm thick SiO_2 lower cladding layer. SiN layer deposition is often followed by a chemical mechanical polishing step to smooth surface roughness and reduce optical loss. Reactive ion etching (RIE) is then used to define a ridge or strip waveguide structure using a photoresist or silicon dioxide hard mask. It has been shown that etching in CHF_3 and O_2 gas mixtures leads to smooth, vertical Si_3N_4 waveguide sidewalls [55]. Additional etch gases which are used include CF_4, CH_2F_2, N_2 and Ar [55,56]. Hydrogen is known to play an important role in Si_3N_4 etch processes [57,58]. For example, the hydrofluorocarbon CHF_3 can provide higher Si_3N_4 etch rates, higher selectivity to SiO_2 and more anisotropic profiles than etching in fluorocarbons such as CF_4 [56–58]. The formation and removal of polymer residue in hydrofluorocarbon chemistries on the etched surface allows for anisotropic etch profiles but can also increase surface roughness and requires systematic optimization of the RIE process [55]. O_2 is typically added for removal of polymer residue and improving etch selectivity to the SiO_2 substrate and/or SiO_2 hard mask [55]. Careful tuning of the various process gases and RIE parameters can lead to good selectivity over SiO_2 or photoresist masks [55,56]. After patterning, a PECVD or LPCVD oxide is typically deposited to act as a top-cladding layer for the waveguides.

Figure 11.3 shows different types of silicon nitride waveguides. Silicon nitride's wide range of applications with different requirements and versatile fabrication methods has led to the development of a variety of waveguide geometries. The three most common waveguide geometries can be classified based on low, moderate or high confinement of light within the Si_3N_4 waveguide core, as shown in Figure 11.3(a), (b) and (c), respectively.

Low-confinement waveguides use a thin waveguide core (~40–100 nm) with wide waveguide widths (~3–6 μm). The weak confinement causes the waveguide mode area to expand into the cladding, resulting in low effective indices and small mode overlap with the etched sidewalls of the waveguide, thus avoiding scattering due to etching induced roughness, a significant source of loss for most waveguides. Limiting the sidewall interaction allows low-confinement silicon nitride waveguides to achieve very low waveguide loss, with losses of <0.1 dB/m having been reported [59]. The minimum achievable loss then becomes highly sensitive to the

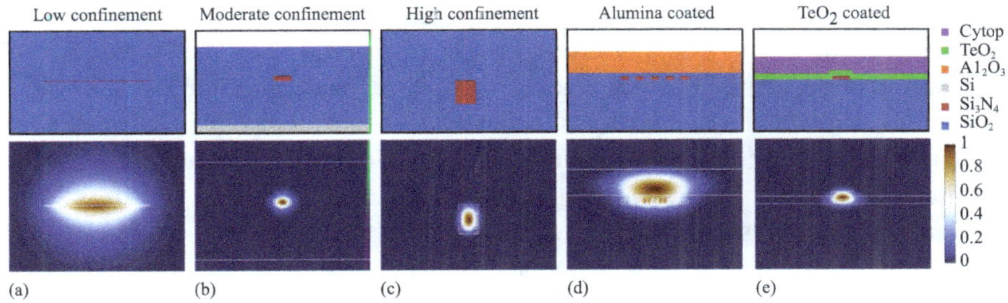

Figure 11.3 Structure (top) and mode profile (bottom) of (a) low-confinement, (b) moderate-confinement and (c) high-confinement silicon nitride waveguides, and hybrid (d) alumina- and (e) TeO_2-coated silicon nitride waveguides for new functionalities, including rare-earth-doped optical amplifiers and lasers on a silicon nitride platform

oxide top-cladding. Waveguides are grown on a thermally oxidized silicon dioxide layer and then top cladded up to 3 μm in a tetraethyl orthosilicate LPCVD process. Further cladding in a PECVD oxide gives waveguides with a loss of ~0.3 dB/m. To achieve <0.1 dB/m loss, the chip must be top cladded by flip chip bonding of a second thermally oxidized wafer, which has lower losses than PECVD oxide. The primary trade-off in such waveguides is a large minimum bend radius, ranging from 1 to several mm at 1,550 nm, leading to large device footprints.

High-confinement waveguides are a more recently developed platform, geared towards non-linear device applications of silicon nitride [60–62]. The large waveguide core leads to high overlap with the silicon nitride, which has higher Kerr non-linearity than the cladding, low or anomalous dispersion and ultra-low losses due to minimal interaction of the optical mode with the waveguide sidewall. The trade-offs of such a waveguide structure are multimode operation and greater fabrication complexity. A long-standing challenge in the fabrication of high-confinement waveguides was the stress-cracking of silicon nitride that occurs for films grown above ~400 nm. Development of the 'photonic damascene process', where silicon nitride films are grown into etched silicon dioxide layers containing stress release patterns, has enabled the ability to grow high-quality silicon nitride layers to thicknesses of greater than 1 μm without stress-cracking. While low and high-confinement waveguide platforms are specifically tailored towards low propagation losses and high non-linearity, respectively, the moderate-confinement waveguides are designed to offer a compromise in optimization for increased design flexibility [9]. Moderate-confinement waveguides use the standard patterning and etching process applied to silicon nitride thin films on the order of ~200–400 μm in thickness. Moderate-confinement waveguides can have broadband, single-mode wavelength operation from the visible to the MIR, and offer applications in sensing, non-linear optics and telecommunications. The moderate and high-confinement designs have mode sizes and profiles determined by the SiN waveguide core size, as can be seen in the mode profiles of Figure 11.3(b) and (c). More complex waveguide geometries are also offered through specific foundries, such as the box-shell and double-stripe TriPleX geometries developed by LioniX [63]. Such waveguides offer distinct advantages, such as mode-matching to fibre modes for low-loss fibre-chip coupling and polarization insensitivity by compensating for the material birefringence with the waveguide structure.

Using post-processing steps after SiN chip fabrication has enabled novel materials and waveguide structures and additional devices and functionalities to be incorporated into silicon nitride platforms. An example of a hybrid waveguide structure is the rare-earth-doped alumina or aluminium oxide (Al_2O_3)-coated silicon nitride waveguides developed for optically pumped monolithic lasers by research groups at the Massachusetts Institute of Technology (MIT) and University of California, Santa Barbara (UCSB) [64,65]. In such Al_2O_3–Si_3N_4 waveguides, the silicon nitride layer is composed of a thin strip or multiple pieces separated by small gaps and is separated from the alumina layer by a thin PECVD SiO_2 layer [64,65] in order to create a wavelength insensitive waveguide design [66]. An ~1-μm-thick rare-earth-doped aluminium oxide layer is blanket deposited over the entire chip outside the silicon nitride foundry in a single post-processing step. The silicon nitride and aluminium oxide layers form a composite waveguide structure with a shared optical mode, guided by the patterning in the silicon nitride layer. An example of the waveguide design and optical mode profile can be found in Figure 11.3(d). A large percentage of the optical intensity overlaps with the doped alumina layer, a requirement for achieving optical gain and lasing. This design has been used to demonstrate rare-earth-doped waveguide lasers by the addition of distributed Bragg reflector (DBR) and distributed feedback (DFB) cavities in the silicon nitride layer. The required grating feature sizes on the order of ~200 nm are enabled by the high-resolution Si_3N_4 patterning possible via stepper or immersion lithography, building on earlier laser interference lithography methods to define grating cavity features in the SiO_2 top-cladding [64–67]. A simpler approach for additive processing of novel materials onto silicon nitride waveguides can be achieved by conformally coating layers directly onto exposed silicon nitride strip waveguides. This approach has recently been explored in TeO_2-coated Si_3N_4 waveguides [68–72] and is motivated by the higher refractive index of the TeO_2 vs. Al_2O_3 (2.08 vs. 1.65 at 1,550 nm), which tends to pull the optical mode up into the TeO_2 layer. As shown in Figure 11.3(e), the mode is highly confined in the high-refractive-index composite Si_3N_3–TeO_2 waveguide core, allowing for small bends and highly compact and efficient optically pumped devices. High-Q microring resonators (MRRs) [68] and waveguides with low propagation losses [69] have been demonstrated in such Si_3N_4 strip waveguides coated in tellurium oxide layers. It has been shown that by doping with rare-earth ions such as erbium and thulium, compact optical amplifiers [70,71] can be realized on a silicon nitride chip. Such waveguides are also highly promising for non-linear device applications, on account of the higher non-linearity of TeO_2 compared to Si_3N_4 [72]. These two examples illustrate the flexibility afforded by silicon nitride platforms and future potential for developing composite waveguide structures with widely varying structural and optical properties.

11.2.3 Photonic integration platforms

In this section, the state of the art in SiN photonic integration platforms is reviewed. The extensive applications now associated with SiN waveguides have been enabled by numerous high-quality fabrication processes developed in either academic labs or foundries or both. Multi-project wafer (MPW) runs, which allow for the chip design space on lithography masks and fabrication expenses to be shared among many users, have significantly increased access and reduced the cost of SiN technology for academic researchers and industry. The state of the art in recent silicon nitride platforms and their properties are summarized in Table 11.1. The waveguide types are distinguished based on the categorization proposed in [8], including single stripe or strip (including low and moderate-confinement designs), double stripe, multisegmented, high confinement (including etched, filled box and damascene versions) and box shell. SiN waveguide technology generally uses thermally oxidized silicon wafers or SOI wafers as substrates and SiO_2 (coated or bonded) as the upper cladding. Table 11.1 also summarizes the applied lithography techniques and waveguide propagation loss of different SiN platforms based on the published literatures. The table is not meant to be an exhaustive list but is instead meant to provide a general overview of platforms available and variations in their properties to highlight the accessibility and diversity of SiN technology while pointing the reader to references for further details.

Several research labs and companies have worked significantly on developing low-loss SiN waveguides for passive devices and other applications. Current platforms build on comprehensive earlier works on thin films, waveguides and devices, including works at the University of Twente [7], MIT [73] and the University of Trento [74]. LioniX International and its acquired companies, based in the Netherlands, have been developing SiN technology and applications since 2001. The TriPleX platform of SiN technology from LioniX [8,10,61,63], developed in part in collaboration with researchers at the University of Twente and Politecnico di Milano, has focused on PICs for tele- and datacom, life science and metrology, operating from the visible to the MIR (405–2,350 nm). Almost all the waveguide geometries are available in TriPleX platform. The waveguide geometries can be selected in the layout design varying for the specific applications. Ultra-low propagation losses of 0.007 dB/cm (single mode) and 0.0005 dB/cm (multimode) have been achieved by using single-stripe geometry [8]. To maintain efficient fibre-to-chip coupling with various waveguide geometries for the broad wavelength operation, monolithic spot-size converter technology has also been developed.

A wide range of applications have been realized or achieved by using TriPleX technology, including telecommunications, microwave photonics, high-performance filters, optical beamforming networks, miniaturized biomedical devices and hybrid light sources [8]. The propagation losses of Si_3N_4 planar waveguides (single stripe) have been systematically investigated by LioniX and

Table 11.1 State of the art of silicon nitride photonic integration platforms[a]

Foundry (Lab(s))	Substrate (SiO_2 thickness)	Silicon nitride (SiN thickness)	Waveguide type(s)	Lithography method(s)	Propagation loss (dB/cm)	Applications/emphasis
A*STAR[b] (SUTD) [12]	Si/SiO_2[c]	ICP-CVD Si-rich SiN_x (780 nm)	Single stripe	Deep-ultraviolet (DUV)	4.6@1,550 nm	Non-linear optics and others
A*STAR and AMF[b] (University of Toronto and Huawei) [16,87,88]	200-mm SOI[d]	PECVD and LPCVD Si_3N_4 or SiN_x	SiN/Si,[e] single stripe, double stripe	DUV	0.8@1,580 nm, 0.24@1,270 nm	Optical switching, routing, microwave photonics and others
AIM Photonics Academy[b] (MIT) [64,66,85,86]	300-mm Si/SiO_2[c] and SOI[d]	PECVD and LPCVD SiN_x (100–200 nm)	SiN/Si,[e] single stripe, multisegmented, double stripe	193 nm immersion	0.4@1,550 nm	Biphotonic applications, WDM, optical computing, quantum communications, sensing, monolithic lasers and others
AMO [77]	150-mm Si/SiO_2[c] (6 μm)	LPCVD Si_3N_4 (50–500 nm)	Single stripe	i-line stepper	~0.1@1,550 nm	Radar, sensors, lasers and others
Applied Nanotools[b] [78]	Si/SiO_2[c]	LPCVD Si_3N_4	Single stripe (~400 nm)	Electron beam (e-beam)		Prototyping, various
CEA-LETI[b] (Université Grenoble Alpes, Ecole Centrale de Lyon and DTU) [89,90]	200-mm SOI[d]	PECVD SiN_x (600 nm) SiN_x or LPCVD Si_3N_4 (370 nm)	SiN/Si,[e] asymmetric double stripe or double rib	DUV	0.8@1,310 nm, 1.0–1.5@940 nm	LIDAR, CWDM transceivers, fibre grating couplers, frequency comb generation and others
Cornell University and Columbia University [11,35,60]	Si/SiO_2[c]	PECVD and LPCVD Si_3N_4	High confinement (etched)	E-beam	0.04@1,550 nm	Non-linear photonics, frequency comb generation, communications and others

(Continues)

Georgia Tech [76,92]	Si/SiO_2[c] (5 or 6 μm)	LPCVD Si_3N_4 (~200 or 400 nm)	Si_3N_4 single stripe on 50-nm pedestal layer or bonded upper SiO_2/SOI structure	E-beam	$Q=3.4\times10^6$@652–660 nm; 0.02@1,550 nm for Si/ Si_3N_4 coupler	Sensing and optical data processing applications in the visible range or optical information processing, RF photonics, optical communication and others
HKUST [82,93]	4-in. Si/SiO_2[c] (~2 μm)	PECVD N-rich SiN_x (~300 nm) or PECVD Si_3N_4 (780 nm)	Single stripe and high confinement (etched)	Contact and e-beam	~17@680 nm or $Q=1.7\times10^6$	Optical biochemical sensing, non-linear optics, quantum optics and others
IMB-CNM, CSIC[b] (UPV and VLC Photonics) [9]	100-mm Si/SiO_2[c] (2.5 μm)	LPCVD Si_3N_4	Strip or rib	Stepper	1.4@1,550 nm	Biophotonics, tele/datacom and sensing applications from VIS to NIR
imec[b] (Ghent University) [13,84]	200-mm Si/HDP-CVD SiO_2 (2 or 2.4 μm)	PECVD Si_3N_4 (230 nm)	Stripe, SiO_2 rib coated with Si_3N_4	DUV	0.65@532 nm, 1.33@700 nm, 0.62@900 nm	Evanescent field-based sensing, visible-VNIR application and others
LIGENTEC[b] (EPFL) [62,79]	Si/SiO_2[c] (0.13–3.5 or 4 μm)	LPCVD Si_3N_4 (up to 1.52 μm)	High confinement (damascene)	Contact and e-beam	0.055@MIR wavelengths	LIDAR, quantum, biomedical and non-linear photonics and others
LioniX[b] (University of Twente, Politecnico di Milano) [8,10,61,63]	100-mm Si/SiO_2[c] (8 μm)	LPCVD Si_3N_4	TriPleX: box shell, stripes, and high confinement (filled box)	Contact, e-beam and stepper	0.007@1.58 μm	Communications, microwave photonics, sensors, visible light applications and others
LioniX[b] (UCSB) [59,75]	100-mm Si/SiO_2[c] (8 μm)	LPCVD Si_3N_4	Single stripe	Contact	0.02–0.09@1,550 nm	Ring resonators, optical reflectometry, monolithic lasers and others
MIT [73,94]	Si/SiO_2[c] (4 μm)	LPCVD, Si-rich SiN_x	Single stripe	Contact and e-beam	2.1@3.7 μm	Communications, filters, MIR light generation, biochemical sensors and others

(Continues)

Table 11.1 (*Continued*)

Foundry (Lab(s))	Substrate (SiO_2 thickness)	Silicon nitride (SiN thickness)	Waveguide type(s)	Lithography method(s)	Propagation loss (dB/cm)	Applications/emphasis
Sandia Labs[b] [80,81,91]	Si/SiO_2[c] (5 μm) and SOI[d]	LPCVD Si_3N_4 (200 nm)	SiN/Si,[e] Single stripe	DUV	0.11–1.45@1,550 nm	Chemical sensing, communications and others
University of Southampton[b] [30,45]	150-mm Si/SiO_2[c] (2 μm)	PECVD Si-rich SiN_x (300 nm)	Single stripe	E-beam	~1.5@1,500 nm	Non-linear signal processing applications and others
STMicro-Electronics[b] (Université de Sherbrooke and Lyon) [83]	300-mm SOI[d]	(600 nm) SiN_x	SiN/Si,[e] single stripe	Standard		Low-cost biosensing and others
University of Trento and FBK [74,95]	4-in. Si/boro phosphosilicate glass (2 μm) or 6-in. Si/SiO_2[c] (5 μm)	LPCVD Si_3N_4 (200 nm) or PECVD SiON (600 nm)	Single stripe	Contact	0.1@632.8 nm or 2@1,550 nm	Non-linear, quantum optical integrated photonic circuits and others
UCSB [65,96]	100-mm Si/SiO_2[c] (15 μm)	LPCVD Si_3N_4 (40–175 nm)	Single stripe	248-nm stepper	<0.25@1,530–1,600 nm or 0.002–0.005@1,550 nm	Optical communications, sensing, navigation and others

[a]Partially adapted from the comprehensive table provided in [9].
[b]Multi-project wafer (MPW) runs offered.
[c]Thermally oxidized Si.
[d]Silicon-on-insulator (SOI).
[e]SiN-on-SOI integrated photonic platform with active functionality.

researchers at UCSB [59,75], and low-loss single-stripe waveguides with a lower residual stress have been realized by introducing a wafer-bonding process for the upper cladding. Such a wafer-bonding process is faster than the PECVD process of upper cladding, and it introduces less stress during manufacturing. Also of note, is the low-loss platform developed by researchers at Georgia Tech, particularly for visible photonic devices, where it is challenging to achieve low losses because of the stronger influence of scattering due to etched waveguide sidewalls [76]. A moderate-confinement SiN platform has recently been developed by Instituto de Microelectrónica de Barcelona (IMB) in Spain, in collaboration with Universitat Politècnica de València (UPV) and VLC Photonics, using a systematic approach [9]. AMO, based in Germany, has also recently developed a flexible LPCVD nitride platform with layer thickness ranging from 50 to 500 nm and metal heaters for switches [77]. Electron-beam-lithography-based MPW services are also emerging from companies such as Applied Nanotools in Canada [78], who are currently developing a silicon nitride platform, and will allow access to smaller feature sizes and new device designs.

A variety of SiN platforms have been developed for specific applications, including non-linear optical devices and quantum optics (which require thicker SiN waveguides) and sensing (which requires low-loss low or moderate-confinement visible and NIR waveguides). Reliable fabrication of thick and stoichiometric Si_3N_4 waveguides with high stress have recently been achieved in the École Polytechnique Fédérale de Lausanne (EPFL) and LIGENTEC [62,79] taking advantage of a novel photonic damascene fabrication process. Microresonators (1.35 μm thick and with Q factors of 3.7×10^6) have been achieved with a close to unity sample yield [62]. Another approach has also been reported to overcome the stress limitations of the Si_3N_4 film by using mechanical trenches to isolate photonic devices from propagating cracks [60]. Ring resonators (910 nm thick and with Q factors of 7×10^6) have been developed by researchers at Cornell University [11,35,60] for non-linear applications. High-Q MRRs have been also developed for liquid-phase chemical sensing application by Sandia National Labs [80,81] and biochemical sensing application by the Hong Kong University of Science and Technology (HKUST) [82], taking advantage of the large refractive index difference between SiN and SiO_2. SiN ring resonator-based devices for biosensing have been developed by researchers at the Université de Sherbrooke and Université de Lyon and manufactured by STMicroelectronics on 300 mm SOI in an industrial environment – the Datacom Advanced PHotonic Nanoscale Environment (DAPHNE) industrial platform [83].

The usage of the SiN platform for sensors can leverage the wide availability of smartphones for sensing in the visible or NIR wavelength [82]. For the visible and very-NIR (VNIR) (500–950 nm) wavelength window, a PECVD SiN BioPIX platform and unique waveguide structures, including an SiO_2 rib coated with Si_3N_4, have been developed by imec and Ghent University [13,84], and the dependence of loss on waveguide width, wavelength and cladding has been systematically investigated. Compared to stoichiometric Si_3N_4, silicon-rich SiN films can enable much larger non-linear parameters (5–500×) [12]. The linear and non-linear characteristics of BEOL Si_xN_y waveguides have been systematically investigated in the University of Southampton [30,45] and by A*STAR in collaboration with the Singapore University of Technology and Design (SUTD) [12]. The non-linear Kerr response of Si_xN_y waveguides can be increased by an order of magnitude by optimizing the fabrication parameters. Moreover, Si_xN_y waveguides can exhibit negligible TPA [12], enabling high power operation.

Several foundries now offer open-access Si_3N_4-SOI platforms, which combine the advantages of passive silicon nitride and passive/active silicon waveguides and devices. Owing to its CMOS-compatibility, silicon nitride waveguide technology can be built into state-of-the-art silicon photonics processes, significantly enhancing the capabilities and expanding the applications of SOI photonic chips. Si_3N_4 waveguides can enable new functionalities, including wider wavelength operation down to the visible range and lower loss passive devices, which complement the highly compact passive and high-speed active functionalities offered by SOI waveguides and devices. There are several examples of silicon nitride waveguides built into SOI platforms developed in recent years. Low-loss Si_3N_4 waveguides and active device integration using an Si_3N_4-on-SOI platform has been developed by the AIM Photonics Academy in collaboration with researchers at MIT [64,66,85,86], including demonstration of monolithic rare-earth-doped aluminium oxide (Al_2O_3:Er^{3+}) DBR, DFB, microring and tuneable lasers with a 300 mm wafer-scale fabrication process. Such a scalable approach allows for a dense integration of Si_3N_4 passive devices, SOI active and passive devices and ultra-narrow linewidth and tuneable light sources into silicon-based microphotonic systems [64,66,85,86]. A CMOS-compatible Si_3N_4-on-SOI platform has been developed by A*STAR and Advanced Micro Foundry (AMF) and researchers at the University of Toronto [87,88] by using monolithic BEOL integration. Moreover, the propagation loss and inter-optical layer transition loss have been systematically investigated, enabling active Si device integration with low-loss passive SiN waveguides [16,87,88]. An SiN-on-SOI platform has also been developed by CEA-LETI [89,90] with collaborators, including the Université Grenoble Alpes, Ecole Centrale de Lyon and Technical University of Denmark (DTU), using either PECVD or LPCVD SiN for applications, including (de-)multiplexers for CWDM optical transceivers and broadband fibre grating couplers and frequency comb generation for telecommunications. Many additional foundries are now offering Si_3N_4-on-SOI integrated optics platforms, including STMicroelectronics [83] and Sandia Labs [91], and the technology has become widely available to researchers and industry, opening up possibilities for many new devices and applications.

11.3 Silicon nitride integrated optical devices

Because of its low losses and amorphous structure, silicon nitride integrated optics technology has traditionally been focused on passive devices, particularly for optical communications and sensing. More recently, there has been a growing focus on non-linear optical devices and systems. Research and development on monolithic and hybrid integration of novel materials with SiN has led to new active functionalities and a growing portfolio of SiN-based devices. In this section, silicon nitride passive components, sensors, non-linear optical devices, active devices, including switches, modulators and detectors, and optical amplifiers and light sources are reviewed.

11.3.1 Passive integrated optical devices

Passive photonic devices consist of photonic circuit elements that require no external input for the device to function, in contrast to active optoelectronic components such as switches and modulators which require electrical input for control. The most basic passive circuit element is the earlier discussed waveguide, which is used to confine light within an optical mode and guide it across the chip between circuit elements. Beyond the waveguide, other basic passive elements used in silicon nitride photonic circuits include mode-size converters, directional couplers, splitters/combiners, ring resonator filters, Mach–Zehnder interferometers (MZIs) and wavelength division multiplexers (WDMs).

Mode-size converters are fundamental PIC building blocks used to adiabatically couple light between different waveguide structures and for low-loss fibre-chip coupling. In the absence of a monolithic or hybrid-integrated light source, light from an external source must be coupled onto the PIC by a fibre. As the optical mode of a fibre is typically significantly larger than that of an SiN waveguide mode, the waveguide core is usually thinned, both vertically and laterally at the outcoupling facets, to reduce the confinement of light within the waveguide creating an expanded mode. The expanded mode then has greater overlap with the mode of the fibre, which results in a significantly improved coupling efficiency. The taper then gradually transitions from the reduced waveguide size to its normal size, over lengths ranging from ~0 1 to 1.0 mm, converting from the expanded waveguide mode with minimal waveguide loss. Highly efficient spot-size converters have been developed on SiN platforms, allowing for low-loss fibre-chip coupling [10].

Couplers and splitters/combiners are basic elements which are used for routing light in PICs. Different types of couplers and splitters/combiners and their trade-offs in terms of versatility, footprint, fabrication robustness, wavelength sensitivity, polarization sensitivity and excess loss are outlined in [10]. Directional couplers are based on coupled mode theory and involve two closely spaced waveguides, such that their evanescent fields overlap and a combined mode between the two structures exists. Light coupling from one waveguide to the other is achieved over a distance related to the effective refractive index difference between the symmetric and antisymmetric modes. The coupling length is given by $L_{\text{coupling}} = \lambda/2\Delta n_{\text{eff}}$, where λ is the wavelength and Δn_{eff} is the effective index difference, and L_{coupling} is proportional to the gap between the waveguides [97]. The coupling ratio can be adjusted by changing the length or gap of the coupler. Because of silicon nitride's high-resolution patterning and high refractive index, allowing for small waveguide gaps and high-confinement waveguides with tight bends, respectively, highly compact directional couplers can be realized. In addition, densely integrated 50/50 and $M \times N$ SiN splitters/combiners based on Y-junctions and multimode interferometers (MMIs), respectively, can be designed. The maturity of silicon nitride technology has enabled important building blocks such as directional couplers and splitters/combiners to be established in various platforms [10].

Interferometric devices are widely used fundamentally important elements in PICs. They have applications in filtering and wavelength division multiplexing and are the base component of most integrated photonic sensors and modulators, in which the interference pattern of the device is modified by altering the optical path length. Two common integrated optics interferometric devices are the MZI and the MRR. An MZI [98] splits incoming light into two separate arms which are later recombined after some propagation length, where optical path length differences between the two arms create an interference pattern at the output. Alternatively, an MRR evanescently couples a portion of light from a bus waveguide into a waveguide loop, typically in the shape of a ring or racetrack, which experiences constructive interference within the loop when the wavelength matches the optical path length of the ring. When this resonance condition occurs, light intensity builds up in the ring and destructively interferes when coupling back into the bus waveguide, creating a dip in the transmitted intensity through the bus waveguide. The effect of light intensity build-up within the cavity also gives ring resonators applications in active devices that require high-intensity light, such as non-linear frequency comb generators and lasers. The circular layout of MRRs allows for a compact design, which, along with their interferometric and active device applications, makes MRRs one of the most commonly used building blocks in PICs for a wide array of applications.

MRRs [99] are characterized based on their extinction ratio (ER), free spectral range (FSR) and quality factor (Q factor). The ER is the level of transmitted signal drop that occurs when the device is on resonance and is at a maximum when the intensity build-up within the ring is equal to the intensity in the bus waveguide, creating perfect destructive interference. The ER can be controlled by adjusting the size of the evanescent coupling gap between the ring and bus waveguides. The FSR is the spectral distance between resonant wavelengths, which decreases as the optical path length of the resonator becomes larger. Silicon nitride's high refractive index allows for SiN ring resonators to achieve tight bending radii (as small as ~10 μm in high-confinement waveguides) resulting in short optical path lengths and FSRs as high as 20 nm. In weaker confinement SiN waveguides, the larger minimum bend radius (>100 μm) results in FSRs on the order of 2 nm and below. In general, large FSRs are more desirable for filtering and multiplexing to reduce signal pickup from nearby resonances.

The Q factor is a metric describing the spectral linewidth of the resonant modes of a cavity and the light intensity build-up within the cavity, determined by the losses and photon lifetime in the resonator. The loaded Q factor of the cavity, Q_l, is given by $1/Q_l = 1/Q_i + 1/Q_e$, where Q_i is the internal Q factor of the ring, determined by losses in the ring, and Q_e is the external Q factor, determined by the coupling strength to the bus waveguide. The highest quality factor an MRR can reach is limited by the optical propagation loss of the ring waveguide, which is determined by material-related absorption, scattering at imperfections and scattering at roughened etched waveguide sidewalls. Since SiN absorption is intrinsically low and thin-film quality is very high, much effort on SiN MRR Q-factor enhancement has focused on the optimization of etch processes and waveguide designs to reduce waveguide sidewall roughness and minimize the sidewall interaction of the optical mode, respectively. The propagation loss can be determined from the measured quality factor in the unloaded or weak coupling case, using the equation $Q_i = 2\pi n_g/a\lambda_0$ [100], where n_g is the group index, related to the effective index and dispersion by $n_g = n_{\mathrm{eff}} + \lambda(dn_{\mathrm{eff}}/d\lambda)$, a is the attenuation coefficient (in cm^{-1}), converted to propagation loss (in dB/cm) by the relation $\alpha = 10 \cdot \log e \cdot a \approx 4.34 \cdot a$ [97], and λ_0 is the wavelength of the resonant mode. Silicon nitride ring resonators are capable of achieving Q factors $>10^7$, leading to very narrow resonant linewidths. Filtering and multiplexing applications generally require large ERs, over narrow linewidths/high-Q factors. However, non-linear and laser applications that require a large intensity build-up within the cavity and/or ultra-narrow spectral linewidths need quality factors $>\sim10^6$, motivating the use of high-Q silicon nitride MRRs and resonant structures for such applications. Figure 11.4(a) shows how the MRR design for low- to moderate-confinement silicon nitride waveguide platforms changes with waveguide thickness, where thin waveguides have high-Q factors and small FSRs, while thicker waveguides allow for more compact rings with larger FSRs, but at the cost of lower Q factors.

WDMs and demultiplexers are key elements to take advantage of an optical waveguide's ability to transmit multiple signals simultaneously at different wavelengths. By adding a second drop waveguide on the opposite side of ring resonator, light at resonant wavelengths can transfer their power from the bus to drop waveguide. An optical filter can be readily designed using such a ring or multiple coupled rings with through and drop ports [101]. Figure 11.4(b) shows an example of a third-order ring resonator filter and its transmission spectra [102]. A series of ring resonators with drop waveguides acting as filters at different resonant wavelengths can form a WDM. An alternative option is to use star couplers, to split different wavelengths into different waveguides through free propagation in a multimode region in order to diverge light into an array of different waveguides [103,104]. Because of silicon nitride's low losses, it is also ideal for long passive structures such as delay lines, used in applications such as optical beamforming networks. Further details regarding these and other passive building blocks can be found in [10].

Si_3N_4 passive device fabrication is traditionally carried out via masked processes such as contact lithography or deep-ultraviolet immersion lithography and etching, or damascene processing (see references in Table 11.1). For finer feature resolution and device proximity, electron-beam (e-beam) writing is also used, which allows the process to be carried out without the predefinition of a physical mask, as is needed in traditional lithography. For research and prototyping, this is an attractive advantage which allows changing designs from run-to-run with short turnaround. However, e-beam is an expensive technology and is therefore less

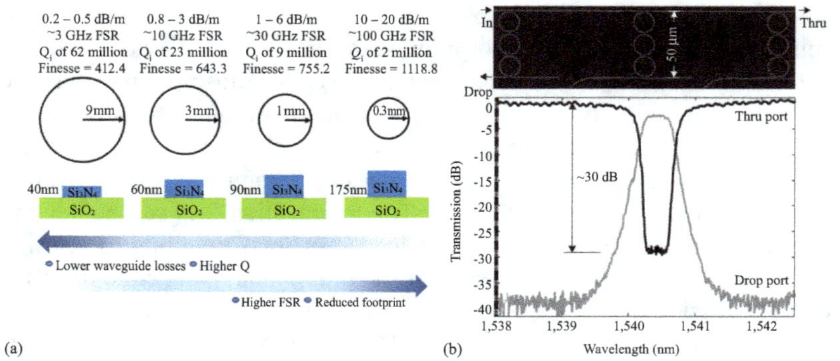

Figure 11.4 (a) Illustration showing how metrics for silicon nitride ring resonators change with silicon nitride waveguide thickness. In low- to moderate-confinement designs, thin waveguides are able to achieve higher Q factors; however, thicker waveguides have larger FSRs and more compact device sizes [96]. (b) A third-order silicon nitride microring resonator filter and its spectral response. (a) © 2018 IEEE. Reprinted, with permission, from Reference [96] and (b) © 2006 The Optical Society. Reprinted, with permission, from Reference [102]

accessible for prototyping to many researchers, and many low- to moderate-confinement SiN waveguide designs allow for larger feature sizes.

Instead, laser direct writing (LDW) can be considered as an alternative [105], as its cost advantages in terms of facilities and equipment are in-line with research and prototyping constraints. LDW uses an optical source to induce photorefractive changes in a material enabling waveguide fabrication without the use of an etch step. Here, typically, pulsed ultraviolet light is used to cause refractive index changes on the order of 10^{-3} refractive index units (RIU) in polymers [106] and doped glass [107]. Due to these limitations on material selection and small refractive index changes, typical devices made using LDW suffer from large bend radii, and relatively large waveguide widths. The latter is due to optical scattering during the exposure process from higher energies required to induce index changes [108]. Recently, an alternative fabrication process has been developed, which enables the fast turnaround of Si_3N_4 waveguide devices for prototyping and research development [109]. In contrast to LDW, lower energy continuous wave (CW) UV laser light can be used to expose a photoresist mask layer enabling feature sizes down to ~1.0 μm. After development, the resist layer can act as an etch mask, allowing pattern transfer to the layer below the resist. This combination of maskless laser writing and standard photoresists enables a low-cost process which allows for prototyping of various waveguide designs in different optical materials.

A demonstration of this UV laser resist mask writing technique is shown in Figure 11.5 from the Centre for Emerging Device Technologies at McMaster University, where the writing process is characterized, and passive Si_3N_4 waveguide devices are fabricated and tested [109]. Figure 11.5(a) shows a diagram of the

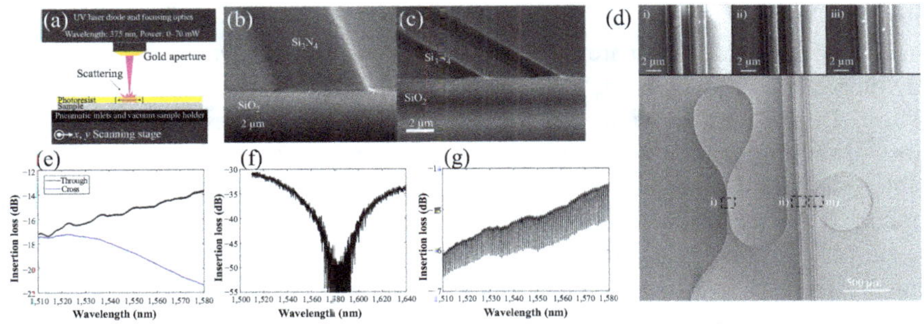

Figure 11.5 (a) Diagram of a UV-laser-written system showing lateral UV light scattering during exposure of a negative photoresist etch mask layer on an Si_3N_4 film. (b) SEM image of an Si_3N_4 waveguide fabricated using 70-mW UV laser power and a development step to define the negative resist mask followed by reactive-ion etching. (c) Parallel fabricated waveguides in a coupler region. (d) SEM images of the total footprint (bottom) and coupling regions (top) of proof-of-concept Si_3N_4 integrated optical devices fabricated using UV laser resist mask patterning. (e–g) Transmission spectra of (e) a directional coupler showing 50/50 coupling at 1,510 nm for 40 μm coupling length, (f) a 300-μm-radius Sagnac loop interferometer with 40 μm coupler length, demonstrating a 20 dB extinction ratio at ~1,580 nm and (g) a 300-μm-radius point coupled-ring resonator with multiple resonances shown. © 2019 The Optical Society. Reprinted, with permission, from Reference [109]

set-up used to expose the photoresist, which is a commercial patterning system consisting of a 375 nm laser source and focusing optics above an air bearing controlled stage. A negative photoresist is used, and exposure settings were varied to optimize the sidewall definition, which is dominated by scattering of the UV light. The technique was shown to produce smooth sidewalls with higher exposure energies and widths suitable for single-mode low-confinement Si_3N_4 waveguides. Figure 11.5(b) and (c) displays waveguides fabricated after RIE and adjacent features characterized for gaps in structures such as directional couplers and ring resonators. A Sagnac loop, directional coupler and ring resonator were fabricated with widths and gaps of 1.9 and 1.1 μm, 1.1 and 2.1 μm, and 1.6 and 1.5 μm, respectively, as shown in Figure 11.5(d). Edge coupling was used to demonstrate 50/50 coupling at 1,510 nm for the directional coupler, a 20-dB ER at 1,580 nm for the Sagnac loop and a Q factor of 13,000 at 1,576 nm for the ring resonator, as displayed in Figure 11.5(e), (f) and (g), respectively. Details regarding the experimental procedure and fabrication can be found in [109].

11.3.2 Silicon nitride devices for sensing

Silicon nitride's wide transparency, ultra-low losses, flexible waveguide structures and volume PIC production methods make SiN platforms particularly attractive for sensing. This section provides an overview of SiN waveguides and their applicability to sensing, particularly focusing on their advantages in terms of access to the visible spectrum for biosensor applications, and high optical power operation with minimal TPA for Raman spectroscopy.

Modern health diagnostics and environmental sensing are benefitting substantially from various lab-on-chip (LOC)-integrated optical devices. These devices enable compact, sensitive and rapid assessment and, in comparison to electrical devices, are insensitive to electromagnetic interference and electrical conduction of the fluid under test. Optical sensing devices are based on compact optical circuits and high-quality resonator devices which are readily achievable with SiN [9,15]. Although the optical sensing space is dominated by demonstrations in the SOI material platform [110], SiN has a unique and irreplaceable role to play in the sensor market of the future.

Integrated optical sensing comes in two common forms: absorption-based refractive index sensing where the evanescent tails of waveguide modes interact with surface events such as microfluidics [82], gases [9] or binding analytes [111] and Raman spectroscopy where inelastic scattering of laser light is measured by a sensitive detector and used to fingerprint molecular bonds [112,113]. The broad transparency of Si_3N_4 waveguides allows for both of these measurements to take place in the visible and short NIR spectral range, where a wide variety of biologically and chemically relevant substances can be detected, including antigens [113], proteins [114] and viruses [115].

To increase the interaction between the evanescent tail or light for Raman scattering and the top sensing medium, many approaches have been attempted, including fabricating slot or channel waveguides [113], interference-based structures [111] or resonator-based devices [116]. Figure 11.6 highlights one of these approaches from the Photonics Research Group, INTEC Department – imec and Center for Nano- and Biophotonics from Ghent University, Belgium [113]. Figure 11.6(a) shows a cross section schematic and optical mode profile of a 220-nm-thick and 700-nm-wide Si_xN_y waveguide on a thermally oxidized silicon substrate. Optimization of the waveguide parameters was carried out to maximize the Raman conversion efficiency and ensure a maximum amount of light interactions with the surrounding analyte in order to excite the molecules while maintaining sufficient refractive index contrast to collect the scattered light back into the waveguide. Details regarding the design parameters can be found in [113]. To simulate an analyte, isopropanol (IPA) was used, and 50-mW optical power at an excitation wavelength of 785 nm was launched into the waveguide. Figure 11.6(b) demonstrates the results comparing Raman spectra collected from a Raman microscope to the IPA cladded strip waveguide structure, demonstrating an improvement of two orders of magnitude after background waveguide transmission spectrum subtraction. Significant challenges in Raman spectroscopy are subtraction

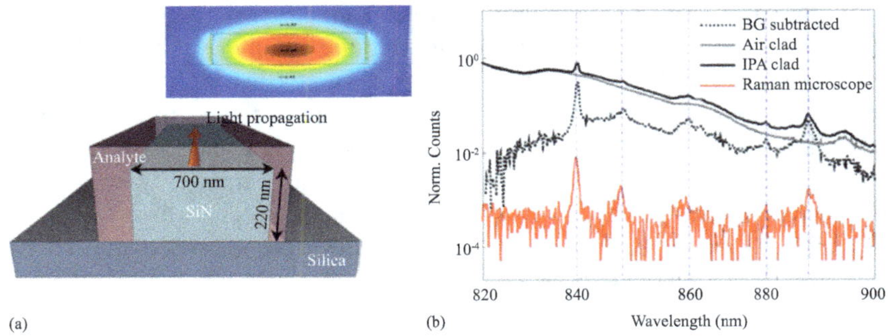

Figure 11.6 (a) SiN Raman waveguide sensor cross section shown with surrounding analyte with $n_{SiN} = 1.89$, $n_{silica} = 1.45$ and $n_{IPA} = 1.37$. Inset: modal field profile for 785 nm. (b) Waveguide transmission when covered with IPA (black), air (grey) and resulting background subtraction (dotted) showing Raman spectra. Control Raman spectra (red) taken with conventional confocal microscope. © 2017 SPIE. Reprinted, with permissions, from Reference [113]

of the background waveguide effects and increasing the sensitivity of scattered light for detection. To circumvent these issues, the waveguide structure can be designed such that the majority of the optical intensity interacts with the analyte. This can be achieved with subwavelength waveguide designs, or slot waveguides can be used to situate the mode in between two closely spaced Si_3N_4 waveguides. In [113], a direct comparison was carried out for the Raman spectra collected for a 1.6-cm-long strip vs. slot waveguide for 7 s collection time with similar excitation power of 50 mW at 785 nm. By decreasing the waveguide influence through geometric design, background spectrum subtraction was shown to lead to an increase in signal-to-noise ratio by a factor of 5 in the slot waveguide.

As state-of-the-art silicon nitride integrated optical sensors continue to emerge, so will new applications and the demands on sensitivity, resolution and limits of detection will not cease to rise. However, certain limitations can be noticed when one considers continuously increasing the light overlap with the sensing medium, which is for the most part of an aqueous nature for biological applications. Performance metrics which are important for biosensors include the refractive index sensitivity (*S*), which is defined as the change in resonance wavelength ($\Delta\lambda_{res}$) per RIU change of the surface (top cladding) medium (Δn_{clad}), $S = \Delta\lambda_{res}/\Delta n_{clad}$, as well as the intrinsic limit of detection (iLoD), iLoD $= \lambda_{res}/QS$, which has units in RIU and can be understood as the minimum index change needed to shift any one resonance by a full-linewidth [110]. The latter equation shows that quality factor (*Q*), sensitivity (*S*) and iLoD all exist with trade-offs. Advances in silicon nitride waveguide fabrication and ring resonators with higher *Q* factors are enabling improved limits of detection in sensors. Yet in such high-*Q* resonators, the mode either overlaps strongly with a low-loss cladding (low-confinement case) or is strongly confined to the SiN waveguide core (high-confinement case). When considering increasing the sensitivity by exposing more of the waveguide mode to the analyte, which acts as a lossy cladding layer, the trade-off becomes higher absorption, decreased photon lifetime in the cavity and lower Q, limiting the iLoD. This motivates the need for new measurement schemes where integrated active components such as lasers with narrower linewidth [115] can be used to interrogate sensing regions rather than absorption-based techniques solely. Hybrid approaches with light-emitting materials in SiN platforms have shown promise in this direction [117,118].

11.3.3 Non-linear optical devices

Non-linear optical devices have recently emerged as enabling technologies for a range of applications in silicon-based photonics, from all-optical signal processing, ultra-low power all-optical switching to quantum photonics. Silicon nitride has proven to be an ideal medium for non-linear integrated photonics, on account of its ultra-low linear losses, low TPA, high Kerr index and mature fabrication methods. Optical non-linearities arise when the intensity of light propagating in a dielectric medium is high such that the response of the material can no longer be described by liner estimations and the induced polarization becomes non-linear. The polarization terms can be expressed by the Taylor series expansion [119], $P = \varepsilon_0(\chi^1 \cdot E + \chi^2 \cdot EE + \chi^3 \cdot EEE + \cdots)$, where E is electric field, ε_0 is permittivity of free space and χ^n is nth order susceptibility. The first term (χ^1) is the dominant response of a material where its real part is related to the linear refractive index (n_0) and its imaginary part is related to the linear absorption (a). The second term (χ^2) is observed in non-centrosymmetric materials and is responsible for non-linear phenomena such as second-harmonic generation. Bulk Si_3N_4 or amorphous Si_3N_4 thin films do not possess intrinsic χ^2; however, in practical applications, it can be induced in a waveguide by stress through material growth or by ion migration [120]. The third term (χ^3) gives rise to what is called the non-linear refractive index or Kerr index (n_2) which is responsible for phenomena such as self-phase modulation (SPM) and four-wave mixing (FWM). The non-linear index can be calculated from χ^3 by $n_2 = (3/4{n_0}^2\varepsilon_0 c) \cdot \chi^3$, where c is the speed of light in a vacuum. The extent of non-linear effects in a particular waveguide is described by what is referred to as the non-linear parameter (γ), which depends on n_2, the effective mode area (A_{eff}) and the wavelength of incident light (c/ω) according to $\gamma = (n_2 \cdot \omega / A_{\text{eff}} \cdot c)$. The n_2 value of LPCVD Si_3N_4 is 2.4×10^{-19} m^2/W at 1,550 nm, which is ~7.5 times that of silica, and the small mode area in high-confinement SiN waveguides allows for high γ [12]. Silicon-rich silicon nitride is also a promising platform and has been shown to have higher n_2 and γ than stoichiometric Si_3N_4 while still maintaining negligible TPA [12]. Crystalline materials such as silicon have significantly higher n_2, where $n_{2,\text{Si}} = 1.9 \times 10^2 \times n_{2,\text{Si}_3}\text{N}_4$. However, SiN has been established as the most suitable platform for on-chip non-linear Kerr devices because of its lower linear absorption and lower non-linear losses due to negligible TPA, leading to higher non-linear figures of merit [11,12].

Important considerations in Kerr non-linear devices are SPM, FWM and group velocity dispersion (GVD). SPM is a common non-linear effect resulting from the third-order susceptibility. It is a process whereby an optical pulse modifies its own phase leading to a pulse broadening effect. Observation of SPM in an Si_3N_4 waveguide was reported in [121]. When spectral broadening becomes extensive with the combination of other linear and non-linear processes, it can give rise to SCG. FWM is another non-linear effect arising from the third-order susceptibility. In a traditional sense, FWM occurs when two waves interact in a Kerr medium and generate non-linear sum and difference frequencies. This is a nondegenerate case of FWM; however, one strong pump can also incite FWM leading to two new frequencies, a case of degenerate FWM. FWM is used in applications, including wavelength conversion, broadband frequency generation and parametric amplification. However, due to the phase shift introduced by non-linearities and GVD, the efficiencies of both processes are highly dependent on achieving a phase-matching condition. Many demonstrations of these processes have been reported in SiN platforms, including frequency conversion in stoichiometric Si_3N_4 in [122,123] and parametric amplification in silicon-rich nitride platforms [124,125].

Dispersion engineering has become a crucial step for the design of non-linear optical devices. This is due to the requirement for anomalous and low GVD over a wide range of wavelengths to achieve phase matching in non-linear waveguides and optical resonators. In waveguides, the GVD is often represented by the dispersion parameter D, which describes the pulse temporal spreading per unit wavelength per unit propagation distance, where $D \approx -(\lambda/c) \cdot (d^2n/d\lambda^2)$ (ps/nm/km) [119]. The GVD is limited by the choice of material since it depends on both the material dispersion and waveguide geometry. For stoichiometric Si_3N_4, the GVD is normal over the telecommunication band unless thicker waveguides of at least ~700 nm are employed. Fabricating such waveguides through conventional LPCVD deposition and etching is challenging because of cracks induced by stress. Different procedures have been proposed to overcome this problem such as the damascene reflow process [79]. In addition, techniques have been proposed to engineer GVD by geometrical or refractive index optimization [126,127]. Recently, tellurium oxide (TeO_2)-coated Si_3N_4 waveguides have been studied that promise a simple way of achieving anomalous GVD on lower height LPCVD Si_3N_4 [72]. The deposited TeO_2 layer serves as the extra waveguide core height required for anomalous GVD and can be deposited on standard Si_3N_4 waveguide by single post-processing step described in [69]. In addition to removing the necessity of thicker Si_3N_4, TeO_2 helps to optimize the non-linear parameter since it has significantly higher n_2, more than 50 times that of silica [128]. Figure 11.7(a) and (b) shows a comparison of the GVD parameter calculated using a finite element method modesolver in thicker stoichiometric Si_3N_4 [129] and TeO_2-coated Si_3N_4 waveguides of ~half the height, respectively.

Optical frequency combs (OFCs) are the most prominent application of non-linear effects in integrated photonics. OFCs can be generated through a variety of processes, including SCG or Kerr comb generation (KCG). A thorough analysis, distinction and review of Kerr combs have been provided in several references,

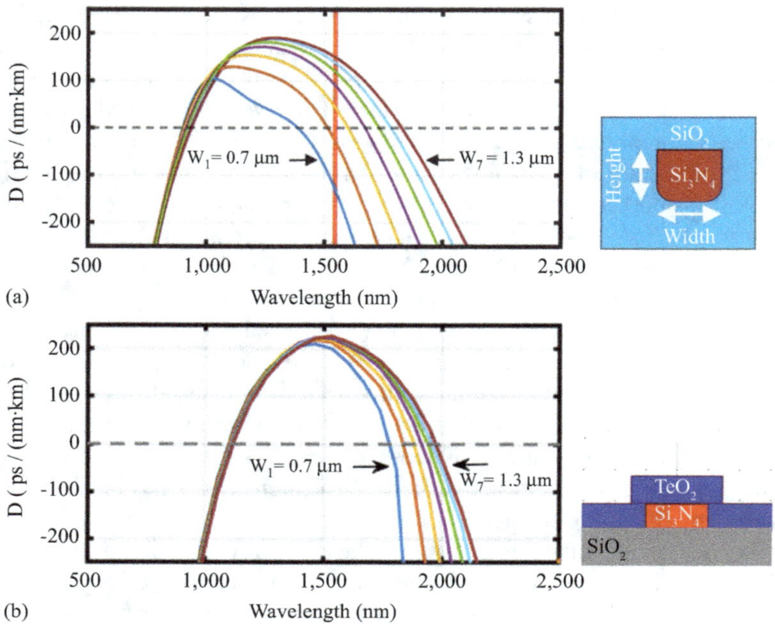

Figure 11.7 (a) Numerically calculated GVD coefficient D, for stoichiometric Si_3N_4 of height 0.9 μm and width varied from 0.7 to 1.3 μm in 0.1-μm steps at a wavelength of 1,560 nm, for the fundamental TM mode. (b) Similar calculation repeated for TeO_2-coated Si_3N_4 of half height (i.e. 0.45 μm), with the same variation in width and 0.45-μm-thick TeO_2 coating, showing the possibility for engineered anomalous dispersion using significantly thinner Si_3N_4 waveguide structures. (a) © 2017 The Optical Society. Reprinted/adapted, with permission, from Reference [129]

including [120]. Comb generation through SCG is achieved by pumping non-linear waveguides with a pulsed laser source which leads to the generation of new frequencies by combinations of processes such as cascaded FWM, SPM and dispersive waves (DWs) generation. With proper dispersion engineering, wider supercontinuum than that obtained from SPM can be achieved. The frequency spacing of the resulting comb is determined by the repetition rate of the pump pulses. For Si_3N_4, SCG extending across the visible and into the NIR wavelength range was demonstrated in [130], and two-octave SCG has been demonstrated in [129], as shown in Figure 11.8(a) and (b). On the other hand, OFC generation through KCG is achieved in microresonators with CW pumping where comb spacing depends on the resonator FSR as shown in Figures 11.8(c), (d) and (e) [131]. Demonstration of very high-Q microresonators in Si_3N_4 has enabled KCG and ultimately generation of broadband OFCs. Through proper tuning of a single CW

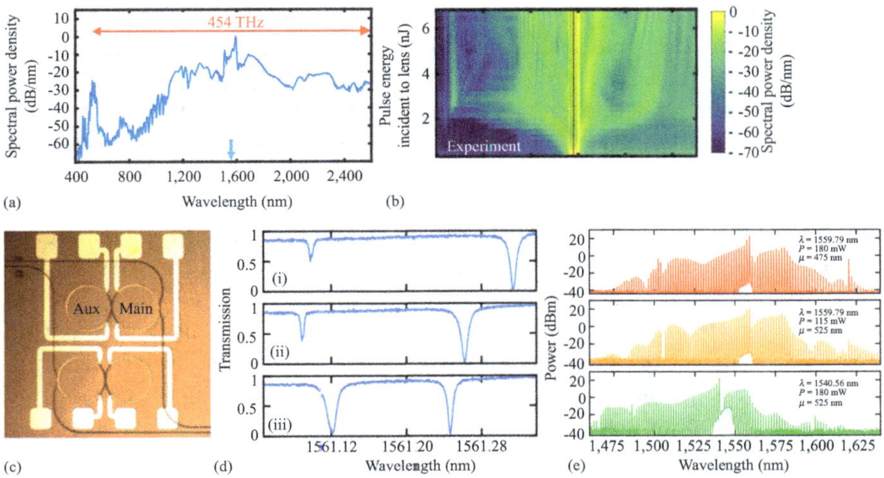

Figure 11.8 *(a) Supercontinuum spectrum generated in a 1-µm-wide and 0.9-µm-high Si_3N_4 waveguide using a launched pump pulse energy of 1.4 nJ. The pump wavelength of 1,560 nm is indicated with an arrow. (b) The corresponding measured spectral power density (color-coded). (c, d and e) Demonstration of Kerr comb generation in a normal dispersion regime with a coupled-ring geometry in Si_3N_4. (c) Image of the fabricated coupled-ring resonators with integrated platinum resistive heaters. (d) Transmission spectra of the device using a counter-propagating probe one free spectral range from the pump wavelength (1,559.7 nm) measured (i) as the main ring heater is tuned, (ii) before parametric oscillation and (iii) during comb generation. (e) Measured comb spectra for different configurations of the coupled-ring geometry with varying coupling gaps, pump wavelength and power as indicated in the legends. (a and b) © 2017 The Optical Society. Reprinted, with permissions, from Reference [129] and (c–e) © 2019 The Optical Society. Reprinted, with permission, from Reference [131]*

pump's frequency and/or amplitude, dissipative Kerr solitons [132] can be excited to facilitate generation of ultra-broadband and highly coherent OFCs. Such OFCs have been reported in [133,134]. OFCs have wide applications in areas such as tele- and data communications, spectroscopy, timing and clocks, frequency synthesis and microwave photonics.

11.3.4 Active devices: switches, modulators and photodetectors

Silicon nitride PICs have traditionally focused on passive optical functionalities, including waveguides and metal electrodes for thermo-optic tuning as basic building blocks. However, applications of PICs in high-speed communications require semiconductor optoelectronic devices for key active functions, including switching, signal modulation and photodetection. Recently, there have been significant developments in integrating optoelectronic materials and devices on silicon nitride platforms using hybrid approaches. Here, SiN-based switches, modulators and detectors are briefly surveyed.

Waveguide phase modulators are typically based on interference structures such as MZIs and MRRs. Electrical signals are used to alter the refractive index and shift the phase of light in the waveguide to switch the output of the device from constructive interference, where the signal transmits across the modulator, to destructive interference, which turns off the signal, representing the on and off states of the device, respectively. In silicon and III–V semiconductor platforms, refractive index modulation is readily achieved by the plasma dispersion effect, where refractive index changes are induced by changing the electron/hole carrier concentration in doped regions (e.g. by electrically biasing a p–n junction). Unlike silicon and III–V materials such as InP, silicon nitride is an electrical insulator. Therefore, it cannot be electrically doped effectively, and the plasma dispersion effect is not a viable modulation technique for silicon nitride waveguides. Instead, alternative methods of achieving an electrically induced refractive index change must be applied in silicon nitride. The simplest approach is to apply the thermo-optic effect [135], which changes the refractive index of a medium by changing its temperature through localized integrated heaters. Thermo-optic modulation techniques are generally only able to achieve low switching speeds (~1 kHz) and require large amounts of power for efficient heat dissipation. More effective modulation schemes rely on the integration of new materials onto the silicon nitride platform to achieve modulation via stress-optic, electro-optic or acousto-optic effects. Lithium niobate is a particularly effective and well-used material for electro-optic modulation, and heterogeneous integration of lithium niobate with silicon nitride waveguides has been developed by researchers at the University of Massachusetts and others [136–138]. III–V modulators have also been demonstrated, including Mach–Zehnder modulators integrated with SiN waveguides on silicon achieving 40 Gbit/s operation developed by NTT [139]. An alternative material option, which is compatible with silicon nitride and can be used in modulator designs, is lead zirconate titanate (PZT), which can be grown onto silicon nitride waveguides by a chemical solution deposition process by cyclic spin coating and annealing of PZT on a lanthanide seed layer. This process has been used to demonstrate phase modulators on a silicon nitride platform using stress-optic, electro-optic and acousto-optic techniques. PZT has been used to create electro-optic modulators using the Pockels effect, where the silicon nitride waveguide shares an optical mode with the PZT thin film [140], as displayed in Figure 11.9(a). Additionally, PZT layers located significantly above the silicon nitride waveguiding layer can be compressed and expanded by an applied electric field from two electrodes, creating stress-optic effects in the underlying silicon nitride and SiO_2 [141,142], as shown in Figure 11.9(b). Reconfigurable switching has also been

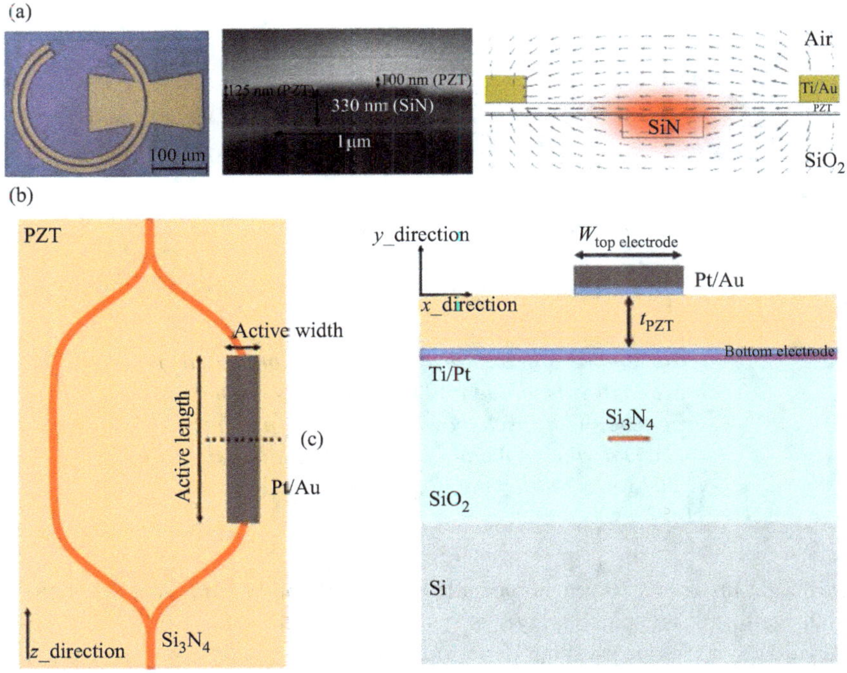

Figure 11.9 *(a) A PZT Pockels effect modulator, using an SiN microring resonator and a PZT film deposited closely over the Si_3N_4 waveguide to interact with the optical mode. (b) A stress-optic Mach–Zehnder modulator, where a PZT film on the silicon dioxide top cladding is compressed/expanded by electrical impulses to stress the underlying Si_3N_4 waveguide and cladding. (a) Reprinted from [140] with the permission under Creative Commons License (https://creativecommons.org/licenses/by/4.0/) and (b) © 2015 The Optical Society. Reprinted, with permission, from Reference [141]*

demonstrated using phase-change materials such as Ge–Sb–Te on silicon nitride ring resonator structures via photothermal or electrothermal control [143].

Photodetectors are used to convert optical signals into electrical signals, in order to interface PICs with electrical circuits and measurement equipment. Photogeneration of electrical charge carriers by the absorption of signal light in a semiconducting material is the most commonly used detection technique. Telecommunication applications use wavelengths in the C-band (1,530–1,565 nm), which lies above the bandgap of silicon. Silicon is only able to absorb light at wavelengths up to ~1,100 nm, making it a nonideal material for photodetection for many waveguide applications. Instead, germanium, which is able to efficiently

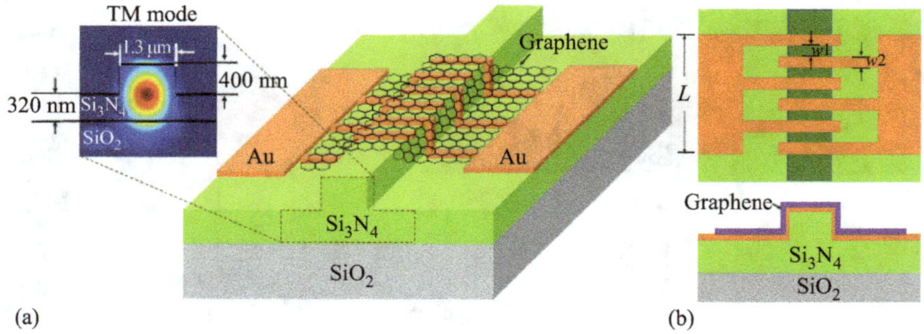

Figure 11.10 Graphene on silicon nitride integrated photodetector. (a) Silicon nitride rib waveguide and gold contacts coated in a layer of graphene, with inset showing the mode profile of the waveguide. (b) Aerial view of metal contact pattern, and detector cross section. © 2018 AIP Publishing. Reprinted, with permission, from Reference [148]

absorb light up to ~1,700 nm, is typically used in silicon-based photonics platforms. In SOI platforms, an epitaxial layer of germanium is grown onto an exposed silicon waveguide by a chemical vapour deposition process. Signal light guided towards the waveguide photodetector is then absorbed in the germanium layer, and the generated electrical carriers are swept out by a reverse biased p–n junction formed by doping either the germanium or silicon layer. Growth of high-quality germanium films on silicon waveguides is enabled by the engineering the lattice mismatch between germanium and silicon using a silicon/germanium (SiGe) buffer layer. Integration of germanium photodetectors with silicon nitride waveguide platforms and high-speed detection has been demonstrated by researchers at MIT, IHP and others [144,145]. Photodetection has also been demonstrated using III–V–SiN hybrid integration [146]. As illustrated in Figure 11.10, monolithic approaches to the integration of photodetectors on silicon nitride have been successfully demonstrated by the integration of graphene deposited onto waveguides by chemical vapour deposition processes. Successful photodetection at C-band wavelengths using this method, with 33 GHz bandwidth, and 2.36 A/W responsivity, has been demonstrated [147,148].

Silicon photonics platforms can include a combination of silicon nitride and silicon waveguides, in multilevel PIC technology, to combine the low losses of silicon nitride waveguides with the already well-developed optoelectronic capabilities of silicon devices [77,83,86–92,149]. Silicon nitride can be integrated as either a front or back end of line process, above or below the silicon waveguides. Si/SiN circuits rely on vertical interlayer light coupling between silicon and silicon nitride waveguides using nano-taper designs. End of line processing, including via etching and metallization, is used to supply the electrical contacts for heaters and the buried silicon modulators and detectors. Optoelectronic functions can be carried out in the silicon layer, and light can be coupled into and out of the low-loss silicon nitride layer for passive and non-linear functions such as optical delay lines, WDMs and OFCs.

11.3.5 Amplifiers and lasers

Optical amplification and lasing has been a challenge in silicon nitride integrated optics, because SiN is a transparent dielectric and not an optoelectronic light-emitting material. Therefore, these functions were previously carried out in separate active components or photonic chips. However, the significant advances in SiN fabrication and introduction of hybrid and monolithically integrated light-emitting materials have led to new possibilities for SiN on-chip amplifiers and light sources. Parametric amplifiers and light sources have been developed by exploiting the non-linear Kerr effect, as described in Section 11.3.3. In addition, recent developments building on silicon nitride's ultra-low losses and high-resolution patterning for low-loss waveguides and high-Q cavity structures have led to new hybrid III–V and monolithic rare-earth-doped amplifiers and lasers. In this section, a survey of optical amplifiers and lasers on silicon nitride platforms is reviewed.

An optical amplifier is one of the key components in optical systems, used to increase the power of a degraded signal after many sources of loss in the system, including propagation losses and device insertion losses. Common optical amplifiers include erbium-doped fibre amplifiers (EDFAs), semiconductor optical amplifiers (SOAs) and Raman fibre amplifiers. Silicon nitride waveguides are a promising route to on-chip amplifiers because of their ultra-low losses. However, aside from non-linear parametric amplification, which is less ideal for amplification of high-speed signals due to its lower efficiencies and requirement for phase matching, optical gain directly in SiN, e.g., by doping or stimulated Raman emission, is difficult because of SiN's low rare-earth solubility [48–52] and negligible Raman gain coefficient [69], respectively. As with amplifiers and lasers in silicon photonics platforms, a natural approach is to use heterogeneously or hybrid-integrated SOAs because of their compact size [150]. Nevertheless, an attractive lower cost and scalable alternative is developing monolithic amplifiers, in order to bring the low noise, broadband and efficient performance of an EDFA onto the chip, made possible by SiN's low losses. Aluminium oxide (Al_2O_3) is known to be a good rare-earth host material with high rare-earth solubility, allowing high gain per unit length and high total gain possible in a compact footprint in rare-earth-doped spiral waveguides [151,152]. Researchers at the University of Twente recently developed a hybrid active–passive Al_2O_3–Si_3N_4 platform, showing vertical coupling using SiN taper structures and gain in an Er-doped spiral waveguide [153,154]. Furthermore, ultra-high gain per unit length was demonstrated in an $Al_2O_3{:}Er^{3+}$/SiN slot waveguide structure using an $Al_2O_3{:}Er^{3+}$ coating deposited by atomic layer deposition by researchers at Aalto University, CNRS – Université Paris-Sud/Saclay and colleagues [155].

Tellurium oxide (TeO_2) is another promising candidate for on-chip rare-earth-doped amplifiers and lasers because of its high rare-earth solubility, wide emission spectrum, low phonon energy and high refractive index (n=2.1), enabling compact

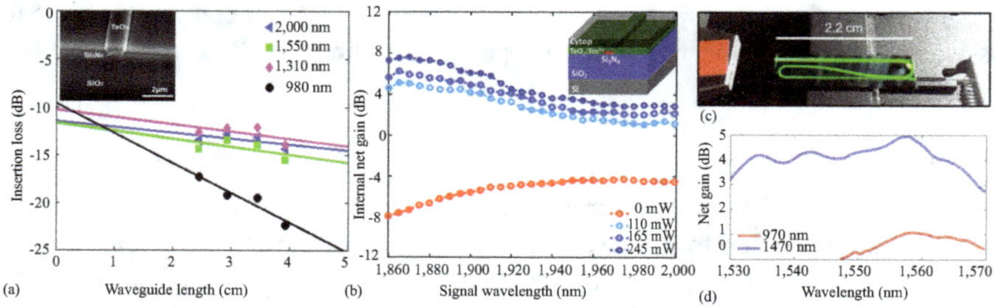

Figure 11.11 (a) Insertion loss measurements at different wavelengths in TeO_2-coated Si_3N_4 waveguide platform. Inset: SEM cross section image of an Si_3N_4 waveguide coated with a 0.3-μm-thick TeO_2 film [69]. (b) Internal net gain vs. wavelength measured in a 6.7-cm-long paperclip Tm-doped waveguide amplifier at different launched 1,620 nm pump powers. Inset: cross section profile of the TeO_2: Tm^{3+}-coated Si_3N_4 waveguide structure [70]. (c) Image of the green light emission in TeO_2:Er^{3+}-coated Si_3N_4 amplifier [71]. (d) Internal net gain vs. wavelength in Er-doped amplifier under 970 and 1,470 nm pumping [71].

devices [128]. As shown in Figure 11.11, researchers at McMaster University in collaboration with LioniX developed a high-quality tellurium-oxide-coated silicon nitride waveguide platform [69]. Cut-back measurements in Figure 11.11(a) show that the propagation losses are as low as 0.6 dB/cm at 2-μm wavelength, which makes this platform appropriate for compact linear, non-linear and active tellurite glass devices in silicon nitride PICs. As displayed in Figure 11.11(b), a TeO_2:Tm^{3+} waveguide amplifier on a low-loss Si_3N_4 chip has been demonstrated with 7.6 dB peak internal net gain at 1,870 nm wavelength [70]. In addition, 5 dB net gain has been demonstrated in a 6.7-cm-long erbium-doped tellurium-oxide-coated silicon nitride waveguide for 35 mW of launched 1,470 nm pump power with low background propagation losses of 0.25 dB/cm (Figure 11.11(c) and (d)) [71].

Recently, there has been significant attention drawn to hybrid integration of III–V gain materials and lasers with low-loss silicon nitride platforms. Si_3N_4 is CMOS compatible and has a seven times lower thermo-optic coefficient [156], negligible free-carrier absorption and lower non-linearity compared to Si, offering excellent thermal stability and high-*Q* cavity features for laser operation, particularly narrow-linewidth semiconductor lasers [157–163]. An optically integrated InP–Si_3N_4 single-mode laser was reported in [157] by a combination of a reflective SOA (RSOA) with an Si_3N_4 waveguide feedback circuit on separate submounts with a tuning range of more than 43 nm centred near 1,550 nm and a linewidth as low as 90 kHz. As shown in Figure 11.12(a)–(c), a compact Si_3N_4 hybrid single-mode laser cavity consisting of a III–V RSOA gain chip and low-loss Si_3N_4 ring

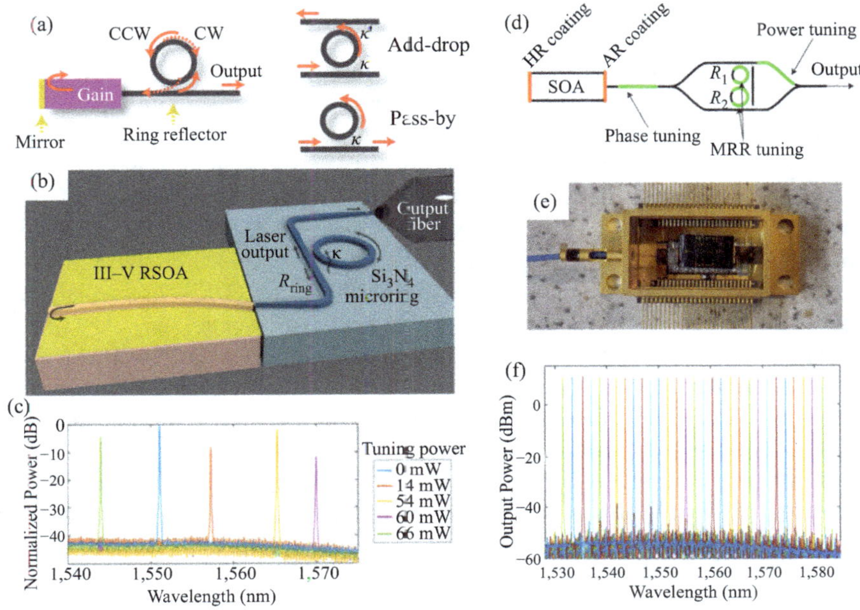

Figure 11.12 (a) Schematic and (b) 3D rendition of hybrid III–V/Si_3N_4 laser cavity design. (c) Output spectrum of the laser for different tuning powers applied to the Si_3N_4 ring. (d) Schematic, (e) top view image and (f) laser output spectra of packaged hybrid III–V/Si_3N_4 laser using a Vernier ring design. (a–c) © 2017 The Optical Society. Reprinted/adapted, with permission, from Reference [158] and (d–f) reprinted from [161] with the permission under Creative Commons License (https://creativecommons.org/licenses/by/4.0/)

resonator chip was demonstrated in [158] with lasing at discrete wavelengths across a 27 nm range, an output power up to 1.7 mW and a linewidth down to 13 kHz. Other hybrid III–V–Si_3N_4 laser structures have been shown with silicon nitride Bragg gratings for feedback [159], with a Lorentzian linewidth of ~320 Hz demonstrated in [160]. As shown in Figure 11.12(d)–(f), a packaged and fibre-pigtailed hybrid laser was demonstrated with a tuning range of around 50 nm in the C-band with high output power (~10 dB m), side-mode suppression ratio more than 50 dB, narrow linewidth of under 80 kHz across the whole tuning range and μs switching speed [161]. These and numerous other recent works have proven the distinct advantage of Si_3N_4 waveguides for hybrid tuneable and ultra-narrow-linewidth chip-scale lasers for applications in advanced communications, light detection and ranging (LIDAR) and biosensing, among others [162,163].

Silicon nitride waveguides and cavity features have also been demonstrated as a path to high-performance monolithic lasers, including rare-earth-doped aluminium oxide ($Al_2O_3:RE^{3+}$) waveguide lasers [67]. Optically pumped Al_2O_3: RE^{3+} lasers can achieve wide wavelength operation, narrow intrinsic linewidths, high output powers and CW or ultra-short-pulsed emission. The $Al_2O_3:RE^{3+}$ gain layer is deposited on the silicon nitride chip in a straightforward post-processing step by reactive co-sputtering [164]. Grating features are defined in the Si_3N_4 layer using high-resolution stepper lithography. The laser waveguide design is highly wavelength insensitive, allowing for good pump-signal overlap and large mode overlap with the $Al_2O_3:RE^{3+}$ gain layer [66]. Researchers at MIT and UCSB demonstrated Er-doped DBR and DFB lasers on wafer-scale SiN platforms [64–66,165], based on the ultra-narrow-line-width $Al_2O_3:Er^{3+}$ DFB lasers that were demonstrated earlier by researchers at the University of Twente [166]. $Al_2O_3:Er^{3+}$ DFB lasers with high output powers up to 75 mW at 1,563 nm [167] and ultra-narrow linewidths were demonstrated by the MIT researchers using silicon nitride cavities [66]. Using a similar design, a thulium waveguide laser was realized with record high output power for Si-based lasers of up to 387 mW at 1,881 nm [168], as shown in Figure 11.13(a) and (b). A holmium-doped Al_2O_3 laser was also shown, which provides broadband emission from 1,930 to 2,130 nm [169]. Using the same

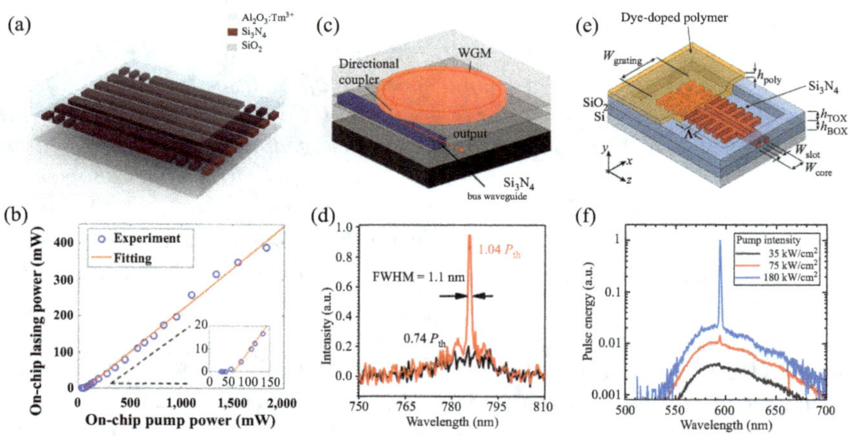

Figure 11.13 (a) Schematic and (b) laser output power vs. pump power of an $Al_2O_3:Tm^{3+}$ DBR laser on a silicon nitride chip showing up to 387 mW output power at 1,881 nm. (c) Schematic and (d) emission spectrum of perovskite disc laser integrated on a silicon nitride photonic chip. (e) Schematic and (f) output spectrum of a silicon nitride organic hybrid slot waveguide DFB laser. (a and b) © 2017 The Optical Society. Reprinted/adapted, with permission, from Reference [168]; (c and d) © 2017 ACS Publications. Reprinted/adapted, with permission, from Reference [176]; and (e and f) reprinted/adapted from [177] with the permission under Creative Commons License (https://creativecommons.org/licenses/by/4.0/)

multilayer silicon nitride platform, ultra-compact rare-earth-doped microring lasers were demonstrated by depositing the Al_2O_3:RE^{3+} gain material into a microring-trench etched into the SiO_2 top cladding next to an Si_3N_4 bus waveguide. Ytterbium-, erbium-, and thulium-doped microcavity lasers have been reported with sub-mW thresholds and emission at 1.0, 1.5 and 1.9 μm, respectively [85,170]. Further, 1.5-μm Er and 1.9-μm Tm Q-switched mode-locked lasers with a novel integrated saturable absorber structure were shown in [171,172]. Building on these designs and exploiting the advanced photonic circuitry possible in the silicon nitride platform, a four-wavelength WDM laser source [173], a tunable C-band laser [174] and an optical frequency synthesizer based on an Er-doped tunable waveguide laser [175] were demonstrated. Other light-emitting materials have been applied to demonstrate monolithic SiN-based lasers using similar methods. As displayed in Figure 11.13(c) and (d), a methylammonium lead iodide ($MAPbI_3$) perovskite microdisc laser integrated on a silicon nitride photonic chip was achieved by a team at AMO and colleagues with a lasing threshold of 4.7 μJ/cm^2 and infrared emission at 785 nm [176]. As displayed in Figure 11.13(e) and (f), an organic single-mode visible DFB laser based on SiN slot waveguides was also shown by investigators at the Austrian Institute of Technology, TU Wien and colleagues with emission at 600 nm [177]. Recently, a sub-hertz linewidth (~0.7 Hz) stimulated Brillouin scattering laser was demonstrated in silicon nitride led by UCSB researchers [178]. The laser is based on non-linear photon–phonon coupling in a high-Q silicon nitride microring resonator and opens up new applications in optical gyroscopes and low-phase-noise photonic oscillators.

11.4 Photonic integrated circuits and applications

The various passive, active and non-linear silicon nitride building block devices described in this chapter have been used to realize PICs for a wide range of applications. With the growing access to SiN platforms and MPW services, SiN PICs are becoming increasingly more widespread. Here, a selection of recent SiN PIC demonstrations in the fields of sensing, communications and LIDAR are reviewed to highlight the versatility of the material platform and significant potential for future developments.

As discussed in Section 11.3.2, SiN platforms offer many advantages for realizing on-chip waveguide sensors, leading to many demonstrations in the space [111]. In Figure 11.14, two SiN waveguide sensors developed by researchers at Ghent University and imec are shown [111]. Figure 11.14(a) displays a refractive index sensor PIC operating in the NIR, where water absorption is negligible, based on the combination of an MZI and an AWG which can be used for the detection of tuberculosis antigen in urine. A broadband source was sampled by the AWG arm of the MZI which consists of 30-channels with 1 nm spacing. This design enabled refractive index sensing when phosphate-buffered solution was flowed across the MZI sensing arm, resulting in a minimal detectable shift of 4 pm and a sensitivity of 5,440 nm/RIU with a final limit of detection of 7×10^{-7} RIU. Sensogram data

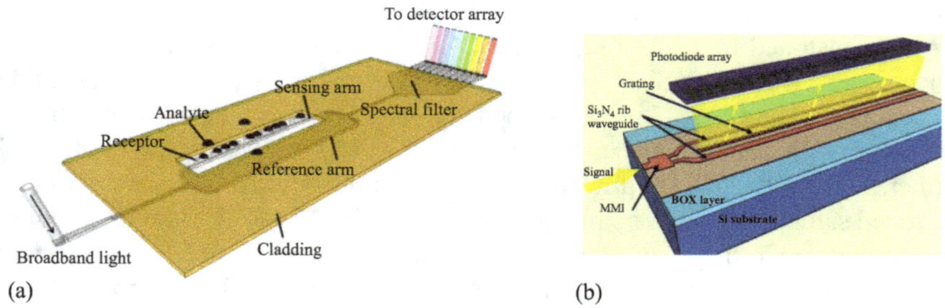

Figure 11.14 (a) Diagram of a refractive index sensor with an MZI and AWG on-chip spectral filter, including input and output light. (b) Diagram of the chip used for stationary Fourier transform spectroscopy with integrated photodiode array. © 2016 SPIE. Reprinted, with permission, from Reference [111]

and AWG details can be found in [111]. Figure 11.14(b) shows a design for a Fourier transform spectrometer (FTS), where improvements upon classical stationary wave integrated FTS are proposed. An MMI splits the signal into two waveguides, which share a beat mode separated by a thin grating, directing light to a photodiode array above. This co-propagating FTS scheme is noted by the authors to be well suited towards Raman spectroscopy due to the high bandwidth and small resolution achievable [111].

Communication bandwidth and network traffic have been increasing exponentially for the last few decades such that it has long been anticipated that copper-based electronic interconnects will reach their theoretical limit and fail to satisfy such high demands in the near future. Based on advances in silicon photonics and silicon nitride platforms, photonic integrated devices are now playing a key role in addressing the bandwidth limit and offering solutions for rapidly evolving data, tele- and wireless communications networks. Silicon-based PICs leverage mature CMOS technology and offer promising advantages of low power, low cost and higher bandwidth through wavelength division multiplexing. PICs for communications include WDMs, transceivers, optical beamformers and microwave photonic circuits, to name a few. WDMs are of particular interest in optical communication as they play a key role in scaling up the capacity of optical interconnects and communication links. In Si_3N_4, WDMs can be realized by arrayed-waveguide gratings, rings and star couplers. A detailed review of demonstrated silicon photonics WDMs based on monolithically integrated Si_3N_4 arrayed-waveguide gratings and thermally tuneable silicon microring filters is given in [179]. Figure 11.15 shows an example of an advanced on-chip microwave photonics signal processor based on an optical beamformer chip for phase arrayed antennas [135]. Optical beamformers have recently been extensively developed for applications in wireless communication. Beamformers allow focusing a microwave signal from antennas into specific directions hence delivering high-quality signals

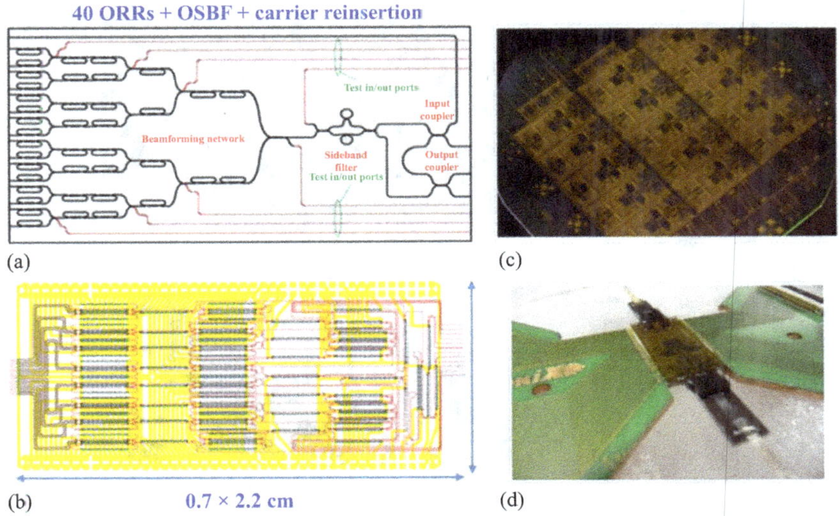

Figure 11.15 An Si_3N_4-based microwave photonic beamformer circuit, (a) schematic of a designed chip, (b) chip mask layout and (c and d) photos of the processed wafer and an individual chip with electrical interconnects and fibre pigtailing. © 2013 The Optical Society. Reprinted, with permission, from Reference [135]

to the receivers. Besides the beamformer integrated circuit shown in Figure 11.15, an overview of microwave photonic circuits based on the high-index-contrast TriPleX silicon nitride platform is given in [135].

Silicon nitride's ability to handle high optical powers by avoiding TPA, and low phase noise and scattering losses make it an ideal candidate for transmitting and receiving high-power free-space optical signals using phased arrays for LIDAR. Figure 11.16 demonstrates a phased array fabricated in Si_3N_4 from the Research Laboratory of Electronics at MIT and highlights its promise as a solution to long-range sources operating in the visible and infrared [180]. The chip shown in Figure 11.16(a) consists of 1,024 antennas making up a 4 mm^2 aperture of Si_3N_4-graded waveguides with a 4 μm pitch, as shown in Figure 11.16(b). A binary network of MMI splitters was used to distribute light which was coupled to the chip via lensed fibre. With 9.1 W, 1,550 nm input power at the facet, a 400 mW peak power was measured 0.5 m from the aperture, resulting in an overall efficiency of −13.6 dB. Due to the broadband transmission of Si_3N_4, the same approach was used for operation at 635 nm, where an antenna circuit with 2 μm pitch and 0.5 mm^2 footprint was fabricated. As displayed in Figure 11.16(c), similar binary MMI schemes were used to distribute the visible light to the array. Figure 11.16(d) shows the chip during measurement and the visible array output on a white card. Details about the phased arrays and their emission characteristics can be found in [180].

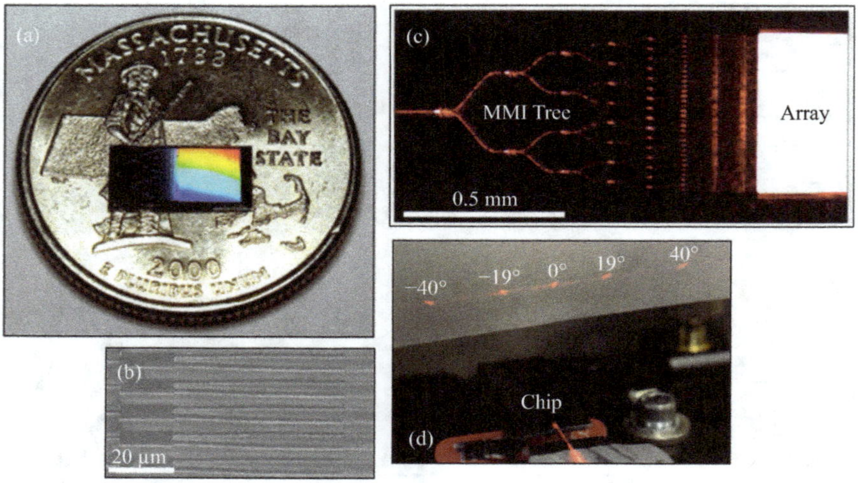

Figure 11.16 (a) Photograph of a 4 mm×4 mm Si_3N_4 infrared phased array chip. (b) SEM image of the last stage of the splitter tree and antennas. (c) Image of a 0.5 mm×0.5 mm phased array designed for visible light. (d) Photograph of output beams with 635-nm operation. © 2017 The Optical Society. Reprinted/adapted, with permission, from Reference [180]

The advanced functionalities available on SiN PIC platforms are motivating their use for many other applications in sensing, communications and ranging [181–184], and in emerging fields such as quantum information processing [185–187] and photonic neural networks [188,189]. Notably, monolithic and hybrid integration of silicon nitride on silicon photonics platforms and with InP chips, respectively, have opened the door to new active functionalities and microsystems. As an example, the recent development of rare-earth lasers on silicon nitride platforms has enabled the integration of erbium-doped waveguide lasers in full active SOI silicon photonics platforms for the first time, and they have been investigated in LIDAR [190], communications [191] and applications. Figure 11.17(a) shows an erbium DBR laser monolithically integrated with a silicon modulator and a germanium photodetector on an SOI chip, while Figure 11.17(b) displays the laser output spectrum and modulation and photodetection of a 2 Gbit/s signal on the same chip. Figure 11.17(c) shows a schematic of the multilayer Al_2O_3:Er^{3+}/Si_3N_4/SOI integration platform developed by researchers at MIT and SUNY Polytechnic Institute [191]. The integration of such rare-earth lasers and other novel devices into SiN PICs is allowing for a wide range of passive, active and non-linear optical functions to be carried out on a single low-cost photonic chip.

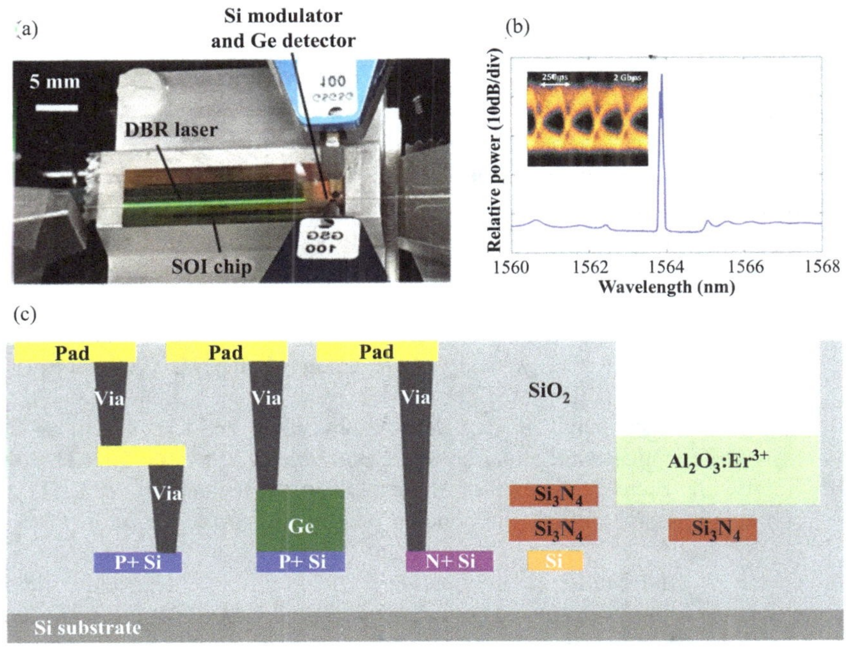

Figure 11.17 *(a) Image of a monolithic Er-doped laser on an SOI chip. (b) Laser output spectrum and eye diagram for a modulated and detected 2 Gbps signal measured on the same chip. (c) Schematic showing the layers in the 3D-integrated Al_2O_3:Er^{3+}/Si_3N_4/active SOI platform. Reprinted/adapted with from [191] under Creative Commons License*

11.5 Conclusion

Silicon nitride is a highly versatile waveguide material and a large range of passive, active and non-linear photonic SiN devices and PICs has been demonstrated in recent years. It offers a low-cost, ultra-low-loss waveguide platform that is both complementary to and enables new functionalities within III–V semiconductor and active silicon photonics platforms. Many silicon nitride fabrication platforms and MPW services are now available to researchers and industry which can be exploited to develop new devices and components operating from the visible to MIR for applications, including communications, ranging, metrology and sensing. Post-processing and hybrid integration of materials, including rare-earth-doped oxides, III–V semiconductors, phase-change materials and 2D materials using silicon nitride waveguides as a base, is enabling new high-performance integrated optical devices and functionalities. In the coming years, building new devices onto versatile silicon nitride platforms will allow photonics engineers to leverage its low losses and mature, low-cost and scalable chip manufacturing to transition new integrated optical components and microsystems from research lab to commercialization.

References

[1] Philipp H.R. 'Optical properties of silicon nitride'. *Journal of the Electrochemical Society*. 1973;**120**(2):295–300.

[2] Stutius W. and Streifer W. 'Silicon nitride films on silicon for optical waveguides'. *Applied Optics*. 1977;**16**(12):3218–3222.

[3] Hickernell F.S., Davis R.L., and Richard F.V. 'The acousto-optic properties of thin-film Si_3N_4, Ta_2O_5, ZnO and 7059 glass on oxidized silicon substrates'. *1978 Ultrasonics Symposium*; Cherry Hill, NJ, USA, Sep 1978 (New York, NY, IEEE, 1978), pp. 60–63.

[4] Henry C.H., Kazarinov R.F., Lee H.J., Orlowsky K.J., and Katz L.E. 'Low loss Si_3N_4–SiO_2 optical waveguides on Si'. *Applied Optics*. 1987;**26**(13):2621–2624.

[5] Valette S. 'State of the art of integrated optics technology at LETI for achieving passive optical components'. *Journal of Modern Optics*. 1988;**35**(6):993–1005.

[6] Gleine W. and Müller J. 'Low-pressure chemical vapor deposition silicon-oxynitride films for integrated optics'. *Applied Optics*. 1992;**31**(12):2036–2040.

[7] de Ridder R.M., Wörhoff K., Driessen A., Lambeck P.V., and Albers H. 'Silicon oxynitride planar waveguiding structures for application in optical communication'. *IEEE Journal of Selected Topics in Quantum Electronics*. 1998;**4**(6):930–937.

[8] Wörhoff K., Heideman R.G., Leinse A., and Hoekman M. 'TriPleX: a versatile dielectric photonic platform'. *Advanced Optical Technologies*. 2015;**4**(2): 189–207.

[9] Munoz P., Mico G., Bru L.A., *et al.* 'Silicon nitride photonic integration platforms for visible, near-infrared and mid-infrared applications'. *Sensors*. 2017;**17**(9):2088.

[10] Roeloffzen C.G.H., Hoekman M., Klein E.J., *et al.* 'Low-loss Si_3N_4 TriPleX optical waveguides: technology and applications overview'. *IEEE Journal of Selected Topics in Quantum Electronics*. 2018;**24**(4):4400321.

[11] Moss D.J., Morandotti R., Gaeta A.L., and Lipson M. 'New CMOS-compatible platforms based on silicon nitride and Hydex for nonlinear optics'. *Nature Photonics*. 2013;**7**(8):597–607.

[12] Tan D.T.H., Ooi K.J.A., and Ng D.K.T. 'Nonlinear optics on silicon-rich nitride—a high nonlinear figure of merit CMOS platform [Invited]'. *Photonics Research*. 2018;**6**(5):B50–B66.

[13] Baets R., Subramanian A.Z., Clemmen S., *et al.* 'Silicon photonics: silicon nitride versus silicon-on-insulator'. *OSA Technical Digest (online), Optical Fiber Communication Conference*; Anaheim, CA, USA, Mar 2016 (Washington, DC, Optical Society of America, 2016), paper Th3J.1.

[14] Rahim A., Ryckeboer E., Subramanian A.Z., *et al.* 'Expanding the silicon photonics portfolio with silicon nitride photonic integrated circuits'. *Journal of Lightwave Technology*. 2017;**35**(4):639–649.

[15] Blumenthal D.J., Heideman R., Geuzebroek D., Leinse A., and Roeloffzen C. 'Silicon nitride in silicon photonics'. *Proceedings of the IEEE*. 2018; **106**(12):2209–2231.

[16] Sacher W.D., Mikkelsen J.C., Huang Y., *et al.* 'Monolithically integrated multilayer silicon nitride-on-silicon waveguide platforms for 3-D photonic circuits and devices'. *Proceedings of the IEEE*. 2018;**106**(12):2232–2245.

[17] Ma T.P. 'Making silicon nitride film a viable gate dielectric'. *IEEE Transactions on Electron Devices*. 1998;**45**(3):680–690.

[18] Wörhoff K., Lambeck P.V., Albers H., Noordman O., van Hulst N.F., and Popma T.J.A. 'Optimization of LPCVD silicon oxynitride growth to large refractive index homogeneity and layer thickness uniformity'. *Proceedings of the SPIE 3099, Micro-Optical Technologies for Measurement, Sensors, and Microsystems II and Optical Fiber Sensor Technologies and Applications*; Munich, Germany, Sep 1997 (SPIE Digital Library, 1997).

[19] Bonnotte E., Gorecki C., Toshiyoshi H., *et al.* 'Guided-wave acoustooptic interaction with phase modulation in a ZnO thin-film transducer on an Si-based integrated Mach–Zehnder interferometer'. *Journal of Lightwave Technology*. 1999;**17**(1):35–42.

[20] Mao S.C., Tao S.H., Xu Y.L., *et al.* 'Low propagation loss SiN optical waveguide prepared by optimal low-hydrogen module'. *Optics Express*. 2008;**16**(25):20809–20816.

[21] Ay F. and Aydinli A. 'Comparative investigation of hydrogen bonding in silicon based PECVD grown dielectrics for optical waveguides'. *Optical Materials*. 2004;**26**(1):33–46.

[22] Mao S.C., Xu Y.L., and Lu G. 'Correspondence relation between [N-H]/[Si-H] ratio and their optical loss properties in silicon nitride thin films'. *4th IEEE Conference on Industrial Electronics and Applications*; Xi'an, China, 2009, pp. 3319–3322.

[23] Gorin A., Jaouad A., Grondin E., Aimez V., and Charette P. 'Fabrication of silicon nitride waveguides for visible-light using PECVD: a study of the effect of plasma frequency on optical properties'. *Optics Express*. 2008;**16**(18):13509–13516.

[24] Herth E., Legrand B., Buchaillot L., Rolland N., and Lasri T. 'Optimization of SiN_x:H films deposited by PECVD for reliability of electronic, microsystems and optical applications'. *Microelectronics Reliability*. 2010;**50**(8):1103–1106.

[25] Yota J., Hander J., and Saleh A.A. 'A comparative study on inductively-coupled plasma high-density plasma, plasma-enhanced, and low pressure chemical vapor deposition silicon nitride films'. *Journal of Vacuum Science & Technology A*. 2000;**18**(2):372–376.

[26] Yang C. and Pham J. 'Characteristic study of silicon nitride films deposited by LPCVD and PECVD'. *Silicon*. 2018;**10**(6):2561–2567.

[27] Wörhoff K., Driessen A., Lambeck P.V., Hilderink L.T.H., Linders P.W.C., and Popma T.J.A. 'Plasma enhanced chemical vapor deposition silicon oxynitride optimized for application in integrated optics'. *Sensors and Actuators A: Physical*. 1999;**74**(1–3):9–12.

[28] Wörhoff K., Lambeck P.V., and Driessen A. 'Design, tolerance analysis, and fabrication of silicon oxynitride based planar optical waveguide for communication devices'. *Journal of Lightwave Technology*. 1999;**17**(8):1401–1407.

[29] Esposito E.M., Mercaldo L.V., Delli Veneri P., Lancellotti L., and Privato C. 'Annealing effects on PECVD-grown Si rich SiN_x thin films'. *Energy Procedia*. 2010;**2**(1):159–164.

[30] Domínguez Bucio T., Khokhar A.Z., Lacava C., *et al.* 'Material and optical properties of low-temperature NH_3-free PECVD SiN_x layers for photonic applications'. *Journal of Physics D: Applied Physics*. 2017;**50**(2):025106.

[31] Hiraki T., Aihara T., Nishi H., and Tsuchizawa T. 'Deuterated SiN/SiON waveguides on Si platform and their application to C-band WDM filters'. *IEEE Photonics Journal*. 2017;**9**(5):2500207.

[32] Chiles J., Nader N., Hickstein D.D., *et al.* 'Deuterated silicon nitride photonic devices for broadband optical frequency comb generation'. *Optics Letters*. 2018;**43**(7):1527–1530.

[33] Roschuk T., Wojcik J., Tan X., Davies J.A., and Mascher P. 'Optical and compositional characterization of SiO_xN_y and SiO_x thin films deposited by electron cyclotron resonance plasma enhanced chemical vapor deposition'. *Journal of Vacuum Science & Technology A*. 2004;**22**(3):883–886.

[34] Bååk T. 'Silicon oxynitride; a material for GRIN optics'. *Applied Optics*. 1982;**21**(6):1069–1072.

[35] Luke K., Okawachi Y., Lamont M.R., Gaeta A.L., and Lipson M. 'Broadband mid-infrared frequency comb generation in a Si_3N_4 microresonator'. *Optics Letters*. 2015;**40**(21):4823–4826.

[36] Liu Y., Lin I.K., and Zhang X. 'Mechanical properties of sputtered silicon oxynitride films by nanoindentation'. *Materials Science and Engineering: A*. 2008;**489**(1–2):294–301.

[37] Frigg A., Boes A., Ren G., *et al.* 'Low loss CMOS-compatible silicon nitride photonics utilizing reactive sputtered thin films'. *Optics Express*. 2019; **27**(26):37795–37805.

[38] Colman P., Husko C., Combrié S., Sagnes I., Wong C.W., and De Rossi A. 'Temporal solitons and pulse compression in photonic crystal waveguides'. *Nature Photonics*. 2010;**4**:862–868.

[39] Safioui J., Leo F., Kuyken B., *et al.* 'Supercontinuum generation in hydrogenated amorphous silicon waveguides at telecommunication wavelengths'. *Optics Express*. 2014;**22**(3):3089–3097.

[40] Shoji Y., Ogasawara T., Kamei T., *et al.* 'Ultrafast nonlinear effects in hydrogenated amorphous silicon wire waveguide'. *Optics Express*. 2010; **18**(6):5668–5673.

[41] Miller R.C. 'Optical second harmonic generation in piezoelectric crystals'. *Applied Physics Letters*. 1964;**5**(1):17–19.

[42] Marin-Palomo P., Kemal J.N., Karpov M., *et al.* 'Microresonator-based solitons for massively parallel coherent optical communications'. *Nature*. 2017;**546**:274–279.

[43] Krückel C.J., Fülöp A., Klintberg T., Bengtsson J., Andrekson P.A., and Torres-Company V. 'Linear and nonlinear characterization of low-stress high-confinement silicon-rich nitride waveguides'. *Optics Express*. 2015; **23**(20):25827–25837.

[44] Lin Q., Painter O.J., and Agrawal G.P. 'Nonlinear optical phenomena in silicon waveguides: modeling and applications'. *Optics Express*. 2007; **15**(25):16604–16644.

[45] Lacava C., Stankovic S., Khokhar A.Z., *et al.* 'Si-rich silicon nitride for nonlinear signal processing applications'. *Scientific Reports*. 2017;**7**(1):22.

[46] Krückel C.J., Fülöp A., Ye Z., Andrekson P.A., and Torres-Company V. 'Optical bandgap engineering in nonlinear silicon nitride waveguides'. *Optics Express*. 2017;**25**(13):15370–15380.

[47] Gondarenko A., Levy J.S., and Lipson M. 'High confinement micron-scale silicon nitride high Q ring resonator'. *Optics Express*. 2009;**17**(14): 11366–11370.

[48] Chang J.S., Eom S.C., Sung G.Y., and Shin J.H. 'On-chip, planar integration of Er doped silicon-rich silicon nitride microdisk with SU-8 waveguide with sub-micron gap control'. *Optics Express*. 2009;**17**(25):22918–22924.

[49] Pirasteh P., Charrier J., Dumeige Y., *et al.* 'Guided photoluminescence study of Nd-doped silicon rich silicon oxide and silicon rich silicon nitride waveguides'. *Journal of Applied Physics*. 2013;**114**(1):014906.

[50] Li R., Yerci S., Kucheyev S.O., van Buuren T., and Dal Negro L. 'Energy transfer and stimulated emission dynamics at 1.1 μm in Nd-doped SiN_x'. *Optics Express*. 2011;**19**(6):5379–5385.

[51] Dal Negro L., Li R., Warga J., and Basu S.N. 'Sensitized erbium emission from silicon-rich nitride/silicon superlattice structures'. *Applied Physics Letters*. 2008;**92**(18):181105.

[52] Yerci S., Li R., Kucheyev S.O., van Buuren T., Basu S.N., and Dal Negro L. 'Energy transfer and 1.54 μm emission in amorphous silicon nitride films'. *Applied Physics Letters*. 2009;**95**(3):031107.

[53] Sandland J.G. Sputtered Silicon Oxynitride for Microphotonics: A Materials Study. Ph.D. Thesis. Boston, MA: Massachusetts Institute of Technology; 2005.

[54] Labbe C., An Y.T., Zatryb G., *et al.* 'Structural and emission properties of Tb^{3+}-doped nitrogen-rich silicon oxynitride films'. *Nanotechnology*. 2017;**28**(11):115710.

[55] Ji X., Barbosa F.A.S., Roberts S.P., *et al.* 'Ultra-low-loss on-chip resonators with sub-milliwatt parametric oscillation threshold'. *Optica*. 2017;**4**(6): 619–624.

[56] El Dirani H., Youssef L., Petit-Etienne C., *et al.* 'Ultralow-loss tightly confining Si_3N_4 waveguides and high-Q microresonators'. *Optics Express*. 2019;**27**(21):30726–30740.

[57] Chen L., Xu L., Li D., and Lin B. 'Mechanism of selective Si_3N_4 etching over SiO_2 in hydrogen-containing fluorocarbon plasma'. *Microelectronic Engineering*. 2009;**86**(11):2354–2357.

[58] Ito T., Karahashi K., Fukasawa M., Tatsumi T., and Hamaguchi S. 'Hydrogen effects in hydrofluorocarbon plasma etching of silicon nitride: beam study with CF^+, $CF_2{}^+$, $CHF_2{}^+$, and CH_2F^+ ions'. *Journal of Vacuum Science and Technology A*. 2011;**29**(5):*050601*.

[59] Bauters J.F., Heck M.J.R., John D.D., *et al.* 'Planar waveguides with less than 0.1 dB/m propagation loss fabricated with wafer bonding'. *Optics Express*. 2011;**19**(24):24090–24101.

[60] Luke K., Dutt A., Poitras C.B., and Lipson M. 'Overcoming Si_3N_4 film stress limitations for high quality factor ring resonators'. *Optics Express*. 2013; **21**(19):22829–22833.

[61] Epping J.P., Hoekman M., Mateman R., *et al.* 'High confinement, high yield Si_3N_4 waveguides for nonlinear optical applications'. *Optics Express*. 2015;**23**(2):642–648.

[62] Pfeiffer M.H.P., Kordts A., Brasch V., *et al.* 'Photonic damascene process for integrated high-Q microresonator based nonlinear photonics'. *Optica*. 2016;**3**(1):20–25.

[63] Morichetti F., Melloni A., Martinelli M., *et al.* 'Box-shaped dielectric waveguides: a new concept in integrated optics?'. *Journal of Lightwave Technology*. 2007; **25**(9):2579–2589.

[64] Purnawirman, Sun J., Adam T.N., *et al.* 'C- and L-band erbium-doped waveguide lasers with wafer-scale silicon nitride cavities'. *Optics Letters*. 2013;**38**(11):1760–1762.

[65] Belt M., Huffman T., Davenport M.L., Li W., Barton J.S., and Blumenthal D.J. 'Arrayed narrow linewidth erbium-doped waveguide-distributed feedback lasers on an ultra-low-loss silicon-nitride platform'. *Optics Letters*. 2013;**38**(22):4825–4828.

[66] Purnawirman, Li N., Magden E.S., *et al.* 'Ultra-narrow-linewidth Al_2O_3: Er^{3+} lasers with a wavelength-insensitive waveguide design on a wafer-scale silicon nitride platform'. *Optics Express*. 2017;**25**(12):13705–13713.

[67] Pollnau M. and Bradley J.D.B. 'Optically pumped rare-earth-doped Al_2O_3 distributed-feedback lasers on silicon [Invited]'. *Optics Express*. 2018;**26**(18):24164–24189.

[68] Frankis H.C., Miarabbas Kiani K., Su D., Mateman R., Leinse A., and Bradley J.D.B. 'High-Q tellurium-oxide-coated silicon nitride microring resonators'. *Optics Letters*. 2019;**44**(1):118–121.

[69] Frankis H.C., Miarabbas Kiani K., Bonneville D.B., *et al.* 'Low-loss TeO_2-coated Si_3N_4 waveguides for application in photonic integrated circuits.' *Optics Express*. 2019;**27**(9):12529–12540.

[70] Miarabbas Kiani K., Frankis H.C., Mbonde H.M., *et al.* 'Thulium-doped tellurium oxide waveguide amplifier with 7.6 dB net gain on a silicon nitride chip'. *Optics Letters*. 2019;**44**(23):5788–5791.

[71] Frankis H.C., Mbonde H.M., Bonneville D.B., *et al.* 'Erbium-doped TeO_2-coated Si_3N_4 waveguide amplifiers with 5 dB net gain'. *Photonics Research.* 2020;**8**(2):127–134.

[72] Mbonde H.M., Frankis H.C., and Bradley J.D.B. 'Enhanced nonlinearity and engineered anomalous dispersion in TeO_2-coated Si_3N_4 waveguides'. *IEEE Photonics Journal.* 2020;**12**(2):2200210.

[73] Barwicz T., Popović M.A., Rakich P.T., *et al.* 'Microring-resonator-based add-drop filters in SiN: fabrication and analysis'. *Optics Express.* 2004; **12**(7):1437–1442.

[74] Daldosso N., Melchiorri M., Riboli F., *et al.* 'Comparison among various Si_3N_4 waveguide geometries grown within a CMOS fabrication pilot line'. *Journal of Lightwave Technology.* 2004;**22**(7):1734–1740.

[75] Bauters J.F., Heck M.J.R., John D., *et al.* 'Ultra-low-loss high-aspect-ratio Si_3N_4 waveguides'. *Optics Express.* 2011;**19**(4):3163–3174.

[76] Shah Hosseini E., Yegnanarayanan S., Hossein Atabaki A., Soltani M., and Adibi A. 'High quality planar silicon nitride microdisk resonators for integrated photonics in the visible wavelength range'. *Optics Express.* 2009; **17**(17):14543–14551.

[77] Rahim A., Goyvaerts J., Szelag B., *et al.* 'Open-access silicon photonics platforms in Europe'. *IEEE Journal of Selected Topics in Quantum Electronics.* 2019;**25**(5):8200818.

[78] Chrostowski L., Shoman H., Hammood M., *et al.* 'Silicon photonic circuit design using rapid prototyping foundry process design kits'. *IEEE Journal of Selected Topics in Quantum Electronics.* 2019;**25**(5):8201326.

[79] Pfeiffer M.H.P., Herkommer C., Liu J., *et al.* 'Photonic damascene process for low-loss high-confinement silicon nitride waveguides [Invited]'. *IEEE Journal of Selected Topics in Quantum Electronics.* 2018;**24**(4):6101411.

[80] Shaw M.J., Guo J., Vawter G.A., Habermehl S., and Sullivan C.T. 'Fabrication techniques for low-loss silicon nitride waveguides'. *Proceedings of the SPIE 5720, Micromachining Technology for Micro-Optics and Nano-Optics III*; San Jose, CA, USA, Jan 2005 (Bellingham, WA, SPIE, 2005), pp. 109–118.

[81] Guo J., Shaw M.J., Vawter G.A., Hadley G.R., Esherick P., and Sullivan C. T. 'High-Q microring resonator for biochemical sensors'. *Proceedings of the SPIE 5728, Integrated Optics: Devices, Materials, and Technologies*; San Jose, CA, USA, Jan 2005 (Bellingham, WA, SPIE, 2005), pp. 83–92.

[82] Wang J., Yao Z., and Poon A.W. 'Silicon-nitride-based integrated optofluidic biochemical sensors using a coupled-resonator optical waveguide'. *Frontiers in Materials.* 2015;**2**:34.

[83] Calvo M., Guerber S., Beaudin G., *et al.* 'Ring resonator designed for biosensing applications manufactured on 300 mm SOI in an industrial environment'. *Japanese Journal of Applied Physics.* 2019;**58**(SB):SBBE02.

[84] Subramanian A.Z., Neutens P., Dhakal A., *et al.* 'Low-loss singlemode PECVD silicon nitride photonic wire waveguides for 532–900 nm

wavelength window fabricated within a CMOS pilot line'. *IEEE Photonics Journal*. 2013;**5**(6):2202809.

[85] Bradley J.D.B., Shah Hosseini E., Purnawirman, *et al.* 'Monolithic erbium- and ytterbium-doped microring lasers on silicon chips'. *Optics Express*. 2014;**22**(10):12226–12237.

[86] Coolbaugh D., Adam T.N., Leake G.L., *et al.* 'Integrated silicon photonics fabrication on a 300mm platform.' *Chip Scale Review*. 2013;**17**(3):20–23.

[87] Huang Y., Song J., Luo X., Liow T.Y., and Lo G.Q. 'CMOS compatible monolithic multi-layer Si_3N_4-on-SOI platform for low-loss high performance silicon photonics dense integration'. *Optics Express*. 2014;**22**(18):21859–21865.

[88] Sacher W.D., Huang Y., Lo G.-Q., and Poon J.K.S. 'Multilayer silicon nitride-on-silicon integrated photonic platforms and devices'. *Journal of Lightwave Technology*. 2015;**33**(4):901–910.

[89] Wilmart Q., El Dirani H., Tyler N., *et al.* 'A versatile silicon-silicon nitride photonics platform for enhanced functionalities and applications'. *Applied Sciences*. 2019;**9**(2):255.

[90] Mak J.C.C., Wilmart Q., Olivier S., Menezo S., and Poon J.K.S. 'Silicon nitride-on-silicon bi-layer grating couplers designed by a global optimization method'. *Optics Express*. 2018;**26**(10):13656–13665.

[91] Lentine A.L., DeRose C.T., Davids P.S., *et al.* 'Silicon photonics platform for national security applications'. *2015 IEEE Aerospace Symposium*; Big Sky, MT, USA, Mar 2015 (New York, NY, IEEE, 1978), pp. 1–9.

[92] Hosseinnia A.H., Atabaki A.H., Eftekhar A.A., and Adibi A. 'High-quality silicon on silicon nitride integrated optical platform with an octave-spanning adiabatic interlayer coupler'. *Optics Express*. 2015;**23**(23):30297–30307.

[93] Wu K. and Poon A.W. 'Waveguide-coupled disk resonators on a crack-free Si_3N_4 film with a dense stress release pattern'. *OSA Technical Digest (online), Conference on Lasers and Electro-Optics*; San Jose, CA, USA, May 2019 (Washington, DC, Optical Society of America, 2019), paper SF2I.4.

[94] Lin P.T., Sing V., Kimerling L., and Agarwal A.M. 'Planar silicon nitride mid-infrared devices'. *Applied Physics Letters*. 2013;**102**(25):251121.

[95] Trenti A., Borghi M., Biasi S., *et al.* 'Thermo-optic coefficient and nonlinear refractive index of silicon oxynitride waveguides'. *AIP Advances*. 2018;**8**(2):025311.

[96] Huffman T.A., Brodnik G.M., Pinho C., Gundavarapu S., Baney D., and Blumenthal D.J. 'Integrated resonators in an ultralow loss Si_3N_4/SiO_2 platform for multifunction applications'. *IEEE Journal of Selected Topics in Quantum Electronics*. 2018;**24**(4):5900209.

[97] Dekker R. All-Optical Processes in Integrated Optical Devices using Materials with Large Third-Order Nonlinearities and Gain. Ph.D. Thesis. Enschede: University of Twente; 2006.

[98] Sepúlveda B., Sánchez del Río J., Moreno M., *et al.* 'Optical biosensor microsystems based on the integration of highly sensitive Mach–Zehnder

interferometer devices'. *Journal of Optics A: Pure and Applied Optics*. 2006;**8**:S561–S566.

[99] Bogaerts W., De Heyn P., Van Vaerenbergh T., *et al.* 'Silicon microring resonators'. *Laser & Photonics Reviews*. 2012;**6**(1):47–73.

[100] Rabiei P., Steier W.H., Zhang C., and Dalton L.R. 'Polymer micro-ring filters and modulators'. *Journal of Lightwave Technology*. 2002;**20**(11): 1968–1975.

[101] Little B.E., Chu S.T., Haus H.A., Foresi J., and Laine J.-P. 'Microring resonator channel dropping filters'. *Journal of Lightwave Technology*. 1997;**15**(6):998–1005.

[102] Watts M.R., Barwicz T., Popović M.A., *et al.* 'High-index-contrast microphotonics, from concept to implementation'. *OSA Technical Digest (online), Conference on Lasers and Electro-Optics*; Long Beach, CA, USA, May 2006 (Washington, DC, Optical Society of America, 2006), paper CTuM3.

[103] Martens D., Subramanian A.Z., Pathak S., *et al.* 'Compact silicon nitride arrayed waveguide gratings for very near-infrared wavelengths'. *IEEE Photonics Technology Letters*. 2015;**27**(2):137–140.

[104] Shang K., Pathak S., Qin C., and Ben Yoo S.J. 'Low-loss compact silicon nitride arrayed waveguide gratings for photonic integrated circuits'. *IEEE Photonics Journal*. 2017;**9**(5):6601805.

[105] Gattass R.R. and Mazur E. 'Femtosecond laser micromachining in transparent materials'. *Nature Photonics*. 2008;**2**(4):219–225.

[106] Eldada L., Xu C., Stengel K.M.T., Shacklette L.W., and Yardley J.T. 'Laser-fabricated low-loss single-mode raised-rib waveguiding devices in polymers'. *Journal of Lightwave Technology*. 1996;**14**(7):1704–1713.

[107] Pelli S., Righini G.C., Scaglione A., Guglielmi M., and Martucci A. 'Direct laser writing of ridge optical waveguides in silica-titania glass sol-gel films'. *Optical Materials*. 1996;**5**(1–2):119–126.

[108] Kiani A., Venkatakrishnan K., Tan B., and Venkataramanan V. 'Maskless lithography using silicon oxide etch- stop layer induced by megahertz repetition femtosecond laser pulses'. *Optics Express*. 2011;**19**(11): 10834–10842.

[109] Bonneville D.B., Méndez-Rosales M.A., Frankis H.C., Gonçalves L.M., Kleiman R.N., and Bradley J.D.B. 'Flexible and low-cost fabrication of optical waveguides by UV laser resist-mask writing'. *Optical Materials Express*. 2019;**9**(4):1728–1737.

[110] Schmidt S., Flueckiger J., Wu W., *et al.* 'Improving the performance of silicon photonic rings, disks, and Bragg gratings for use in label-free biosensing'. *Proceedings of the SPIE 91660M, Biosensing and Nanomedicine VII*; San Diego, CA, USA, Sep 2014 (SPIE Digital Library, 2014).

[111] Ryckeboer E., Nie X., Subramanian A., *et al.* 'CMOS-compatible silicon nitride spectrometers for lab-on-a-chip spectral sensing'. *Proceedings of the SPIE 9891, Silicon Photonics and Photonic Integrated Circuits V*; Brussels, Belgium, May 2016 (SPIE Digital Library, 2016).

[112] Dhakal A., Peyskens F., Clemmen S., *et al.* 'Single mode waveguide platform for spontaneous and surface-enhanced on-chip Raman spectroscopy'. *Interface Focus*. 2016;**6**(4):*20160015*.

[113] Clemmen S., Raza A., Dhakal A., *et al.* 'Spectroscopic sensing with silicon nitride photonic integrated circuits'. *Proceedings of the SPIE* 10106, *Integrated Optics: Devices, Materials, and Technologies XXI*; San Francisco, CA, USA, Feb 2017 (SPIE Digital Library, 2017).

[114] Ksendzov A. and Lin Y. 'Integrated optics ring-resonator sensors for protein detection'. *Optics Letters*. 2005;**30**(24):3344.

[115] He L., Özdemir Ş.K., Zhu J., Kim W., and Yang L. 'Detecting single viruses and nanoparticles using whispering gallery microlasers'. *Nature Nanotechnology*. 2011;**6**(7):428–432.

[116] Goykhman I., Desiatov B., and Levy U. 'Ultrathin silicon nitride microring resonator for biophotonic applications at 970 nm wavelength'. *Applied Physics Letters*. 2010;**97**(8):081108.

[117] Su Z., Li N., Frankis H.C., *et al.* High-Q-factor Al_2O_3 micro-trench cavities integrated with silicon nitride waveguides on silicon'. *Optics Express*. 2018;**26**(9):11161–11170.

[118] Frankis H.C., Su D., Bonneville D.B., and Bradley J.D.B. 'A tellurium oxide microcavity resonator sensor integrated on-chip with a silicon waveguide'. *Sensors*. 2018;**18**(11):4061.

[119] Agrawal G., *Nonlinear Fiber Optics (Fifth Edition)*. Cambridge: Academic Press; 2013.

[120] Gaeta A.L., Lipson M., and Kippenberg T.J. 'Photonic-chip-based frequency combs'. *Nature Photonics*. 2019;**13**(3):158–169.

[121] Tan D.T.H., Ikeda K., Sun P.C., and Fainman Y. 'Group velocity dispersion and self-phase modulation in silicon nitride waveguides'. *Applied Physics Letters*. 2010;**96**(6):061101.

[122] Krückel C.J., Torres-Company V., Andrekson P.A., *et al.* 'Continuous wave-pumped wavelength conversion in low-loss silicon nitride waveguides'. *Optics Letters*. 2015;**40**(6):875–878.

[123] Grassani D., Pfeiffer M.H.P., Kippenberg T.J., and Brès C.-S. 'Second- and third-order nonlinear wavelength conversion in an all-optically poled Si_3N_4 waveguide'. *Optics Letters*. 2019;**44**(1):106–109.

[124] Yang M., Xu L., Wang J., *et al.* 'An octave-spanning optical parametric amplifier based on a low-dispersion silicon-rich nitride waveguide'. *IEEE Journal of Selected Topics in Quantum Electronics*. 2018;**24**(6):1–7.

[125] Ooi K.J., Ng D.K., Choi J., *et al.* 'High gain optical parametric amplification in ultra-silicon-rich nitride (USRN) waveguides'. *Proceedings of the SPIE 10684, Nonlinear Optics and its Applications*; Strasbourg, France, May 2018 (SPIE Digital Library, 2018).

[126] Epping J.P. Dispersion Engineering Silicon Nitride Waveguides for Broadband Nonlinear Frequency Conversion. Ph.D. Thesis. Enschede: University of Twente; 2015.

[127] Chavez Boggio J.M., Bodenmüller D., Fremberg T., *et al.* 'Dispersion engineered silicon nitride waveguides by geometrical and refractive-index optimization'. *Journal of the Optical Society of America B*. 2014;**31**(11): 2846–2857.

[128] Madden S.J. and Vu K.T. 'Very low loss reactively ion etched tellurium dioxide planar rib waveguides for linear and non-linear optics'. *Optics Express*. 2009;**17**(20):17645–17651.

[129] Porcel M.A.G., Schepers F., Epping J.P., *et al.* 'Two-octave spanning supercontinuum generation in stoichiometric silicon nitride waveguides pumped at telecom wavelengths'. *Optics Express*. 2017;**25**(2):1542–1554.

[130] Zhao H., Kuyken B., Clemmen S., *et al.* 'Visible-to-near-infrared octave spanning supercontinuum generation in a silicon nitride waveguide'. *Optics Letters*. 2015;**40**(10):2177–2180.

[131] Kim B.Y., Okawachi Y., Jang J.K., *et al.* 'Turn-key, high-efficiency Kerr comb source'. *Optics Letters*. 2019;**44**(18):4475–4478.

[132] Kippenberg T.J., Gaeta A.L., Lipson M., and Gorodetsky M.L. 'Dissipative Kerr solitons in optical microresonators'. *Science*. 2018;**361**(6402): eaan8083.

[133] Yu S.-P., Briles T.C., Moille G.T., *et al.* 'Tuning Kerr-soliton frequency combs to atomic resonances'. *Physical Review Applied*. 2019;**11**(4): 044017.

[134] Xue X., Weiner A.M., and Qi M. 'Kerr combs from normal and anomalous dispersion silicon nitride microresonators'. *Proceedings of the SPIE 10090, Laser Resonators, Microresonators, and Beam Control XIX*; San Francisco, CA, USA, Feb 2017 (SPIE Digital Library, 2017).

[135] Roeloffzen C.G.H., Zhuang L., Taddei C., *et al.* 'Silicon nitride microwave photonic circuits'. *Optics Express*. 2013;**21**(19):22937–22961.

[136] Jin S., Xu L., Zhang H., and Li Y. '$LiNbO_3$ thin-film modulators using silicon nitride surface ridge waveguides'. *IEEE Photonics Technology Letters*. 2016;**28**(7):736–739.

[137] Chang L., Pfeiffer M.H.P., Volet N., *et al.* 'Heterogeneous integration of lithium niobite and silicon nitride waveguides for wafer-scale photonic integrated circuits on silicon'. *Optics Letters*. 2017;**42**(4):803–806.

[138] Ahmed A.N.R., Shi S., Zablocki M., Yao P., and Prather D.W. 'Tunable hybrid silicon nitride and thin-film lithium niobite electro-optic micro-resonator'. *Optics Letters*. 2017;**44**(3):618–621.

[139] Hiraki T., Aihara T., Takeda K., *et al.* 'Membrane InGaAsP Mach–Zehnder modulator with SiN:D waveguides on Si platform'. *Optics Express*. 2019;**27**(13):18612–18619.

[140] Alexander K., George J.P., Verbist J., *et al.* 'Nanophotonic Pockels modulators on a silicon nitride platform'. *Nature Communications*. 2018;**9**(1): 3444.

[141] Hosseini N., Dekker R., Hoekman M., *et al.* 'Stress-optic modulator in TriPleX platform using a piezoelectric lead zirconate titanate (PZT) thin film'. *Optics Express*. 2015;**23**(11):14018–14026.

[142] Munk D., Katzman M., Hen M., *et al.* 'Surface acoustic wave photonic devices in silicon on insulator'. *Nature Communications*. 2019;**10**(1):4214.

[143] Wu C., Yu H., Li H., Zhang X., Takeuchi I., and Li M. 'Low-loss integrated photonic switch using subwavelength patterned phase change material'. *ACS Photonics*. 2019;**6**(1):87–92.

[144] Ahn D., Hong C.-Y., Liu J., *et al.* 'High performance, waveguide integrated Ge photodetectors'. *Optics Express*. 2007;**15**(7):3916–3921.

[145] Lischke S., Knoll D., Mai C., *et al.* 'Silicon nitride waveguide coupled 67+ GHz Ge photodiode for non-SOI PIC and ePIC platforms'. *2019 IEEE International Electron Devices Meeting (IEDM)*; San Francisco, CA, USA, Dec 2019 (New York, NY, IEEE, 2019), pp. 33.2.1–33.2.4.

[146] Piels M., Bauters J.F., Davenport M.L., Heck M.J.R., and Bowers J.E. 'Low-loss silicon nitride AWG demultiplexer heterogeneously integrated with hybrid III–V/silicon photodetectors'. *Journal of Lightwave Technology*. 2014;**32**(4):817–823.

[147] Wang J., Cheng Z., Chen Z., Xu J.-B., Tsang H.K., and Shu C. 'Graphene photodetector integrated on silicon nitride waveguide'. *Journal of Applied Physics*. 2015;**117**(14):144504.

[148] Gao Y., Tao L., Tsang H.K., and Shu C. 'Graphene-on-silicon nitride waveguide photodetector with interdigital contacts'. *Applied Physics Letters*. 2018;**112**(21):211107.

[149] Bauters J.F., Davenport M.L., Heck M.J.R., *et al.* 'Silicon on ultra-low-loss waveguide photonic integration platform'. *Optics Express*. 2013;**21**(1): 544–555.

[150] Pintus P., Andriolli N., Di Pasquale F., and Bowers J.E. 'Integrated bidirectional optical amplifier for crosstalk-free WDM communication'. *2013 Optical Interconnects Conference*; Santa Fe, NM, USA, May 2013 (New York, NY, IEEE, 2013), pp. 106–107.

[151] Bradley J.D.B., Agazzi L., Geskus D., Ay F., Wörhoff K., and Pollnau M. 'Gain bandwidth of 80 nm and 2 dB/cm peak gain in Al_2O_3:Er^{3+} optical amplifiers on silicon'. *Journal of the Optical Society of America B*. 2010;**27** (2):187–196.

[152] Vázquez-Córdova S.A., Dijkstra M., Bernhardi E.H., *et al.* 'Erbium-doped spiral amplifiers with 20 dB of net gain on silicon'. *Optics Express*. 2014;**22**(21):25993–26004.

[153] Mu J., Dijkstra M., and García-Blanco S.M. 'Resonant coupling for active-passive monolithic integration of Al_2O_3 and Si_3N_4'. *IEEE Photonics Technology Letters*. 2019;**31**(10):771–774.

[154] García-Blanco S.M., van Emmerik C.I., Mu J., de Goede M., Dijkstra M., and Chang L. 'On-chip amplifiers and lasers on the Al_2O_3 integrated photonics platform'. *Proceedings of the SPIE* 10923, *Silicon Photonics XIV*; San Francisco, CA, USA, Feb 2019 (SPIE Digital Library, 2019).

[155] Rönn J., Zhang W., Autere A., *et al.* 'Ultra-high on-chip optical gain in erbium-based hybrid slot waveguides'. *Nature Communications*. 2019;**10**:432.

[156] Arbabi A. and Goddard L. 'Measurements of the refractive indices and thermo-optic coefficients of Si_3N_4 and SiO_x using microring resonances'. *Optics Letters* 2013;**38**: 3878–3381.

[157] Fan Y., Epping J.P., Oldenbeuving R.M., *et al.* 'Optically integrated InP–Si_3N_4 hybrid laser'. *IEEE Photonics Journal*. 2016;**8**(6):1–11.

[158] Stern B., Ji X., Dutt A., and Lipson M. 'Compact narrow-linewidth integrated laser based on a low-loss silicon nitride ring resonator'. *Optics Letters*. 2017;**42**(21):4541–4544.

[159] Iadanza S., Bakoz A.P., Singaravelu P.K.J., *et al.* 'Thermally stable hybrid cavity laser based on silicon nitride gratings'. *Applied Optics*. 2018;**57**(22): E218–E223.

[160] Xiang C., Morton P.A., and Bowers J.E. 'Ultra-narrow linewidth laser based on a semiconductor gain chip and extended Si_3N_4 Bragg grating'. *Optics Letters*. 2019;**44**(15):3825–3828.

[161] Lin Y., Browning C., Timens R.B., *et al.* 'Characterization of hybrid InP-TriPleX photonic integrated tunable lasers based on silicon nitride (Si_3N_4/SiO_2) microring resonators for optical coherent system'. *IEEE Photonics Journal*. 2018;**10**(3):1–8.

[162] Tran M.A., Huang D., and Bowers J.E. 'Tutorial on narrow linewidth tunable semiconductor lasers using Si/III-V heterogeneous integration'. *APL Photonics*. 2019;**4**(11):111101.

[163] Boller K.-J., van Rees A., Fan Y., *et al.* 'Hybrid integrated semiconductor lasers with silicon nitride feedback circuits'. *Photonics*. 2020;**7**(1):4.

[164] Wörhoff K., Bradley J.D.B., Ay F., Geskus D., Blauwendraat T.P., and Pollnau M. 'Reliable low-cost fabrication of low-loss Al_2O_3:Er^{3+} waveguides with 5.4-dB optical gain'. *IEEE Journal of Quantum Electronics*. 2009;**45**(5):454–461.

[165] Belt M. and Blumenthal D.J. 'Erbium-doped waveguide DBR and DFB laser arrays integrated within an ultra-low-loss Si_3N_4 platform'. *Optics Express*. 2014;**22**(9):10655–10660.

[166] Bernhardi E.H., van Wolferen H.A.G.M., Agazzi L., *et al.* 'Ultra-narrow-linewidth, single-frequency distributed feedback waveguide laser in Al_2O_3:Er^{3+} on silicon'. *Optics Letters*. 2010;**35**(14):2394–2396.

[167] Shah Hosseini E., Purnawirman, Sun J., *et al.* 'CMOS compatible 75 mW erbium-doped distributed feedback laser'. *Optics Letters*. 2014;**39**(11): 3106–3109.

[168] Li N., Purnawirman, Su Z., *et al.* 'High-power thulium lasers on a silicon photonics platform'. *Optics Letters*. 2017;**42**(6):1181–1184.

[169] Li N., Magden E.S., Su Z., *et al.* 'Broadband 2-μm emission on silicon chips: monolithically integrated holmium lasers'. *Optics Express*. 2018; **26**(3):2220–2230.

[170] Su Z., Li N., Magden E.S., *et al.* 'Ultra-compact and low-threshold thulium microcavity laser monolithically integrated on silicon'. *Optics Letters*. 2016;**41**(24):5708–5711.

[171] Kärtner F.X., Callahan P.T., Shtyrkova K., *et al.* 'Integrated rare-earth doped mode-locked lasers on a CMOS platform'. *Proceedings of the SPIE 10686, Silicon Photonics: From Fundamental Research to Manufacturing*; Strasbourg, France, May 2018 (SPIE Digital Library, 2018).

[172] Shtyrkova K., Callahan P.T., Li N., *et al.* 'Integrated CMOS-compatible Q-switched mode-locked lasers at 1900nm with an on-chip artificial saturable absorber'. *Optics Express*. 2019;**27**(3):3542–3556.

[173] Purnawirman, Magden E.S., Singh G., *et al.* 'Wavelength division multiplexed light source monolithically integrated on a silicon photonics platform'. *Optics Letters*. 2017;**42**(9):1772–1775.

[174] Li N., Vermeulen D., Su Z., *et al.* 'Monolithically integrated erbium-doped tunable laser on a CMOS-compatible silicon photonics platform'. *Optics Express*. 2018;**26**(13):16200–16211.

[175] Xin M., Li N., Singh N., *et al.* 'Optical frequency synthesizer with an integrated erbium tunable laser'. *Light: Science & Applications*. 2019;**8**:122.

[176] Cegielski P.J., Giesecke A.L., Neutzner S., *et al.* 'Monolithically integrated perovskite semiconductor lasers on silicon photonic chips by scalable top-down fabrication'. *Nano Letters*. 2018;**18**(11):6915–6923.

[177] Vogelbacher F., Sagmeister M., Kraft J., *et al.* 'Slot-waveguide silicon nitride organic hybrid distributed feedback laser'. *Scientific Reports*. 2019;**9**:18438.

[178] Gundavarapu S., Brodnik G.M., Puckett M., *et al.* 'Sub-hertz fundamental linewidth photonic integrated Brillouin laser'. *Nature Photonics*. 2019;**13**:60–67.

[179] Dong P. 'Silicon photonic integrated circuits for wavelength division multiplexing applications'. *IEEE Journal of Selected Topics in Quantum Electronics*. 2016;**22**(6):370–378.

[180] Poulton C.V., Byrd M.J., Raval M., *et al.* 'Large-scale silicon nitride nanophotonic phased arrays at infrared and visible wavelengths'. *Optics Letters*. 2017;**42**(1):21–24.

[181] Heideman R., Hoekman M., and Schreuder E. 'TriPleX-based integrated optical ring resonators for lab-on-a-chip and environmental detection'. *IEEE Journal of Selected Topics in Quantum Electronics*. 2012;**18**(5):1583–1596.

[182] Yurtsever G., Považay B., Alex A., Zabihian B., Drexler W., and Baets R. 'Photonic integrated Mach–Zehnder interferometer with an on-chip reference arm for optical coherence tomography'. *Biomedical Optics Express*. 2014;**5**(4):1050–1061.

[183] Moreira R.L., Garcia J., Li W., *et al.* 'Integrated ultra-low-loss 4-bit tunable delay for broadband phased array antenna applications'. *IEEE Photonics Technology Letters*. 2013;**25**(12):1165–1168.

[184] Li N., Su Z., Purnawirman, *et al.* 'Athermal synchronization of laser source with WDM filter in a silicon photonics platform'. *Applied Physics Letters*. 2017;**110**(21):21105.

[185] Ramelow S., Farsi A., Clemmen S., *et al.* 'Silicon-nitride platform for narrowband entangled photon generation'. *ArXiv*. 2015;**1508**:04358.

[186] Zhang X., Zhang Y., Xiong C., and Eggleton B.J. 'Correlated photon pair generation in low-loss double-stripe silicon nitride waveguides'. *Journal of Optics*. 2016;**18**(7):074016.

[187] Taballione C., Wolterink T.A.W., Lugani J., *et al.* '8×8 reconfigurable quantum photonic processor based on silicon nitride waveguides'. *Optics Express*. 2019;**27**(19):26842–26857.

[188] Cheng Z., Ríos C., Pernice W.H.P., Wright D., and Bhaskaran H. 'On-chip photonic synapse'. *Science Advances*. 2017;**3**(9):e1700160.

[189] Chiles J., Buckley S.M., Nam S.W., Mirin R.P., and Shainline J.M. 'Design, fabrication, and metrology of 10 × 100 multi-planar integrated photonic routing manifolds for neural networks'. *APL Photonics*. 2018; **3**(10):106101.

[190] Notaros J., Li N., Poulton C.V., *et al.* 'CMOS-compatible optical phased array powered by a monolithically-integrated erbium laser'. *Journal of Lightwave Technology*. 2019;**37**(24):5982–5987.

[191] Li N., Xin M., Su Z., *et al.* 'A silicon photonic data link with a monolithic erbium-doped laser.' *Scientific Reports*. 2020;**10**:1114.

Chapter 12

Femtosecond laser writing of integrated optical structures in glasses

Shane M. Eaton[1], *Belén Sotillo*[2], *Toney T. Fernandez*[3], *Vibhav Bharadwaj*[1], *Argyro N. Giakoumaki*[1], *Thien Le Phu*[1], *María Ramos Vázquez*[4], *Antonio Ancona*[5], *Roberto Osellame*[1] *and Roberta Ramponi*[1]

Femtosecond (fs) laser microprocessing is a direct, maskless technique capable of inducing a permanent refractive index increase buried beneath the surface of transparent glasses, enabling photonic circuit fabrication in three-dimensional (3D) geometries. We describe how among the many variables in fs laser waveguide writing, the repetition rate has the most important role as it influences the heat accumulation between laser pulses, which determines the regime of modification and the resulting morphological change. In most silicate and phosphate glasses, higher repetition rates are shown to be beneficial for driving increased heat accumulation and ion migration, leading to rapid fabrication of low-loss optical waveguides with tuneable size and refractive index contrast. In pure silica, which has low absorption due to its high bandgap, heat accumulation effects are reduced and higher fluences provided by the second harmonic visible wavelength from Yb-based fs lasers are required to form highly confining optical waveguides. We discuss several key 3D photonic components realized by fs laser waveguide writing technology for applications in astronomy, telecommunications and sensing.

12.1 Introduction

Focused fs laser pulses yield peak intensities greater than 10 TW/cm^2, which cause strong nonlinear absorption and localized energy deposition in the bulk of

[1]Istituto di Fotonica e Nanotecnologie (IFN)-CNR, Milano, Italy
[2]Departamento de Física de Materiales, Facultad de Ciencias Físicas, Universidad Complutense de Madrid, Ciudad Universitaria, Madrid, Spain
[3]MQ Photonics Research Centre, Department of Physics and Astronomy, Macquarie University, Sydney, Australia
[4]ICRM, Interdisciplinary Graduate School, Nanyang Technological University, Singapore, Singapore
[5]Istituto di Fotonica e Nanotecnologie (IFN)-CNR, Bari, Italy

transparent glasses. After several picoseconds, the laser-excited electrons transfer their energy to the lattice, leading to a permanent modification. Depending on the laser and material properties, this modification may result in damaged and irregular scattering centres, or smooth structures with a positive refractive index alteration. Such smooth structures of positive refractive index change (Δn) can confine light at visible and infrared (IR) wavelengths, and such waveguides may be formed along arbitrary 3D paths by translating the sample relative to the laser using computer-controlled motion stages. Additional advantages of fs laser direct writing include rapid, maskless single-step fabrication, and its versatility in producing waveguides in many different glasses, both passive and active. In this chapter, the fundamentals in the fs laser–glass interaction are first described. We discuss the relevant exposure parameters to produce low loss, high Δn optical waveguides in a variety of glasses at low and high repetition rate regimes. We then describe how this fs laser writing can produce novel devices for use in important fields such as telecommunications, quantum information, astrophotonics, mechanical sensing and lab-on-a-chip, exploiting building blocks such as directional couplers, Bragg grating waveguides (BGWs), active-doped waveguides and buried microfluidic channels.

12.2 Fundamentals of buried medication of glasses with focused femtosecond laser pulses

12.2.1 Nonlinear absorption

Focused fs laser pulses, with wavelengths typically in the visible or near-IR, do not have enough photon energy to be linearly absorbed in glasses. Instead, valence electrons may be promoted to the conduction band through nonlinear photoionization, which proceeds by multiphoton ionization and/or tunnelling photoionization pathways, depending on the laser and glass properties [1]. In addition to nonlinear photoionization, avalanche photoionization also occurs, explaining the small variation in threshold intensity for breakdown with bandgap [2]. Because of this low dependence of the breakdown threshold on the bandgap energy, fs laser nanofabrication can be applied to a wide range of glasses and other transparent materials.

Multiphoton absorption occurs due to the simultaneous absorption of several photons by an electron in the valence band. The number of photons m required to bridge the bandgap must satisfy $mh\nu > E_{gap}$, where E_{gap} is the bandgap energy, and ν is the laser frequency. For high laser intensity and low frequency, nonlinear ionization instead proceeds via quantum tunnelling. The strong laser field distorts the band structure and reduces the energetic barrier between the valence and conduction bands, allowing for direct band-to-band transitions by tunnelling. Keldysh showed that multiphoton and tunnelling photoionization can be described by the same theory, with a transition between the two processes defined by the Keldysh parameter:

$$\gamma = \frac{\omega}{e}\sqrt{\frac{m_e c n \varepsilon_0 E_{gap}}{I}} \tag{12.1}$$

where ω is the laser frequency, I is the laser intensity at the focus, m_e is the effective electron mass, e is the fundamental electron charge, c is the speed of light, n is the refractive index, and ε_0 is the permittivity of free space. For γ much less (greater) than 1.5, tunnelling (multiphoton) ionization is the dominant process. However, for $\gamma \sim 1.5$, nonlinear photoionization is a combination of both tunnelling and multiphoton ionization, which is the case for typical fs laser waveguide writing conditions in glass [3].

Electrons already present in the conduction band, provided by thermally excited impurities or nonlinear photoionization, may absorb laser radiation by free carrier absorption. After the sequential linear absorption of several photons, a conduction band electron's energy exceeds the conduction band minimum, allowing the hot electron to impact ionize a bound electron in the valence band, leading to two excited electrons at the conduction band minimum. These two electrons can then undergo free carrier absorption and impact ionization, and the process can repeat itself so long as the laser field is present and sufficiently strong. This process is referred to as avalanche ionization.

For sub-picosecond laser pulses, absorption is faster than energy coupling to the lattice, decoupling the absorption and lattice heating processes [3]. Seeded by nonlinear photoionization, the electron density in the conduction band increases via avalanche ionization until the plasma frequency approaches the laser frequency, at which point the plasma becomes strongly absorbing. For a typical fs laser with 1 μm wavelength, the plasma frequency equals the laser frequency when the free carrier density $\sim 10^{21}$ cm^{-3}, known as the critical density, commonly used to define optical breakdown. In terms of intensity, this breakdown threshold is $\sim 10^{13}$ W/cm^2 in glasses.

Since the electron–phonon lattice coupling time is about 10 ps, the absorbed laser energy is transferred to the lattice well after the laser pulse is gone. As short pulses require less energy to achieve the intensity for breakdown and because the absorption is decoupled from the lattice heating, more precise laser nanofabrication is possible relative to longer pulses. Another advantage of using fs laser pulses is a deterministic breakdown, since nonlinear photoionization can seed the electron avalanche. This is in contrast to the stochastic breakdown with longer pulses which rely on the low concentration of impurities (about one impurity electron in conduction band per focal volume), randomly distributed in the substrate to seed an electron avalanche [4].

12.2.2 Relaxation and material modification

Although it is well accepted that nonlinear photoionization and avalanche ionization are responsible for the creation of a free electron plasma, the physics are less clear once the electrons have transferred their energy to the lattice and the material is modified. In the nearly 3,000 published articles on optical waveguide writing citing the first work by the Hirao group [5], the reported morphological changes can be generally classified into three types of modifications: a smooth refractive index change [6], a form birefringent refractive index modification [7–10] and microexplosions resulting in empty voids [11]. The type of modification depends not only on many exposure parameters such as energy, pulse duration, repetition rate, wavelength, polarization, focal length, and scan velocity, but also on material properties such as bandgap and thermal conductivity. However, in pure fused silica glass, the most commonly processed material for waveguide writing, these three different morphologies can be observed by simply changing the incident laser energy [12].

12.2.2.1 Pulse energy

An isotropic regime of modification is useful for optical waveguides, where smooth and uniform refractive index modification is required for low propagation loss. At low pulse energies just above the modification threshold (~100 nJ for typical fs laser focusing conditions), a smooth refractive index modification has been observed in fused silica [12], which Krol's group has attributed to densification from rapid quenching of the melted glass in the focal volume [13]. In fused silica, the density and hence refractive index increases when glass is quickly cooled from a higher temperature [14]. Micro-Raman spectroscopy experimentally confirmed an increase in the concentration of three-member and four-member rings in the silica network in the laser-exposed region, signalling a densification of the glass [12]. Shock waves generated by focused fs laser pulses giving rise to stress have been shown to play a role in driving densification under certain conditions [15].

Another possible contribution to morphological change produced by focused fs laser pulses is the formation of colour centres, which alter the absorption spectra and hence the refractive index via a Kramers–Kronig mechanism [16]. Although laser-induced colour centres have been reported in glasses exposed to fs laser radiation [17,18], to date only a weak connection between colour centre formation and refractive index change has been shown. Waveguides formed in fused silica [19] were found to exhibit photo-induced absorption peaks at 213 and 260 nm, corresponding to positively charged oxygen vacancies and non-bridging oxygen hole centres defects, respectively. However, the colour centres were completely erased after annealing at 400 °C, even though waveguide behaviour was observed up to an annealing temperature of 900 °C. It is therefore unlikely that colour centres played a significant role in the refractive index change. Other research in borosilicate glasses has supported this claim [18]. In Yb-doped-phosphate glasses, Withford's group has shown that laser-induced colour centres contribute about 15% to the observed refractive index increase [20]. Using integrated BGWs, the authors were able to accurately study the photobleaching and thermal annealing of the fs-laser-induced colour centres. The colour centres were stable for temperatures below 70 °C, which is below the operating point during the waveguide laser application. However, the green luminescence generated by the Yb ions resulted in a photobleaching of the colour centres during laser operation, which must be corrected by pre-ageing techniques.

It is clear that both densification and colour centre formation play a role in the fs-laser-induced refractive index change, but their contributions will vary

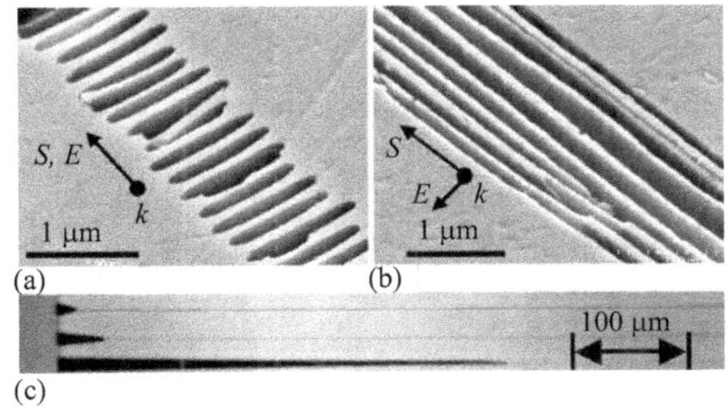

Figure 12.1 Scanning electron microscopy image of buried nanogratings formed after etching (sample cleaved and polished at writing depth) with polarization parallel (a) and perpendicular (b) to the scan direction. Overhead view (c) of etched microchannels demonstrating polarization selective etching with parallel (top), 45° (middle) and perpendicular (bottom) linear polarization. © 2006 Springer Nature/ Applied Physics. Reproduced, with permission, from Reference [24]

depending on the glass composition and the fs laser exposure conditions, adding to the complexity in modelling fs laser waveguide writing.

In multicomponent glasses, significant research has been recently devoted to understanding the mechanism of ultrafast laser-induced migration of ions/elements and its effect on the resulting refractive index change. The fundamental idea behind this was to pinpoint the physical mechanism at its smallest dimension, which in this case would be at the atomic level, with the goal of predesigning the glass to achieve both high refractive index change and low loss waveguides via ultrafast laser inscription.

For higher pulse energies (~150–500 nJ for typical fs laser focusing conditions), birefringent refractive index changes have been observed in fused silica glass, as first reported by Sudrie *et al.* [8]. Kazansky's group argued that the birefringence was due to periodic nanostructures caused by interference of the laser field and the induced electron plasma wave [9]. These nanogratings develop after multiple laser pulses [7,21] and are always oriented perpendicularly to the laser polarization [22] as shown in Figure 12.1. Their structural properties can be controlled with the laser-processing parameters, allowing for precise tuning of their birefringent properties [23]. However, the mechanisms responsible for the self-organization of nanogratings are not yet fully understood.

In 2008, Taylor's group discovered that nanogratings consist of self-aligned nanocracks [24]. They proposed that inhomogeneous dielectric breakdown results in the formation of a nanoplasma resulting in Si–O bonds deformation or breaking, which leads to growth and self-organization of nanoplanes [10]. The model was found to accurately predict the experimentally measured nanograting period for a certain range of experimental conditions in fused silica.

The Nolte group recently published an important review paper, giving further experimental insight into nanogratings [7]. They applied non-destructive small-angle X-ray scattering and found that characteristic sizes of the smallest features were nanocavities of dimensions 30×200×300 nm^3. The dimensions of these nanocavities were independent of exposure parameters, whereas exposure to multiple laser pulses leads to an increase in their total number. They then applied focused ion beam milling to remove a portion of the sample and found that hollow cavities are the primary constituents of nanogratings and that their sheet-like arrangement gives rise to their periodicity. Nolte's group proposed that the dangling bonds in the glass matrix exhibit enhanced absorption which can explain the accumulative behaviour for multiple-pulse irradiation in the formation of nanogratings [25].

A systematic study of the laser writing parameters in the orientation and periodicity of the nanogratings was performed by Stankevič *et al.* [26]. The orientation and the directionality have been explained based on anisotropic heat diffusion model. The heat diffusion along the direction of writing is enhanced by the interaction of the hot electrons in the plasma with the laser electric field leading to a tilt in the nanograting wave vector.

The large index contrast between the cavities and the surrounding material was found to be the cause of the high birefringence of nanogratings [27] despite the small feature size. Continuous grating planes emerge as adjacent cavities link due to their close proximity, whereas the material in between remains devoid of pores and is therefore more resilient. The formation of continuous, parallel nanocavities facilitates the preferential etching of these areas with HF acid due to the creation of vulnerable sites and the enabling of the acid penetration [22,28].

The selective etching of laser-modified glass can be exploited to fabricate buried microchannels for microfluidic applications (Figure 12.1). These nanogratings are not usually suitable for waveguide devices as birefringence is often seen as a detriment. However, there are optical applications where nanogratings are useful such as rewritable optical memory [22], birefringent waveplates [29] and integrated polarization beam splitters [30].

Until recently, it was thought that nanogratings could only be formed in pure fused silica glass. Nolte's group [31] applied a tuneable pulse duration fs laser to show that it is indeed possible to form buried nanostructures in two other glasses: Corning ULE TiO_2-doped silicate glass and Schott Borofloat 33 borosilicate glass (Figure 12.2). The birefringence of nanogratings in ULE is comparable to those in fused silica, while the nanostructures in borosilicate glass show much lower birefringence. Interestingly, the period of the nanogratings is also dependent on the type of the glass, being 250 nm for ULE (similar to fused silica) but only 60 nm in the case of Borofloat 33. As the properties of nanogratings in ULE and borosilicate differ significantly from those in fused silica, a more general model of nanograting formation must account for this differing behaviour amongst glasses.

At even higher pulse energies (>500 nJ for typical fs laser focusing conditions), pressures greater than Young's modulus are generated in the focal volume, creating a shockwave after the electrons have coupled their energy to the ions (~10 ps) [12]. The shockwave leaves behind a less dense or even hollow core, depending on the laser and material properties [32]. By conservation of mass, this

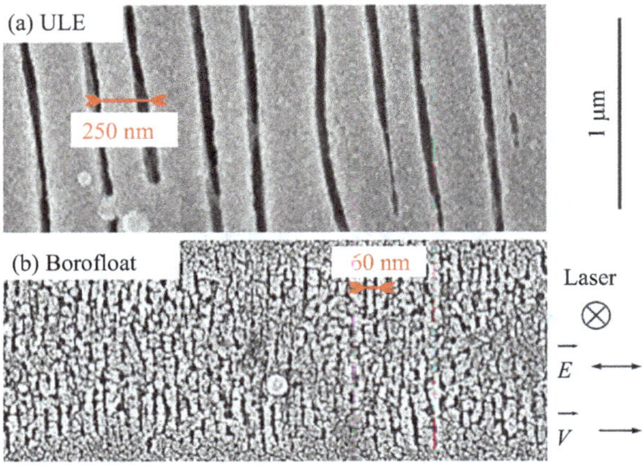

Figure 12.2 SEM images of periodic nanostructures of laser-induced birefringent structures in (a) ULE (200-nJ, 120-fs pulses) and (b) Borofloat (400-nJ, 400-fs pulses). © 2013 The Optical Society. Reproduced, with permission, from Reference [31]

core is surrounded by a shell of higher refractive index. Such voids may be exploited for 3D memory storage [33] or photonic bandgap materials [34].

12.2.2.2 Repetition rate

The above interpretations for the structural changes induced by focused fs lasers typically assumed single-pulse interactions but can likely be extended to explain modification from multipulse interactions during waveguide writing, assuming the repetition rate is low enough that thermal diffusion has carried the heat away from the focus before the next pulse arrives [12]. In this case, the following pulses may add to the overall modification but still act independently of one another.

For high repetition rates (>100 kHz), the time between laser pulses is less than the time for heat to diffuse away, giving rise to a build-up of temperature in the focal volume. For sufficiently high pulse energy, the glass near the focus is melted, and as more laser pulses are absorbed, this melted volume continues to expand until the laser is removed, at which point rapid cooling produces a region of altered refractive index. For scanned waveguide exposures, the size of the melted volume is determined by the effective number of pulses in the laser spot size, $N=2w_0R/v$, where $2w_0$ is the spot size ($1/e^2$), R is the repetition rate and v is the scan velocity.

Figure 12.3 shows microscope images of borosilicate glass modified by static laser exposure of 400-nJ pulse energy with varied repetition rate and number of pulses. Spherical-laser-modified zones were observed for all static exposures tested and arise from the 3D symmetry of heat diffusion from a small laser absorption volume of a ~2-µm diameter. These refractive index structures are due to localized melting within a cumulative heating zone that is built up over many laser pulses, which then cools rapidly to resolidify after the exposure. Evidence of heat accumulation is noted at

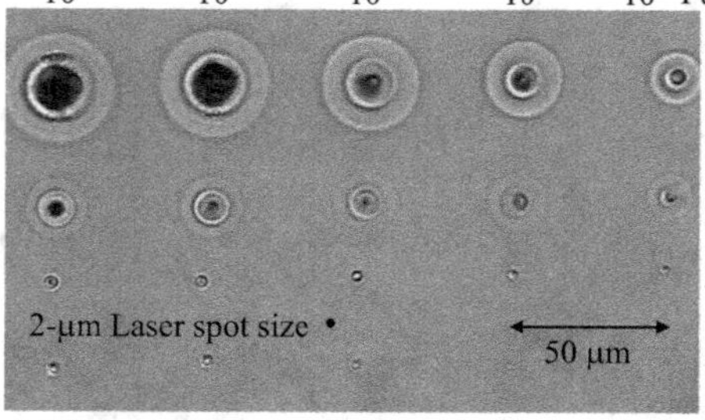

Figure 12.3 Optical microscope images showing modified volumes created in borosilicate glass with 400-nJ pulse energy from a femtosecond laser. Total pulse (top) and net fluence (bottom) are shown for each column and the repetition rate is indicated for each row. Laser direction is normal to page. © 2005 The Optical Society. Reproduced, with permission, from Reference [35]

repetition rates above 200 kHz, where the diameter of the modified volume significantly exceeds the ~2-μm laser spot size. Within each row (constant repetition rate) in Figure 12.3, one can see a modest increase in the diameter of the heat-affected zone despite a four order-of-magnitude increase in exposure. More striking is the 10-fold increase in modified zone size when the repetition rate is increased from 0.1 to 1 MHz in each column. Since the total laser exposure is identical within any column, a 200-kHz repetition rate represents the onset for cumulative heating effects above which thermal diffusion controls the properties of optical circuits formed by the fs laser. One can also appreciate that the size of the modification zone grows more quickly with the number of pulses when in the cumulative heating regime.

12.2.2.3 Focusing

Linear effects such as dispersion, diffraction, aberration and nonlinear effects such as self-focusing, plasma defocusing and energy depletion influence the propagation of focused fs laser pulses in glasses, resulting in an altered energy distribution at the focus, distorting the final refractive index modification.

Neglecting spherical aberration and nonlinear effects, the spatial intensity profile of a focused fs laser beam can be well represented by the paraxial wave equation and Gaussian optics. The diffraction-limited minimum waist radius w_0 (1/2 the spot size) for a collimated Gaussian beam focused in glass is given by the following equation:

$$w_0 = \frac{M^2 \lambda}{\pi \mathrm{NA}} \tag{12.2}$$

where M^2 is the Gaussian beam propagation factor [36], NA is the numerical aperture of the focusing objective and λ is the free-space wavelength. The Rayleigh range z_0 (1/2 the depth of focus) inside a glass of refractive index n is given by the following equation:

$$z_0 = \frac{M^2 n\lambda}{\pi \mathrm{NA}^2} \tag{12.3}$$

Chromatic and spherical aberration alter the intensity distribution near the focus so that (12.2) and (12.3) are no longer valid approximations. Chromatic aberration as the result of dispersion in the lens can be corrected by using chromatic aberration-corrected microscope objectives for the wavelength spectrum of interest. For lenses made with easily formed spherical shapes, light rays that are parallel to the optic axis but at different distances from the optic axis do not converge to the same point, resulting in spherical aberration. This can be addressed by using multiple lenses such as those found in microscope objectives or using an aspheric focusing lens. In waveguide writing where light is focused inside glass, the index mismatch at the air–glass interface introduces additional spherical aberration. As a result, there is a strong depth dependence for fs-laser-written buried structures [37,38], which is even more pronounced for higher NA objectives [3] except for oil-immersion lenses [39] or dry objectives with collars that can correct for spherical aberration at different focusing depths [37].

Dispersion from mirror reflection and transmission through materials can broaden the pulse width, which can reduce the peak intensity and alter the energy dissipation at the focus. However, it is only for short pulse <40-fs oscillators with large bandwidths that dispersion becomes an issue. In this case, precompensation of the dispersion through the microscope objective is required to obtain the shortest pulse at the focus [40].

The spatially varying intensity of a Gaussian laser beam can create a spatially varying refractive index in glasses. As the nonlinear refractive index n_2 is positive in glasses, the refractive index is higher at the centre of the beam compared to the wings. This variation in refractive index creates a positive lens which focuses the beam inside a dielectric with a strength that is dependent on the peak power. If the peak power of the fs laser pulses exceeds the critical power P_0 for self-focusing [3]

$$P_0 = \frac{3.77\,\lambda^2}{8\pi n_0 n_2} \tag{12.4}$$

a collapse of the pulse to a focal point is predicted. However, as the beam self focuses, the increased intensity drives nonlinear ionization which creates a free-electron plasma, which acts as a diverging lens that counters self-focusing. A balance between self-focusing and plasma defocusing leads to filamentary propagation, which results in vertically elongated refractive index structures, which are undesirable for transversely written waveguide structures (the standard geometry for waveguide fabrication). Filaments can be exploited to fabricate waveguide devices by scanning the sample vertically along the optical axis. Self-focusing can be avoided in waveguide fabrication by tightly focusing the laser beam with a microscope objective to reach the intensity for optical breakdown without exceeding the critical power.

In fused silica, $n_0 = 1.45$ and $n_2 = 3.5\times10^{20}$ m^2/W [41] so that for $\lambda =$ 1,030 nm, the critical power is ~3 MW. From (12.4), the critical power is proportional to the square of the laser wavelength; therefore, lower critical powers result when working with the second harmonic wavelength of fs lasers. Also, the critical power is inversely related to the nonlinear (and linear) refractive index, making it difficult to form waveguides in heavy metal oxide glasses with $n_0 \sim 2$, $n_2 \sim 10^{-18}$ m^2/W [42] and chalcogenide glasses (CHGs) with $n_0 \sim 2.5$, $n_2 \sim 10^{-17}$ m^2/W [43].

12.3 Femtosecond laser waveguide writing in glasses

The properties of fs-laser-written waveguides in glass depend strongly on laser exposure conditions such as scan speed, average power, and NA. These parameters influence the net fluence, NF:

$$\mathrm{NF} = \frac{2w_0 R F_p}{v} \tag{12.5}$$

where w_0 is the waist radius, R is the repetition rate, F_p is the per-pulse fluence (pulse energy per area) and v is the scan speed. Related to the dwell time, the effective number of pulses per spot size diameter ($2w_0$) during scanning is given by $N = 2w_0R/v$. Refractive index modification in fs laser waveguide writing occurs above a bulk modification threshold intensity, typically ~10^{13} W/cm^2 in glass [3]. Above this threshold, increasing NF via decreased scan speed or increased fluence usually results in increased refractive index and, at sufficiently high NF, irregular and overly damaged modification tracks.

12.3.1 Low repetition rate fabrication

At low repetition rates, the time between pulses is long enough so that thermal diffusion has carried the heat away from the focus before the following pulse arrives. The threshold repetition rate for heat accumulation is not a fixed value but also depends on several factors, including the glass properties, the laser pulse energy and the focusing NA, and will be discussed in further detail in the next section. As described previously, in the low repetition rate waveguide writing regime, the ensuing pulses may add to the overall modification but still act independently of one another. Most results in the field of fs laser microfabrication have been carried out at low repetition rates, and usually at 1 kHz, due to the common availability of 800-nm amplified Ti:sapphire fs lasers. One limitation of waveguide writing in the single-pulse interaction regime is that waveguide cross sections take on a shape similar to the asymmetric focal volume from a depth of focus that exceeds the transverse spot size. The resulting waveguides written with the transverse writing scheme, where the sample is translated transversely relative to the incident laser, are elliptical-like, giving modes that couple poorly to standard optical fibre. Several methods have been proposed to produce a more symmetric focal volume including astigmatic focusing with a cylindrical lens telescope [44], slit reshaping [45], spatial light modulator [46] or multiscan writing [47,48].

12.3.1.1 Fused silica glass

The best result in fused silica to date was obtained with a multiscan writing procedure (NA = 0.4, pulse energy 180 nJ, speed 10 μm/s) with 20 scans separated transversely by 0.4 μm to form waveguides with nearly square cross sections (~8 μm × 8 μm) with a refractive index change of 4×10^{-3}, suitable for low-loss coupling to single-mode (SM) fibre (SMF) [47]. The waveguides written in planar lightwave circuits (PLC) glass were nearly identical to the waveguides written in fused silica. A propagation loss of ~0.1 dB/cm at 1,550 nm was reported, which is the lowest reported in the field, and is attributed to the gentle modification enabled by the low-fluence, multiscan fabrication method.

The effect of laser writing polarization on waveguide properties was studied using a 1-kHz, 100-fs Ti:sapphire laser with a 0.5-mm slit placed before the 0.5-NA focusing objective [49] to obtain symmetric waveguide cross sections. With 3-μJ pulse energy measured after the slit and 25-μm/s writing speed, a refractive index change of 2×10^{-3} was measured for circular polarization, about twice as high as that obtained with linear polarizations. The enhancement was attributed to the higher photoionization rate for circular polarization compared to linear polarization in the range of intensities studied (40–50 TW/cm^2). Nanogratings also influence the transmission properties of fs-laser-written waveguides in fused silica. Using a broadband laser source, Bragg stop bands in the UV have recently been discovered [50], due to the ~100-nm pitch of the nanogratings formed by fs laser writing. For more details on nanogratings and their applications, the reader is referred to the excellent reviews by Taylor *et al.* [22] and Richter *et al.* [7].

12.3.1.2 Silicate and phosphate glasses

Although fused silica is the most commonly used glass for photonics due to its high transmission, excellent temperature resistance and biocompatibility, silicate and phosphate glasses offer similar characteristics at lower cost. Further, silicate and phosphates may be doped with active ions for lasing applications.

The most common silicate glasses are boro-aluminosilicate glasses, which along with silica (SiO_2) contain boron trioxide (B_2O_3), aluminium oxide (Al_2O_3) and sodium oxide (Na_2O). Waveguides have been successfully fabricated by low repetition rate fs lasers in several borosilicates, including Schott Duran [51], Corning 7890 [18] and Corning 1737. In Corning EAGLE2000, a common glass used primarily in displays owing to its low density, Zhang *et al.* explored a wide range of processing conditions with a 1-kHz Ti:sapphire fs laser. By tuning the pulse durations from 50 fs to 2 ps, promising windows for waveguide writing at both 100 fs and 1 ps were identified [52], disproving the common belief at the time that sub-picosecond pulses were needed to form optical waveguides in glass. The most important consequence of this work was a fortuitous discovery of BGWs. By scanning the sample at 0.5 mm/s so that successive pulses only partially overlapped, a periodic refractive index modulation along the waveguide was created. The partially overlapped refractive index voxels produced segmented waveguides that showed strong and narrowband Bragg reflection yet still maintaining high optical confinement waveguiding at 1,550-nm wavelength.

High-quality optical waveguides have been demonstrated in phosphate glass, commonly doped with Erbium and Ytterbium ions for waveguide laser applications. Osellame *et al.* applied an astigmatic writing method with a cylindrical lens telescope to form symmetric waveguides with low damping loss (0.25 dB/cm) at 1,550-nm wavelength with a 1-kHz, 150-fs Ti:sapphire laser with 0.3-NA objective, 5-μJ pulse energy and 20-μm/s writing speed [44]. This was the first time a beam-shaping method was applied to correct for the asymmetric intensity distribution in transverse waveguide writing. The slit shaping method was later applied by Ams *et al.* to produce symmetric waveguides (Figure 12.4) with similar propagation loss in the same active glass [45].

12.3.1.3 Chalcogenide glasses

CHGs have attracted much attention due to their high non-linearity, high photosensitivity and a wide IR transmission window [53,54]. The potential applications of these glasses are in all-optical switching [55,56], mid-IR photonics [57–59] or continuum generation [60,61], among others. However, the high linear and refractive index of CHGs poses a challenge for ultrafast laser processing. Initial studies of fs laser inscription were focused on arsenic-sulphides-based glasses [62,63]. Many of these glasses show a negative Δn when exposed to the fs laser pulses, or show a weak positive index contrast, which is highly dependent on the processing parameters [57,64]. For example, D'Amico *et al.* [64], using an amplified Ti:sapphire ultrafast laser delivering 800-nm pulses (130 fs) of 500-mW power with repetition rate varying between 10 and 100 kHz, showed that a weak positive Δn can be produced in As_2S_3 at low accumulation doses. An increase in the accumulation doses produces a flip in the sign of the refractive index change from positive to negative. As a solution, they proposed Ge doping to obtain a high contrast positive Δn, up to ~10^{-3}. On the other hand, for CHGs exhibiting a negative Δn, several waveguide geometries can be used, as suggested by Arriola *et al.* in [57].

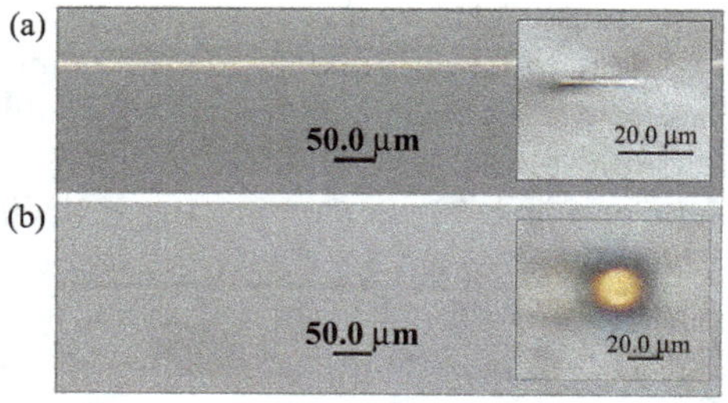

Figure 12.4 Microscope images of waveguides fabricated in phosphate glass (a) without and (b) with a 500-μm slit. © 2005 The Optical Society. Reproduced, with permission, from Reference [45]

Among the CHGs, gallium lanthanum sulphides (GLSs) are attractive because they are non-toxic and are transparent from 0.55 to 10 μm. Moreover, GLS has the highest nonlinear figure of merit among CHGs [65] and the interaction with fs laser pulses yields a positive refractive index change. Due to the application in IR photonics, the waveguides reported in the literature are typically designed for this wavelength range.

The few studies performed at low repetition rate in GLS applied slit shaping and multiscan techniques to obtain more symmetrical waveguides [57,66]. Gretzinger *et al.* studied the formation of waveguides in GLS glass using a Ti: sapphire laser with a repetition rate of 1 kHz [66]. In order to obtain a material modification strong enough to support optical modes at 1,550 nm, they stretched the pulse duration to 1.5 ps. Applying the slit beam shaping method and focusing the laser with a 40× microscope objective, they were able to write symmetric waveguides (Figure 12.5) [66]. The lowest propagation loss was 2.0 dB/cm for waveguides inscribed with a slit of 320 μm, 200-nJ pulse energy, scanning speed of 1.5 mm/min and four overpasses.

In comparison, high repetition rate fabrication has proved to produce better waveguides in GLS in terms of losses and mode field diameter (MFD) [56,66–68]. For this reason, most of the fs laser inscription studies done in GLS have been performed using high repetition rate fabrications and will be described in the next section.

12.3.2 High repetition rate fabrication

The recent development of high repetition rate, high-power fs lasers opens new avenues for manipulating thermal relaxation effects that control the properties of

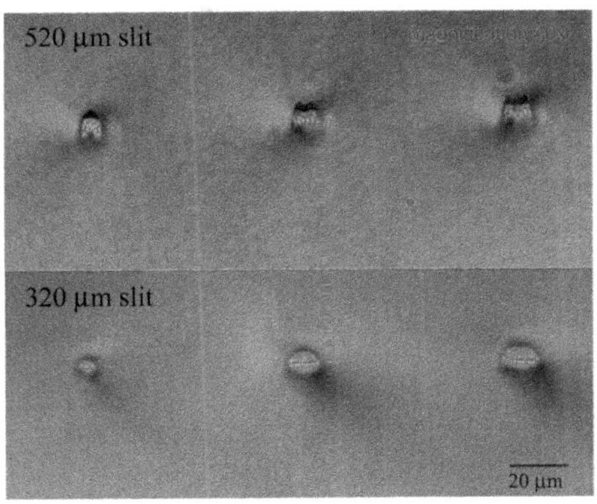

Figure 12.5 Microscope images of the side view of waveguides fabricated in GLS with a 1-kHz laser delivering pulses of 1.5 ps and using a slit of (top) 520 μm and (bottom) 320 μm. The pulse energy is 200, 350 and 500 nJ (from left to right), with a translation speed of 1.5 mm/min and four overpasses. © 2015 The Optical Society. Reproduced, with permission, from Reference [66]

optical waveguides formed when ultrashort laser pulses are focused inside glasses. At low-to-moderate repetition rates (1–100 kHz), an increase in laser pulse energy leads to the formation of larger modification structures as thermal diffusion extends the laser-heated region far outside the focal volume. As repetition rate increases, the time between laser pulses becomes shorter than the time for the absorbed laser radiation to diffuse out of the focal volume and heat builds up in the focal volume. Schaffer *et al.* first reported heat accumulation in the bulk of glass using a 25-MHz fs laser oscillator [69]. With increased dwell time, a dramatic increase in the size of laser-modified structures was observed compared to structures formed with single-pulse interactions, where no variation in modification size with dwell time was observed. The combination of high repetition rate and heat accumulation offers fast writing speeds and cylindrically symmetric waveguides together with benefits of annealing and decreased thermal cycling that are associated with low propagation and coupling loss to standard optical fibre.

12.3.2.1 Fused silica glass

Unlike silicate glasses [38], it is more difficult to drive heat accumulation in fused silica due to its higher bandgap (lower absorption) and higher melting point [70]. By taking advantage of the higher average power offered by the latest fs laser technology, cumulative heating has recently been demonstrated in fused silica, but, unlike other glasses [35] where low propagation loss waveguides were reported, the structures written in fused silica were nonuniform and unable to guide light [40]. To produce low loss smooth waveguides in fused silica at high repetition rates, the second harmonic wavelength 520 nm was found to be beneficial [71], while keeping the energy below the threshold for heat accumulation to avoid irregular damage tracks.

As one cannot take advantage of heat accumulation to produce symmetric waveguides in fused silica, a recent study employed a high 1.25-NA oil-immersion lens along with the optimum green wavelength [70]. Compared to 0.55-NA, the waveguide formed by 1.25-NA focusing was significantly less elongated (Figure 12.6), which was attributed to a more symmetric focal volume and reduced self-focusing. The peak $\Delta n = 0.022$ represented the highest refractive index increase ever reported for a fs-laser-written waveguide in fused silica.

The mode profiles at 1,550 nm for the optimum waveguides written by 0.55- and 1.25-NA focusing are shown in Figure 12.7, along with the mode for SMF with 10.5-μm MFD. The waveguide mode produced by 0.55-NA focusing (MFD = 10 μm × 12 μm) becomes significantly smaller and more symmetric (MFD = 7 μm × 7 μm) with 1.25-NA focusing, defining the smallest every reported mode for a laser-written waveguide in fused silica. Theoretical simulations predict that small $R = 15$ mm bends are now possible with these processing conditions, allowing for more dense integration of photonic circuits. Future efforts will be directed to applying the latest high-power, high-repetition rate fs laser technology offering even shorter (sub 100 fs) pulse durations, with the goal of producing uniform positive refractive index change structures in fused silica.

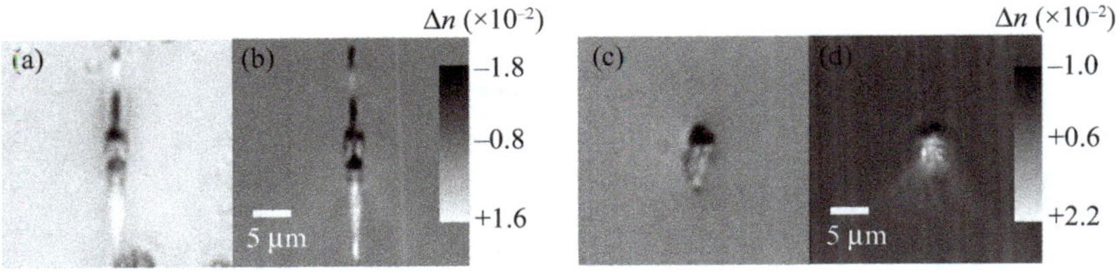

Figure 12.6 Cross-sectional microscope images (a and c) and RNF profiles (b and d) for waveguides fabricated with 0.55-NA (a and b) and 1.25-NA (c and d) lenses. © 2011 Elsevier. Reproduced, with permission, from Reference [70]

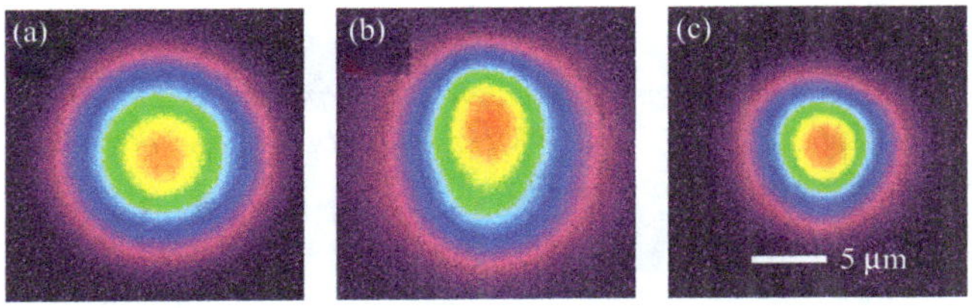

Figure 12.7 Mode profile at 1,550-nm wavelength for (a) SMF, waveguide writing with 0.55-NA (b) and 1.25-NA (c) objectives. © 2011 Elsevier. Reproduced, with permission, from Reference [70]

12.3.2.2 Silicate and phosphate glasses

Following early work by Schaffer [3], further insight into heat accumulation effects in borosilicate glass was provided by Eaton *et al.* [38]. Using a finite difference thermal diffusion model, the temperature in the focal volume was calculated as a function of pulse number (dwell time) for repetition rates of 0.1–1 MHz in EAGLE2000 borosilicate glass. As shown in Figure 12.8, at a 100-kHz repetition rate, the temperature relaxes to below the softening point (985 °C) before the following pulse arrives, resulting in minimal heat accumulation and large temperature cycling during waveguide writing. At 0.5 and 1-MHz repetition rates, heat accumulation is more evident, leading to a larger melted volume which increases with pulse number and repetition rate.

To better understand the contributions of thermal diffusion and heat accumulation to the resulting waveguide morphology, a variable repetition rate (0.1–5 MHz) 1,045 nm fs fibre laser was applied to waveguide writing in EAGLE2000 borosilicate glass [38] with a 0.55-NA focusing lens. The threshold for heat accumulation was found by comparing waveguide diameters produced by scanned exposures with that of single-pulse exposures, where the only contribution is from thermal diffusion. The threshold was defined as the minimum pulse energy to increase the waveguide diameter 2-fold over the diameter produced by diffusion in a single-pulse interaction.

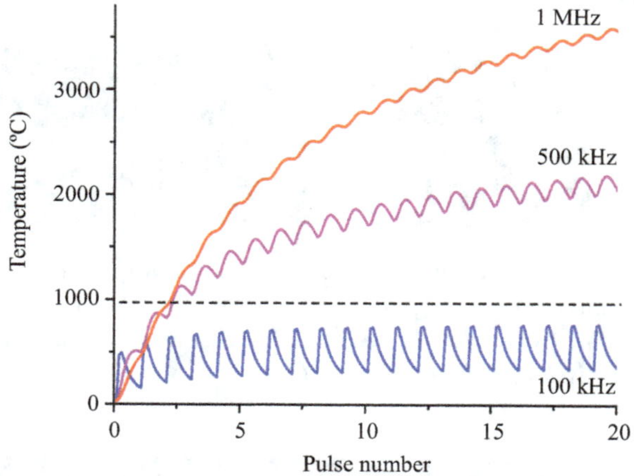

Figure 12.8 Model of glass temperature versus exposure at repetition rates of 100 kHz, 500 kHz and 1 MHz. The absorbed energy was 200 nJ at each repetition rate [72]

The single-pulse modification diameter increased from 2 to 6 μm over a pulse energy range of 100–1,200 nJ.

The threshold pulse energy for driving heat accumulation is shown in Figure 12.9 as a function of repetition rate (200 kHz to 2 MHz). The single-pulse modification threshold was 50 nJ (2.5 J/cm^2) and invariant with repetition rate. In contrast, the energy threshold for heat accumulation decreased from 900 nJ at 200 kHz to 80 nJ at 2 MHz and was only weakly dependent on scan speed. One can note that the threshold average power for driving heat accumulation is nearly constant, at about 160 mW.

In this 200 kHz to 2 MHz range of repetition rates, the thermal diffusion scale length $\sqrt{D/R}$ decreases from 1.6 to 0.5 μm, where D is the thermal diffusivity and R is the repetition rate. This decrease in the characteristic thermal diffusion length means that the effective laser heating volume decreases with increasing repetition rate, thereby reducing the threshold pulse energy for heat accumulation. For $R >$ 1 MHz, the thermal diffusion scale length of 0.7 μm falls inside the laser waist radius of $w_0 = 0.8$ μm, resulting in an asymptotic limit of the heat accumulation threshold energy to a minimum value of 80 nJ at 2-MHz repetition rate in Figure 12.9. Beyond 2 MHz, the available laser pulse energy was below this value, preventing the observation of cumulative heating in EAGLE2000. Similarly, when the laser was operated at the lowest 100-kHz repetition rate, the 2-μJ maximum pulse energy available was too low to drive heat accumulation effects beyond a larger thermal diffusion diameter of 8 μm.

In the same study, a new laser-processing window was discovered for producing low-loss waveguides across the range of repetition rates tested. By holding the average power constant and delivering the same NF, similar waveguide morphology and low propagation loss (~0.3 dB/cm at 1,550 nm wavelength) resulted across

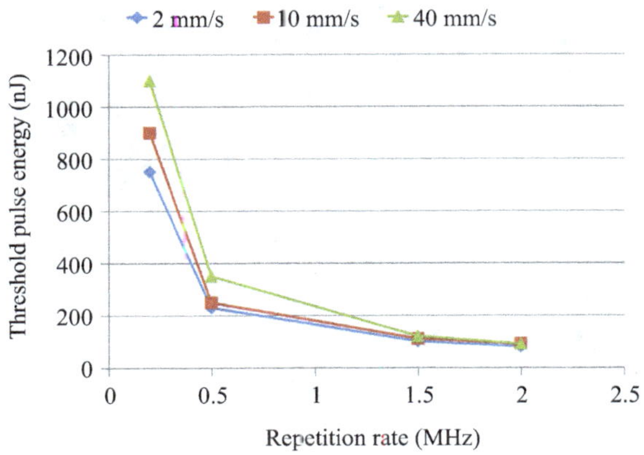

Figure 12.9 Experimental values of threshold pulse energy for driving heat accumulation in borosilicate glass as a function of laser repetition rate for scan speeds of 2, 10 and 40 mm/s and 0.55-NA focusing. © 2008 The Optical Society. Reproduced, with permission, from Reference [38]

this wide range of repetition rates. As the repetition rate increases, the pulse energy and hence the strength of thermal diffusion is decreased, but this is counteracted by the increasing pulse delivery rate to produce the similar waveguide shapes shown in Figure 12.10. The constant 200-mW average power to form optimum waveguides is consistent with the constant average power (160 mW) heat accumulation threshold over the same repetition rate range. Other researchers have also identified that average power is the key parameter in heat accumulation defined waveguides: Osellame *et al.* found that a constant 250-mW power resulted in low-loss waveguides in phosphate glass from repetition rates of 500–900 kHz [39]. Richter *et al.* found a 200-mW average power threshold for driving melting in fused silica from 1- to 9-MHz repetition rate, which erased nanogratings [73].

The core of the waveguides shown in the refracted near field (RNF) profiles in Figure 12.10 is attributed to the high-temperature spikes induced within the laser spot size by each laser pulse, while the outer lower contrast cladding is formed by a more slowly evolving near-Gaussian temperature distribution, with the overall size determined by the maximum diameter where the temperature exceeds the melting point.

Figure 12.11 plots the insertion loss and MFD versus repetition, at this optimum processing window of 200 mW and 25 mm/s. The decreasing insertion loss (IL) with an increasing repetition rate is associated with increasingly stronger heat accumulation that results in higher refractive change and smaller MFD for best coupling to optical fibres at 1.5 MHz. The increased IL and MFD from 1.5 to 2 MHz is attributed to insufficient laser pulse energy (100 nJ) at 2 MHz for driving significant heat accumulation above the ~90-nJ threshold from Figure 12.10. Beyond 2-MHz repetition rate, only narrow ~2-μm diameter waveguides were formed that showed no evidence of heat accumulation.

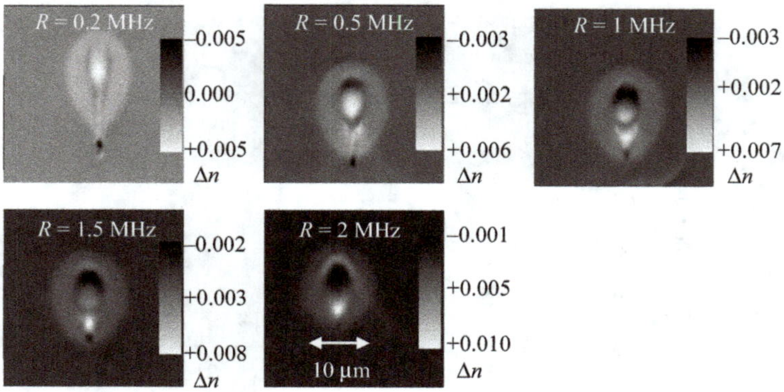

Figure 12.10 Cross-sectional refractive index profiles of waveguides written with 200-mW average power, 25-mm/s scan speed and repetition rates of 0.2, 0.5, 1, 1.5 and 2 MHz. © 2008 The Optical Society. Reproduced, with permission, from Reference [38]

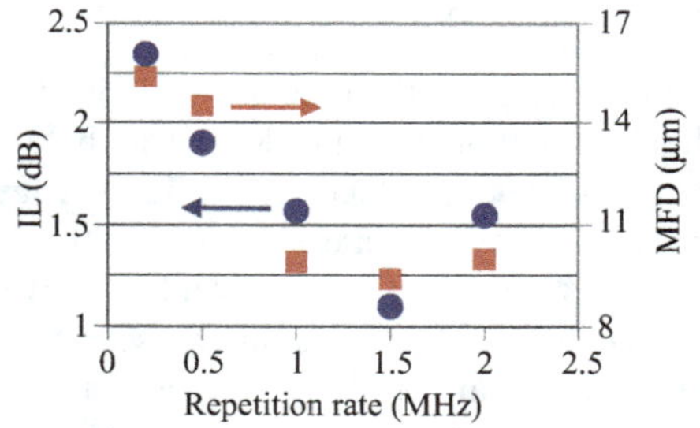

Figure 12.11 IL and MFD versus repetition rate for waveguides formed with 200-mW average power and 15-mm/s scan speed. © 2008 The Optical Society. Reproduced, with permission, from Reference [38]

The refractive index profile for the optimum waveguide at 15-mm/s scan speed was used to confirm the accuracy of the RNF measurements, as shown in Figure 12.12. The RNF data was imported into a mode solving routine (Lumerical MODE Solutions), with the resulting mode profile showing excellent agreement with the experimentally measured mode profile, confirming the accuracy of the RNF measurements.

12.3.2.3 Chalcogenide glasses

Several CHGs have been selected as a platform to fabricate waveguides using high repetition rate ultrafast laser fabrication [57,63,74] due to their suitable properties to develop nonlinear photonic circuits within a wide transmission window. Among them, the best reported results in terms of propagation losses have been obtained in GeS_4 binary glasses using moderate repetition rate of 100 kHz (1.1 dB/cm at

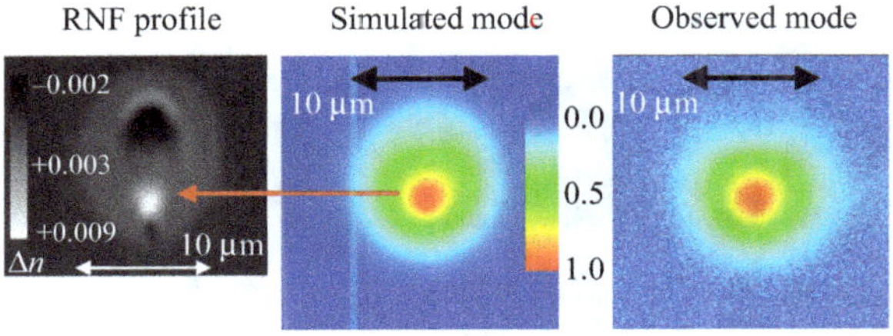

Figure 12.12 Waveguide fabricated with 1.5-MHz repetition rate, 200-mW power and 15-mm/s scan speed: cross-sectional refractive index profile (left), simulated mode profile (middle) and measured mode profile (right). Red arrow indicates position of mode relative to waveguide.

633 nm [59]) or, for high repetition rate, in GLS glass for different operating wavelengths. GLS waveguides with low propagation losses of 1.5 dB/cm at 633-nm wavelength [68] and 0.65 dB/cm at 1,560-nm wavelength [67] have previously been demonstrated with fs laser writing. In addition, fs inscribed mid-IR photonic circuits for astronomical applications were reported with ~1-dB/cm propagation loss at 3.39-µm wavelength [58,66] and 0.22 dB/cm at 4-µm wavelength [75].

As described earlier, GLS is the CHG with the highest nonlinear figure of merit, so it is useful to study the effect of laser writing on the nonlinear refractive index (n_2) of the material. Recently, Vázquez *et al.* [56] reported waveguides in GLS with propagation losses of 0.1 dB/cm at 800 nm. The fs laser used was a Yb: KGW regenerative amplified system with 230-fs pulse duration, 1,030-nm wavelength and a repetition rate of 500 kHz, focused in the sample with a 0.42-NA microscope. Waveguides were laser inscribed using the multiscan shaping approach, thus generating a gentle modification and avoiding nonlinear propagation effects to obtain a symmetric refractive index profile. X-ray microanalysis and Raman spectroscopy confirmed the gentle modification produced in the material. The microscope images and MFD of the waveguides are shown in Figure 12.13(a)–(c). Directional couplers with a maximum coupling ratio of 54% were also fabricated to exploit them as all-optical switches and thus to obtain an evaluation of the n_2 value from the switching parameters (Figure 12.13(d)).

When applying high optical intensities, all-optical switching in the directional coupler is manifested as an intensity-dependent refractive index change induced in the nonlinear material, known as the optical Kerr effect ($n = n_0 + n_2 I$, where n_0 and n_2 correspond to the linear and nonlinear refractive indices, respectively, and I is the intensity of the control light).

Then, for obtaining the n_2 value, ultrafast all-optical switching was performed with fs laser pulses using laser-written directional couplers. The nonlinear refractive index in the waveguides ($n_2 = 9.0 \times 10^{-19}$ m^2/W) was found to be slightly reduced but on the same order of magnitude as the pristine material ($n_2 = 2.16 \times 10^{-18}$ m^2/W

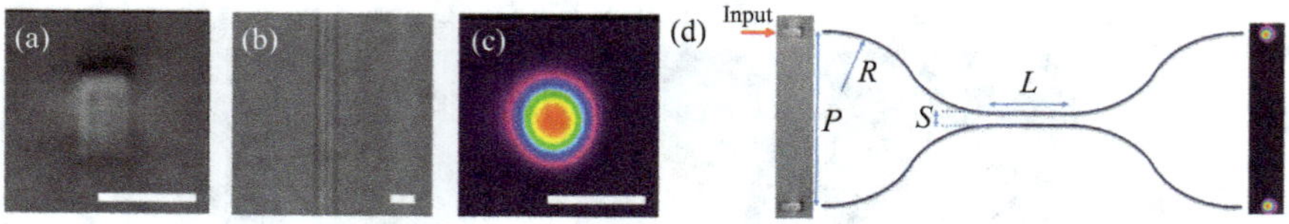

Figure 12.13 Transverse (a) and overhead (b) microscope view of the waveguide written at 70 nJ and 5 mm/s. (c) The corresponding guided mode at 808-nm wavelength. The scale bar corresponds to 10 μm. (d) Sketch of a symmetric directional coupler showing the various parameters defining its radius of curvature, R, interaction length, L, centre-to-centre separation between the waveguides in the interaction region, s, and separation between the two ports, p. On the left, a side view optical microscope image of the two input ports is shown, and, on the right, the output modes measured for a directional coupler with 50% coupling ratio is shown. © 2018 The Author(s). Published by IOP Publishing Ltd. Reproduced, with permission, from Reference [56] under Creative Commons 3.0 License

[76]). Similar behaviour was reported by Demetriou *et al.* [77]. These results evidence that GLS is a promising platform for laser-written integrated nonlinear photonics [78] since the laser writing technique applied preserves the particular nonlinearity of the glass.

12.3.3 Ion migration in high repetition rate modification of multicomponent glasses

The migration of ions/elements upon irradiation of an fs laser towards the densified or rarefied zones became a recent interest in the laser waveguide writing community as it was demonstrated that tuning the glass properties could increase the refractive index change by an order of magnitude [79,80]. In initial work, the Miura and Hirao group found that the ring-shaped refractive index profile during static fs laser irradiation was the result of ion exchange between network formers and network modifiers [81]. They explained this by the Soret effect or thermodiffusion which is a temperature-gradient-driven diffusion in a multicomponent system.

As shown in Figure 12.14, the network former SiO_2 concentrated in the central region and modifiers like CaO and Na_2O concentrated at the periphery of the molten region. This initial observation of ion migration during the static fs laser exposure was based on the characterization of migration perpendicular to the incoming laser beam.

However, it is important to also consider the direction parallel to incident laser beam, as the laser-induced modification is not symmetric. Recent studies considering both directions and other multicomponent glasses found a different behaviour of ion migration compared to the initial static irradiation study. It was found that in most cases, the densified/high refractive index zones were enriched by

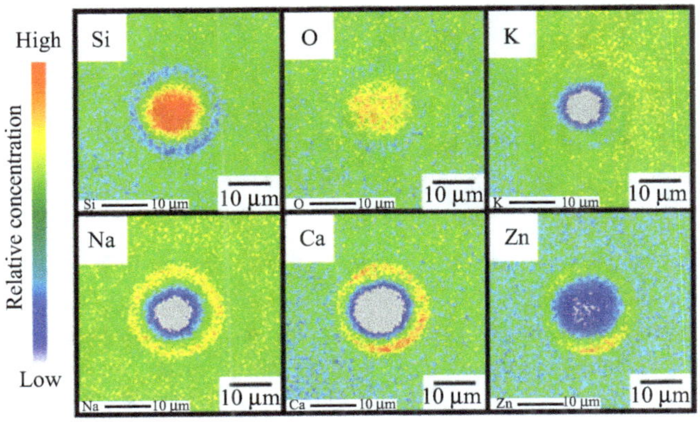

Figure 12.14 The EPMA mapping of migration of ions in a silicate glass upon fs laser irradiation reported by Kanehira et al. © 2008 AIP Publishing. Reprinted, with permission, from Reference [81]

multivalent ions, whereas the rarefied/low refractive index zones were enriched by single-valent ions (Figure 12.15) [82]. But there are cases where one observes silicon or aluminium reversing its role and migrating towards the low refractive index zones such as in borosilicate glasses [83]. Very recently this role reversal, which was a deviation from the general Soret effect, was also explained by Miura and Hirao's group using simulations [84]. They demonstrated that SiO_2 concentrates on the hot side in an SiO_2 rich melt and on the cold side in an SiO_2 poor melt through nonequilibrium molecular dynamics and concluded that the partial molar enthalpy is one of the dominant factors in the fs-laser-induced Soret effect responsible for ion migrations. Most recently, the migration of anions to form laser-written waveguides was reported for the first time in CHGs [75].

A strong initial effort to identify the stimulus for ion migration was carried out using in situ plasma emission microscopy [85]. The work found that the transient plasma distribution during waveguide writing drives the ions in specific directions that eventually determine the final refractive index distribution. The paper also demonstrates that by modulating the spherical aberration upon waveguide writing can tune the transient plasma shape and thereby the refractive index profile.

Figure 12.16 gives an overview of the general trend for elements migrating due to fs laser irradiation: when observed perpendicular to the laser direction (Figure 12.16(a)), the glass formers segregate in the epicentre whereas the intermediate and the network modifiers segregate in outer concentric regions with the molten region. This elemental redistribution can be partially explained from the transient temperature distribution and Soret effect, but the situation is more complicated when there is a strong flow of glass melt at high repetition rates and NFs, requiring more sophisticated elemental and molecular dynamic simulations. The trend of elements migrating along the parallel direction during high repetition rate fs laser irradiation (Figure 12.16(b)) is for multivalent ions to segregate in the

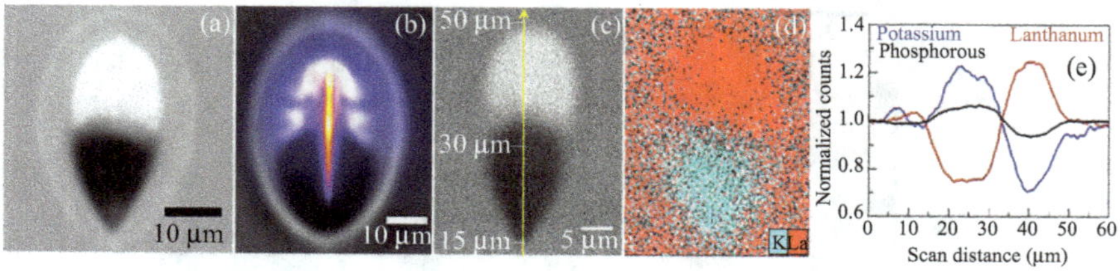

Figure 12.15 *(a) Differential interference contrast image of a lanthanum phosphate glass waveguide. (b) In situ image of the relative position of laser-induced plasma with respect to the waveguide as it is being formed. (c) Secondary electron image of the waveguide. (d) EDX mapping showing the distribution of La and K. (e) EDX line profile across the waveguide (yellow arrow in (c)). © 2015 IoP Publishing, Ltd. Reprinted, with permission, from Reference [85]*

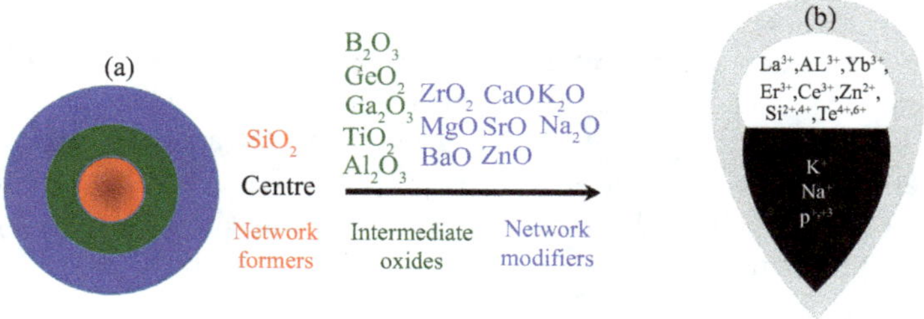

Figure 12.16 *Migrating elements observed in the perpendicular (a) and parallel direction (b). In both cases, ion migration only occurs inside the molten zone. In (a) the colour-coded concentric circular zones are within the central molten region with the outer zone not shown. In (b), white and black regions are within the molten zone, with the outer zone indicated by grey. © 2018 Elsevier. Reproduced, with permission, from Reference [79]*

densified zone with the single-valent ions migrating to the rarefied zone. For both directions, no evidence of ion migration has been observed in the cladding zone outside the inter-molten region.

The application of tuned/tailored ion migration can be found in space selective phase separation for the fabrication of microfluidic channels for lab-on-a-chip [86], tuning the V-number of waveguides to adapt to a predesigned MFD, thereby reducing the coupling losses [87], optimizing the performance of optical devices [88] or designing an optical micropipette that could draw elements out or inject elements into an area of particular interest. A detailed review of ion migrations in both high and low repetition rates published to date can be found in [79].

12.3.4 Comparison of low and high repetition rate processing

In addition to pulse energy, scan speed and focusing, several other exposure parameters have been found to influence the resulting properties of fs-laser-written waveguides. These include pulse duration [52,89], polarization [38,49], direction [90], wavelength [70] and migration of ions [79].

In fused silica, where heat accumulation is not driven, waveguide properties were found to vary with incident laser polarization, across a repetition rate range of 1 kHz [49] to 1 MHz [38]. In contrast, no detectable difference in waveguide transmission properties was found when waveguides were formed with different polarizations in borosilicate glass within the heat-accumulation regime at MHz repetition rates [38]. In addition, the waveguide properties in borosilicate glass were invariant to pulse duration when varied 300 to 700 fs, in contrast to fused silica, where pulse duration was observed to strongly affect waveguide mode size and loss [49]. The sensitivity to pulse duration and polarization in fused silica is associated with form birefringence arising from nanogratings formed within the laser-modified volume [91]. In cumulative heating regime, melting erases the nanogratings [73] as shown in Figure 12.17, which likely reduces the polarization and pulse duration dependence.

Due to energy depletion, self-focusing and plasma defocusing, pulse duration influences the spatial distribution of the energy density at the focus [92]. At a 1-kHz repetition rate, where heat accumulation does not occur, the dependence of waveguide properties on pulse duration in fused silica glass [93] was attributed to nonlinear pulse propagation. However, in the heat accumulation regime, the

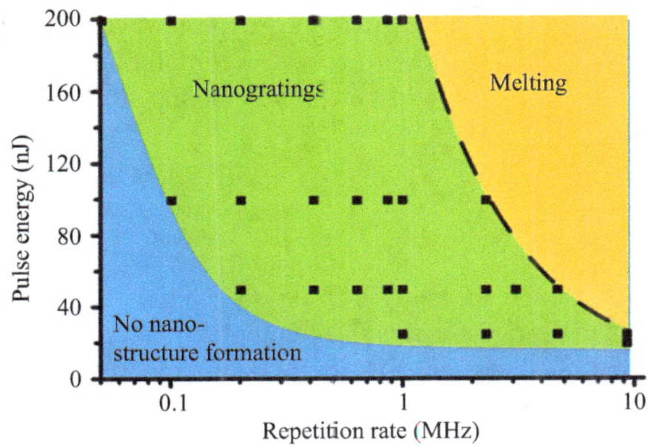

Figure 12.17 Map of nanograting formation in fused silica as a function of repetition rate and pulse energy. The dashed line represents 230 mW, the heat accumulation threshold. Above this power, the nanogratings are erased by melting of the glass. © 2011 Springer Nature/Applied Physics A. Reproduced, with permission, from Reference [73]

spherically symmetric thermal diffusion washes out the elliptical distribution of energy in the focal volume to produce waveguides with nearly circular cross sections. Therefore, one would expect pulse duration, despite its effect on the energy distribution at the focus, to play a smaller role in forming waveguides in the heat accumulation regime.

Kazansky *et al.* recently discovered the quill effect [90], in which laser material modification is influenced by the writing direction due to a pulse front tilt in the ultrafast laser beam. Although any material should show a direction dependence due to a pulse front tilt, the effect was found to be almost negligible when processing borosilicate glass within the heat accumulation regime as evidenced by a directional coupler formed by arms written in opposite direction but showing a remarkably high peak coupling ratio of 99% [94].

In borosilicate glass, a constant average power of about 160 and 230 mW was needed for driving heat accumulation and for producing the optimum waveguides, respectively [38]. Waveguide writing within the heat accumulation regime is somewhat similar to continuous wave (CW) writing of waveguides in photosensitive glass with UV lasers [95], where the average power dictates the NF and refractive index contrast. In a study of fused silica over a wide repetition rate range and where heat accumulation was not driven, a constant pulse energy was required to achieve the lowest loss waveguides [70]. In pure fused silica, the individual laser pulse each contributed to the modification, in much the same way as fabrication with 1-kHz (1-ms pulse separation) repetition rate Ti:sapphire lasers, but with advantage of faster speeds owing to a faster pulse delivery rate.

Finally, the behaviour of ion migration strongly depends on the laser repetition rate: at repetition rates higher than 200 kHz, heat accumulation can drive the ions near the focal volume [81], and, moreover, there exist specific fluence regimes where light and heavy element migration can form waveguides [87]. At lower repetition rates, even though there is no heat accumulation, it was demonstrated that spatial distribution of heat build-up can be modulated to drive the ions to and from the heat sources [96].

12.4 Applications

Researchers have exploited the recipes to produce highly confining and low loss optical waveguides to fabricate a wealth of complex photonic devices using the fs laser writing method. Such devices find use in many important technological fields including passive and active devices for telecommunications, sensing, lab-on-a-chip, astronomy and quantum optics.

12.4.1 Photonic devices

In this section, we highlight four breakthrough applications in photonics made possible by fs laser writing. These include BGWs for sensing and filtering, active-doped glass waveguides for laser applications, 3D passive light routing devices for astronomy and cascaded directional couplers for quantum information.

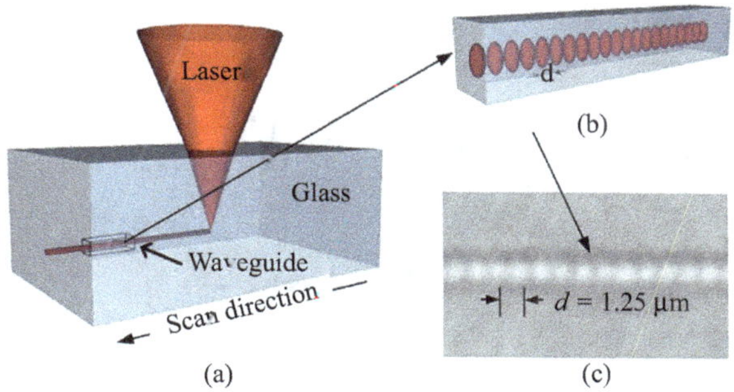

Figure 12.18 Schematic illustration (a) of the waveguide fabrication process with the sample translated perpendicularly to the focused laser beam direction, leading to an array of isolated focal volumes (b) under high (>1 mm/s) scan velocities. An overhead microscope image (c) of a waveguide showing isolated modification volumes written with 1.25-mm/s scan speed. © 2007 The Optical Society. Reproduced, with permission, from Reference [98]

12.4.1.1 Bragg grating waveguides for telecom/sensing

The BGW was discovered accidentally: in searching for the waveguide processing window in a borosilicate glass with a 1-kHz, 800-nm Ti:sapphire laser, researchers in Toronto explored every possible exposure parameter, including those well outside the optimum values previously reported in the literature. Although typical scan speeds reported in the dozens of previous works on waveguide writing in glass were about 50 μm/s, Zhang attempted scan speeds greater than 1 mm/s, and in doing so serendipitously discovered the BGW, which has opened up many new applications for fs laser writing [97].

When translating a sample with a velocity v relative to the focused fs laser beam, the number of pulses overlapped in a focal spot size diameter of $2w_0$ is given by $N = 2w_0R/v$, where R is the laser repetition rate. For a 1-kHz repetition rate fs laser and a typical focus spot diameter of 1 μm, the number of pulses/spot is $N = 1/v$, where v is the speed in mm/s. At speeds greater than 1 mm/s, the number of pulses overlapping will be a fraction of 1, resulting in segmented voxels as shown in Figure 12.18. The separation between these segmented voxels is given by $\Lambda = v/R$. Although such structures appear unsuitable for waveguide applications, they guide light surprisingly well and additionally offer a very strong Bragg reflection at a wavelength of $\lambda_B = 2n_{\mathrm{eff}}\Lambda$

This novel BGW device was demonstrated to have both low-loss (1-dB/cm propagation loss) light guiding and high-strength Bragg resonance (30-dB transmission dip) in borosilicate glass at 1,550-nm wavelength [98]. However, for materials like fused silica, this single-pulse interaction failed to generate sufficiently large refractive index change for guiding at telecom wavelengths.

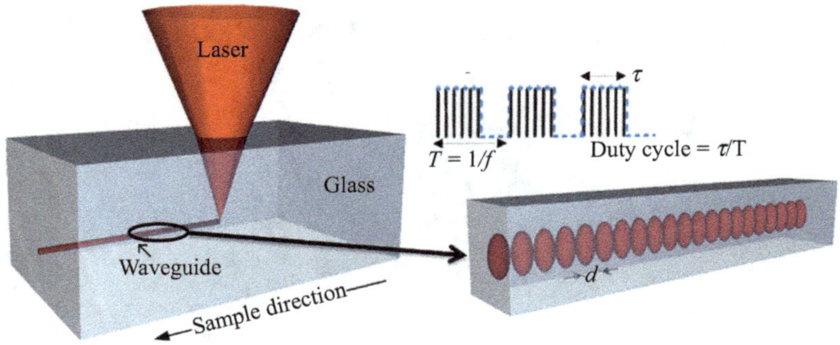

Figure 12.19 Burst method for writing Bragg grating waveguides with focused femtosecond laser pulses: the pulse train is modulated with frequency f. © 2007 The Optical Society. Reproduced, with permission, from Reference [99]

Unlike the single-pulse interaction used to form BGWs with the previously discussed method, a multipulse method is preferred for waveguide writing in glasses, where hundreds to thousands of laser pulses overlap in the laser focal volume to induce large refractive index changes. Zhang *et al.* invented another technique to address the limitations of the single-pulse BGW fabrication method. Using the scan speed and pulse energy required to produce optimum continuous waveguides, the laser intensity was modulated during waveguide inscription with a fast shutter to form a segmented BGW (Figure 12.19). In this case, the bursts of laser pulses from the modulated pulse train are responsible for the voxels that make up the segmented waveguide. In the burst method, the separation between voxels is independent of the repetition rate and is instead tuned with the modulation frequency of the shutter f so that $\Lambda = v/f$. For a Bragg wavelength $\lambda_B = 1{,}550$-nm ($\Lambda =$ 0.53 μm for fused silica) and a typical waveguide scan speed of 10 mm/s, modulation rates of about 20,000 Hz are required. Since the response time of mechanical shutters is on the order of 1 ms, they are too slow to be applied to the burst method of BGW fabrication. Therefore, acousto-optic modulators (AOMs) are often applied to BGW fabrication at the expense of reduced transmission (~70% diffraction efficiency) and added complexity since the beam must be reduced to about 1-mm width to achieve fast switching times [99].

The end result is a BGW structure similar to the one presented in Figure 12.19, but each voxel is now formed by multiple laser pulses to ensure sufficient refractive index change for near-IR waveguiding. The voxel size and separation can be further optimized by adjusting the modulation duty cycle (Figure 12.19) for high refractive index contrasts targeting strong grating response.

The burst method was optimized for fused silica, where heat accumulation effects are less pronounced, yielding a 35-dB transmission dip in the Bragg transmission spectrum near 1,550-nm wavelength and well-confined modes well matched to SMF (Figure 12.20), with 0.5-dB/cm propagation loss. It was assumed that

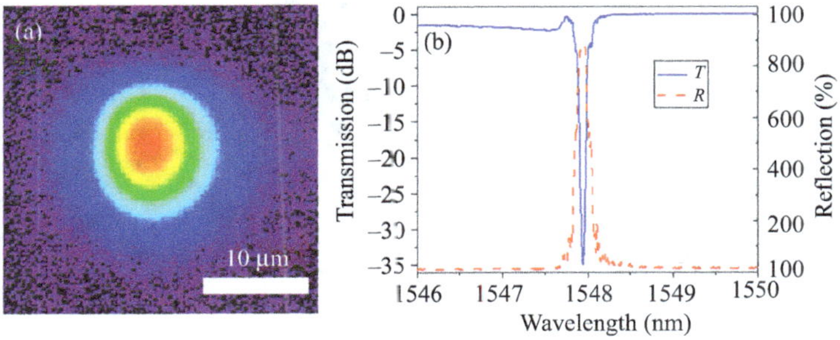

Figure 12.20 (a) Mode profile and (b) transmission/reflection spectra of BGW written with burst method in fused silica. © 2007 The Optical Society. Reproduced, with permission, from Reference [99]

in borosilicate glass where heat accumulation is present at high repetition rates, the periodicity of the segmented voxels would be washed out since the heat-affected cladding of waveguide zones tapers away slowly when the laser is shuttered off [72]. However, the Withford group demonstrated that BGW response is still possible even when cumulative heating effects are present [100]. They showed that when a high repetition rate pulse train is modulated with an AOM, the transient build-up of heat accumulation within the sample can lead to the formation of a permanent nano-void, which can be exploited to fabricate BGWs.

The fs laser BGW writing method has been applied to form several novel devices in bulk glass, including narrowband multi-wavelength filters [101] and chirped BGWs having 20-nm bandwidths [102]. An important application was demonstrated using BGWs in an optical sensor network inside bulk glass that directly measures the 3D distribution of temperature and strain (Figure 12.21) [103].

The Bragg reflection peak is sensitive to the chip temperature and strain and so several BGWs were fabricated in different positions inside a fused silica glass substrate. In this way, it was possible to have a 3D map of the temperature and strain in the chip with high spatial resolution. The thermo-optic and strain-optic responses of BGWs devices were reported and found to be similar to traditional fibre Bragg gratings (FBGs) in silica fibre. Further, thermal annealing studies revealed a high stability of both waveguides and Bragg resonances for these devices, showing that the sensor is suitable for a high-temperature environment. Such 3D grating networks open new directions for strain and temperature sensing in optical circuits, actuators, and micro-electro-mechanical systems (MEMS) and civil structures.

The fs laser BGW writing method has also been extended to use inside optical fibres. Herman's group demonstrated temperature-compensated 3D fibre shape sensing using BGWs that were distributed inside a single coreless optical fibre (Figure 12.22). Efficient coupling between the laser-written photonic elements and a standard SMF was obtained by 3D laser writing of a 1×3 directional coupler to meet with the core waveguide in the fusion-spliced SMF.

Simultaneous interrogation of nine Bragg gratings, distributed along three laterally offset waveguides, was achieved through a single waveguide port to

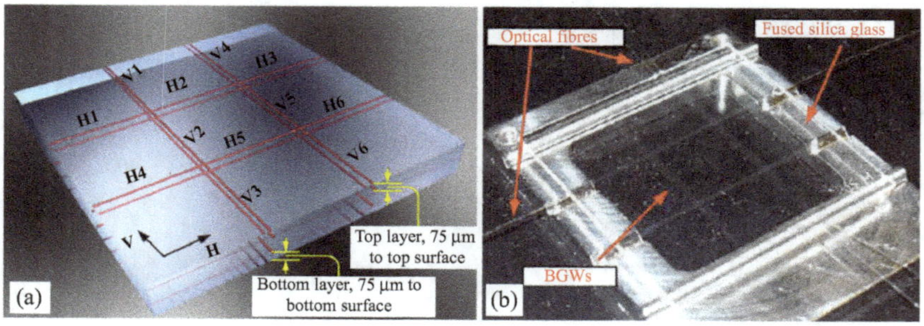

Figure 12.21 A 3D-sensing network in a fused silica plate (50 mm×50 mm×1 mm) consisting of multi-wavelength BGW segments (red lines) with a geometry as shown in the drawing (a) and the photograph (b). Single-mode fibres are shown epoxied for butt-coupling to six of the BGW segments. © 2008 The Optical Society. Reproduced, with permission, from Reference [103]

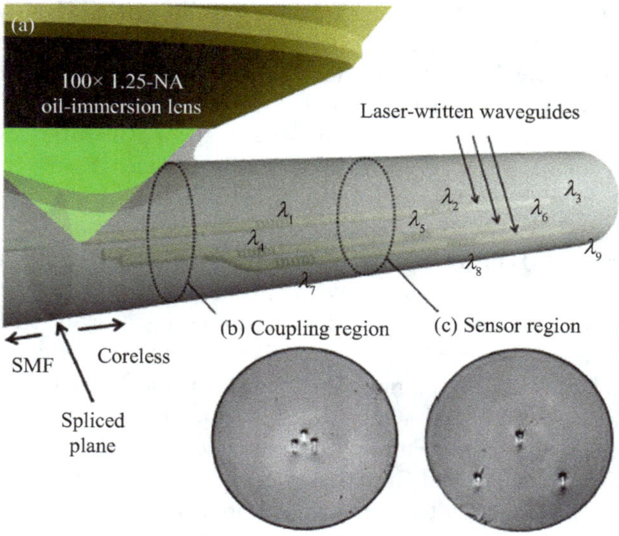

Figure 12.22 (a) Schematic of a 3D distributed shape and thermal sensor written in coreless fused silica fibre by a femtosecond laser focused with an oil-immersion lens. Microscope images of the fibre cross section (125-μm diameter) at the (b) coupling and (c) sensor regions, showing the arrangement of the laser-written waveguides. © 2013 The Optical Society. Reproduced, with permission, from Reference [104]

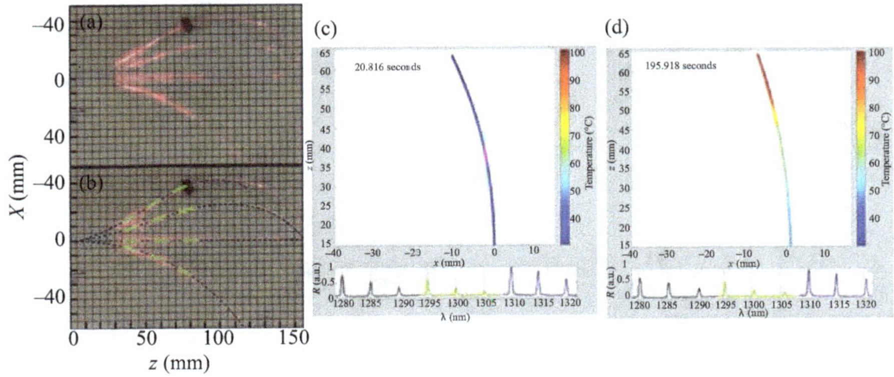

Figure 12.23 (a) Superimposed photographs of the fibre sensor bent in various shapes showing scattered light when end-fired with a red laser. (b) Calculated fibre profiles in dashed lines superimposed onto the fibre shapes. The green lines indicate the BGW triplet locations; fibre bending under (c) uniform and (d) graded temperature: single frames from a time sequence of fibre shape and temperature profile (top) calculated from the Bragg wavelength shifts in the reflection spectrum of nine BGWs (bottom), where red lines represent the Bragg wavelengths of unstrained gratings at room temperature. © 2013 The Optical Society. Reproduced, with permission, from Reference [104]

temporally monitor the Bragg wavelength shifts to infer both the shape and temperature profile along the fibre length (Figure 12.23). The fibre sensor was calibrated for bend sensing up to 2.5×10^{-2}-mm^{-1} curvature, temperature sensing up to 250 °C, and real-time dynamic 3D fibre shape sensing with positional error of 0.6 mm. The compact size and high-temperature tolerance make the 3D fibre sensor attractive for structural, industrial, reactor and pipeline applications and also integration into biomedical and catheter devices.

In another application, Bragg gratings waveguides (FBGs) were laser written in bulk fused silica (the core of SMF) along with adjacent stress-inducing continuous tracks [105]. The fs-laser-generated stress tracks induced a birefringence causing a spectral splitting of the transverse electric (TE) and transverse magnetic (TM) Bragg resonances. By optimization of the writing energy and placement of the adjacent stress inducing tracks, a birefringence of up to 2×10^{-3} was achieved, giving a beat length of 775 μm. Such a short polarization mode beat length is attractive compared to commercial polarization-maintaining fibres that have typical beat lengths ranging from 1.5 to 5 mm.

12.4.1.2 Active devices

Femtosecond laser micromachining has been applied to several active glasses to produce both waveguide amplifiers and lasers. Typical active ions are ytterbium (Yb^{3+}) and erbium (Er^{3+}) due to their compatibility with the telecommunication

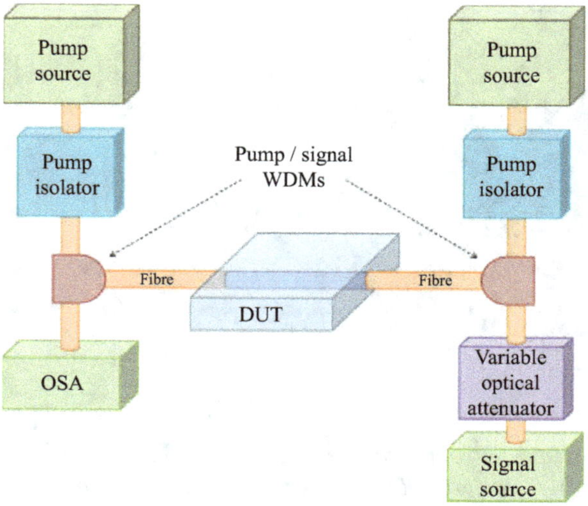

Figure 12.24 Active waveguide device configuration and characterization set-up. DUT, device under test; OSA, optical spectrum analyser. © 2009 John Wiley and Sons. Reproduced, with permission, from Reference [107]

band. In particular, Er-doped and Er/Yb-doped glasses have allowed the demonstration of waveguide amplifiers operating in the whole C-band (1,530–1,565 nm), as well as tuneable lasers and SM or mode-locked laser operation.

Optical amplifiers are achieved by doping an active material inside a host passive transparent glass. The active elements are typically atoms which absorb photons at the pump wavelength to invert the electronic populations, followed by electron decay and photon emission at the signal wavelength to be amplified. A phosphate glass (Kigre QX) doped with Er and Yb was found to be ideal for fs laser waveguide writing with a low-repetition-rate system, leading to the first demonstration of an active waveguide, exhibiting internal gain of 0.7 dB at 1,550 nm [106]. The waveguide was 25-mm long and diode-pumped at 980 nm under a bi-propagating scheme (Figure 12.24) [107].

The SM-guided field was approximately Gaussian but quite large and asymmetric compared to the mode supported by SMF, resulting in high coupling losses which prevented net gain. A significant improvement was demonstrated by exploiting a cavity-dumped Yb oscillator with a high repetition rate (500 kHz) which drove heat accumulation and thermal diffusion effects during waveguide inscription [39]. The same phosphate glass was used, but the doping concentrations were optimized to obtain higher gain per unit length and longer devices. The cavity-dumped system led to high-quality optical waveguides with low coupling losses of 0.2 dB/facet and propagation losses of 0.3 dB/cm, enabling net gain over the entire C-band as shown in Figure 12.25. The peak internal gain per length was 2.5 dB/cm, comparable to the

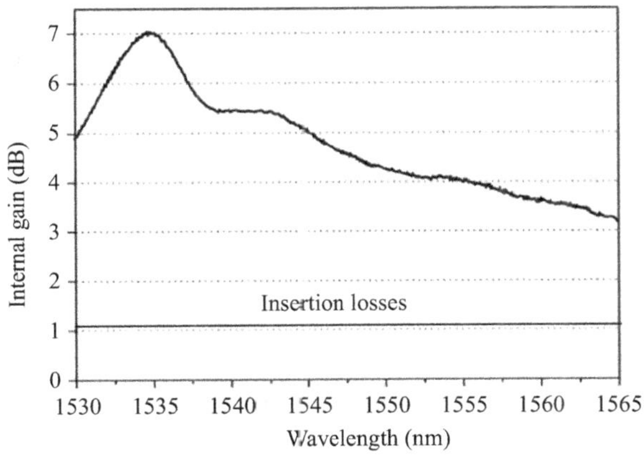

Figure 12.25 Internal gain of the Kigre waveguide amplifier as a function of wavelength. © 2006 IEEE. Reproduced, with permission, from Reference [39]

3-dB/cm value reported in state-of-the-art Er/Yb-doped waveguide amplifiers fabricated by ion exchange in a similar phosphate glass.

Waveguide lasers are waveguide amplifiers but with a feedback element for the signal wavelength. To achieve lasing, the amplifier must also provide net gain. The feedback may be formed by three different approaches: (i) external dielectric mirrors, (ii) external Bragg reflectors such as FBGs and (iii) distributed feedback achieved by the inscription of BGWs.

Using the second approach, several waveguide lasers have been reported by fs laser writing [107], demonstrating low lasing threshold powers and reasonably high slope efficiencies. The first fs-laser-written waveguide laser was fabricated in an Er/Yb-doped phosphate glass (Kigre QX) by means of a frequency doubled fs laser, operated at 166-kHz repetition rate [108]. The waveguides provided 2.3-dB peak net gain at 1,534 nm with 420-mW pump power at 980-nm wavelength in a bi-propagating pumping scheme. The laser cavity was formed using a 99.8% reflectivity FBG on one facet and a 68% reflectivity FBG on the other facet, acting as output coupler. A maximum output power of 1.7 mW with 2% slope efficiency was reported. Improvement in waveguide quality was obtained using a 500-kHz repetition rate fs laser which led to significantly better laser properties: a maximum output power of 30 mW with 8.4% slope efficiency was obtained at 1,534 nm, with laser action and wavelength tunability demonstrated across the entire C-band [39].

The passive mode-locked operation of an fs-laser-written waveguide laser was demonstrated using a carbon-nanotube-based saturable absorber [109]. When assembled in a ring cavity configuration, continuous-wave laser action and self-starting, stable mode-locking were observed (Figure 12.26). The duration of transform-limited sech^2 pulses was 1.6 ps at a repetition rate of 16.7 MHz. Exploiting the same ring cavity configuration, a mode-locked Er-doped bismuthate

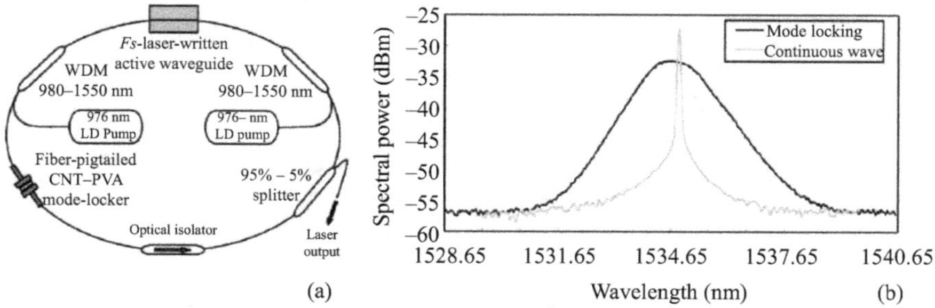

Figure 12.26 Carbon-nanotube-mode-locked fs-written waveguide laser in Er/Yb-doped phosphate glass: (a) schematic set-up of the ring cavity and (b) output spectrum in continuous wave and mode-locked lasing regimes. © 2006 AIP. Reproduced, with permission, from Reference [109]

glass waveguide laser was demonstrated, having 320-fs duration pulses, 40-MHz repetition rate with 1.25-mW output power at 1.56-μm wavelength [110].

An improvement in stability of fs-laser-written waveguide lasers was achieved using the third method for producing feedback, by fabricating the Bragg gratings directly in the waveguides. Withford's group demonstrated such a device in an erbium/ytterbium-doped phosphate glass [111]. A 500-nm pitch produced a strong first-order reflection at 1,535 nm. When pumped at 980 nm, lasing occurred with a slope efficiency of 17%.

12.4.1.3 3D architectures for astronomy

Another application of an fs laser writing is the burgeoning field of astrophotonics, where photonic lanterns are needed to provide an efficient interface between multimode (MM) 'light-bucket' fibres and high-fidelity, SM photonic devices. Narrowband spectral filtering of MM light on a 3D-integrated photonic chip using photonic lanterns and BGWs was recently demonstrated by the Withford group in a boro-aluminosilicate glass (Corning EAGLE 2000) [112]. The fabricated chip comprised back-to-back lanterns with a horizontal array of straight waveguides in the middle as shown in Figure 12.27. Each photonic lantern was composed of 19 individual SM waveguides, which are brought close to each other to produce the MM waveguide input. These 19 waveguides fan out to a horizontal array of straight waveguides. The waveguides contained either one, two, three or four gratings, with different Bragg resonances written next to each other in the SM waveguide core.

Pulses from an fs laser oscillator (800 nm, 5.1 MHz, 50 fs) were focused with a 1.25-NA oil immersion objective 300 μm below the surface of the glass sample. The photonic lantern structures were written in the cumulative-heating regime at a repetition rate of 5 MHz, a pulse energy of 35 nJ and a translation speed of 12.5 mm/s. In a second step after the waveguide inscription, the laser repetition rate was reduced to 52 kHz with a pulse picker to place point-by-point modifications for

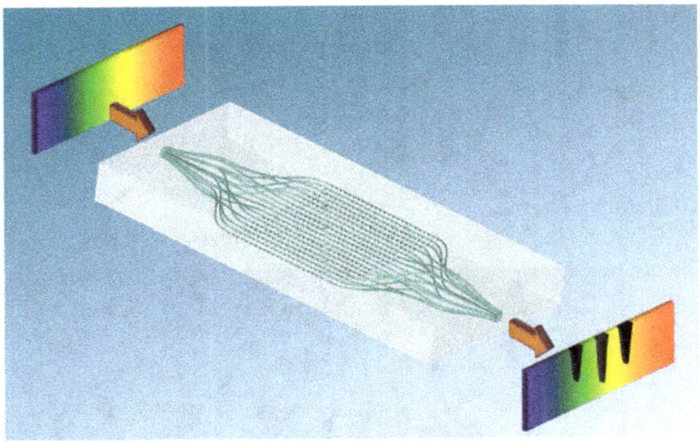

Figure 12.27 Sketch of a photonic lantern for astrophotonics. © 2009 John Wiley and Sons. Reproduced, with permission, from Reference [112]

the gratings into the SM waveguides. Due to the high-focusing NA, the point-by-point features were only ~1 × 4 μm in cross section, enabling the placement of multiple gratings side-by-side and spaced by 1 μm within the SM waveguide (Figure 12.28).

The period and therefore the Bragg wavelength (λ_B) of the gratings was tuned by changing the translation speed according to $v = \lambda_B R/(2n_{\mathrm{eff}})$. In this way, four photonic lantern structures were written with single (λ_B = 1,545 nm), double (λ_B = 1,545 and 1,552 nm), triple (λ_B = 1,545, 1,552 and 1,559 nm) and quadruple (λ_B = 1,545, 1,552, 1,559 and 1,563 nm) BGWs, with a scan speed of about 27 mm/s. Transmission dips of up to 5 dB were measured in both photonic lanterns and reference SM waveguides with 10 mm long gratings. The result demonstrates efficient and uniform performance of the gratings in the photonic lantern; thus, such devices could be employed for space-division multiplexed communication systems or astronomical systems for suppression of the telluric absorption lines.

12.4.1.4 Directional couplers for quantum information

An exciting new application of fs laser writing is the on-chip integration of quantum information experiments conducted with single photons. Quantum information technologies are based on encoding information with quantum systems [113], such as trapped ions, quantum dots or photons, and on harnessing non-classical phenomena to increase the computational speed of certain algorithms. Photons are ideally suited to quantum information systems because of their good coherence and it has been demonstrated that universal quantum computation can be achieved with linear optical networks and a single photon detector [114]. Most quantum information experiments use bulk optics, but phase stability issues limit the development of large and complex quantum circuits. Femtosecond laser writing is an ideal candidate for creating a more stable quantum information platform in an integrated optical chip.

Marshall *et al.* demonstrated how fs-laser-inscribed directional couplers (Figure 12.29) could be exploited for quantum information systems [115].

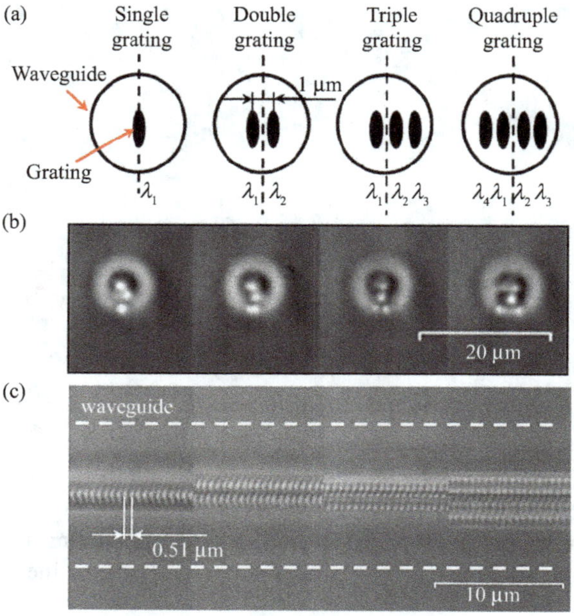

Figure 12.28 (a)Cross-sectional diagram of multiple Bragg gratings inscribed within fs-laser-written waveguides. Cross-sectional (b) and overhead (c) optical microscope images of BGWs. © 2009 John Wiley and Sons. Reproduced, with permission, from Reference [112]

Directional couplers, an important building block for photonic integrated circuits, have been previously fabricated by fs laser writing for use at telecom wavelengths [94,116,117]. However, for quantum optics experiments, optical waveguides were designed to be SM and low loss at 800-nm wavelength [118]. Experiments showing two and three photons non-classical interference were conducted in the laser-fabricated couplers. The two-photon (three-photon) experiment used a directional coupler with a 1:1 (1:2) splitting ratio and showed an interference visibility as high as 96% (84%). In these experiments, the polarization of the photons was held fixed, with the qubits encoded in the path of the photons. However, many quantum information processes and sources of entangled photon states are based on the polarization degree of freedom, with information encoded in the polarization state of single photons.

To address this point, Sansoni *et al.* demonstrated the first integrated optical chip able to support polarization encoded qubits [119]. The directional coupler was realized in borosilicate glass chip, written by a 300-fs, 1,030-nm Yb:KYW (potassium gadolinium tungstate) cavity-dumped oscillator. Heat accumulation effects driven by the 1-MHz repetition rate yielded nearly circular guided modes at 800 nm, without any additional beam shaping required. The most important

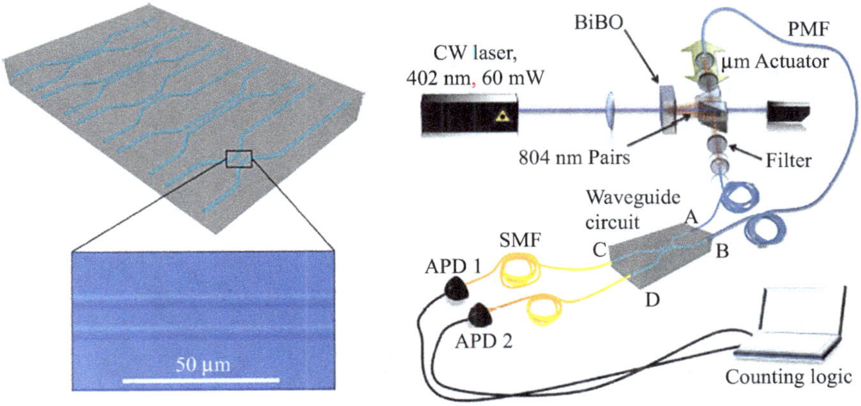

Figure 12.29 (Left) Schematic of an array of directional couplers fabricated by fs direct writing in a fused silica chip and an optical microscope image showing the central coupling region where the waveguides were separated by 10 μm. (Right) Measurement set-up based on spontaneous parametric down conversion of a continuous wave 402-nm laser diode. © 2009 The Optical Society. Reproduced, with permission, from Reference [115]

challenge for polarization encoded qubits is to provide low birefringence waveguides. A large birefringence will not only rotate the polarization of incoming photons, but also it can destroy the coherence of the large-bandwidth photons used in quantum experiments. Laser fabrication parameters in fs laser waveguide writing were optimized to achieve a waveguide birefringence as low as 7×10^{-5}, lower than the ~10^{-4} value obtained by waveguides produced by conventional silica-on-silicon photolithography. Hong–Ou–Mandel experiments with entangled two-photon states were performed with the integrated directional coupler and visibilities of the non-classical interference dips and peaks higher than 90% were observed, indicating the high quality of the device produced.

According to theory, a quantum computer can outperform a classical computer in several tasks, but to date, there has been no empirical proof that such tasks can be achieved in practice. The 'boson-sampling problem' may be the best opportunity for such a demonstration [120], where the probability distribution for a particular arrangement of bosons to appear at the output of an energy-conserving linear unitary transformation is sampled, assuming a particular input arrangement. However, the evolution of bosons undergoing arbitrary linear unitary transformations quickly becomes more difficult to predict using classical computers as the number of particles and modes are increased [121]. Photons propagating in a multiport interferometer solve the boson sampling problem, which has motivated the development of technologies that enable precise control of multiphoton interference in large interferometers. Crespi *et al.* showed how fs laser writing could be used to achieve

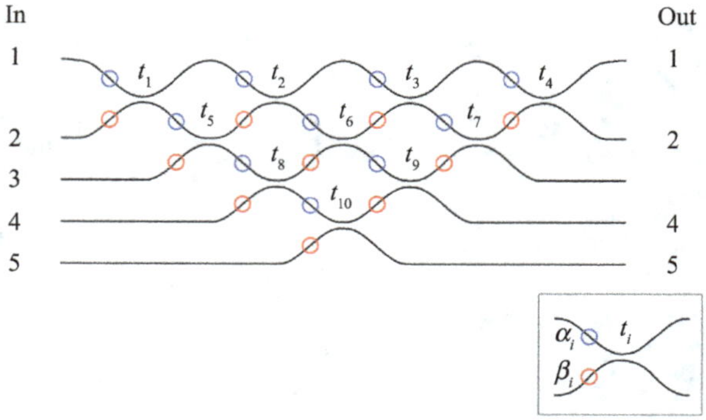

Figure 12.30 Layout of multimode interferometers for an arbitrary 5×5 mode transformation using a network of beam splitters (evanescently coupled directional couplers) with different transmissivities t_i. The blue and red boxes indicate different phase shifters. © 2013 Springer Nature/Nature Photonics. Reproduced, with permission, from Reference [121]

simultaneous control of all the parameters describing an arbitrary interferometer [121]. They implemented a small instance of the boson sampling problem by studying three-photon interference in a five-mode integrated interferometer using evanescent coupling of directional couplers (Figure 12.30) in place of the beam splitters used in bulk optics.

The main challenge in implementing the integrated layout is to independently control each of the 10 transmissivities t_i and 15 phase shifts a_i, β_i of an arbitrary, five-mode chip. This is because in a typical photonic circuit (Figure 12.31(a)), changes in the coupler geometry to modulate the transmissivity will modify the optical path (and the phase shifts), and vice versa.

Phase shifters were implemented by deforming the S-bend waveguides at the input of each directional coupler to stretch the optical path without affecting the transmissivity of the surrounding couplers (Figure 12.31(b)). Achieving independent control of the transmission of each directional coupler is more challenging, which is usually controlled by tuning the interaction length and/or the waveguide spacing. However, changing either parameter induces a variation in the optical path leading to a phase shift. Here the 3D advantage of fs laser writing truly shines: the researchers cleverly rotated one arm of the directional coupler out of the main circuit plane (Figure 12.31(c)), which alters the waveguide separation in the coupling region without affecting the path lengths. With this photonic device, the researchers were able to experimentally confirm that the permanent formula that governs the quantum-mechanical behaviour of non-interacting bosons holds for up to three photons interfering in a randomly chosen, five-mode interferometer.

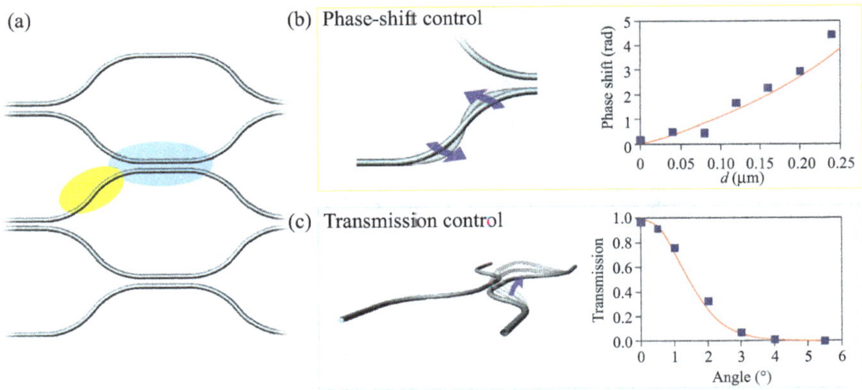

Figure 12.31 *(a)Controlled deformation of the S-bend at the input of each directional coupler and the coupling geometry allow independent control over the phase shift (b) and transmission (c). A calibration of the experimental transmission dependence on the rotation angle at 806-nm wavelength is given (squares), compared to theory (solid line). © 2013 Springer Nature/Nature Photonics. Reproduced, with permission, from Reference [121]*

12.4.2 Microfluidic devices

Initial demonstrations of the fs laser irradiation and chemical etching (FLICE) technique, aqueous solutions of HF in low concentrations (0.5%–5%) were used to create hollow channels in fused silica [122–124]. The nanogratings that are formed upon laser irradiation create vulnerable sites for the dissolution of SiO_2 which locally speed up the material's etching rate, thus enabling the creation of hollow structures in the bulk of fused silica. In this case, the etched silica microchannels are quite uniform in cross-sectional shape, but the lengths are limited to about 500 μm due to the poor selectivity of the etchant [124]. Solutions with higher content of HF (up to 20%) were found to promote the dissolution of the exposed areas at a higher rate, and longer structures were achieved with approximately 2-mm length, but, due to the aggressive etching process, the channels were tapered in shape [125]. To counteract this phenomenon, a spiralled conical pattern can be written with the opposite profile of the tapered channel, with an increasing diameter along the initial etching location. The resulting channel is uniform in cross-sectional shape, as shown in Figure 12.32. Furthermore, longer channels of 4 mm have been demonstrated with this method, with an aspect ratio of 20 [126].

Recently, the use of concentrated solutions of KOH has been proposed for the etching of modification tracks in fused silica. Researchers have demonstrated a more selective etching to yield longer channels with a higher aspect ratio at the expense of longer etching time [127]. A novel hybrid approach [128] that combines HF acid and KOH aqueous solutions in subsequent steps enables even more control in etching microchannels written in fused silica with focused fs laser pulses.

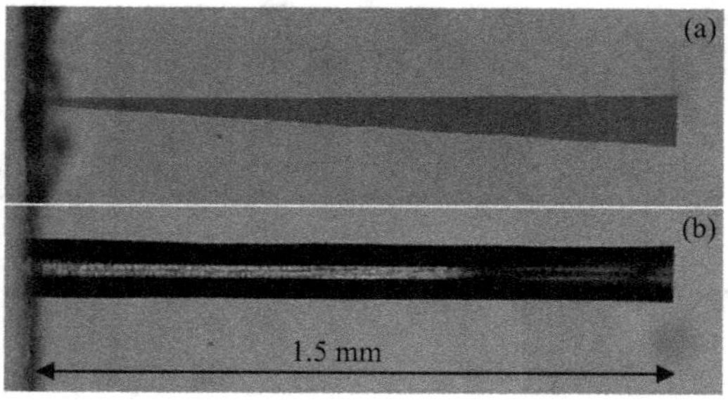

Figure 12.32 Microscope image of (a) conical spiral inscribed in fused silica glass and (b) the HF-etched microchannel showing a uniform cylindrical shape. © 2009 The Optical Society. Reproduced, with permission, from Reference [126]

Referred to as hybrid FLICE (H-FLICE), the method combines the fast and larger volume etching of HF with the slower but more accurate and selective etching of KOH.

12.4.2.1 Hybrid FLICE for particle filter

One of the key functions in microfluidics is the microfilter, which allows the passage of materials of certain dimensions, while blocking the rest. Embedded microfilters have been demonstrated [129,130] in fused silica by FLICE and in polymer by two-photon polymerization, the glass platform having the advantage of being inert to many chemical reagents and lacking autofluorescence. The most impressive microparticle filter was recently demonstrated in fused silica exploiting the versatile H-FLICE fabrication method. The micro-filter was directly fabricated inside a square channel with 80 μm × 80 μm cross section and 2.25-mm length [128]. Filtering was accomplished by a 15 × 15 grid of 2-μm diameter, 100-μm-long microchannels, spaced by 5 μm (Figure 12.33). The device was written with an oil-immersion 1.25-NA objective lens at a central depth of 160 μm.

As shown in Figure 12.33(b) and (c), there are three separate zones that can be observed after inscription: (A) reverse tapered access channels (to counteract the tapering effect described during HF etching), (B) square microchannels consisting of a parallelepipeds of 80 μm × 80 μm completely written in the volume with 1-μm spaced lines and (C) the filter itself. Zone A is the longest part of the channel and was etched with HF; zone B can be etched with KOH or HF and acts as a safety zone to prevent the exposure of the ultrafine filter pillars to the HF and lastly, zone C was etched with KOH, to preserve the initial 2-μm diameter of the fs-laser-inscribed lines.

To test the filtering capability, the channel was filled with water-containing polystyrene beads of 1- and 3-μm diameter. Figure 12.33(f) demonstrates that the 3-μm beads are blocked by the filter, whereas the smaller 1-μm beads pass through.

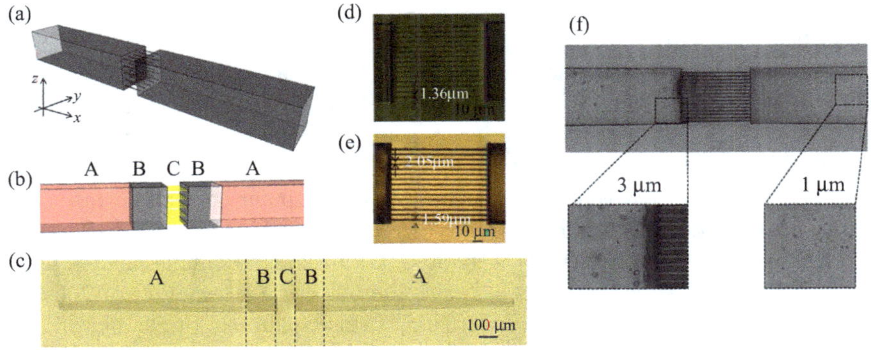

Figure 12.33 Schematic diagram (a) the device in 3D, two arms of the channel separated by the filtering matrix, (b) difference zone, where HF or KOH is used for etching (see the text); the microscope image of (c) the imprint of the device in fused silica by laser irradiation, (d) the filter structure before etching and (e) the filter structure after etching. (f) Mixture of 1- and 3-μm beads inserted in the channel's input arm: the 3-μm beads are stopped by the filter, whereas the smaller ones flow through. © IoP Publishing, Ltd. Reproduced, with permission, from Reference [128]

12.4.2.2 Liquid–liquid dynamic interfaces

Lab-on-a-chip is now a commonly known concept, and significant efforts have been made for the realization of multifunctional integrated systems for chemical analysis [131], cell culture and biochemical systems investigation [132], but most importantly for multiphase chemical reactions even for miscible solutions. Microfluidic reactors have already been proven valuable due to their high surface-to-volume ratio, the scale-out capabilities for industrial applications, the higher yield over batch reactors and the versatility of the microfluidic chip set-ups [133]. By manufacturing a microfluidic chip with a suitable geometry, it is possible to manage simultaneously two or more fluids and create dynamic interfaces between them while avoiding active mixing due to laminar flow [133].

Fused silica is compatible with a wide variety of organic solvents as well as water, and since it is not gas-permeable, it can be used as a material for the fabrication of microreactors for a wide variety of reactions, including water splitting for hydrogen production. FLICE was exploited for the fabrication and characterization of a double Y-branch fused silica microfluidic device for the introduction, interaction and separation of two miscible solutions characterized by laminar flow [134]. The laser-fabricated, fused silica microfluidic device was used for the study of the diffusion of Rhodamine 6G (R6G) in the ethanol–ethanol interface. The angle between the two inlets and the height of the chamber were varied, and the diffusion was qualitatively determined for different flow rates. R6G was used as a colorant for one of the two streams due to its optical properties and its well-established diffusivity in ethanol.

The fs laser used for device fabrication in fused silica glass was a regeneratively amplified Yb:KGW system (Pharos, Light Conversion) with 230-fs pulse duration, 515-nm wavelength (frequency doubled) and 500-kHz repetition rate focused with a 0.42-NA microscope objective (M Plan Apo SL50X Ultra-Long Working Distance Plan-Apochromat, Mitutoyo). Computer-controlled, three-axis motion stages (ABL-1000, Aerotech) interfaced by CAD-based software (ScaBase, Altechna) with an integrated AOM were used to translate the sample relative to the laser to form the desired structures. An average power (pulse energy) of 200 mW (400 nJ) and a scan speed of 5 mm/s were used to laser pattern a single microfluidic channel and the microfluidic double Y device shown in Figure 12.34(a). A multi-scan writing procedure with 7-μm spacing between scans was adopted to form the microfluidic device. The sample was etched in a sonication bath of 20% HF aqueous solution, with a temperature of 37 °C. The resulting chamber's internal dimensions were 2 mm × 100 μm (length × width) with a height h of 100 μm (Figure 12.34(b)).

To complete the characterization of the chip, the behaviour of diffusive mass transfer was studied by varying the separation angle between the inlet/outlet channels, θ, and the pumping pressure of the fluids, p. To avoid turbulence, the inlet tubes and the interaction chamber were designed considering the continuity of the fabrication process and the equality in resistance for both the inlets and outlets of the chip. A microfluidic pump (OB1, Elveflow, Paris, France) was connected to

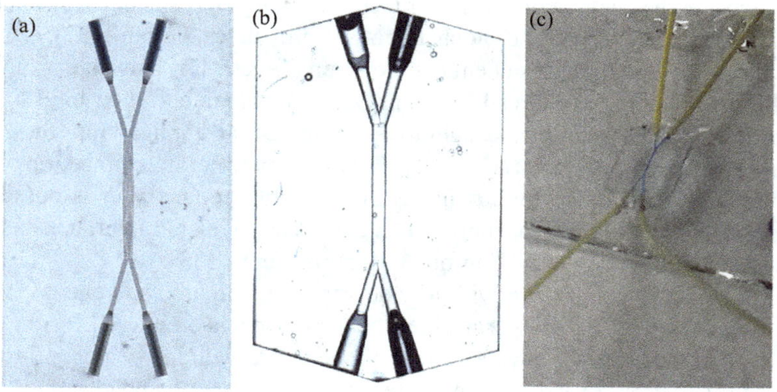

Figure 12.34 Steps of the creation of a buried glass microfluidic chip. (a) Optical microscope image of the modifications created by the laser writing for the fabrication of the double Y microfluidic chip, (b) the hollowed out microfluidic chip after etching and (c) image of a ready to use double Y microfluidic chip, with mounted tubing of 360/150 OD/ID. In the chip, blue dye solution is flowing for better visualization

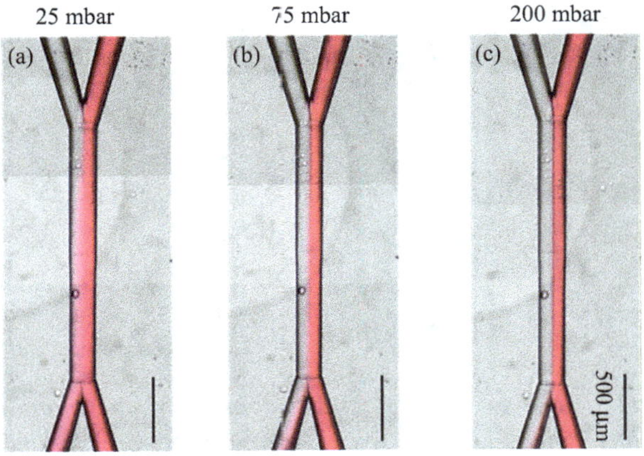

Figure 12.35 Diffusion behaviour inside a microfluidic chip with 30° incident angle and h = 500 μm, at increasing pumping pressure (a) ΔP = 25 mbar; (b) ΔP = 75 mbar; (c) ΔP = 200 mbar. Scale bars correspond to 500 μm. Arrow indicates the flow direction

the reservoirs of the solutions. Polytetrafluoroethylene tubing was inserted into the reservoirs and drove the fluids into the chips by polyether ether ketone tubing with an outer (inner) diameter of 360 μm (150 μm) by using an appropriate adapter. The latter tubing was connected to the glass chip using UV glue (Figure 12.34(c)).

The flow velocity is directly dependent on the pumping pressure according to the Hagen–Poiseuille equation, and it is the only parameter other than the geometrical characteristics of the chip that can affect the diffusion of R6G in ethanol. Diffusion is a time-dependent process, and it is obvious that a slowly flowing solution (i.e., in the case of $\Delta P = 25$ mbar) exhibits greater diffusion, as seen in Figure 12.35.

The effect of the angle ($\theta = 30°$, 60°, and 80°) between the two inlets was studied for microfluidic chips. Angles greater than 80° exhibited significant diffusion due to the trajectory of the fluids and were excluded for the purposes of the work. For each angle value, increased pressure resulted in an increased value of the Peclet coefficient and reduced diffusive mixing. For a given pumping pressure, a smaller separation angle resulted in reduced mixing between the two parallel flowing fluids. A separation angle of $\theta = 30°$ was thus found to be most suitable for minimum diffusion. In future work, chemical reactions at the interface of the parallel laminar flows in the laser-inscribed buried branching network will be performed.

12.4.2.3 Refractive index sensor

The most impressive 3D integrated optofluidic device formed by fs laser micromachining is a refractive index sensor which employs an unbalanced Mach–Zehnder interferometer with a sensing arm directly crossing a microfluidic channel

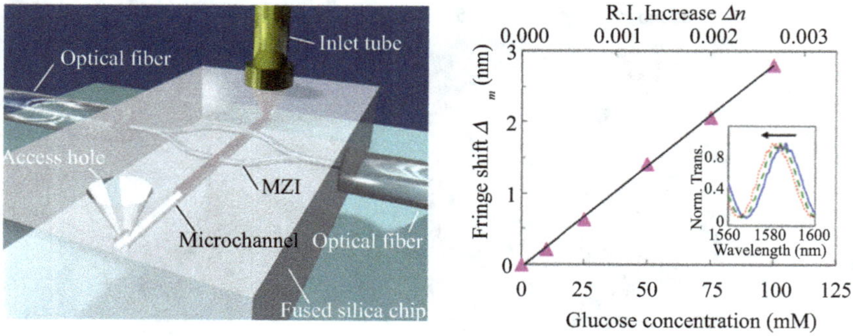

Figure 12.36 Left: schematic of the fs-laser-fabricated microfluidic channel and integrated MZI. The sensing arm crosses the channel orthogonally, while the reference arm passes over it. Right: measured fringe shift for different concentrations of glucose (inset: 0 mM solid; 50 mM dashed and 100 mM dotted); the corresponding refractive index increase is shown along the upper x-axis. © 2010 The Royal Society of Chemistry. Reproduced, with permission, from Reference [135]

and the reference arm passing above it (Figure 12.36) [135]. The device geometry, lying on a tilted plane, is impossible to obtain with other conventional methods.

Integration of the optical device was shown both in a commercial microfluidic chip, where lithographically defined channels were already present, and in a home-made chip, where microchannels were fabricated by FLICE. The interferometer's waveguides were inscribed in fused silica, using the second harmonic (515-nm wavelength) of 350-fs pulses from a cavity-dumped Yb:KYW (potassium yttrium tungstate) laser oscillator with 1-MHz repetition rate. The 90-nJ pulses were focused through a 0.6-NA objective, with the sample translated at 100-μm/s. The interferometer was successfully used to sense the refractive index variations of the content of the microchannel, with a sensitivity of 10^{-4} RIU (Figure 12.36). Potential applications include label-free sensing in electrophoresis, where one can take further advantage of the spatial resolution in the microchannel, given by the direct-crossing geometry.

12.5 Conclusions

Over the years, fs laser waveguide writing has evolved from a laboratory peculiarity into a viable solution for producing high-quality photonic devices for use in diverse industrially relevant fields. This direct writing technique allows fine adjustment and tuning of device parameters from one device to the next, aided by characterization feedback, thus shortening the optimization process and reducing costs compared to photolithographic techniques where a new mask is required, and several devices are fabricated at every run. The wide array of photonic components that have been demonstrated by this technology, such as couplers, interferometers, Bragg gratings, waveguide lasers and amplifiers, has enabled the facile fabrication of complex microsystems that find application in several different fields such as telecommunications, lab-on-a-chip and integrated quantum optics. The potential of this technology is so great that several unforeseen fields will certainly benefit from this versatile 3D photonic device fabrication technique in the future.

References

[1] Stuart BC, Feit MD, Herman S, Rubenchik AM, Shore BW, and Perry MD. Nanosecond-to-femtosecond laser-induced breakdown in dielectrics. Physical Review B. 1996;53(4):1749.

[2] Gattass RR and Mazur E. Femtosecond laser micromachining in transparent materials. Nature Photonics. 2008;2(4):219–25.

[3] Schaffer CB, Brodeur A, and Mazur E. Laser-induced breakdown and damage in bulk transparent materials induced by tightly focused femtosecond laser pulses. Measurement Science and Technology. 2001;12(11):1784–94.

[4] Du D, Liu X, Korn G, Squier J, and Mourou G. Laser-induced breakdown by impact ionization in fused silica with pulse widths from 7 ns to 150 fs. Applied Physics Letters. 1994;64(23):3071–3.

[5] Davis KM, Miura K, Sugimoto N, and Hirao K. Writing waveguides in glass with a femtosecond laser. Optics Letters. 1996;21(21):1729–31.

[6] Miura K, Qiu JR, Inouye H, Mitsuyu T, and Hirao K. Photowritten optical waveguides in various glasses with ultrashort pulse laser. Applied Physics Letters. 1997;71(23):3329–31.

[7] Richter S, Plech A, Steinert M, *et al.* On the fundamental structure of femtosecond laser-induced nanogratings. Laser & Photonics Reviews. 2012; 6(6):787–92.

[8] Sudrie L, Franco M, Prade B, and Mysyrewicz A. Writing of permanent birefringent microlayers in bulk fused silica with femtosecond laser pulses. Optics Communications. 1999;171(4–6):279–84.

[9] Shimotsuma Y, Kazansky PG, Qiu J, and Hirao K. Self-organized nanogratings in glass irradiated by ultrashort light pulses. Physical Review Letters. 2003;91(24):247405.

[10] Bhardwaj VR, Simova E, Rajeev PP, *et al.* Optically produced arrays of planar nanostructures inside fused silica. Physical Review Letters. 2006; 96(5):057404–4.

[11] Glezer EN and Mazur E. Ultrafast-laser driven micro-explosions in transparent materials. Applied Physics Letters. 1997;71(7):882–4.

[12] Itoh K, Watanabe W, Nolte S, and Schaffer CB. Ultrafast processes for bulk modification of transparent materials. MRS Bulletin. 2006;31(8):620–5.

[13] Chan JW, Huser T, Risbud S, and Krol DM. Structural changes in fused silica after exposure to focused femtosecond laser pulses. Optics Letters. 2001;26(21):1726–8.

[14] Bruckner R. Properties and structure of vitreous silica. I. Journal of Non-Crystalline Solids. 1970;5(2):123–75.
[15] Sakakura M, Shimizu M, Shimotsuma Y, Miura K, and Hirao K. Temperature distribution and modification mechanism inside glass with heat accumulation during 250 kHz irradiation of femtosecond laser pulses. Applied Physics Letters. 2008;93(23):3.
[16] Hirao K and Miura K. Writing waveguides and gratings in silica and related materials by a femtosecond laser. Journal of Non-Crystalline Solids. 1998;239(1–3):91–5.
[17] Chan JW, Huser TR, Risbud SH, and Krol DM. Modification of the fused silica glass network associated with waveguide fabrication using femtosecond laser pulses. Applied Physics A. 2003;A76(3):367–72.
[18] Streltsov AM and Borrelli NF. Study of femtosecond-laser-written waveguides in glasses. Journal of the Optical Society of America B (Optical Physics). 2002;19(10):2496–504.
[19] Saliminia A, Vallee R, and Chin SL. Waveguide writing in silica glass with femtosecond pulses from an optical parametric amplifier at 1.5 microns. Optics Communications. 2005;256(4–6):422–7.
[20] Dekker P, Ams M, Marshall GD, Little DJ, and Withford MJ. Annealing dynamics of waveguide Bragg gratings: evidence of femtosecond laser induced colour centres. Optics Express. 2010;18(4):3274–83.
[21] Sharma S, Oliveira V, Herrero P, and Vilar R. Internal structure of the nanogratings generated inside bulk fused silica by ultrafast laser direct writing. Journal of Applied Physics. 2014;116(5):053106.
[22] Taylor R, Hnatovsky C, and Simova E. Applications of femtosecond laser induced self-organized planar nanocracks inside fused silica glass. Laser & Photonics Reviews. 2008;2(1–2):26–46.
[23] Ramirez LPR, Heinrich M, Richter S, *et al.* Tuning the structural properties of femtosecond-laser-induced nanogratings. Applied Physics A. 2010; 100(1):1–6.
[24] Hnatovsky C, Taylor RS, Simova E, *et al.* Fabrication of microchannels in glass using focused femtosecond laser radiation and selective chemical etching. Applied Physics A. 2006;84(1–2):47–61.
[25] Richter S, Jia F, Heinrich M, *et al.* The role of self-trapped excitons and defects in the formation of nanogratings in fused silica. Optics Letters. 2012;37(4):482–4.
[26] Stankevič V, Račiukaitis G, Bragheri F, *et al.* Laser printed nano-gratings: orientation and period peculiarities. Scientific Reports. 2017;7(1):1–8.
[27] Tamaki T, Watanabe W, Nagai H, Yoshida M, Nishii J, and Itoh K. Structural modification in fused silica by a femtosecond fiber laser at 1558 nm. Optics Express. 2006;14(15):6971–80.
[28] Qi J, Wang Z, Xu J, *et al.* Femtosecond laser induced selective etching in fused silica: optimization of the inscription conditions with a high-repetition-rate laser source. Optics Express. 2018;26(23):29669–78.

[29] Fernandes LA, Grenier JR, Herman PR, Aitchison JS, and Marques PV. Femtosecond laser writing of waveguide retarders in fused silica for polarization control in optical circuits. Optics Express. 2011;19:18294–301.

[30] Corrielli G, Crespi A, Geremia R, *et al.* Rotated waveplates in integrated waveguide optics. Nature Communications. 2014;5:4249.

[31] Richter S, Miese C, Döring S, *et al.* Laser induced nanogratings beyond fused silica-periodic nanostructures in borosilicate glasses and ULE^{TM}. Optical Materials Express. 2013;3(8):1161–6.

[32] Juodkazis S, Nishimura K, Tanaka S, *et al.* Laser-induced microexplosion confined in the bulk of a sapphire crystal: evidence of multimegabar pressures. Physical Review Letters. 2006;96(16):166101.

[33] Glezer EN, Milosavljevic M, Huang L, *et al.* Three-dimensional optical storage inside transparent materials. Optics Letters. 1996;21(24):2023–5.

[34] Juodkazis S, Matsuo S, Misawa H, *et al.* Application of femtosecond laser pulses for microfabrication of transparent media. Applied Surface Science. 2002;197–198:705–9.

[35] Eaton SM, Zhang H, Herman PR, *et al.* Heat accumulation effects in femtosecond laser-written waveguides with variable repetition rate. Optics Express. 2005;13(12):4708–16.

[36] Johnston TF. Beam propagation (M2) measurement made as easy as it gets: the four-cuts method. Applied Optics. 1998;37(21):4840–50.

[37] Hnatovsky C, Taylor RS, Simova E, Bhardwaj VR, Rayner DM, and Corkum PB. High-resolution study of photoinduced modification in fused silica produced by a tightly focused femtosecond laser beam in the presence of aberrations. Journal of Applied Physics. 2005;98(1):013517.

[38] Eaton SM, Zhang H, Ng ML, *et al.* Transition from thermal diffusion to heat accumulation in high repetition rate femtosecond laser writing of buried optical waveguides. Optics Express. 2008;16(13):9443–58.

[39] Osellame R, Chiodo N, Della Valle G, *et al.* Waveguide lasers in the C-band fabricated by laser inscription with a compact femtosecond oscillator. IEEE Journal of Selected Topics in Quantum. 2006;12(2):277–85.

[40] Osellame R, Chiodo N, Maselli V, *et al.* Optical properties of waveguides written by a 26 MHz stretched cavity Ti:sapphire femtosecond oscillator. Optics Express. 2005;13(2):612–20.

[41] Sudrie L, Couairon A, Franco M, *et al.* Femtosecond laser-induced damage and filamentary propagation in fused silica. Physical Review Letters. 2002;89(18).

[42] Siegel J, Fernandez-Navarro JM, Garcia-Navarro A, *et al.* Waveguide structures in heavy metal oxide glass written with femtosecond laser pulses above the critical self-focusing threshold. Applied Physics Letters. 2005;86(12).

[43] Ta'eed V, Baker NJ, Fu L, *et al.* Ultrafast all-optical chalcogenide glass photonic circuits. Optics Express. 2007;15(15):9205–21.

[44] Osellame R, Taccheo S, Marangoni M, *et al.* Femtosecond writing of active optical waveguides with astigmatically shaped beams. Journal of the Optical Society of America B (Optical Physics). 2003;20(7):1559–67.

[45] Ams M, Marshall GD, Spence DJ, and Withford MJ. Slit beam shaping method for femtosecond laser direct-write fabrication of symmetric waveguides in bulk glasses. Optics Express. 2005;13(15):5676–81.

[46] Sakakura M, Sawano T, Shimotsuma Y, Miura K, and Hirao K. Fabrication of three-dimensional 1×4 splitter waveguides inside a glass substrate with spatially phase modulated laser beam. Optics Express. 2010;18(12):12136–43.

[47] Nasu Y, Kohtoku M, and Hibino Y. Low-loss waveguides written with a femtosecond laser for flexible interconnection in a planar light-wave circuit. Optics Letters. 2005;30(7):723–5.

[48] Thomson RR, Campbell S, Blewett IJ, *et al.* Active waveguide fabrication in erbium-doped oxyfluoride silicate glass using femtosecond pulses. Applied Physics Letters. 2005;87(12):121102–3.

[49] Little DJ, Ams M, Dekker P, Marshall GD, Dawes JM, and Withford MJ. Femtosecond laser modification of fused silica: the effect of writing polarization on Si-O ring structure. Optics Express. 2008;16(24):20029–37.

[50] Li J, Ho S, Haque M, and Herman PR. Nanograting Bragg responses of femtosecond laser written optical waveguides in fused silica glass. Optical Materials Express. 2012;2(11):1562–70.

[51] Ehrt D, Kittel T, Will M, Nolte S, and Tunnermann A. Femtosecond-laser-writing in various glasses. Journal of Non-Crystalline Solids. 2004;345–46:332–7.

[52] Zhang H, Eaton SM, Li J, Nejadmalayeri AH, and Herman PR. Type II high-strength Bragg grating waveguides photowritten with ultrashort laser pulses. Optics Express. 2007;15(7):4182–91.

[53] Zakery A and Elliott SR. Optical properties and applications of chalcogenide glasses: a review. Journal of Non-Crystalline Solids. 2003;330(1):1–12.

[54] Eggleton BJ, Luther-Davies B, and Richardson K. Chalcogenide photonics. Nature Photonics. 2011;5:141.

[55] Willner AE, Khaleghi S, Chitgarha MR, and Yilmaz OF. All-optical signal processing. Journal of Lightwave Technology. 2014;32(4):660–80.

[56] Vázquez MR, Sotillo B, Rampini S, *et al.* Femtosecond laser inscription of nonlinear photonic circuits in gallium lanthanum sulphide glass. Journal of Physics: Photonics. 2018;1(1):015006.

[57] Arriola A, Gross S, Ams M, *et al.* Mid-infrared astrophotonics: study of ultrafast laser induced index change in compatible materials. Optical Materials Express. 2017;7(3):698–711.

[58] Arriola A, Mukherjee S, Choudhury D, Labadie L, and Thomson RR. Ultrafast laser inscription of mid-IR directional couplers for stellar interferometry. Optics Letters. 2014;39(16):4820–2.

[59] Bérubé J-P, Messaddeq S-H, Bernier M, Skripachev I, Messaddeq Y, and Vallée R. Tailoring the refractive index of Ge-S based glass for 3D embedded waveguides operating in the mid-IR region. Optics Express. 2014;22(21):26103–16.

[60] Psaila ND, Thomson RR, Bookey HT, *et al.* Supercontinuum generation in an ultrafast laser inscribed chalcogenide glass waveguide. Optics Express. 2007;15(24):15776–81.

[61] Yu Y, Gai X, Wang T, *et al.* Mid-infrared supercontinuum generation in chalcogenides. Optical Materials Express. 2013;3(8):1075–86.
[62] Zoubir A, Richardson M, Rivero C, *et al.* Direct femtosecond laser writing of waveguides in As2S3 thin films. Optics Letters. 2004;29(7):748–50.
[63] Efimov O, Glebov L, Richardson K, *et al.* Waveguide writing in chalcogenide glasses by a train of femtosecond laser pulses. Optical Materials. 2001;17(3):379–86.
[64] D'Amico C, Cheng G, Mauclair C, *et al.* Large-mode-area infrared guiding in ultrafast laser written waveguides in sulfur-based chalcogenide glasses. Optics Express. 2014;22(11):13091–101.
[65] Requejo-Isidro J, Mairaj A, Pruneri V, Hewak D, Netti M, and Baumberg J. Self-refractive non-linearities in chalcogenide based glasses. Journal of Non-Crystalline Solids. 2003;317(3):241–6.
[66] Gretzinger T, Gross S, Ams M, Arriola A, and Withford MJ. Ultrafast laser inscription in chalcogenide glass: thermal versus athermal fabrication. Optical Materials Express. 2015;5(12):2862–77.
[67] McMillen B, Zhang B, Chen KP, Benayas A, and Jaque D. Ultrafast laser fabrication of low-loss waveguides in chalcogenide glass with 0.65 dB/cm loss. Optics Letters. 2012;37(9):1418–20.
[68] Hughes M, Yang W, and Hewak D. Fabrication and characterization of femtosecond laser written waveguides in chalcogenide glass. Applied Physics Letters. 2007;90(13):131113.
[69] Schaffer CB, García JF, and Mazur E. Bulk heating of transparent materials using a high-repetition-rate femtosecond laser. Applied Physics A. 2003; 76(3):351–4.
[70] Eaton SM, Ng ML, Osellame R, and Herman PR. High refractive index contrast in fused silica waveguides by tightly focused, high-repetition rate femtosecond laser. Journal of Non-Crystalline Solids. 2011;357(11):2387–91.
[71] Shah L, Arai A, Eaton SM, and Herman PR. Waveguide writing in fused silica with a femtosecond fiber laser at 522 nm and 1-MHz repetition rate. Optics Express. 2005;13(6):1999–2006.
[72] Eaton SM. Contrasts in thermal diffusion and heat accumulation effects in the fabrication of waveguides in glasses using variable repetition rate femtosecond laser; 2008; Toronto: University of Toronto.
[73] Richter S, Heinrich M, Döring S, Tünnermann A, and Nolte S. Formation of femtosecond laser-induced nanogratings at high repetition rates. Applied Physics A. 2011;104(2):503–7.
[74] Sabapathy T, Kiran M, Ayiriveetil A, Kar AK, Ramamurty U, and Asokan S. Nanoindentation studies on waveguides inscribed in chalcogenide glasses using ultrafast laser. Optical Materials Express. 2013;3(6):684–90.
[75] Gretzinger T, Fernandez TT, Gross S, Arriola A, and Withford M. Boson band mapping: revealing ultrafast laser induced structural modifications in chalcogenide glass. Optics Letters. 2020;45(13):3369–3372.
[76] Hewak DW, Brady D, Curry RJ, *et al.* Chalcogenide glasses for photonics device applications. Photonic Glasses and Glass-Ceramics. 2010:29–102.

[77] Demetriou G, Hewak DW, Ravagli A, Craig C, and Kar A. Nonlinear refractive index of ultrafast laser inscribed waveguides in gallium lanthanum sulphide. Applied Optics. 2017;56(19):5407–11.
[78] Vázquez MR, Sotillo B, Rampini S, *et al.* Femtosecond laser inscription of nonlinear photonic circuits in Gallium Lanthanum Sulphide glass. Journal of Physics: Photonics. 2018;1(1):015006.
[79] Fernandez T, Sakakura M, Eaton S, *et al.* Bespoke photonic devices using ultrafast laser driven ion migration in glasses. Progress in Materials Science. 2018;94:68–113.
[80] Moreno-Zarate P, Gonzalez A, Funke S, *et al.* Imaging ellipsometry determination of the refractive index contrast and dispersion of channel waveguides inscribed by fs-laser induced ion-migration. Physica Status Solidi (A). 2018;215(19):1800258.
[81] Kanehira S, Miura K, and Hirao K. Ion exchange in glass using femtosecond laser irradiation. Applied Physics Letters. 2008;93(2):3.
[82] Toney Fernandez T, Haro-González P, Sotillo B, *et al.* Ion migration assisted inscription of high refractive index contrast waveguides by femtosecond laser pulses in phosphate glass. Optics Letters. 2013;38(24):5248–51.
[83] Fuerbach A, Gross S, Little D, *et al.*, editors. Refractive index change mechanisms in different glasses induced by femtosecond laser irradiation. Pacific Rim Laser Damage 2016: Optical Materials for High-Power Lasers; 2016; Yokohama, Japan: International Society for Optics and Photonics.
[84] Shimizu M, Matsuoka J, Kato H, *et al.* Role of partial molar enthalpy of oxides on Soret effect in high-temperature CaO–SiO_2 melts. Scientific Reports. 2018;8(1):15489.
[85] Fernandez TT, Siegel J, Hoyo J, Sotillo B, Fernandez P, and Solis J. Controlling plasma distributions as driving forces for ion migration during fs laser writing. Journal of Physics D: Applied Physics. 2015;48(15):155101.
[86] Yu Y, Chen Y, Chen J, *et al.* Fabrication of microchannels by space-selective control of phase separation in glass. Optics Letters. 2016;41(14):3371–4.
[87] Fernandez TT, Sotillo B, del Hoyo J, *et al.* Dual regimes of ion migration in high repetition rate femtosecond laser inscribed waveguides. IEEE Photonics Technology Letters. 2015;27(10):1068–71.
[88] del Hoyo J, Moreno-Zárate P, Escalante G, Vallés JA, Fernández P, and Solís J. High-efficiency waveguide optical amplifiers and lasers via fs-laser induced local modification of the glass composition. Journal of Lightwave Technology. 2017;35(14):2955–9.
[89] Fukuda T, Ishikawa S, Fujii T, Sakuma K, and Hosoya H, editors. Low-loss optical waveguides written by femtosecond laser pulses for three-dimensional photonic devices. Photon Processing in Microelectronics and Photonics III; 2004; San Jose, CA: SPIE.
[90] Yang W, Kazansky PG, Shimotsuma Y, Sakakura M, Miura K, and Hirao K. Ultrashort-pulse laser calligraphy. Applied Physics Letters. 2008;93(17): 171109–3.

[91] Hnatovsky C, Taylor RS, Rajeev PP, *et al.* Pulse duration dependence of femtosecond-laser-fabricated nanogratings in fused silica. Applied Physics Letters. 2005;87(1).

[92] Rayner DM, Naumov A, and Corkum PB. Ultrashort pulse non-linear optical absorption in transparent media. Optics Express. 2005;13(9):3208–17.

[93] Zhang H, Eaton SM, and Herman PR. Low-loss Type II waveguide writing in fused silica with single picosecond laser pulses. Optics Express. 2006; 14(11):4826–34.

[94] Eaton SM, Chen W, Zhang L, *et al.* Telecom-band directional coupler written with femtosecond fiber laser. IEEE Photonics Technology Letters. 2006;18(17–20):2174–6.

[95] Svalgaard M, Poulsen CV, Bjarklev A, and Poulsen O. Direct UV writing of buried singlemode channel waveguides in Ge-doped silica films. Electronics Letters. 1994;30(17):1401–3.

[96] Sakakura M, Kurita T, Shimizu M, *et al.* Shape control of elemental distributions inside a glass by simultaneous femtosecond laser irradiation at multiple spots. Optics Letters. 2013;38(23):4939–42.

[97] Zhang H, Eaton SM, and Herman PR. Low-loss Type II waveguide writing in fused silica with single picosecond laser pulses. Optics Express. 2006; 14(11):4826–34.

[98] Zhang H, Eaton SM, Li J, Nejadmalayeri AH, and Herman PR. Type II high-strength Bragg grating waveguides photowritten with ultrashort laser pulses. Optics Express. 2007;15(7):4182–91.

[99] Zhang H, Eaton SM, and Herman PR. Single-step writing of Bragg grating waveguides in fused silica with an externally modulated femtosecond fiber laser. Optics Letters. 2007;32(17):2559–61.

[100] Miese C, Withford MJ, and Fuerbach A. Femtosecond laser direct-writing of waveguide Bragg gratings in a quasi-cumulative heating regime. Optics Express. 2011;19(20):19542–50.

[101] Zhang HB, Eaton SM, Li JZ, and Herman PR. Femtosecond laser direct writing of multiwavelength Bragg grating waveguides in glass. Optics Letters. 2006;31(23):3495–7.

[102] Zhang H. Chirped Bragg grating waveguides directly written inside fused silica glass with an externally modulated ultrashort fiber laser. IEEE Photonics Technology Letters. 2009;21(5):277–9.

[103] Zhang H, Ho S, Eaton SM, Li J, and Herman PR. Three-dimensional optical sensing network written in fused silica glass with femtosecond laser. Optics Express. 2008;16(18):14015–23.

[104] Lee KKC, Mariampillai A, Haque M, Standish BA, Yang VXD, and Herman PR. Temperature-compensated fiber-optic 3D shape sensor based on femtosecond laser direct-written Bragg grating waveguides. Optics Express. 2013;21(20):24076–86.

[105] Fernandes LA, Grenier JR, Marques PV, Aitchison JS, and Herman PR. Strong birefringence tuning of optical waveguides with femtosecond laser

irradiation of bulk fused silica and single mode fibers. Journal of Lightwave Technology. 2013;31(22):3563–9.

[106] Osellame R, Taccheo S, Cerullo G, *et al.* Optical gain in Er-Yb doped waveguides fabricated by femtosecond laser pulses. Electronics Letters. 2002;38(17):964–5.

[107] Ams M, Marshall GD, Dekker P, Piper JA, and Withford MJ. Ultrafast laser written active devices. Laser & Photonics Reviews. 2009;3(6):535–44.

[108] Taccheo S, Della Valle G, Osellame R, *et al.* Er:Yb-doped waveguide laser fabricated by femtosecond laser pulses. Optics Letters. 2004;29(22):2626–8.

[109] Della Valle G, Osellame R, Galzerano G, *et al.* Passive mode locking by carbon nanotubes in a femtosecond laser written waveguide laser. Applied Physics Letters. 2006;89(23).

[110] Beecher SJ, Thomson RR, Psaila ND, *et al.* 320 fs pulse generation from an ultrafast laser inscribed waveguide laser mode-locked by a nanotube saturable absorber. Applied Physics Letters. 2010;97(11).

[111] Marshall GD, Dekker P, Ams M, Piper JA, and Withford MJ. Directly written monolithic waveguide laser incorporating a distributed feedback waveguide-Bragg grating. Optics Letters. 2008;33(9):956–8.

[112] Spaleniak I, Gross S, Jovanovic N, *et al.* Multiband processing of multimode light: combining 3D photonic lanterns with waveguide Bragg gratings. Laser & Photonics Reviews. 2014;8(1):L1–5.

[113] Ladd TD, Jelezko F, Laflamme R, Nakamura Y, Monroe C, and O'Brien JL. Quantum computers. Nature. 2010;464(7285):45–53.

[114] Knill E, Laflamme R, and Milburn GJ. A scheme for efficient quantum computation with linear optics. Nature. 2001;409(6816):46–52.

[115] Marshall GD, Politi A, Matthews JCF, *et al.* Laser written waveguide photonic quantum circuits. Optics Express. 2009;17(15):12546–54.

[116] Eaton SM, Chen W-J, Zhang H, *et al.* Spectral loss characterization of femtosecond laser written waveguides in glass with application to demultiplexing of 1300 and 1550 nm wavelengths. Journal of Lightwave Technology. 2009;27(9):1079–85.

[117] Chen W-J, Eaton SM, Zhang H, and Herman PR. Broadband directional couplers fabricated in bulk glass with high repetition rate femtosecond laser pulses. Optics Express. 2008;16(15):11470–80.

[118] Meany T, Gross S, Jovanovic N, Arriola A, Steel M, and Withford MJ. Towards low-loss lightwave circuits for non-classical optics at 800 and 1,550 nm. Applied Physics A. 2014;114(1):113–8.

[119] Sansoni L, Sciarrino F, Vallone G, *et al.* Polarization entangled state measurement on a chip. Physical Review Letters. 2010;105(20):200503.

[120] Ralph T. Quantum computation: boson sampling on a chip. Nature Photonics. 2013;7(7):514–5.

[121] Crespi A, Osellame R, Ramponi R, *et al.* Integrated multimode interferometers with arbitrary designs for photonic boson sampling. Nature Photonics. 2013;7(7):545–9.

[122] Marcinkevicius A, Juodkazis S, Watanabe M, *et al.* Femtosecond laser-assisted three-dimensional microfabrication in silica. Optics Letters. 2001;26(23):1912–14.

[123] Bellouard Y, Said A, Dugan M, and Bado P. Fabrication of high-aspect ratio, micro-fluidic channels and tunnels using femtosecond laser pulses and chemical etching. Optics Express. 2004;12(10):2120–9.

[124] Hnatovsky C, Taylor RS, Simova E, Bhardwaj VR, Rayner DM, and Corkum PB. Polarization-selective etching in femtosecond laser-assisted microfluidic channel fabrication in fused silica. Optics Letters. 2005; 30(14):1867–9.

[125] Maselli V, Osellame R, Cerullo G, *et al.* Fabrication of long microchannels with circular cross section using astigmatically shaped femtosecond laser pulses and chemical etching. Applied Physics Letters. 2006;88 (19):191107–3.

[126] Vishnubhatla KC, Bellini N, Ramponi R, Cerullo G, and Osellame R. Shape control of microchannels fabricated in fused silica by femtosecond laser irradiation and chemical etching. Optics Express. 2009;17(10):8685–95.

[127] Kiyama S, Matsuo S, Hashimoto S, and Morihira Y. Examination of etching agent and etching mechanism on femtosecond laser microfabrication of channels inside vitreous silica substrates†. The Journal of Physical Chemistry C. 2009;113(27):11560–6.

[128] LoTurco S, Osellame R, Ramponi R, and Vishnubhatla K. Hybrid chemical etching of femtosecond laser irradiated structures for engineered microfluidic devices. Journal of Micromechanics and Microengineering. 2013; 23(8):085002.

[129] Choudhury D, Ramsay WT, Kiss R, Willoughby NA, Paterson L, and Kar AK. A 3D mammalian cell separator biochip. Lab on a Chip. 2012;12 (5):948–53.

[130] Amato L, Gu Y, Bellini N, Eaton SM, Cerullo G, and Osellame R. Integrated three-dimensional filter separates nanoscale from microscale elements in a microfluidic chip. Lab on a Chip. 2012;12(6).

[131] Weigl BH and Yager P. Microfluidic diffusion-based separation and detection. Science. 1999;283(5400):346–7.

[132] Kuo JS and Chiu DT. Controlling mass transport in microfluidic devices. Annual Review of Analytical Chemistry. 2011;4:275–96.

[133] McMullen JP and Jensen KF. Integrated microreactors for reaction automation: new approaches to reaction development. Annual Review of Analytical Chemistry. 2010;3:19–42.

[134] Italia V, Giakoumaki AN, Bonfadini S, *et al.* Laser-inscribed glass microfluidic device for non-mixing flow of miscible solvents. Micromachines. 2019;10(1):23.

[135] Crespi A, Gu Y, Ngamsom B, *et al.*. Three-dimensional Mach-Zehnder interferometer in a microfluidic chip for spatially-resolved label-free detection. Lab on a Chip. 2010;10(9):1167–73.

Chapter 13

Optical waveguides produced by ion beams

Feng Chen[1]

Ions of selected elements are accelerated and implanted into solids, which thereby modifies the related properties of the targets. This process is normally called 'ion implantation' or 'ion irradiation'. By using these energetic ion beams, the physical, chemical, electrical, and optical properties of the target materials could be changed to a certain extent; therefore, new applications correlated to these material modifications would be expected towards diverse purposes [1,2]. In the areas of optics, the implantation of metallic ions (e.g., Au or Ag) into dielectrics may be used to synthesize embedded nanoparticles to realize plasmonic effects due to the localized interaction of light fields with the external medium [3]. In *integrated optics*, ion beams with diverse parameters (i.e., ion species, fluences, energies, and beam scales) could be utilized to fabricate optical waveguides with tailored geometries in a broad variety of optical materials, including glass, crystals, ceramics, and polymers [4]. There have been several books and review articles demonstrating the research progress in this topic [4–7]. Although some physical mechanisms require further investigation for detailed understanding, ion beam technology as a technique for waveguide fabrication seems to be somehow mature to the scientific community.

In this chapter, the ion beam techniques for waveguide fabrication will be overviewed in Section 13.1. In Section 13.2, the typical refractive index profiles of the ion-beam-produced waveguides will be introduced briefly. Section 13.3 demonstrates the fabrication for two dimensionally confined waveguides, i.e., channel waveguides, by using diverse ion beam solutions. Finally, the selected applications of ion-beam-produced waveguides will be highlighted in Section 13.4.

13.1 Ion beam techniques for waveguide fabrication

Ion beam techniques can be roughly classified into a few regimes according to the species, energies, fluences, and beam scales of the used ions. A number of ion beam facilities are applied to generate ion beams with diverse parameters. The incident energetic ions lose their energies through the interactions with the target atoms.

[1]School of Physics, State Key Laboratory of Crystal Materials, Shandong University, Jinan, P.R. China

There are two major energy deposition processes for the ion-beam implantation/irradiation. The elastic collisions between the ions and nuclei of targets result in nuclear energy deposition (correlated to nuclear stopping power S_n), while inelastic collisions between the ions and electrons for ionization are determined by the electronic stopping power S_e [1,2]. The damage induced by S_n and S_e plays crucial roles on the changes of physical properties of the dielectrics and further the alternation of the refractive indices; therefore, the behaviours of S_n and S_e are dominating for the waveguide geometries [4–7].

For ion beams with diverse parameters, the S_n and S_e for dielectrics are of significant difference. It is reasonable to define a few typical regimes of ion beam technology for waveguide fabrication, i.e., light-ion implantation, heavy-ion implantation, swift heavy-ion irradiation, and proton beam writing. These four techniques are different for the ion beam parameters, and usually with different S_n and S_e behaviours.

13.1.1 Light-ion implantation

Light ions for the implanted waveguides in optical materials are mainly referring to H and He. Tandem accelerators or ion implanters can be used to generate light-ion beams. The first ion-implanted waveguide was reported in 1968, i.e., a proton-implanted fused silica waveguide [8]. In crystalline materials, Townsend, who was considered as the 'Father of ion-implanted waveguides' later, and his coworkers reported on the first waveguide in $LiNbO_3$ crystal by He^+ ion implantation [9]. The typical energy of H^+ ions is of 500 keV to 2 MeV, while for He^+ ions it is 1–3 MeV. The thickness of light-ion-implanted waveguides is from 3 to 10 μm, depending on the ion energies, ion species, and materials. In the regime of light-ion implantation, the dominant damage effect of the energy deposition is the nuclear one (S_n), which could be accumulated linearly (below the value for amorphization). The required fluences are of 1×10^{16} to 1×10^{17} ions/cm^2 for H^+ and 5×10^{15} to 5×10^{16} ions/cm^2 for He^+ ions. In addition, the electronic energy deposition of the H and He ions is mainly inducing point defects, which can be removed by a thermal annealing at 200°C–300°C in air [4–9].

Figure 13.1(a) shows the typical curves of S_n and S_e for light-ion implantation in dielectrics (e.g., 2 MeV He^+ ion implanted into $LiNbO_3$ crystal) calculated by SRIM (stopping and range of ions in matter) code, a well-known computer programme [10]. As one can see, the nuclear damage S_n profile is with a peak mainly located at the end of ion range (at depth of ~4 μm), while the S_e is relatively so small that can be negligible. By using the post-implantation annealing process, the point defects or colour centres induced by the electronic damage may be removed significantly.

13.1.2 Heavy-ion implantation

Heavy-ion implantation is considered as an intermediate state between light-ion implantation and swift heavy-ion irradiation [6]. For waveguide fabrication, the so-called heavy ions refer to those with moderate atomic mass (e.g., O, F, Si, or Ar)

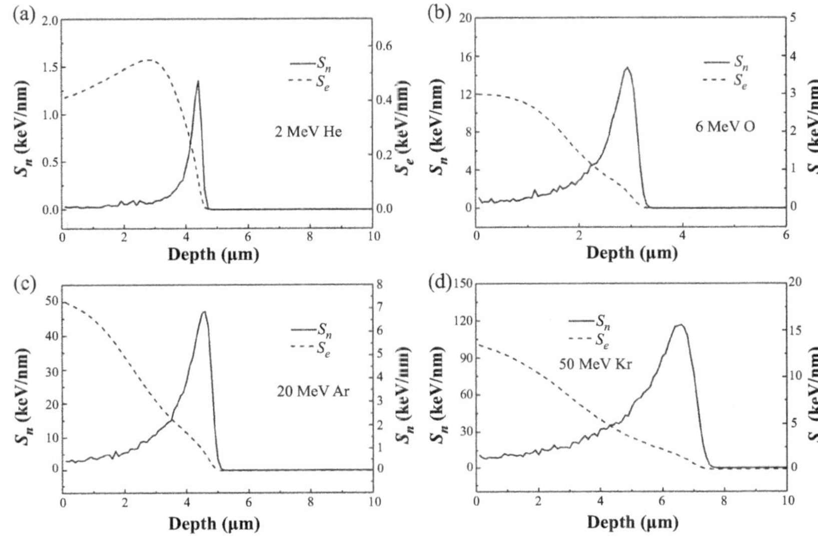

Figure 13.1 The calculated S_e (dashed line) and S_n (solid line) curves of ion implanted into $LiNbO_3$ crystal based on SRIM code: (a) 2 MeV He, (b) 6 MeV O, (c) 20 MeV Ar, and (d) 50 MeV Kr

instead of the very heavy ions such as Au or Pt. To obtain such heavy-ion beams, tandem accelerators operating at higher voltages are often used. To ensure an adequate thickness of the waveguide layer, the energy of heavy-ion implantation is typically of 3–10 MeV. The fluence typically ranges from 1×10^{14} to 1×10^{15} ions/cm^2 to induce adequate refractive index changes. Nevertheless, the energy of heavy-ion implantation may have some overlaps with the swift heavy-ion irradiation. In this regime, the refractive index changes may be considered as the synergetic effects of both S_e and S_n [11,12]. There were numerous works published for heavy-ion-implanted waveguides in crystals and glass [13–16].

Figure 13.1(b) shows S_e and S_n curves of 6 MeV O implanted into $LiNbO_3$ crystal by SRIM, as a typical example for heavy-ion-implanted waveguides. As is indicated, both the S_e and S_n play roles for the damage formation. In addition, the electronic damage region is mainly located in the path of the ion range, while the nuclear damage is located in the region at the end of the ion track. Unlike the case of light-ion implantation, the impact of S_e cannot be ignored. Although the electronic damage is not a linear accumulation of those from single ions, there may be a threshold value of S_e for sufficient impact on the damage formation to generate refractive index changes. For example, for $LiNbO_3$ crystal, the threshold value of S_e is about 2.2 keV/nm [17], that is, as $S_e > 2.2$ keV/nm, the accumulation of single ion's electronic damage would be practical.

13.1.3 Swift heavy-ion irradiation

The heavy ions with energies more than 1 MeV/amu are defined as swift heavy ions for waveguide formation [18,19]. Due to the velocity effect, the electronic energy deposition S_e is dominant over the nuclear contribution S_n during the most path of the ion trajectory [20]. There are two main mechanisms for swift heavy-ion irradiation. The first refers to the case that S_e is high enough, above a threshold for single ion forming an amorphous track. Below this threshold, i.e., for the second case, the electronic damage could be the overlapping impact of several highly disordered nano-tracks [21]. For refractive index changes to build waveguide layer, the fluence of the above-threshold case is of order of 10^{11}–10^{12} ions/cm^2, while the below-threshold case requires 10^{13}–10^{14} ions/cm^2. The threshold value for S_e in $LiNbO_3$ is ~5.5 keV/nm [21].

Figure 13.1(c) and (d) depicts the S_e and S_n curves for swift heavy-ion irradiation in $LiNbO_3$ crystal under 20 MeV Ar and 50 MeV Kr ion beams, respectively. As one can see, the electronic damage peak (S_e) positions are quite different from those of the light- and heavy-ion implantation. The electronic damage, in the swift heavy-ion irradiation regime, is associated with the formation of nano-tracks [18,19,21], which plays dominant roles on the structure and refractive index modifications.

13.1.4 Proton beam writing

Proton beam writing is a direct-write technique for micro- or nanoscale patterning [22]. As a waveguide fabrication method, it has been applied to write channel waveguides in a number of materials, such as polymers, glass, Nd:YAG, and Nd: GGG crystals/ceramics [23–29]. In this regime, the proton beams are focused down to diameters of less than 1 μm (according to the system), and the main modification region in the materials is located at the end of ion range, i.e., in the nuclear damage region. Therefore, the waveguide location inside the materials is determined by the ion energy (typically at 1–2 MeV), since the ion specie is H. The S_e and S_n curves of proton beam writing are similar to those of light-ion implantation. However, ion fluence plays significant roles on the refractive index changes of the nuclear damage region. To ensure positive index changes to form the waveguide core, the fluence of the proton-beam writing is usually lower than that of light-ion implantation. In addition, focused He^+ ion beams (at energy of 2–3 MeV) can also be used to write waveguides in materials, which was successfully applied in KTP crystal [30] and Nd:YAG ceramic [28].

13.2 Refractive index profiles

Ion beams modify the refractive indices of the materials through structural modification, which is correlated to the 'damage'. In Section 13.1, it has been demonstrated that the incident energetic ions lose their energies through nuclear collisions and electronic excitation, which are mainly determined by S_n and S_e, respectively. The damage accumulation to modify the refractive index mainly depends on the ion fluence, and in some cases, also on the single ion's impact or ion-beam scales. For optical waveguides produced by energetic ion beams, one can basically classify the refractive index profiles of the waveguide structures into five typical configurations. These index distributions may not be applicable to all the materials; accordingly, the refractive index changes are dependent on both ion beam parameters and material natures [4–7]. For single crystals, the symmetry of the materials is of great significance for the index changes under the processing of diverse ion beam conditions. For crystals belonging to the same crystal systems (reflecting the symmetry of single crystals), the birefringence is another critical factor for index changes. In addition, the refractive index profiles of the ion-implanted waveguides are normally relying on index reconstruction technique (such as reconstruction calculation method) [31] instead of direct measurement. Reasonable assumptions of physical models are to a great extent to fit the index changes.

13.2.1 Optical barrier (S_n)-type profile

The most well-known profile of ion-implanted waveguides is so-called 'optical barrier' type, which is particularly suitable for a light-ion implantation regime [4]. As the incident ions penetrate the materials, the majority will stop at the end of ion range, i.e., in the nuclear damage region. Following the S_n curve, the nuclear damage will generally induce volume expansion in this region, and the refractive index will be reduced accordingly in the buried layer. Since in the main path of the ions' trajectory, the nuclear damage is very low (see Figure 13.1(a)); therefore, the refractive index in the region between the surface (usually air) and buried nuclear damage layer is relatively higher, forming a sandwich index structure. The nuclear damage layer with low index was named after 'optical barrier' by Townsend [4], which became the most well-known model for ion-implanted waveguides. Early works on ion-implanted waveguides were mostly performed by He ion implantation at ~2 MeV energy. In these cases, the S_n plays dominant roles while the S_e mainly induces point defects that could be removed by thermal annealing [4].

Figure 13.2(a) shows a typical profile of optical-barrier-type index. One of the shortcomings of the S_n barrier type is that the thickness of the low-index layer for single-energy implantation is less than 1 μm, and the refractive index contrast (Δn) between the waveguide and barrier is of 10^{-3}. In this case, the tunnelling effect of the mode field may result in leaky modes, resulting in higher losses. In practice, multiple-energy light or low-mass ion implantation (e.g., 1.8+2.0+2.2 MeV He) could be used to broaden the barrier [4,32–34]. In addition, the multiple-energy implantation can also be utilized to fabricate multilayer waveguides. For example, a two-waveguide-layer structure can be produced by 2.0+2.8 MeV He ion implantation into $LiNbO_3$ crystal.

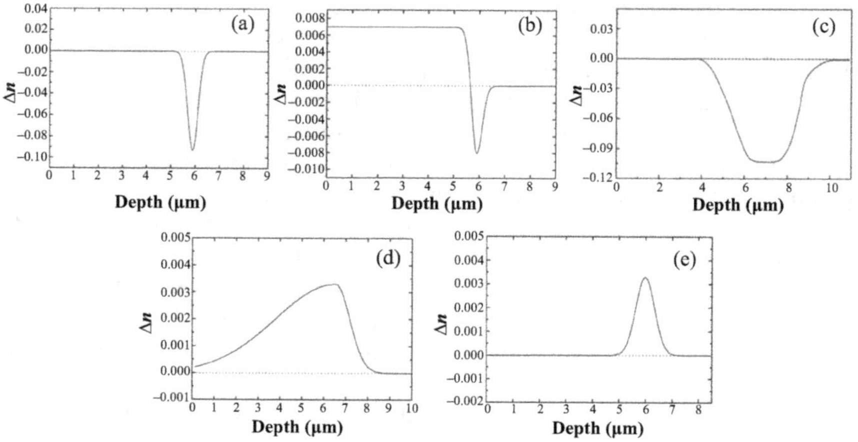

Figure 13.2 Typical refractive index profiles of ion-beam-produced waveguides of configurations of (a) optical barrier (S_n), (b) enhanced well (S_e) plus optical barrier (S_n), (c) optical barrier (S_e), (d) enhanced well (S_e), and (e) enhanced well (S_n) types

13.2.2 Enhanced well (S_e) plus optical barrier (S_n)-type profile

The enhanced well refers to a positive change of refractive index. In some crystals, the index may be increased somehow in an electronic damage region, which often happens in the regime of heavy-ion implantation [12,13,15,16]. Generally, in this case, the impact of the electronic damage cannot be ignored, and meanwhile, the nuclear damage may also create index decrease at the end of ion range (Figure 13.2(b)). Therefore, an enhanced well plus barrier-type index profile is formed due to the synergetic effect of S_e and S_n. Nevertheless, the nature of the materials is another crucial factor, e.g., the birefringence and crystal system (symmetry configuration). For crystals with large birefringence, it may be possible to obtain the increase of one main refractive index with a lower value by ion-beam-induced damage. For example, under ion implantation, in $LiNbO_3$ there will be a positive index change of extraordinary n_e in the main trajectory, while in YVO_4 the enhanced index well occurs for ordinary n_o. The refractive index change is also depending on the ion beam conditions. Usually, higher fluence induces larger index changes of materials [35]. In addition, such kind of profiles with missing modes may also be realizable in He-ion-implanted $LiNbO_3$, in which the damage is associated with crystal properties, e.g., alternation of spontaneous polarization [7].

13.2.3 Optical barrier (S_e)-type profile

In the regime of swift heavy-ion irradiation, the single ion's impact or synergetic effect of a few ions may form a thick buried barrier inside the materials [36,37].

The barrier formation is due to the electronic damage. The maximum index change could be as large as 0.1, which is much larger than that in regimes of ion implantation. Figure 13.2(c) shows the typical profile of S_e-based optical-barrier-type index. It has been well investigated for $LiNbO_3$ crystals under diverse heavy-ion irradiation conditions. In fact, the high-index change of S_e barrier reflects the fact that the electronic damaged region experiences phase transition from the crystalline to the amorphous. Similar effect can be confirmed in other crystals, such as KGW [38]. Nevertheless, the S_e barrier depends on the electronic excitation, while the lattice structures of the crystals also play an important role. This phenomenon reveals the complicated behaviours of the different lattices under the irradiation of ion beams.

13.2.4 Enhanced well (S_e)-type profile

Figure 13.2(d) depicts a typical enhanced-well-type profile induced by electronic excitation. For crystals like Nd:YAG, the swift heavy-ion irradiation at low fluences will bring out positive index changes during most path of the ion tracks. In this case, the waveguide layer could be thick, while the general index change remains less than 10^{-3} [39–41]. Since the fluence is very low (~10^{12} cm^{-2}), the nuclear damage is therefore negligible. The refractive index profile is an enhanced-well type that is correlated to the electronic damage. Another feature of this kind of swift heavy-ion-irradiated waveguide is that the low-fluence irradiation does not allow the overlap of ion tracks; as a result, the waveguide core region still remains crystalline in the track-free zones.

13.2.5 Enhanced well (S_n)-type profile

The enhanced well (see Figure 13.2(e)) may be buried at the nuclear damage region in the case of proton beam writing. Usually, the fluence used in the proton-beam-written waveguides is lower than proton implantation to form a low-index barrier. For example, for Nd:YAG, the buried channel waveguides for proton beam writing require 5×10^{15}–2×10^{16} ions/cm^2, while for proton-implanted barrier it may be as high as 1×10^{17} ions/cm^2 [27]. a slight modification of low-fluence protons on materials may happen with a partial compression of lattice [42]. When the fluence is high enough, a lattice expansion, instead of compression, will happen, which results in the negative index changes at the nuclear damage region. Nevertheless, it is a rough explanation and may be applicable only in some crystals.

13.3 Channel waveguide fabrication

The ion-beam implantation/irradiation (excluding proton beam writing) usually forms a guiding layer, i.e., a planar waveguide that restricts light diffraction along the vertical direction. For more practical applications of photonic chips, channel waveguides are desired, which allow the 2D confinement of light fields. People have developed a number of techniques to produce channel waveguides in optical materials, including mask-assisted implantation, laser ablation or diamond saw dicing, focused beam writing, and optical illumination [43]. These solutions are applied either on a planar waveguide substrate for further processing or to directly write channel waveguides inside the bulks.

13.3.1 Mask-assisted implantation of stripe waveguides

During the implantation or irradiation, a suitable mask is used to block the ions in the mask-free region of the samples. Therefore, some regions of samples are not affected by the ion-induced damages, remaining the original properties of the bulks. In the ion-implanted regions, the refractive index is modified, and the waveguides may be formed in the ion-modified channels. The optical fibre, photoresist mask, and metallic mask are utilized to realize selective ion implantation [4,44]. Figure 13.3 depicts the schematic plots of the mask-assisted implantation of stripe waveguides typically used in the fabrication. It should also be indicated that the

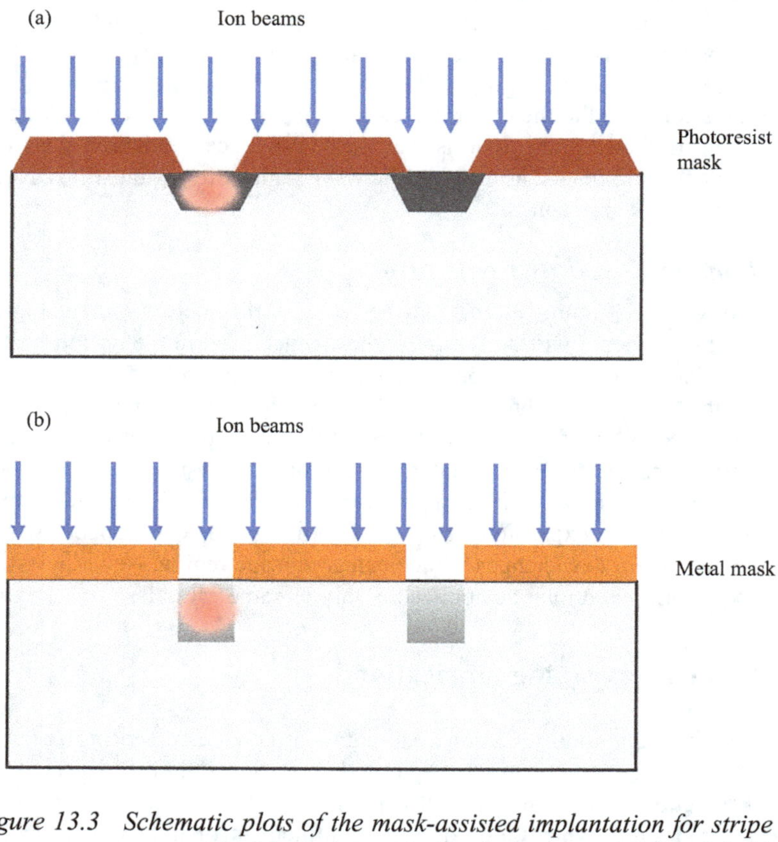

Figure 13.3 Schematic plots of the mask-assisted implantation for stripe waveguide fabrication with (a) photoresist masking and (b) metal masking

thickness, shapes, and used materials of masks are crucial for stripe waveguide formation. For example, the photoresists are typically with trapezoidal configuration of cross section, and thickness should be no less than 5 μm [45,46]. This design was significantly important for the completeness of the side walls of the refractive index barrier, and to avoid complicated processing of the side walls by multi-energy implants in the case of rectangular cross-sectional masks. For swift heavy-ion irradiation, metal masking is more suitable because the penetration depth of the high-energy ions is more than 10 μm. Such metal masks are now commercially available and can be produced by the modern micromachining techniques. A successful example can be seen in [44].

13.3.2 Lateral patterning of ridge waveguides

In addition to the stripe waveguides, one can use so-called ridge waveguides to realize 2D confinement of light propagation. One of the most interesting features of the ridge waveguides is that the lateral confinement is often achieved by the side walls of air. This allows additional deposition of optical elements (such as electrodes) on the side walls and a high refractive index contrast between the waveguide cores and the surroundings. As of yet, ion-beam etching, femtosecond laser ablation, and diamond saw dicing have been utilized to form air grooves of the ridges [47–50].

Figure 13.4 shows the schematic plots of the ridge waveguide formation. The femtosecond laser ablation of air grooves in dielectric materials is often with relatively high roughness (more than 0.5 μm), while for diamond saw dicing the roughness could be as low as 10 nm. Nevertheless, the advantage of the femtosecond laser ablation is the 3D micromachining capability, which can be used to produce curved structures, e.g., Y-branches [48]. And the post-ablation chemical etching could be used to considerably reduce the roughness of the surface of ablated grooves.

13.3.3 Focused beam writing of buried waveguides

Focused beam writing has become a powerful tool to achieve 3D lithography of submicron scale patterns. By using focused protons or He ion beams, it is realizable to directly write channel waveguides that are buried inside the optical materials (see Figure 13.5 for schematic). As of yet, the proton beam writing has been applied onto a number of dielectrics, including glasses (fused silica, ZBLAN, etc.), crystals (e.g., Nd:YAG and Nd:GGG), and ceramics (Nd:YAG), and the focused He^+ ion beams are used to fabricate channel waveguides in Nd:YAG and KTP [23–30,51]. The waveguide cores are located at the end of ion range, which are correlated to the positive well of refractive index due to the ion beams. The micro-photoluminescence maps of the waveguide cross sections clearly indicate the micro-modification at the nuclear damage region. Another feature of focused ion beam writing is that the depth of the waveguide cores could be changed by using ions with different energies. It should be also pointed out that, for gallium lanthanum sulfide (GLS) glass [26], the proton beam writing fabricates a surface channel

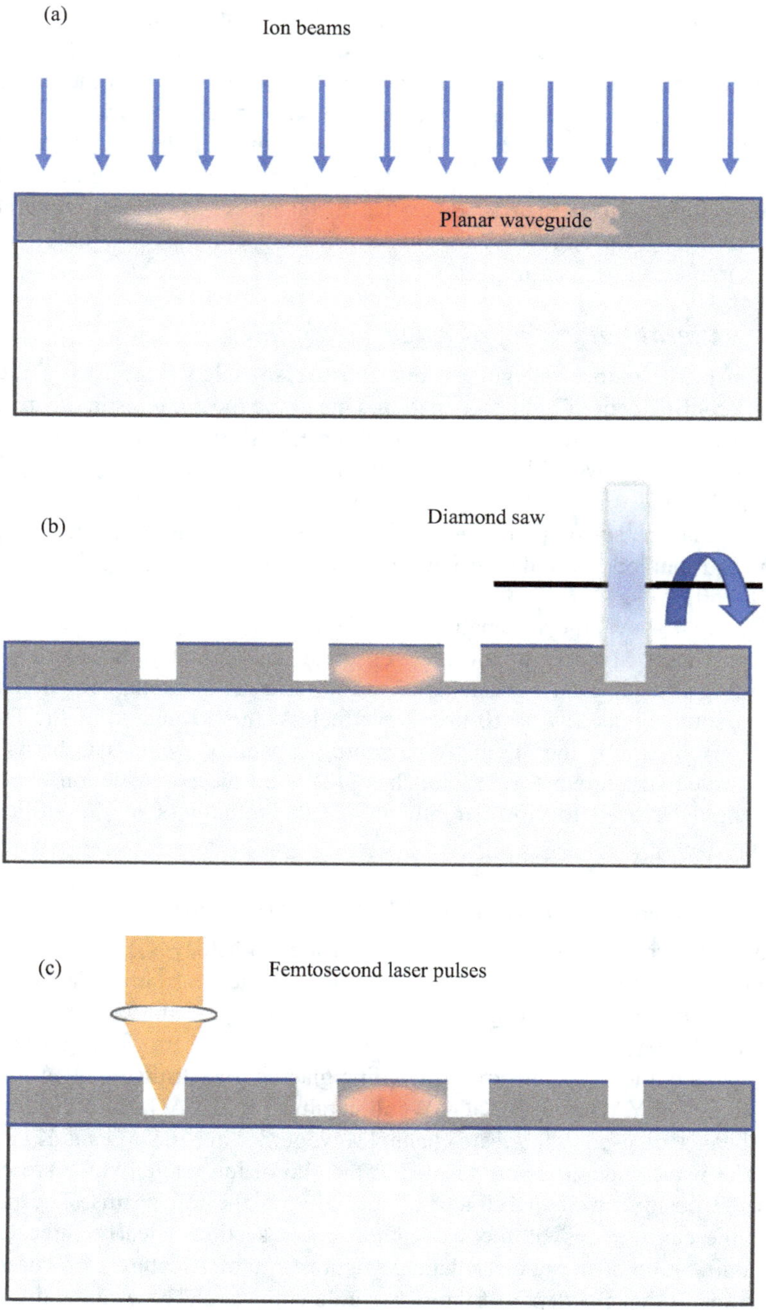

Figure 13.4 Schematic plots of the ridge waveguide formation: (a) formation of planar waveguide layer on the surface , followed by (b) precise diamond saw dicing or (c) femtosecond laser ablation

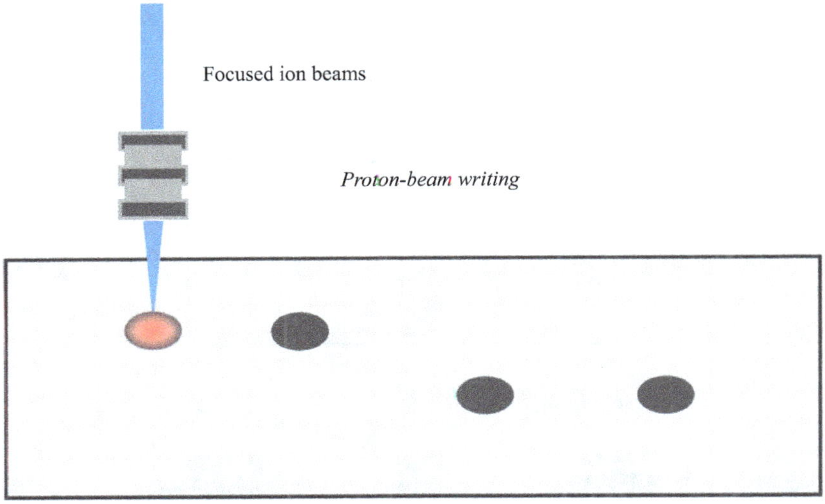

Figure 13.5 *Schematic plot of the buried channel waveguides produced by proton beam writing*

waveguide instead of a buried one, which shows a different effect on the working region of focused protons on dielectrics.

13.3.4 Selective illumination of reconfigurable waveguides

In crystals that exhibit special properties, one may also produce reconfigurable waveguides. $LiNbO_3$ crystal possesses excellent photorefractive properties, particularly doped by Fe or Cu ions [52]. Owing to the electrical field rearrangement under the light illumination, the refractive index of the irradiated region may be reduced. The typical design for this kind of waveguide is to use photomasked illumination (with stripes or patterns) on an ion-implanted $LiNbO_3$ planar waveguide (Figure 13.6). The selective illumination can be performed by either coherent light (e.g., green laser) or incoherent light (e.g., white light) [53,54]. In the open stripes without photomask, the illumination induces negative index changes, while in the photomask-covered regions the index remains unchanged. Therefore, the waveguide cores can be formed in the regions covered by mask stripes. As the unmasked crystals are under homogenous white light illumination, the electrical fields will be redistributed due to the electrons' movement, resulting in the index recovery. This white light illumination works as erasing of channels. In this way, the channel waveguides can be constructed, erased, and reconstructed by successive steps of light-illumination. The reconfigurable waveguides and beam splitters have been realized in $Fe:LiNbO_3$ crystal by using either green laser or white light irradiation [53,54].

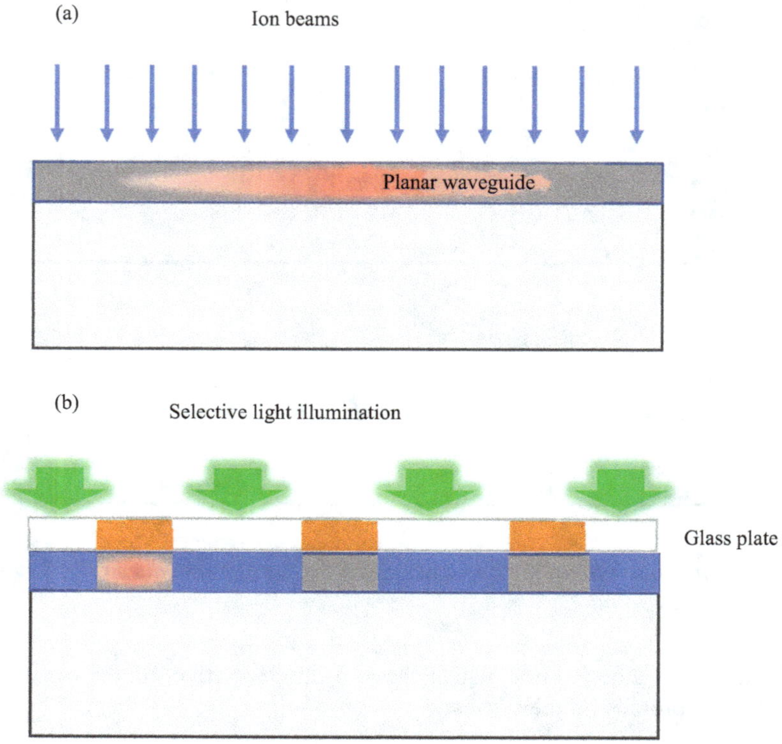

Figure 13.6 Schematic plots of the reconfigurable waveguides: (a) formation of planar waveguide layer on the surface, followed by (b) selective illumination of coherent or incoherence light through a glass plate with stripe patterns

13.4 Selected applications

Ion-beam-produced optical waveguides receive a number of applications in optics and photonics. Depending on the properties of the materials, some integrated devices, such as electro-optic modulators, frequency/wavelength converters, waveguide amplifiers, waveguide lasers, and waveguide sensors, have been fabricated. In this subsection, several successful examples will be demonstrated. More works on the applications could be referenced in literatures.

13.4.1 Electro-optic modulators

The electro-optic modulators, which are used to modulate the light-beam propagation (powers, phases, and polarizations), receive wide applications in optical systems. In ion-implanted waveguides, electro-optic modulators have been

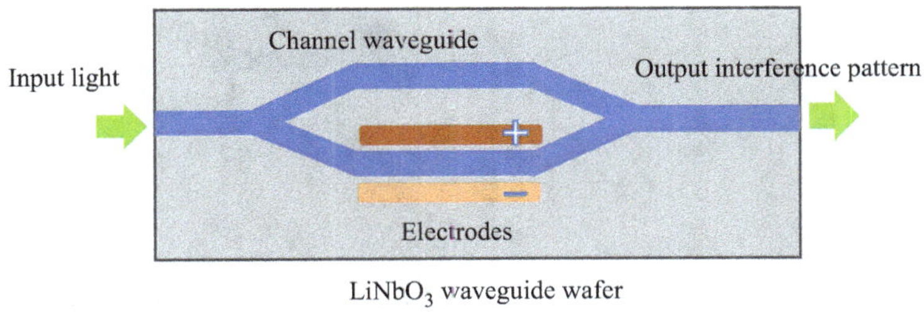

Figure 13.7 Schematic plot of an ion-implanted $LiNbO_3$ electro-optic modulator

implemented in electro-optic crystals, such as $LiNbO_3$ and β-BBO [55,56]. Owing to the additionally applied electrical field on the devices, the light propagation in the waveguides could be modulated through phase or polarization; therefore, functions like optical switches could be realized. The prototype design is typically with Mach–Zehnder geometry. Depending on the bulk properties, the $LiNbO_3$ modulators work in visible and near-infrared bands, while the β-BBO devices operate in blue or ultraviolet light band. Figure 13.7 shows the schematic plot of an ion-implanted $LiNbO_3$ electro-optic modulator. Majkic *et al.* reported on electro-optically tuneable microring resonators in F^{3+}-ion-implanted $LiNbO_3$ waveguide [57]. The 14.5 MeV irradiation creates a high refractive index contrast (Δn=0.17). With the typical microring plus bus waveguide geometry, for the microring with an 80-μm radius, the optical resonances are with modulation depth of 11 dB, free spectral range of 2 nm, and an electro-optical tunability of 10 pm/V.

13.4.2 Frequency converters

Based on nonlinear crystals, the ion-beam-produced waveguides could be utilized to achieve frequency conversion through two mechanisms, i.e., phase matching (PM) or quasi PM (QPM). The most common frequency conversion is the second harmonic generation (SHG), which is also called frequency doubling. The phase-matched nonlinear waveguides include $KNbO_3$, GdCOB, and KTP, and the second harmonics could be extended to violet band [58–62]. The frequency conversion based on QPM is realized in periodically poled crystals, such as periodically poled lithium niobate (PPLN) [63]. The operation wavelength band of the devices depends on the period of the domain structures. For phase-matched waveguides, it is required to be with guided modes at polarization of both fundamental and second harmonics, since the birefringence of the crystals plays crucial roles to realize frequency conversion. Due to the refractive index modification by using ion beams, only nonlinear crystals with refractive index responses sensitive to both fundamental and second harmonic light polarizations can be used for phase-matched SHG, which also requires high-quality waveguiding properties along these orientations. For $KNbO_3$, the He^+-ion-implanted channel waveguides were with low propagation loss of ~1 dB/cm [58]. For GdCOB, the ion-produced ridge

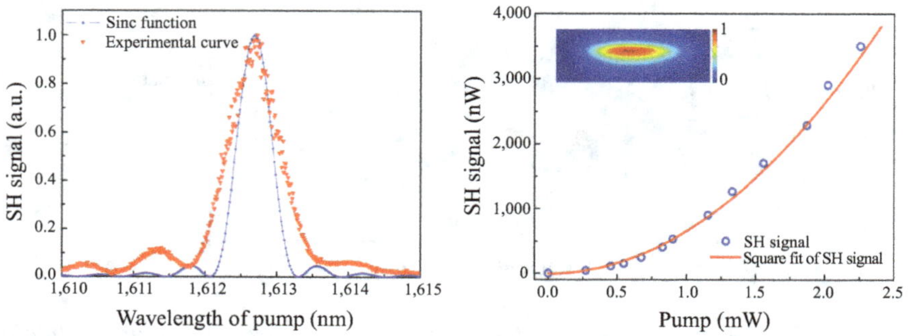

Figure 13.8 The measured second harmonic power as function of pump power at wavelength 1,612.7 nm for the PPMgOLN waveguide. © 2015 Optical Society of America. Reproduced, with permission, from Reference [64]

waveguides present higher losses (~4 dB/cm). The green–blue–violet SHG of $KNbO_3$ waveguides were well investigated by Günter's group. They applied both birefringent and Cerenkov configurations to realize the SHG. For the $KNbO_3$ ridge waveguide, conversion efficiency of 12%/W was achieved for blue light SHG [58]. By using GdCOB ridge waveguides, a conversion efficiency up to 11.4% was obtained for 1,064→532-nm green SHG under pulsed laser pump [60].

The PPLN waveguides produced by ion implantation/irradiation have been successfully applied for SHG in diverse wavelength regimes. The further diamond saw dicing enables normalized second harmonic conversion efficiency of 20.3%/(W cm^2) telecommunication L-band for a high-quality periodically poled MgO:$LiNbO_3$ (PPMgOLN) ridge waveguide, reaching at least 70% of the bulk system [64]. Figure 13.8 shows the measured second harmonic power as the function of pump power at wavelength 1,612.7 nm for the PPMgOLN waveguide.

13.4.3 Waveguide amplifiers

Waveguide optical amplifiers have been produced in both glass and crystals by using ion-beam-techniques [44,65,66]. Erbium (Er)-doped materials are favourite platforms for the operation of amplification of signals at telecommunication bands (i.e., at ~1.5 μm). The Er-doped waveguide amplifiers (EDWA) are used to amplify infrared light at wavelengths in optical communication bands, which is due to the Er^{3+} ions photoluminescence emissions [67]. Other rare-earth ions, such as Nd^{3+}, can also support signal amplification at the corresponding wavelength. The ion-implanted waveguide amplifiers have been realized in Er^{3+}-, Yb^{3+}-codoped phosphate glass, and Nd:YAG ceramics. The proton beam writing enables buried channel waveguides in phosphate glass, allowing C-band amplification. The Nd:YAG can be used for 1.06- or 1.3-μm amplification, which has been realized in a carbon-ion-irradiated ceramic waveguide (high gain of ~24 dB/cm at 1,064 nm and ~6 dB/cm at 1,319 nm) [66]. Figure 13.9 shows the gain measurements as the ratio

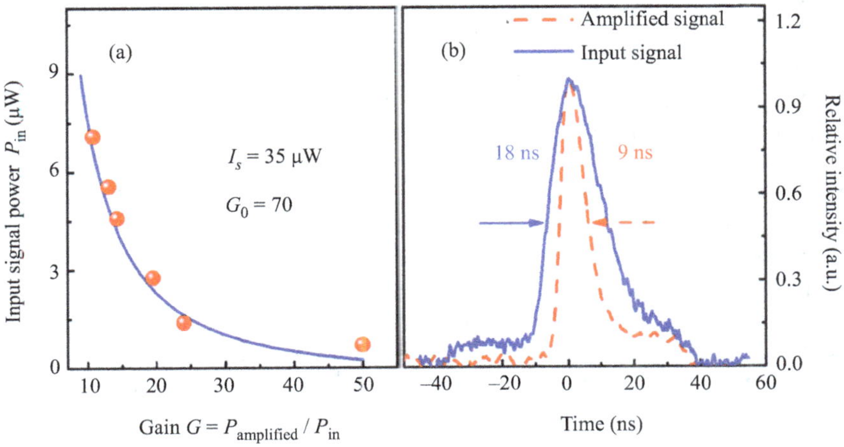

Figure 13.9 *(a) Gain measurements as the ratio of input power (Red dots) and the fitting result (solid line) for a carbon-ion-irradiated Nd:YAG ceramic waveguide; (b) the input (solid line) and amplified (dash line) 1064 nm signal pulse for the waveguide. © 2013 Optical Society of America. Reproduced, with permission, from Reference [44]*

of input power for a carbon-ion-irradiated Nd:YAG ceramic waveguide. The 1,064-nm pulse laser amplification reached a net small-signal gain value of ~18.4 dB for a 7-mm long waveguide [44]. The EDWA produced by proton beam writing achieved a net gain of 1.72 dB/cm at 1.5 μm band [65].

13.4.4 Waveguide lasers

Waveguide lasers serve as miniature light sources in integrated photonics [68,69]. Benefiting from the compact geometries of waveguides, the lasing thresholds of waveguide lasers are much lower than those of bulk lasers. The lasing efficiencies in waveguide systems are comparable to those of the bulks. Research on waveguide lasers receives much more attentions than other applications for ion-implanted waveguides. The early work performed in last century demonstrated continuous wave (CW) waveguide lasers in He-ion-implanted crystals (e.g., Nd:YAG and Nd: $MgO:LiNbO_3$), and the reduced lasing thresholds were as low as a few milliwatts [4]. However, the output powers were also relatively low, less than 20 mW [4,70,71]. As the waveguide technology became more mature, the performance of ion-implanted waveguide lasers obtained great improvement in comparison to the early-reported works. As of yet, the output power of CW waveguide lasers produced by ion irradiation has reached a few hundred mW [40]. In addition, by using proton beam writing, the waveguide laser beam could be as small as possible. With Nd:YAG crystal, Yao *et al.* reported 1.06-μm waveguide lasers with a beam diameter of only ~2 μm, which was considered as the smallest solid-state waveguide laser [72]. The slope efficiency of the ion-irradiated waveguide lasers produced by swift heavy-ion irradiation also reaches high values. For example, the Nd:YCOB planar waveguide lasing at 1.06 μm has a slope efficiency of 68%, which is close to the quantum limit value [73]. The proton beam written Nd:GGG waveguide laser, with a slope efficiency of 66%, exhibits an excellent performance of CW lasing [29]. Figure 13.10 shows the output power of waveguide lasers at 1 μm as the function of absorbed pump power for the 170-MeV Ar-ion-irradiated Nd:YCOB waveguides [73]. The lasing wavelength of the waveguide lasers has also been realized in green (~531 nm) [74] and near-infrared (~800 nm) [75], in addition to the 1-μm wavelength.

In addition to the operation in a CW regime, also pulsed waveguide lasers have been implemented in ion-beam-produced waveguides. As the most important 2D material, graphene is a broadband saturable absorber for Q switching or mode locking to generate laser pulses [76]. Combined with 2D materials, the waveguide lasers can be operated in a pulsed regime. The Q-switched waveguide lasers can be realized by either direct absorption or evanescent field absorption [77,78]. Tan *et al.* reported on the Nd:YAG waveguide laser in a Q-switched regime, reaching a pulse duration of 57 ns, a repetition rate of 4.1 MHz, and an output power of 72 mW [77,78]. For the swift heavy-ion-irradiated Nd:YAG ridge waveguide, the maximum output average power reaches 110 mW [40]. Following works have been performed

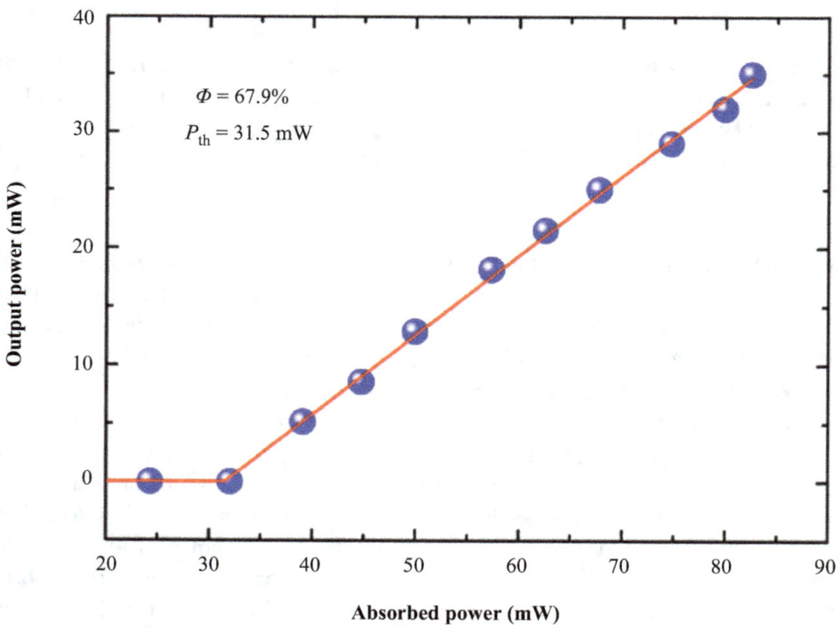

Figure 13.10 The output powers as the function of absorbed pump powers for the 180 MeV Ar ion irradiated waveguides. © 2011 Optical Society of America. Reproduced, with permission, from Reference [73]

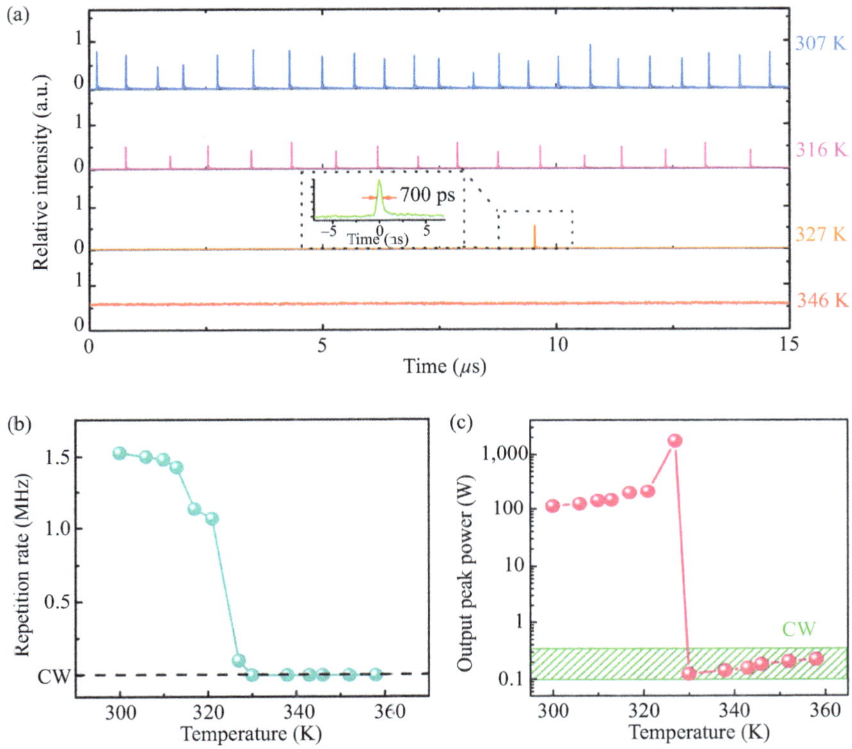

Figure 13.11 (a) Pulse trains of Q-switched waveguide laser at different temperatures. Inset: pulse duration at 327 K. (b) repetition rate and (c) peak power at different temperatures. Reproduced from [80] under CC BY 4.0 License

on other 2D materials, such as black phosphorus and WS_2 [79]. In addition, by using phase-change material VO_2 as a saturable absorber, the pulse duration could be reduced as short as 700 ps [80]. Figure 13.11 shows the pulse trains of a Q-switched VO_2-based Nd:YAG waveguide laser at different temperatures [80].

13.4.5 Waveguide sensors

In addition to the pulsed waveguide lasers, the combination of ion-beam-produced waveguides with 2D materials enables the possibility of sensing, which is due to the additional absorption of 2D materials of the guided light. When covered by 2D materials, the waveguide mode will have interactions with 2D materials layer. In this case, the ultrathin-layered materials usually possess the feature of polarization-sensitive absorption of light waves [81]. Based on this property, Tan *et al.* reported on a novel optical sensor combining 2D materials (MoS_2) thin film and a micro-fluidic structure to achieve the sensitive monitoring of refractive index (see

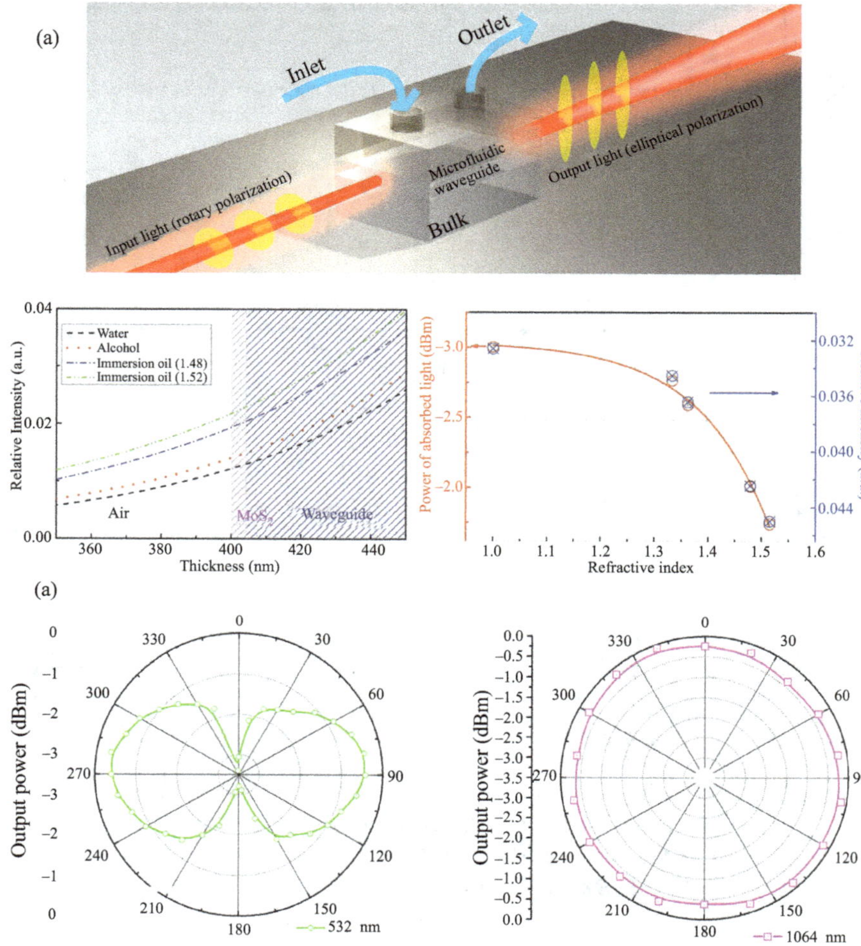

Figure 13.12 (a) The optical sensor based on ion-implanted waveguide combining MoS_2 thin film and a microfluidic structure to achieve the sensitive monitoring of refractive index, and (b) the polarization-dependent absorption of MoS_2 at 532 nm. Reproduced from [82] under CC BY 4.0 License

Figure 13.12(a) for schematic plot) [82]. In Figure 13.12(b), one can see the polarization-dependent absorption of MoS_2 at 532 nm. Based on this feature, in the designed refractive index sensor, as the microfluidic channel was filled with the tested liquid, e.g., water ($n < 1.33$), alcohol ($n < 1.35$), or immersion oil ($n <$ 1.48 or 1.52), there was a sharp variation of absorbed power within the index range of 1.33–1.52 (see Figure 13.12(c)).

Another intriguing sensor design refers to the use of ion-implanted waveguide lasers. Li *et al.* reported on an intracavity biosensor based on the Nd:YAG waveguide laser [83]. Graphene/WSe_2 heterostructure was utilized as the key element for the sensor device, owing to the evanescent field interaction with the waveguide modes or laser modes. The biosensor device can work in either passive or active regime. In the active operation regime (i.e., waveguide lasing was realized, and the detection of probe light signals was based on generated laser powers), the sensitivity of the biosensor reaches 10 mW/RIU, which was successfully applied to distinguish the concentration of the dextrose solution and tumour cells. This work indicates that by using waveguide lasers as probe signals, the biosensor could be constructed for high-resolution sensing.

13.5 Conclusions and outlook

Ion beam technology has been successfully applied to fabricate optical waveguides in numerous (more than 100) optical materials, including amorphous glasses, single crystals, and polycrystalline ceramics. As a physical technique, it offers possibility of fabrication under low temperature, exhibiting a wide applicability of materials. In addition, by choosing diverse solutions of ion beam techniques, waveguides with various geometries can be produced. A number of intriguing applications, including electro-optic modulators, frequency converters, waveguide amplifiers, waveguide lasers, and waveguide sensors, have been realized on the basis of ion-beam-produced waveguide structures. It should be also pointed out that some new techniques have also been developed on the basis of the understanding and innovation of ion-implanted dielectrics. For example, an evolution of ion-beam slicing of $LiNbO_3$ crystals results in a so-called LNOI technique, which enables ultrathin (down to 100 nm) $LiNbO_3$ single crystalline films with wafer size [84]. The large scale (up to 4-in. diameter) of LNOI wafers offers a unique platform for the development of on-chip devices. The He ion implantation plays crucial roles on the thin-film exfoliation from the bulk crystal. The LNOI structures have successfully been applied in a number of areas of photonics, telecommunications, and information science (microcavity, quantum chips, electro-optic devices, filters, data storages, sensors, etc.) [85–91]. In summary, ion beam technology has shown its unique advantages for the fabrication of waveguides and photonic devices, and more potential applications could be expected in various aspects in future.

Acknowledgement

This work was supported by National Nature Science Foundation of China (NSFC) under grant No. 11535008.

References

[1] Wesch W. and Wendler E. (eds.). *Ion Beam Modification of Solids*. Switzerland: Springer; 2016.

[2] Schmidt B. and Wetzig K. *Ion Beams in Materials Processing and Analysis*. Wien: Springer-Verlag; 2013.

[3] Stepanov A. L. 'Nonlinear optical properties of implanted metal nanoparticles in various transparent matrixes: a review'. *Rev. Adv. Mater. Sci.* 2011, vol. 27(2), pp. 115–145.

[4] Townsend P. D., Chandler P. J., and Zhang L. *Optical Effects of Ion Implantation*. Cambridge: Cambridge Univ. Press; 1994.

[5] Chen F., Wang X. L., and Wang K. M. 'Developments of ion implanted optical waveguides in optical materials: a review'. *Opt. Mater.* 2007, vol. 29(11), pp. 1523–1542.

[6] Chen F. 'Micro- and submicrometric waveguiding structures in optical crystals produced by ion beams for photonic applications'. *Laser Photonics Rev.* 2012, vol. 6(5), pp. 622–640.

[7] Chen F. 'Photonic guiding structures in lithium niobate crystals produced by energetic ion beams'. *J. Appl. Phys.* 2009, vol. 106(8), p. 081101.

[8] Schineller E. R., Flam R. P., and Wilmot D. W. 'Optical waveguides formed by proton irradiation of fused silica'. *J. Opt. Soc. Am.* 1968, vol. 58(9), pp. 1171–1176.

[9] Destefanis G. L., Townsend P. D., and Gailliard J. P. 'Optical waveguides in $LiNbO_3$ formed by ion implantation of helium'. *Appl. Phys. Lett.* 1978, vol. 32(5), pp. 293–295.

[10] Ziegler J. F. Computer Code SRIM, 2013. http://www.srim.org.

[11] Bentini G. G., Bianconi M., Chiarini M., *et al.* 'Effect of low dose high energy O^{3+} implantation on refractive index and linear electro-optic properties in X-cut $LiNbO_3$: planar optical waveguide formation and characterization'. *J. Appl. Phys.* 2002, vol. 92(11), pp. 6477–6483.

[12] Hu H., Lu F., Chen F., Shi B. R., Wang K. M., and Shen D. Y. 'Extraordinary refractive-index increase in lithium niobate caused by low-dose ion implantation'. *Appl. Opt.* 2001, vol. 40(22), pp. 3759–3761.

[13] Hu H., Lu F., Chen F., Shi B. R., Wang K. M., and Shen D. Y. 'Monomode optical waveguide in lithium niobate formed by MeV Si^+ ion implantation'. *J. Appl. Phys.* 2001, vol. 89(9), pp. 5224–5226.

[14] Berneschi S., Conti G. N., Banyasz I., *et al.* 'Ion beam irradiated channel waveguides in Er^{3+}-doped tellurite glass'. *Appl. Phys. Lett.* 2007, vol. 90(12), p. 121136.

[15] Chen F., Wang X. L., Wang K. M., Lu Q. M., and Shen D. Y. 'Optical waveguides formed in Nd:YVO_4 by MeV Si^+ implantation'. *Appl. Phys. Lett.* 2007, vol. 80(19), pp. 3473–3475.

[16] Vàzquez G. V., Rickards J., Lifante G., Domench M., and Cantelar E. 'Low dose carbon implanted waveguides in Nd:YAG'. *Opt. Express* 2003, vol. 11 (11), pp. 1291–1296.

[17] Bentini G. G., Bianconi M., Correra L., *et al.* 'Damage effects produced in the near-surface region of x-cut $LiNbO_3$ by low dose, high energy implantation of nitrogen, oxygen, and fluorine ions'. *J. Appl. Phys.* 2004, vol. 96(1), pp. 242–247.

[18] Szenes G., Horvath Z. E., Pecz B., Paszti F., and Toth L. 'Tracks induced by swift heavy ions in semiconductors'. *Phys. Rev. B* 2002, vol. 65(4), p. 045206.

[19] Ridgway M. C., Bierschenk T., Giulian R., *et al.* 'Tracks and voids in amorphous Ge induced by swift heavy-ion irradiation'. *Phys. Rev. Lett.* 2013, vol. 110(24), p. 245502.

[20] Agullo-Lopez F., Garcia G., and Olivares J. 'Lattice preamorphization by ion irradiation: fluence dependence of the electronic stopping power threshold for amorphization'. *J. Appl. Phys.* 2005, vol. 97(9), p. 093514.

[21] Rivera A., Olivares J., Garcia G., Cabrera J. M., Agullo-Rueda F., and Agullo-Lopez F. 'Giant enhancement of material damage associated to electronic excitation during ion irradiation: the case of $LiNbO_3$'. *Phys. Status Solidi A* 2009, vol. 206(6), pp. 1109–1116.

[22] Watt F., Breese M. B. H., Bettiol A. A., and van Kan J. A. 'Proton beam writing'. *Mater. Today* 2007, vol. 10(6), pp. 20–29.

[23] Bettiol A. A., Rao S. V., Teo E. J., van Kan J. A., and Watt F. 'Fabrication of buried channel waveguides in photosensitive glass using proton beam writing'. *Appl. Phys. Lett.* 2006, vol. 88, p. 171106.

[24] Sum T. C., Bettiol A. A., Florea C., and Watt E. 'Proton-beam writing of poly-methylmethacrylate buried channel waveguides'. *J. Lightwave Technol.* 2006, vol. 24(10), pp. 3803–3809.

[25] Rao S. V., Bettiol A. A., and Watt F. 'Characterization of channel waveguides and tunable microlasers in SU8 doped with rhodamine B fabricated using proton beam writing'. *J. Phys. D: Appl. Phys.* 2008, vol. 41(19), p. 192002.

[26] An Q., Cheng C., Vanga S. K., Bettiol A. A., and Chen F. 'Proton beam writing of chalcogenide glass: a new approach for fabrication of channel waveguides at telecommunication O and C bands'. *J. Lightwave Technol.* 2014, vol. 32(22), pp. 4365–4369.

[27] Benayas A., Jaque D., Yao Y., *et al.* 'Microstructuring of Nd:YAG crystals by proton-beam writing'. *Opt. Lett.* 2010, vol. 35(23), pp. 3898–3900.

[28] Yao Y. C., Zhang C., Vanga S. K., Bettiol A. A., and Chen F. 'Proton or helium ion beam written channel waveguides in Nd:YAG ceramics'. *Opt. Mater.* 2013, vol. 35(12), pp. 2257–2260.

[29] Yao Y. C., Dong N. N., Chen F., Vanga S. K., and Bettiol A. A. 'Proton beam writing of Nd:GGG crystals as new waveguide laser sources'. *Opt. Lett.* 2011, vol. 36(21), pp. 4173–4175.

[30] Dong N. N., Yao Y. C., Jia Y. C., *et al.* 'Buried channel waveguides in $KTiOPO_4$ nonlinear crystal fabricated by focused He^+ beam writing'. *Opt. Mater.* 2012, vol. 35(2), pp. 184–186.

[31] Chandler P. J. and Lama F. L. 'A new approach to the determination of planar waveguide profiles by means of a non-stationary mode index calculation'. *Opt. Acta* 1986, vol. 33(2), pp. 127–143.

[32] Wang K. M., Chen F., Hu H., *et al.* 'Double waveguide in $NdLiP_4O_{12}$ laser crystal formed by MeV He^+ ion implantation'. *Nucl. Instrum. Methods Phys. Res., Sect. B* 2005, vol. 227(3), pp. 271–274.

[33] Wang K. M., Lu F., Meng M. Q., *et al.* 'Double barrier structures in $LiNbO_3$ waveguide created by MeV He^+ implantation'. *Opt. Commun.* 1997, vol. 141(3–4), pp. 141–144.

[34] Chen F., Wang X. L., Wang K. M., *et al.* 'Bi-layer optical waveguide formed by double boron ion implantation into strontium barium niobate crystal'. *Nucl. Instrum. Methods Phys. Res., Sect. B* 2005, vol. 234(4), pp. 467–470.

[35] Chen F., Wang X. L., Wang K. M., Lu Q. M., and Shen D. Y. 'Ion-implanted Nd:YVO_4 planar waveguide: refractive-index characterization and propagation mode reduction'. *Opt. Lett.* 2002, vol. 27(13), pp. 1111–1113.

[36] Olivares J., Garcia G., Garcia-Navarro A., Agullo-Lopez F., Caballero O., and Garcia-Cabanes A. 'Generation of high-confinement step-like optical waveguides in $LiNbO_3$ by swift heavy ion-beam irradiation'. *Appl. Phys. Lett.* 2005, vol. 86(18), p. 183501.

[37] Olivares J., Garcia-Navarro A., Garcia G., *et al.* 'Nonlinear optical waveguides generated in lithium niobate by swift-ion irradiation at ultralow fluences'. *Opt. Lett.* 2007, vol. 32(17), pp. 2587–2589.

[38] Garcia-Navarro A., Olivares J., Garcia G., *et al.* 'Fabrication of optical waveguides in KGW by swift heavy ion beam irradiation'. *Nucl. Instrum. Methods Phys. Res., Sect. B* 2006, vol. 249(1–2), pp. 177–180.

[39] Ren Y. Y., Dong N. N., Chen F., *et al.* 'Swift heavy-ion irradiated active waveguides in Nd:YAG crystals: fabrication and laser generation'. *Opt. Lett.* 2010, vol. 35(19), pp. 3276–3278.

[40] Jia Y. C., Tan Y., Cheng C., Vázquez de Aldana J. R., and Chen F. 'Efficient lasing in continuous wave and graphene Q-switched regimes from Nd:YAG ridge waveguides produced by combination of swift heavy ion irradiation and femtosecond laser ablation'. *Opt. Express* 2014, vol. 22(11), pp. 12900–12908.

[41] Ren Y. Y., Dong N. N., Chen F., and Jaque D. 'Swift nitrogen ion irradiated waveguide lasers in Nd:YAG crystal'. *Opt. Express* 2011, vol. 19(6), pp. 5522–5527.

[42] Benayas A., Dong N. N., Yao Y. C., Chen F., Bettiol A. A., and Jaque D. 'Thermal optimization and erasing of Nd:YAG proton beam written waveguides'. *Opt. Lett.* 2011, vol. 38(16), pp. 3278–3280.

[43] Chen F. 'Construction of two-dimensional waveguides in insulating optical materials by means of ion beam implantation for photonic applications: fabrication methods and research progress'. *Crit. Rev. Solid State Mater. Sci.* 2008, vol. 33(3–4), pp. 165–182.

[44] Tan Y., Luan Q. F., Liu F. Q., Akhmadaliev S., Zhou S. Q., and Chen F. 'Swift carbon ion irradiated Nd:YAG ceramic optical waveguide amplifier'. *Opt. Express* 2013, vol. 21(12), pp. 13992–13997.

[45] Fluck D., Pliska T., Günter P., Bauer St., Beckers L., and Buchal Ch. 'Blue-light second-harmonic generation in ion-implanted $KNbO_3$ channel waveguides of new design'. *Appl. Phys. Lett.* 1996, vol. 69, p. 4133.

[46] Tan Y., Chen F., Wang L., Wang X. L., Wang K. M., and Lu Q. M. 'Optical channel waveguides in $KTiOPO_4$ crystal produced by proton implantation'. *J. Lightwave Technol.* 2008, vol. 26(10), pp. 1304–1308.

[47] Tan Y., Chen F., and Zhang H. J. 'Optical ridge waveguides in SBN crystal produced by low-dose carbon ion implantation followed by a sputter etching technique'. *Opt. Express* 2007, vol. 15(25), pp. 16696–16701.

[48] Chen C., Akhmadaliev S., Romero C., *et al.* 'Ridge waveguides and Y-branch beam splitters in $KTiOAsO_4$ crystal by 15 MeV oxygen ion implantation and femtosecond laser ablation'. *J. Lightwave Technol.* 2017, vol. 35(2), pp. 225–229.

[49] Jia Y. C., Rüter C. E., Akhmadaliev S., Zhou S. Q., Chen F., and Kip D. 'Ridge waveguide lasers in Nd:YAG crystals produced by combining swift heavy ion irradiation and precise diamond blade dicing'. *Opt. Mater. Express* 2013, vol. 3(4), pp. 433–438.

[50] Cheng Y. Z., Lv J. M., and Chen F. 'Mid-infrared ZnSe ridge waveguides fabricated by swift Kr8+ ion irradiation combined with precise diamond blade dicing'. *Opt. Mater. Express* 2015, vol. 5(10), pp. 2292–2299.

[51] Sum T. C., Bettiol A. A., van Kan J. A., Watt F., Pun E. Y. B., and Tung K. K. 'Proton beam writing of low-loss polymer optical waveguides'. *Appl. Phys. Lett.* 2003, vol. 83, pp. 1707–1709.

[52] Kip D. 'Photorefractive waveguides in oxide crystals: fabrication, properties, and applications'. *Appl. Phys. B* 1998, vol. 67(2), pp. 131–150.

[53] Tan Y., Chen F., Wang X. L., Wang L., Shandarov V. M., and Kip D. 'Formation of reconfigurable optical channel waveguides and beam splitters on top of proton-implanted lithium niobate crystals using spatial dark soliton-like structures'. *J. Phys. D: Appl. Phys.* 2008, vol. 41(10), p. 102001.

[54] Tan Y., Chen F., Stepic M., Shandarov V., and Kip D. 'Reconfigurable optical channel waveguides in lithium niobate crystals produced by combination of low-dose O^{3+} ion implantation and selective white light illumination'. *Opt. Express* 2008, vol. 16(14), pp. 10465–10470.

[55] Bentini G. G., Bianconi M., Cerutti A., *et al.* 'Integrated Mach-Zehnder micro-interferometer on $LiNbO_3$'. *Opt. Laser Eng.* 2007, vol. 45(3), pp. 368–372.

[56] Degl'Innocenti R., Majkic A., Vorburger P., Poberaj G., Günter P., and Döbeli M. 'Ultraviolet electro-optic amplitude modulation in β-BaB_2O_4 waveguides'. *Appl. Phys. Lett.* 2007, vol. 91, p. 051105.

[57] Majkic A., Koechlin M., Poberaj G., and Günter P. 'Optical microring resonators in fluorine-implanted lithium niobate'. *Opt. Express* 2008, vol. 16 (12), pp. 8769–8779.

[58] Fluck D. and Günter P. 'Second-harmonic generation in potassium niobate waveguides'. *IEEE J. Sel. Top. Quantum Electron.* 2000, vol. 6(1), pp. 122–131.

[59] Ren Y. Y., Jia Y. C., Chen F., Lu Q. M, Akhmadaliev S., and Zhou S. Q. 'Second harmonic generation of swift carbon ion irradiated Nd:GdCOB waveguides'. *Opt. Express* 2011, vol. 19(13), pp. 12490–12495.

[60] Jia Y. C., Chen F., Vázquez de Aldana J. R., Akhmadaliev S., and Zhou, S. Q. 'Femtosecond laser micromachining of Nd:GdCOB ridge waveguides for second harmonic generation'. *Opt. Mater.* 2011, vol. 34(11), pp. 1913–1916.

[61] Chen C., Rueter C. E., and Volk M. F. 'Second harmonic generation of diamond-blade diced $KTiOPO_4$ ridge waveguides'. *Opt. Express* 2016, vol. 24(15), pp. 16434–16439.

[62] Cheng Y. Z., Jia Y. C., Akhmadaliev S., Zhou S. Q., and Chen F. 'Guided-wave phase-matched second-harmonic generation in $KTiOPO_4$ waveguide produced by swift heavy-ion irradiation'. *Opt. Eng.* 2014, vol. 53(11), p. 117102.

[63] Huang Q., Liu P., Liu T., Zhang L., Zhou Y. F., and Wang X. L. 'Second harmonic generation in periodically poled $LiNbO_3$ waveguides formed by oxygen-ion implantation'. *Phys. Status Solidi RRL* 2012, vol. 6(5), pp. 205–207.

[64] Wang L., Haunhorst C. E., Volk M. F., Chen F., and Kip D. 'Quasi-phase-matched frequency conversion in ridge waveguides fabricated by ion implantation and diamond dicing of MgO:$LiNbO_3$ crystals'. *Opt. Express* 2015, vol. 23(23), pp. 30188–30194.

[65] Liu K., Pun E. Y. B., Sum T. C., Bettiol A. A., van Kan J. A., and Watt F. 'Erbium-doped waveguide amplifiers fabricated using focused proton beam writing'. *Appl. Phys. Lett.* 2004, vol. 84, pp. 684–686.

[66] Tan Y., Shang Z., Vanga S. K., Bettiol A. A., and Chen F. 'High-gain optical waveguide amplifier based on proton beam writing of Nd:YAG crystal'. *Opt. Express* 2015, vol. 23(11), pp. 14612–14617.

[67] Zimmerman D. R. and Spiekman L. H. 'Amplifiers for the masses: EDFA, EDWA, and SOA amplets for metro and access applications'. *J. Lightwave Technol.* 2004, vol. 22(1), pp. 63–70.

[68] Grivas C. 'Optically pumped planar waveguide lasers, Part I: Fundamentals and fabrication techniques'. *Prog. Quantum Electron.* 2011, vol. 35(6), pp. 159–239.

[69] Grivas C. 'Optically pumped planar waveguide lasers: Part II: Gain media, laser systems, and applications'. *Prog. Quantum Electron.* 2016, vol. 45–46, pp. 3–160.

[70] Domenech M., Vázquez G. V., Cantelar E., and Lifante G. 'Continuous-wave laser action at λ=1064.3 nm in proton- and carbon-implanted Nd:YAG waveguides'. *Appl. Phys. Lett.* 2003, vol. 83, p. 4110.

[71] Domenech M., Vázquez G. V., Flores-Romero E., Cantelar E., and Lifante G., 'Continuous-wave laser oscillation at 1.3 μm in Nd:YAG proton-implanted planar waveguides'. *Appl. Phys. Lett.* 2005, vol. 86, p. 151108.

[72] Yao Y. C., Tan Y., Dong N. N., Chen F., and Bettiol A. A. 'Continuous wave Nd:YAG channel waveguide laser produced by focused proton beam writing'. *Opt. Express* 2010, vol. 18(24), pp. 24516–24521.

[73] Ren Y. Y., Dong N. N., Jia Y. C., *et al.* 'Efficient laser emissions at 1.06 μm of swift heavy ion irradiated Nd:YCOB waveguides'. *Opt. Lett.* 2011, vol. 36(23), pp. 4521–4523.

[74] Ren Y. Y., Jia Y. C., Dong N. N., *et al.* 'Guided-wave second harmonics in Nd:YCOB optical waveguides for integrated green lasers'. *Opt. Lett.* 2012, vol. 37(2), pp. 244–246.

[75] Pollnau M., Grivas C., Laversenne L., Wilkinson J. S., Eason R. W., and Shepherd D. P. 'Ti:Sapphire waveguide lasers'. *Laser Phys. Lett.* 2007, vol. 4(8), pp. 560–571.

[76] Bao Q. and Loh K. P. 'Graphene photonics, plasmonics, and broadband optoelectronic devices'. *ACS Nano* 2007, vol. 6(5), pp. 3677–3694.

[77] Tan Y., Akhmadaliev S., Zhou S. Q., Sun S. Q., and Chen F. 'Guided continuous-wave and graphene-based Q-switched lasers in carbon ion irradiated Nd:YAG ceramic channel waveguide'. *Opt. Express* 2014, vol. 22(3), pp. 3572–3577.

[78] Tan Y., Cheng C., Akhmadaliev S., Zhou S. Q., and Chen F. 'Nd:YAG waveguide laser Q-switched by evanescent-field interaction with graphene'. *Opt. Express* 2014, vol. 22(8), pp. 9101–9106.

[79] Tan Y., Guo Z. N., Ma L. A., *et al.* 'Q-switched waveguide laser based on two-dimensional semiconducting materials: tungsten disulfide and black phosphorous'. *Opt. Express* 2016, vol. 24(3), pp. 2858–2866.

[80] Tan Y., Chen L. W., Wang D., *et al.* 'Tunable picosecond laser pulses via the contrast of two reverse saturable absorption phases in a waveguide platform'. *Sci. Rep.* 2016, vol. 6, p. 26176.

[81] Bao Q. L., Zhang H., Wang B., *et al.* 'Broadband graphene polarizer'. *Nat. Photonics* 2016, vol. 5(7), pp. 411–415.

[82] Tan Y., He R. Y., Cheng C., Wang D., Chen Y. X., and Chen F. 'Polarization-dependent optical absorption of MoS_2 for refractive index sensing'. *Sci. Rep.* 2016, vol. 4, p. 7523.

[83] Li G. H., Li H. Y., Gong R. M., *et al.* 'Intracavity biosensor based on the Nd: YAG waveguide laser: tumor cells and dextrose solutions'. *Photonics Res.* 2017, vol. 5(6), pp. 728–732.

[84] Mercante A. J., Shi S. Y., Yao P., Xie L. L., Weikle R. M., and Prather D. W. 'Thin film lithium niobate electro-optic modulator with terahertz operating bandwidth'. *Opt. Express* 2018, vol. 26(11), pp. 14810–14816.

[85] Boes A., Corcoran B., Chang L., Bowers J., and Mitchell A. 'Status and potential of lithium niobate on insulator (LNOI) for photonic integrated circuits'. *Laser Photonics Rev.* 2018, vol. 12(4), p. 1700256.

[86] Luo R., Jiang H. W., Rogers S., Liang H. X., He Y., and Lin Q. 'On-chip second-harmonic generation and broadband parametric down-conversion in a lithium niobate microresonator'. *Opt. Express* 2017, vol. 25(20), pp. 24531–24539.

[87] Wang C., Xiong X., Andrade N., *et al.* 'Second harmonic generation in nano-structured thin-film lithium niobate waveguides'. *Opt. Express* 2017, vol. 25(6), pp. 6963–6973.

[88] Chang L., Li Y. F., Volet N., Wang L. R., Peters J., and Bowers J. E. 'Thin film wavelength converters for photonic integrated circuits'. *Optica* 2016, vol. 3(5), pp. 531–535.
[89] Wang C., Li Z. Y., Kim M. H., *et al.* 'Metasurface-assisted phase-matching-free second harmonic generation in lithium niobate waveguides'. *Nat. Commun.* 2017, vol. 8, p. 2098.
[90] Poberaj G., Hu H., Sohler W., and Gunter P. 'Lithium niobate on insulator (LNOI) for micro-photonic devices'. *Laser Photonics Rev.* 2012, vol. 6(4), pp. 488–503.
[91] Li G. L., Jia Y. C., and Chen F. 'Research progress of photonics devices on lithium-niobate-on-insulator thin films'. *Acta Physica Sinica* 2020, vol. 69 (15), p. 157801.

Index